CAMBRIDGE STUDIES IN
ADVANCED MATHEMATICS 52

EDITORS

W. FULTON, D. J. H. GARLING, K. RIBET, P. WALTERS

GEOMETRIC CONTROL THEORY

T0275771

Already published

GEOMETRIC CONTROL THEORY

VELIMIR JURDJEVIC

University of Toronto

CAMBRIDGE
UNIVERSITY PRESS

CAMBRIDGE UNIVERSITY PRESS
Cambridge, New York, Melbourne, Madrid, Cape Town, Singapore, São Paulo

Cambridge University Press
The Edinburgh Building, Cambridge CB2 8RU, UK

Published in the United States of America by Cambridge University Press, New York

www.cambridge.org
Information on this title: www.cambridge.org/9780521495028

First published 1997
This digitally printed version 2008

A catalogue record for this publication is available from the British Library

Library of Congress Cataloguing in Publication data
Jurdjevic, Velimir.
Geometric control theory/Velimir Jurdjevic.
p.   cm. – (Cambridge studies in advanced mathematics; 52)
Includes bibliographical references and index.
ISBN 0-521-49502-4 (hc)
1. Control theory – Congresses. 2. Geometry, Differential –
Congresses. 3. Exterior differential systems – Congresses.
I. Title.   II. Series.
QA402.3.J87   1997
515'.64 – dc20   95-50008
                                                             CIP

ISBN 978-0-521-49502-8 hardback
ISBN 978-0-521-05824-7 paperback

to the memory of my mother,
Ljubica Kontić Djurdjević

Contents

Introduction

Geometric control theory provides the calculus of variations new perspectives that both unify its classic theory and outline new horizons toward which its theory extends. These perspectives grow from the theoretical foundations anchored in two important theorems not available to the classic theory of the calculus of variations.

The more immediate of these two theorems is the "maximum principle" of L. S. Pontryagin and his co-workers, obtained in the late 1950s. The maximum principle, a far-reaching generalization of Weierstrass's necessary conditions for strong minima, provides geometric conditions for a (strong) minimum of an integral criterion, called the "cost," over the trajectories of a differential control system. These conditions are based on the topological fact that an optimal solution must terminate on the boundary of the extended reachable set formed by the competing curves and their integral costs.

An important novelty of Pontryagin's approach to problems of optimal control consists of liberating the variations along the optimal curves from the constricting condition that they must terminate at the given boundary data. Instead, he considers variations that are infinitesimally near the terminal point and that generate a convex cone of directions locally tangent to the reachable set at the terminal point defined by the optimal trajectory. As a consequence of optimality, the direction of decreasing cost cannot be contained in the interior of this cone. This observation leads to the "separation theorem," which can be seen as a generalization of the classic Legendre transform in the calculus of variations, which ultimately produces the appropriate Hamiltonian function. The maximum principle states that the Hamiltonian that corresponds to the optimal trajectory must be maximal relative to the competing directions and that each optimal trajectory is the projection of an integral curve of the corresponding Hamiltonian vector field.

The maximum principle, in its original form, suffers from a serious limitation: It does not provide any information about the optimal trajectories whenever the reachable set formed by the competing curves is contained in a proper submanifold of ambient space, for then the conditions of the maximum principle are automatically satisfied by any competing curve, not because of optimality, but because the reachable set is contained in a manifold whose interior is void relative to the ambient manifold.

Geometric control theory forms a theoretical foundation for extensions of the maximum principle to optimal problems on arbitrary differentiable manifolds and circumvents the aforementioned limitation by providing the natural manifolds that contain the reachable sets and in which the reachable sets have a non-void interior. This theoretical foundation comprises important results concerning the topological and differential properties of the reachable sets and is an essential complement to modern optimal control theory.

The second important theorem, which has not been a part of the classic calculus of variations, but which is at the foundation of geometric control theory, is a theorem of W. L. Chow, published in 1939, concerning the reachable sets of integral curves for an arbitrary family of vector fields. Chow's theorem and the related Hermann-Nagano theorem concerning the existence of integral manifolds lead to a distinguished class of control systems, which I call Lie-determined, whose reachable sets admit easy descriptions in terms of Lie theoretical and algebraic criteria. This class is sufficiently large to include systems described by real analytic vector fields.

The qualitative theory of Lie-determined systems, further enriched with the maximum principle and the associated Hamiltonian formalism, provides a base for modern optimal control. This new subject removes some of the static confines of its classic predecessor, the calculus of variations, and offers a theory that is an exciting blend of differential equations, differential topology, geometry, and analysis. The diversity of problems that fit its theoretical framework accounts for the interdisciplinary character and makes the subject a crossroads for differential geometry, mechanics, and optimal control.

The subject matter is presented in two parts. The first part of this book deals with qualitative properties of the reachable sets defined by an arbitrary family of vector fields and establishes the theoretical foundation required for a study of optimal systems on differentiable manifolds. The second part of the book deals with optimality, and it progresses from linear quadratic optimal problems and time-optimal linear problems to more geometric problems on Lie groups and their homogeneous spaces.

In many respects this division of material parallels C. Carathéodory's treatment of the calculus of variations, published in two volumes in 1935. His first volume deals exclusively with the geometric properties of Pfaffian differential systems, and it was intended to provide the theoretical foundations for the optimal problems with differential constraints, known at the time as the problems of Lagrange. However, for all that those two volumes reflect profound insights into the mathematics of that period, the information contained in the first volume is never properly incorporated into the part on the calculus of variations. Nevertheless, Carathéodory's treatment of the subject, particularly his geometric point of view, considerably influenced the present exposition. In that spirit, I have included several problems treated by Carathéodory, which apart from the beauty of their solutions also illustrate the effectiveness of the new methods.

Among the contemporary books on the subject of the calculus of variations, P. Griffiths's book (1983) has had the strongest impact on this study. In particular, much of the material in this text concerning optimal problems on Lie groups owes its inspiration to Griffiths's book.

The choice of material presented in this book strongly reflects my own views that optimal control is also a natural setting for problems of geometry and mechanics, and therefore the corresponding theory should be a synthesis of the main ideas from all these subjects. For that reason, differential systems on Lie groups and their homogeneous spaces form a very important part of the text.

I have repeatedly defended the foregoing views by treating some of the most classic problems within the control-theory context. Partly for that reason, but mostly on its own merit, I include a derivation of the equations of motion for the "heavy top," a rigid body fixed at one of its points and subjected to the force of gravity. This derivation relies on the maximum principle for the correct Hamiltonian on the cotangent bundle of $SO_3(R)$. The classic equations of the heavy top are the integral curves of the corresponding Hamiltonian vector field expressed in the representation of the cotangent bundle of $SO_3(R)$ as the product $SO_3(R) \times \mathcal{L}^*$, with $\mathcal{L}^*$ equal to the dual of the Lie algebra of $SO_3(R)$.

The same formalism leads to a correct interpretation of a famous theorem of Kirchhoff, obtained in 1859, known as the "elastic kinetic analogue," which relates the equations describing the equilibrium configurations of a thin elastic bar to the equations describing the movements of the heavy top. The extension of this theorem to non-Euclidean spaces of constant curvature leads to new examples of integrable systems sharing the Kowalewski top relations among the coefficients corresponding to the principal moments of inertia.

Apart from looking at classic problems from new perspectives, this book also contains a detailed study of invariant optimal problems on Lie groups, including the use of symmetry based on the generalized Noether theorem and the connections with classic integrability theory.

The main ideas of optimal control are introduced through the quadratic regulator problem and the time-optimal problem, two of the most important problems of linear control theory. Each of these problems is presented in a self-contained manner, with particular emphasis on the phenomena that transcend the linear origins of these problems.

The quadratic regulator problem is treated within the most general class of linear quadratic problems that admit solutions. The resulting theory gives new proofs for the classic inequalities of Wirtinger and Hardy-Littlewood concerning the interpolation formulas for L_2 norms of functions whose derivatives also belong to L_2.

The solutions to the singular case lead to an optimal synthesis consisting of "jumps" and singular solutions. The synthesis is of "turnpike" type: The singular solutions are the turnpikes, and the jumps are its access routes. The geometric methods used to obtain these solutions are based on general geometric notions known as the Lie saturate, and they provide a theoretical base for solving problems with degenerate Legendre conditions. Such problems invariably lead to constrained Hamiltonian systems, and their solutions extend the pioneering work of P. A. M. Dirac in connection with the foundations of quantum mechanics. This class of problems further illustrates the geometric significance of Lie-determined systems.

Linear time-optimal problems, as well as the quadratic problem having bounds on controls, exhibit "chattering" and singular phenomena that also fall outside the classic theory, and so they provide important insights to further developments of the subject matter, while at the same time illustrating the distinctive contributions of optimal control to the classic calculus of variations.

This book is intended for students at the graduate level, although it probably contains more material than can reasonably be covered in one academic year. Some of the material concerning global controllability can be omitted during the first reading, without losing much continuity with the second part of the book on optimality.

My original plan was to write a self-contained presentation of this material accessible to a reader with a good undergraduate education in mathematics. Although I kept such a reader in mind, I soon realized that the paths to the most interesting aspects of the subject demanded of the reader more than I had originally expected. I chose to follow these paths to their natural ends, hoping that well-intentioned readers will still be able to get to the essence of the matter and feel inspired to master some of the mathematical details on their own.

Having understood the mathematical demands that the subject matter imposes, I have tried to make the presentation as accessible as possible by treating each topic in a self-contained manner. This concern accounts for the somewhat independent character of the chapters, particularly those dealing with optimal control. I hope that the pedagogical advantages of this style of presentation outweigh its mathematical inefficiencies.

Acknowledgments

A book exists in the imagination long before it exists in print. As early as 1980, Claude Lobry saw a need for a book on geometric control and challenged me to write it. The early chapters of this book, written much later, grew out of our discussions at the University of Bordeaux and the University of Nice. More important than the original inspiration, however, was the sustained insight and enthusiasm of Ivan Kupka. As a friend, mentor, and a colleague, his influence, which permeates this text all the way from practical details in the proofs to his imagined presence as the reader, has made this book better than it otherwise would have been.

The book was written in several stages spanning the period between two sabbatical leaves. I am grateful to Robbie Gardner and the members of the Department of Mathematics at the University of North Carolina at Chapel Hill for their hospitality and interest during the writing of the first part of the book.

A significant portion of the second part of the book was written at the Institute for Advanced Study at Princeton during the spring of 1994. I benefited from the intellectual intensity of the institute and the freedom to focus on my own work. I am particularly grateful to P. Griffiths for his interest in control theory and in this manuscript during my stay at the institute.

The concluding chapters of the book were written at the University of Bourgogne in Dijon, France, and the Institute National des Sciences Appliqués in Rouen, France. I am thankful to those institutions for their support, and I am indebted to my colleagues B. Bonnard and J. P. Gauthier for making those visits possible.

My thanks also go to the members of my own department at the University of Toronto for the reductions in my teaching responsibilities during the writing of this book. These thanks also extend to our graduate students B. McKay, E. Schippers, J. Mighton, and Q. Yang for their thoughtful reading and constructive criticisms of the earlier drafts. In addition, I am grateful to P. Centore for his assistance with the artwork.

I am especially indebted to my wife Deborah for the verbs and metaphors that gave direction and shape to many contorted mathematical thoughts. Finally, my thanks go to Karin Smith for skillful and patient typing of the manuscript.

Part one

Reachable sets and controllability

For the purposes of this book, a "control system" is any system of differential equations in which control functions appear as parameters. Our qualitative theory of control systems begins with the important geometric observation that each control determines a vector field, and therefore a control system can also be viewed as a family of vector fields parametrized by controls. A trajectory of such a system is a continuous curve made up of finitely many segments of integral curves of vector fields in the family.

This geometric view of control systems fits closely the theoretical framework of Sophus Lie for integration of differential equations and points to the non-commutativity of vector fields as a fundamental issue of control theory. The geometric context quickly reveals the Lie bracket as the basic theoretical tool, and the corresponding theory, known as geometric control theory, becomes a subject intimately connected with the structural properties of the enveloping Lie algebras and their integral manifolds. For this reason, our treatment of the subject begins with differentiable manifolds, rather than with $\mathbb{R}^n$ as is customary in the control-theory literature.

As natural a beginning as it may seem, particularly to the reader already familiar with differential geometry, this point of view is a departure from the usual presentation of control theory, which traditionally has been confined either to linear theory and the use of linear algebra or to control systems in $\mathbb{R}^n$, with an emphasis on optimality. The absence of geometric considerations and explicit mention of the Lie bracket in this literature can be attributed to the historical development of the subject. The following is a brief sketch of the main events:

Control theory, originally developed to satisfy the design needs of servomechanisms, under the name of "automatic control theory," became recognized as a mathematical subject in 1960, with the publication of the early papers of R. Kalman. Kalman challenged the accepted approach to control theory of that period, limited to the use of Laplace transforms and the frequency domain, by

1

showing that the basic control problems could be studied effectively through the notion of a state of the system that evolves in time according to ordinary differential equations in which controls appear as parameters. Aside from drawing attention to the mathematical content of control problems, Kalman's work served as a catalyst for further growth of the subject. Liberated from the confines of the frequency domain and further inspired by the development of computers, automatic control theory became the subject matter of a new science called systems theory.

Systems theory grew out of a desire to merge automata theory, and artificial intelligence, and discrete and continuous control into a single subject concerned with input–output relations parametrized by the states of the system. The level of generality required to keep these subjects together was well beyond the realm of differential equations, and control systems quickly evolved into topological dynamical systems, or polysystems. Systems theory, itself a hybrid of control and automata theory, in its formative period looked to abstract dynamical systems and mathematical logic for its further growth. That initial orientation of systems theory, characteristic of the early 1960s, led away from geometric interpretations of linear theory and was partly responsible for the indifference with which R. Hermann's pioneering work of 1963 (relating Chow's theorem to control theory) was received by the mathematical community.

The significance of the Lie bracket for problems of control became clear around the year 1970 with publication of the papers of R. Brockett, H. Hermes, and C. Lobry, followed by the papers of P. Brunovsky, D. Elliott, A. Krener, H. J. Sussmann, and others. Thanks to that collective effort, differential geometry entered into an exciting partnership with control theory. The mid-1960s theorems of R. Hermann and T. Nagano concerning the existence of integral manifolds for singular distributions, and also the theorem of Chow, from 1939, found applications in problems of control. The control problem, on the other hand, through its distinctive concern for time-forward evolution of systems, led to its own theorems, marking the birth of geometric control theory. These theorems will compose the subject matter for the first part of this book.

1

Basic formalism and typical problems

Although differentiable manifolds have become familiar subjects to modern mathematicians, it still seems appropriate to begin with an introductory discussion of differentiable manifolds. Considering their importance in this exposition, the extra time required to examine their basic properties may well be worthwhile. This choice of introductory material, aside from making the text more accessible, will spare the reader the awkward task of having to translate the notations from other sources into the formalism used here.

1 Differentiable manifolds

Throughout this text, the state space M will be an n-dimensional real differentiable manifold. This means that M is a topological space such that at each point $p \in M$ there exists a neighborhood U of p and a homeomorphism ϕ from U onto an open subset of $\mathbb{R}^n$. It is assumed that n does not vary with the choice of a point p on M. The pair (ϕ, U) is called a *chart* at p, and U is called a coordinate neighborhood. We shall denote $\phi(p)$ by $(x_1(p), \ldots, x_n(p))$, where $x_1, \ldots, x_n$ are called coordinates of p. It follows that each coordinate x_i is a continuous function on U. For any charts (ϕ_1, U_1) and (ϕ_2, U_2) such that $U_1 \cap U_2 \neq \phi$, the restrictions of ϕ_1 and ϕ_2 to $U = U_1 \cap U_2$ are homeomorphisms onto open subsets $\phi_1(U)$ and $\phi_2(U)$ in $\mathbb{R}^n$. Then each of $\phi_2 \circ \phi_1^{-1}$ and $\phi_1 \circ \phi_2^{-1}$ is a mapping from an open set in $\mathbb{R}^n$ into $\mathbb{R}^n$. If we write $(y_1(x_1, \ldots, x_n), \ldots, y_n(x_1, \ldots, x_n))$ for the coordinates of such mappings, then the foregoing charts are said to be *compatible* if each of the functions $y_1(x_1, \ldots, x_n)$ is r times continuously differentiable in each of its arguments.

An atlas on M is a set of compatible charts on M that cover M. Two atlases on M are equivalent if their union is also an atlas on M. M and a class of equivalent atlases is called a C^r manifold. The particular equivalence class of atlases is called a *differentiable structure* on M. A set M may have more than one differentiable structure.

We shall admit only manifolds that satisfy each of the following additional conditions:

(a) For each pair of points p and q, there exist charts (ϕ_1, U_1) and (ϕ_2, U_2), such that $p \in U_1$, $q \in U_2$, and $U_1 \cap U_2 = \phi$. That is, points of M are separated by coordinate neighborhoods (M is Hausdorff).
(b) There exists a countable collection of compatible charts $\{(\phi_i, U_i)\}$ such that M is covered by the union of the sets $\{U_i\}$.

The combination of M together with all of these assumptions and its differentiable structure will be called a C^r *differentiable manifold*, or simply a *differentiable manifold*, with n being the dimension of M. For all practical purposes, we shall be interested in only C^∞ manifolds, which we shall refer to as *smooth*.

A very important class of manifolds in this development is composed of the manifolds that admit an analytic structure. A manifold M will be called analytic if for any compatible charts (ϕ_1, U_1) and (ϕ_2, U_2) the mapping $\phi_1 \circ \phi_2^{-1}$ is analytic as a mapping from an open set in $\mathbb{R}^n$ into $\mathbb{R}^n$. That is, each coordinate function $y_i(x_1, \ldots, x_n)$ is represented by its power-series expansion valid in a neighborhood of each point $(x_1, \ldots, x_n)$ in the domain of y_i.

In general, manifolds are obtained in the following way: Let $f_1(x_1, \ldots, x_n)$, $\ldots, f_m(x_1, \ldots, x_n)$ be any C^r $(r \geq 1)$ functions defined on an open set Ω in $\mathbb{R}^n$. Assume that the Jacobian matrix $(\partial f_i / \partial x_j)$ has constant rank k at each point $x = (x_1, \ldots, x_n)$ in Ω. We may assume, without any loss of generality, that the matrix $(\partial f_i / \partial x_j)$, where $1 \leq i \leq k$, $1 \leq j \leq k$, is nonsingular. Now consider the transformation ψ defined by

$$
\begin{aligned}
y_i &= f_i(x_1, \ldots, x_n), \qquad 1 \leq i \leq k, \\
y_i &= x_i, \qquad i > k.
\end{aligned}
\tag{1}
$$

Because the Jacobian of the foregoing mapping is nonsingular at each $x \in \Omega$, it follows, by the inverse-mapping theorem, that at each $x \in \Omega$, ψ has an inverse ϕ valid in some neighborhood U. So each coordinate x_i is a function of $y_1, \ldots, y_n$. In terms of the new coordinates $y_1, \ldots, y_n$,

(a) $f_i(y_1, \ldots, y_n) = y_i$, $1 \leq i \leq k$, and
(b) $(\partial f_i / \partial y_j)(y_1, \ldots, y_n) = 0$ for i, j such that $i > k$, $j > k$.

Problem 1 Prove condition (b).

It follows from (b) that each function f_i, for $i > k$, is constant as a function of the coordinates $y_{k+1}, \ldots, y_n$. This argument shows that $y_{k+1}, \ldots, y_n$ can be used as coordinates for each point x in the set $M = \{x : f_1(x) = \text{const}, f_2(x) = \text{const}, \ldots, f_m(x) = \text{const}\}$ topologized by the relative topology from $\mathbb{R}^n$. It then follows that each connected component of M is an $(n - k)$-dimensional differentiable manifold, with the system of charts as described earlier. We shall refer to this fact as the "constant-rank theorem."

Example 1 Each n-dimensional real vector space V becomes an n-dimensional smooth (or analytic) manifold under the correspondence $\phi(p) = (x_1, \ldots, x_n)$, with $(x_1, \ldots, x_n)$ the coordinates of p relative to a fixed choice of basis $v_1, \ldots, v_n$ in V. (V, ϕ) is a global chart for V. Any other chart (V, ϕ) is compatible as long as $\psi \circ \phi^{-1}$ is a smooth (or analytic) mapping on $\mathbb{R}^n$. A function f on V is smooth (or analytic) if $f(x_1, \ldots, x_n)$ is a smooth (or analytic) function on an open subset of $\mathbb{R}^n$.

Example 2 Any open subset of a manifold M is a manifold with its coordinate charts and the differentiable structure inherited from M.

Example 3 Let $\text{GL}(V)$ denote the set of all nonsingular linear transformations of an n-dimensional real vector space V. Then $\text{GL}(V)$ is an n^2-dimensional real analytic manifold, for the following reasons:

Use $\text{end}(V)$ to denote the vector space of all linear endomorphisms on V. It follows that $\text{end}(V)$ is an n^2-dimensional vector space, and by Example 1 is a real analytic manifold of the same dimension. Because the determinant of an $n \times n$ matrix is a continuous function of the entries of the matrix, $\text{GL}(V)$ is an open subset of $\text{end}(V)$. By Example 2, it inherits the manifold structure of $\text{end}(V)$. $\text{GL}(V)$ consists of two connected components $\text{GL}^+(V)$ and $\text{GL}^-(V)$. $\text{GL}^+(V)$ is the component that contains the identity.

Example 4 The n-dimensional sphere S^n can be realized as the level surface of $f(x_1, \ldots, x_{n+1}) = x_1^2 + \cdots + x_{n+1}^2 = 1$; f satisfies the constant-rank condition at each point $p = (x_1, \ldots, x_{n+1})$ such that $\sum_{i=1}^{n+1} x_i^2 \neq 0$. On S^n, $\sum_{i=1}^{n+1} x_i^2 = 1$, and at least one of the components x_i is not zero. Hence the remaining components serve as the coordinates on $\mathbb{R}^n$. For instance, the open hemisphere U on S^n consisting of all points $p = (x_1, \ldots, x_{n+1})$, with $x_{n+1} > 0$, is the graph of the mapping $x_{n+1} = \sqrt{1 - (x_1^2 + \cdots + x_n^2)}$ defined on the open disk $x_1^2 + \cdots + x_n^2 < 1$ in $\mathbb{R}^n$.

The stereographic projection of any point p from $p_0 = e_{n+1}$ onto the equatorial plane $x_{n+1} = 0$ is given by $y_i = x_i/(1 - x_{n+1})$, $i = 1, \ldots, n$, where $y_1, \ldots, y_n$ is another system of coordinates on $V = S^n - \{p_0\}$. These two charts are compatible in $U \cap V$ because

$$y_i(x_1, \ldots, x_n) = \cfrac{1}{1 - \sqrt{1 - (x_1^2 + \cdots + x_n^2)}}, \qquad i = 1, \ldots, n,$$

is a smooth mapping from the open annulus $0 < x_1^2 + \cdots + x_n^2 < 1$ onto the open set $y_1^2 + \cdots + y_n^2 > 1$. The inverse map is given by

$$x_i = \frac{2}{1 + y_1^2 + \cdots + y_n^2} y_i, \qquad i = 1, \ldots, n.$$

The sphere S^n cannot be covered by a single coordinate chart because of its compactness (as is well known in cartography) (Figure 1.1).

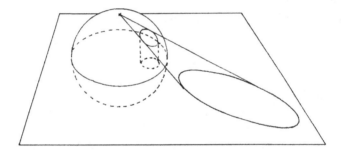

Fig. 1.1.

1.1 Differentiable mappings

Let M be a differentiable manifold, W an open subset of M, and $f : W \to \mathbb{R}^1$. Then f is said to be *differentiable* if for any local chart (ϕ, U) on M, $f \circ \phi^{-1}$ is differentiable as a function from the open subset $\phi(U \cap W)$ of $\mathbb{R}^n$ into $\mathbb{R}^1$.

A continuous mapping $F : M_1 \to M_2$, with both M_1 and M_2 differentiable manifolds, is said to be *differentiable* if for any function f differentiable on M_2, $f \circ F$ is a differentiable function on M_1. We shall write $F^* f$ for the function $f \circ F$, and we shall refer to $F^* f$ as the pull-back of f under F.

Let F be a differentiable mapping from M_1 to M_2, with $\dim M_1 = m$ and $\dim M_2 = n$. Let (ϕ, U) be a chart at p_0 on M_1, and (ψ, V) a chart at $F(p_0)$

on M_2. Write $\phi(p) = (x_1, \ldots, x_m)$ for any $p \in U$, and $\psi(q) = (y_1, \ldots, y_n)$ for any $q \in V$. Then each coordinate function y_i is a differentiable function on M_2. Hence the pull-back $F^* y_i$ is a differentiable function on M_1. But then $F^* y_i \circ \phi^{-1}$ is a differentiable function on an open set in $\mathbb{R}^m$. Therefore, the mapping

$$\psi \circ F \circ \phi^{-1} \text{ is differentiable as a mapping from } \mathbb{R}^m \text{ into } \mathbb{R}^n. \quad (2)$$

Conversely, if F is any continuous mapping from M_1 into M_2 such that F satisfies (2) for any charts (ϕ, U) on M_1 and (ψ, V) on M_2, then F is differentiable.

In particular, any curve $\sigma : I \to M$ is differentiable if $\sigma^* f$ is a differentiable function on the interval I for all differentiable functions f on M. It then follows from (2) that σ is a differentiable curve if and only if $\phi \circ \sigma : I \to \mathbb{R}^n$ is differentiable for any chart (ϕ, U) on M.

1.2 The tangent space

Each point p on a differentiable manifold M defines a real vector space $T_p M$ of the same dimension as M called the tangent space of M at p.

If M were a subset of $\mathbb{R}^n$ such that differentiable curves on M agreed with the restrictions of differentiable curves in $\mathbb{R}^n$ to M, then $T_p M$ could be defined as follows: $v \in T_p M$ if and only if there exists a differentiable curve σ with values in M defined on an interval $(-\varepsilon, \varepsilon)$ such that $\sigma(0) = p$ and such that $(d\sigma/dt)|_{t=0} = v$. However, such a definition would not be intrinsic, for it would depend on the ambient space $\mathbb{R}^n$ as well as M. The intrinsic definition of the tangent space is an abstraction of the preceding idea that does not make any reference to the ambient space, and it goes as follows:

Let $C(p)$ be the space of all differentiable curves σ on M defined on open intervals $(-\varepsilon, \varepsilon)$ in $\mathbb{R}^1$ that satisfy $\sigma(0) = p$. Curves α and β in $C(p)$ are said to be equivalent if $(d/dt)\phi \circ \alpha(t)|_{t=0} = (d/dt)\phi \circ \beta(t)|_{t=0}$ for any chart (ϕ, U) at p. It follows that such equivalence is well defined; that is, if (ϕ, U) and (ψ, V) are two charts at p, then

$$\frac{d}{dt}\phi \circ \alpha(t)|_{t=0} = \frac{d}{dt}\phi \circ \beta(t)|_{t=0}$$

if and only if

$$\frac{d}{dt}\psi \circ \alpha(t)|_{t=0} = \frac{d}{dt}\psi \circ \beta(t)|_{t=0}.$$

Problem 2 Prove the preceding statement.

$T_p M$ is defined as the set of all equivalence classes of $C(p)$. If $\alpha \in C(p)$, let $[\alpha]$ denote its equivalence class. For any chart (ϕ, U) at p, the mapping $\bar{\phi} : T_p M \to \mathbb{R}^n$ defined by $\bar{\phi}([\alpha]) = (d/dt)\phi \circ \alpha(t)|_{t=0}$ is one-to-one and onto $\mathbb{R}^n$. In fact, for any v in $\mathbb{R}^n$, $\alpha(t) = \phi^{-1} \circ (\phi(p) + tv)$ is a curve such that $\bar{\phi}([\alpha]) = v$. We define the linear structure on $T_p M$ so that $\bar{\phi}$ becomes an isomorphism. It remains to show that the linear structure on $T_p M$ is independent of the particular chart.

Now let (ψ, V) be another chart. Let $\bar{\phi}([\alpha]) = v$, and let $\bar{\psi}([\alpha]) = w$. It follows that $w = (d/dt)\psi \circ \alpha(t)|_{t=0} = (d/dt)\psi \circ \phi^{-1} \circ \phi \circ \alpha(t)|_{t=0}$. Therefore the coordinates of v and w transform according to the following formula:

$$ w_i = \sum_{j=1}^{n} \frac{\partial y_i}{\partial x_j} v_j, $$

where $y_i(x_1, \ldots, x_n)$, $i = 1, 2, \ldots, n$, denotes the coordinates of the mapping $\psi \circ \phi^{-1}$. Hence the structure of $T_p M$ is canonical in the sense that it is independent of the choice of coordinates.

1.3 The cotangent space

The cotangent space of M at p, denoted by $T_p^* M$, will be defined in terms of the differentiable functions f, defined in some neighborhood of p, that satisfy $f(p) = 0$. Let $F(p)$ denote such a class of functions. $F(p)$ will have a natural vector-space structure provided that functions that agree on a common domain are regarded as equal. (The domains of elements of $F(p)$ need not all be the same.)

For any f and g in $F(p)$, we shall say that f and g are equivalent if $d(f \circ \phi^{-1})|_{x=\phi(p)} = d(g \circ \phi^{-1})|_{x=\phi(p)}$ for any local chart (ϕ, V) at p. We shall write,

$$ f \circ \phi^{-1}(x_1, \ldots, x_n) = f(x_1, \ldots, x_n) \quad \text{and} \quad d(f \circ \phi^{-1}) = \left(\frac{\partial f}{\partial x_1}, \ldots, \frac{\partial f}{\partial x_n} \right). $$

It follows that if $f \sim \bar{f}$, and if $g \sim \bar{g}$, then $\alpha f \sim \alpha \bar{f}$ and $\alpha g \sim \alpha \bar{g}$ for any real number α, and $f + g \sim \bar{f} + \bar{g}$.

Problem 3 Prove the preceding statements.

Therefore, the space of equivalence classes becomes a vector space with the operations

$$ \alpha[f] + \beta[g] = [\alpha f + \beta g]. $$

The cotangent space T_p^*M is equal to the foregoing vector space of equivalence classes.

For each chart (ϕ, U), the mapping $\bar{\phi} : F(p)/{\sim} \to \mathbb{R}^n$ defined by $\bar{\phi}([f]) = d(f \circ \phi^{-1})$ is a linear isomorphism. If $x_i(q)$ is the ith coordinate function of ϕ, then $f_i(q) = x_i(q) - x_i(p)$, $q \in U$, is an element of $F(p)$, and $\bar{\phi}([f_i]) = e_i$. So $[f_1], \ldots, [f_n]$ is a basis for $F(p)/{\sim}$. Another chart (ψ, V) produces its own basis $[g_1], \ldots, [g_n]$, with $g_i(q) = y_i(q) - y_i(p)$, $q \in V$. Let $[f]$ be an arbitrary element of T_p^*M. Then $[f] = \sum_{i=1}^n v_i[f_i] = \sum_{i=1}^n w_i[g_i]$. It then follows that the coordinates $(w_1, \ldots, w_n)$ are related to the coordinates $(v_1, \ldots, v_n)$ via the following formula:

$$v_i = \sum_{j=1}^n \frac{\partial y_j}{\partial x_i} w_j.$$

We end this section by showing the duality between the elements of T_p^*M and those of $T_p M$. For any $f \in F(p)$ and any $\sigma \in C(p)$, consider the pairing

$$\langle [f], [\sigma] \rangle = \frac{d}{dt} f \circ \sigma|_{t=0}.$$

Because $f \circ \sigma = f \circ \phi^{-1} \circ \phi \circ \sigma$, it follows that the foregoing pairing is well defined and is bilinear. More explicitly,

$$\langle [f], [\sigma] \rangle = \sum_{i=1}^n \frac{\partial f}{\partial x_i} \frac{\partial \sigma_i}{dt},$$

with $d(f \circ \phi^{-1}) = (\partial f/\partial x_1, \ldots, \partial f/\partial x_n)$, and with $(d/dt)\phi \circ \sigma(t)|_{t=0} = (d\sigma_1/dt, \ldots, d\sigma_n/dt)$. Therefore each element of T_p^*M is a linear functional on $T_p M$, and hence

$$T_p^*M = (T_p M)^*.$$

The preceding definitions draw attention to an important distinction between tangent spaces and cotangent spaces. Let M and N be differentiable manifolds, and $\Phi : M \to N$ a differentiable map. We have already mentioned that Φ *pulls back* differentiable functions of N into the differentiable functions on M. However, for differentiable curves the situation is different. For any differentiable curve σ on M, $\Phi \circ \sigma$ is a differentiable curve on N. Thus Φ *pushes forward* curves on M into curves on N. We shall write $\Phi_*\sigma$ for the curve $\Phi \circ \sigma$.

Both the push-forward Φ_* and the pull-back Φ^* induce linear mappings called the tangent map of Φ and the differential of Φ. The tangent map of Φ will be denoted by $T\Phi$ and is a linear mapping from T_pM into $T_{\Phi(p)}N$, whereas the differential $d\Phi$ of Φ is a linear mapping from $T^*_{\Phi(p)}M$ into T^*_pM. These mappings are defined by the following formulas:

$$T\Phi[\alpha] = [\Phi_*\alpha] \quad \text{for any } [\alpha] \text{ in } T_pM, \tag{3a}$$

$$d\Phi[f] = [\Phi^*f] \quad \text{for any } [f] \text{ in } T^*_{\Phi(p)}N. \tag{3b}$$

Problem 4 Show that the mappings in (3) are well defined and that they are both linear.

In terms of the local charts (ϕ, U) at p in M and (ψ, V) at $\Phi(p)$ in N, the preceding formulas translate as follows. Let

$$v = \frac{d}{dt}\phi \circ \alpha(t)|_{t=0}, \qquad w = \frac{d}{dt}\psi \circ \Phi \circ \alpha|_{t=0},$$

and

$$\Phi_i(x_1, \ldots, x_n) = \psi_i \circ \Phi \circ \phi^{-1}(x_1, \ldots, x_n), \qquad i = 1, 2, \ldots, m.$$

Then (3a) is equivalent to

$$w_i = \sum_{j=1}^{n} \frac{\partial \Phi_i}{\partial x_j} v_j, \qquad i = 1, 2, \ldots, m. \tag{4}$$

In order to get an analogous expression for (3b), let f be a differentiable function on N at $\Phi(p)$, and g its pull-back Φ^*f. Denote $g(x_1, \ldots, x_n) = \Phi^*f \circ \phi^{-1}(x_1, \ldots, x_n)$, and $f(y_1, \ldots, y_m) = f \circ \psi^{-1}(y_1, \ldots, y_m)$. Then $g(x_1, \ldots, x_n) = f(\Phi_1(x), \ldots, \Phi_m(x))$, and hence

$$\frac{\partial g}{\partial x_i} = \sum_{j=1}^{m} \frac{\partial \Phi_j}{\partial x_i} \frac{\partial f}{\partial y_j} \quad \text{for each } i = 1, 2, \ldots, n. \tag{5}$$

Equation (4) says that the tangent vectors transform according to the Jacobian matrix $(\partial \Phi_i/\partial x_j)$, and (5) shows that the covectors transform according to its transpose. These transformations are often called the contravariant and covariant, respectively.

2 Vector fields, flows, and differential forms

As a way of introducing the tangent bundle and the cotangent bundle of M, the natural spaces for vector fields and differential forms, it will be convenient first to bring additional ideas into consideration. In particular, it will be useful to think of tangent vectors as objects that act (linearly) on functions and produce directional derivatives. Let us start with the definitions.

2.1 Derivations

Let M be a C^∞, or analytic, manifold. Let $F(p)$ be the space of all differentiable functions defined on open neighborhoods of p. $F(p)$ is a vector space with the usual addition of functions (defined on intersections of domains).

A mapping $L : F(p) \to \mathbb{R}^1$ is called a *derivation* at p if

(a) L is linear, and
(b) $L(f g) = f(p)L(g) + g(p)L(f)$ for any functions f and g in $F(p)$, with fg denoting the product of those functions.

If $f(q) = 1$ for all $q \in M$, then $L(f) = 2L(f)$, and therefore $L(f) = 0$. Thus any derivation of a constant function is zero.

We shall now show that the space of derivations at p is isomorphic with $T_p M$. In general, however, for manifolds that are not smooth, the space of derivations is an infinite-dimensional space and so cannot be equal to $T_p M$.

For each $[\alpha]$ in $T_p M$, and f in $F(p)$, let $\langle f, [\alpha] \rangle = (d/dt)f \circ \alpha(t)|_{t=0}$. Such action is well defined, for if $\alpha \sim \bar{\alpha}$, then $(d/dt)f \circ \bar{\alpha}(t)|_{t=0} = (d/dt)f \circ \phi^{-1} \circ \phi \circ \bar{\alpha}(t)|_{t=0} = (d/dt)(f \circ \phi^{-1}) \circ \phi \circ \alpha(t)|_{t=0} = (d/dt)f \circ \alpha(t)|_{t=0}$. $[\alpha]$ acts linearly on $F(p)$, and it follows immediately by the product rule for the derivatives that such an operation is a derivation. Let $L_{[\alpha]}$ be the derivation induced by the foregoing pairing. We shall now show that for each derivation L at p, there exists an element $[\alpha]$ in $T_p M$ such that $L = L_{[\alpha]}$.

Let (ϕ, U) be a fixed chart at p, and let α_i be the curve defined by $\alpha_i(t) = \phi^{-1}(\phi(p) + te_i)$. Then $[\alpha_1], \ldots, [\alpha_n]$ is a basis for $T_p M$. Now we shall show that $L = \sum_{i=1}^n \lambda_i L_{[\alpha_i]}$ for some numbers $\lambda_1, \ldots, \lambda_n$. Let $\bar{x} = (\bar{x}_1, \ldots, \bar{x}_n)$ denote the coordinates of p, and let $x = (x_1, \ldots, x_n)$ denote an arbitrary point in some neighborhood of $\bar{x}$. Then for any function f in $F(p)$,

$$f \circ \phi^{-1}(x) = f(\bar{x}_1, \ldots, \bar{x}_n) + \sum_{i=1}^n a_i(x_i - \bar{x}_i) + \sum_{i,j}^n (x_i - \bar{x}_i)(x_j - \bar{x}_j) f_{ij}(x),$$

where $a_i = (\partial f/\partial x_i)(x_1, \ldots, x_n)$, with each function f_{ij} a C^∞ function in a neighborhood of $\bar{x}$. (If f were only C^r, then f_{ij} would be C^{r-2}, and that is where the subsequent argument would break down in the C^r case.) So

$$f(q) = f(p) + \sum a_i(x_i(q) - \bar{x}_i) + \sum (x_i(q) - \bar{x}_i)(x_j(q) - \bar{x}_j) f_{ij}(x(q)).$$

Because

$$L((x_i - \bar{x}_i)(x_j - \bar{x}_j) f_{ij}) = (x_i(p) - \bar{x}_i)L((x_j - \bar{x}_j) f_{ij})$$
$$+ f_{ij}(\bar{x})(x_j(p) - \bar{x}_j)L(x_i - \bar{x}_i) = 0,$$

it follows that

$$Lf = \sum_{i=1}^{n} a_i L(x_i - \bar{x}_i) = \sum_{i=1}^{n} L(x_i - \bar{x}_i)L_{[\alpha_i]}(f).$$

The desired coefficients $\lambda_1, \ldots, \lambda_n$ are given by $\lambda_i = L(x_i - \bar{x}_i)$.

Following the usual practice in the literature, we shall write $\partial/\partial x_i$ for $L_{[\alpha_i]}$. Then $\partial/\partial x_i, \ldots, \partial/\partial x_n$ is a basis for the derivations, and each derivation is an expression of the form $\sum_{i=1}^{m} \lambda_i(\partial/\partial x_i)$.

We shall find it convenient to use two notations for the tangent vectors at p, each of which is suggestive in its own way. If we think of $T_p M$ as the equivalence classes of curves at p, then we shall denote its elements by $(d\alpha/dt)|_{t=0}$, and if we think of $T_p M$ as the space of derivations, then we shall denote its elements as $\sum_{i=1}^{n} a_i(\partial/\partial x_i)$, the meaning being that

$$\frac{d\alpha}{dt} = \sum_{i=1}^{n} a_i \frac{\partial}{\partial x_i} \quad \text{if and only if} \quad L_{[\alpha]} = \sum_{i=1}^{n} a_i L_{[\alpha_i]}.$$

We shall adopt a similar convention with the elements of the cotangent space: df is the equivalence class of f in $T_p^* M$, with the understanding that

$$df\left(\frac{d\alpha}{dt}\right) = \langle [f], [\alpha] \rangle = \frac{d}{dt} f \circ \alpha(t)|_{t=0}.$$

In particular, then, $dx_1, \ldots, dx_n$ denotes the dual basis of $\partial/\partial x_1, \ldots, \partial/x_n$. Each dx_i satisfies $dx_i(\partial/\partial x_j) = \delta_{ij}$.

2.2 The tangent bundle and cotangent bundle

The totality of all pairs (p, v), $p \in M$, $v \in T_p M$, is called the tangent bundle and is denoted by TM. The cotangent bundle $T^* M$ is the aggregate of all points

(p, v), where $p \in M$ and $v \in T_p^* M$. Both the tangent bundle and the cotangent bundle are differentiable manifolds, with the following structures:

For each of the foregoing spaces, let π denote the projection onto M given by $\pi(p, v) = p$. Each of TM and T^*M is topologized by the strongest topology under which the projection map is continuous.

For any chart (ϕ, V) on M with coordinates $(x_1, \ldots, x_n)$, $(x_1, \ldots, x_n, y_1, \ldots, y_n)$ defines a system of coordinates at any point $p \in V$, $v \in T_p M$, with $v = \sum_{i=1}^{n} y_i (\partial / \partial x_i)$. With this system of local charts, TM is a $2n$-dimensional manifold. Similarly, T^*M is also a $2n$-dimensional manifold, with charts given by $(x_1, \ldots, x_n, y_1, \ldots, y_n)$, with $v = \sum_{i=1}^{n} y_i dx_i$.

More precisely, when speaking of either the tangent bundle or the cotangent bundle, it is always understood that they are manifolds, with their manifold structure defined as in the preceding paragraph. Notice that if M is a C^k manifold, then both TM and T^*M are C^{k-1} manifolds. But if M is either C^∞ or real-analytic, then both TM and T^*M are C^∞ or real-analytic. As a mapping between smooth manifolds, the projection map π is smooth, with $\pi^{-1}(p)$ equal to $T_p M$ or $T_p^* M$, depending on the case.

Example 5 If an n-dimensional vector space V is regarded as a manifold described by Example 1, then $T(V)$ is isomorphic with $V \times V$, and $T^*(V)$ is isomorphic with $V \times V^*$, with V^* equal to the dual space of V.

Example 6 The tangent bundle of an n-sphere S^n is the totality of all points (p, v) in $\mathbb{R}^{n+1} \times \mathbb{R}^{n+1}$ such that $\sum_{i=1}^{n+1} p_i^2 = 1$ and $\sum_{i=1}^{n+1} p_i v_i = 0$. For $n = 1$, $T(S^1)$ is the same as $S^1 \times \mathbb{R}^1$, but $T(S^2)$ is not equal to $S^2 \times \mathbb{R}^2$ with the product structure of the manifolds S^2 and $\mathbb{R}^2$ (Figure 1.2).

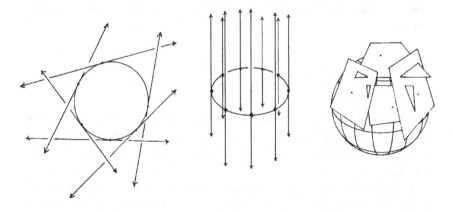

Fig. 1.2.

2.3 Vector fields and differential forms

Definition 1 Let M be a differentiable manifold. A *vector field* X is a mapping from M into TM such that for each $p \in M$ the natural projection π projects $X(p)$ to p. We say that X is a *smooth* vector field if X is a smooth mapping from M (as a smooth manifold) into TM (another smooth manifold).

Definition 2 Any mapping $w : M \to T^*M$ such that the natural projection projects $w(p)$ to p is called a *differential form*. Such a form is said to be smooth whenever w as a mapping is smooth.

Having defined these objects in an "intrinsic" way, let us now examine their meaning in a more intuitive way. It is well known in physics that the position of a particle is a scalar-like quantity, and its velocity is a vector quantity. Therefore, if $p(t)$ is a curve that describes the position, then the velocity dp/dt is a different object, since it is a vector. These two objects "live" in different spaces. The manifold formalism clarifies this issue, and it provides a natural point of view from which differential systems should be studied.

If $dx/dt = F(x)$ is a differential system in $\mathbb{R}^n$, then F *cannot* be viewed as a mapping from $\mathbb{R}^n$ into $\mathbb{R}^n$. Rather, it must be viewed as a mapping from $\mathbb{R}^n$ into the tangent space of $\mathbb{R}^n$, since $F(x(t))$ is equal to the tangent vector of the curve x at $x(t)$. For systems in $\mathbb{R}^n$, it is easy to confuse mappings and vector fields (in much the same way as it is easy to confuse vectors with their duals). It is only on arbitrary manifolds that the genuine differences of these objects become apparent.

Each vector field X in local coordinates $(x_1, \ldots, x_n)$ becomes an expression of the form

$$X_1(x_1, \ldots, x_n)\frac{\partial}{\partial x_1} + X_2(x_1, \ldots, x_n)\frac{\partial}{\partial x_2} + \cdots + X_n(x_1, \ldots, x_n)\frac{\partial}{\partial x_n}.$$

The functions $X_1, \ldots, X_n$ are called coordinates of the vector field X. Strictly speaking X should be expressed in terms of $2n$ coordinates. However, because the first n coordinates contain redundant information, they are suppressed.

A differentiable curve x on M is an *integral curve* of a vector field X if $dx/dt = X \circ x$ for each t in the domain of x. Regarding tangent vectors as derivations, $dx/dt = X \circ x$ means that for each f in a neighborhood of

$x(t)$, $(dx/dt)(f) = X(x)f$. In particular, if (ϕ, U) is a chart at $x(t)$ and if $x_i(t) = \phi_i \circ x(t)$, then

$$\frac{dx}{dt} = \sum_{i=1}^{n} \frac{dx_i}{dt}(t) \frac{\partial}{\partial x_i} = \sum_{i=1}^{n} X_i(x_1(t), \ldots, x_n(t)) \frac{\partial}{\partial x_i}.$$

The particular choice of f equal to x_j leads to

$$\frac{dx_j}{dt} = X_j(x_1(t), \ldots, x_n(t)) \quad \text{for each} \quad j = 1, 2, \ldots, n.$$

This system of differential equations admits solution curves in the open set $\phi(U)$. That is, through each point x^0 in $\phi(U)$ there exist an interval $I = (a, b)$ that contains 0 and a solution curve $x(t) = (x_1(t), \ldots, x_n(t))$ defined on I that passes through x^0 at $t = 0$. Any two such solution curves agree for values of t for which they are both defined. It follows from the theory of differential equations (Coddington and Levinson, 1955) that for each x^0 there exists a maximal open interval $I = (a, b)$ that contains 0, and there is a unique solution curve $x(t)$ defined on I such that $x(0) = x^0$. We shall refer to such a solution curve as the solution curve through x^0.

Any solution curve $x(t)$ in $\phi(U)$ defines an integral curve $p(t) = \phi^{-1}(x_1(t), \ldots, x_n(t))$ on M. Consider now another chart (ψ, V) on M such that $p(t_0) \in U \cap V$ for some t_0. We denote by $(y_1, \ldots, y_n)$ the coordinates on V, and by $Y_1, \ldots, Y_n$ the coordinates of X relative to ψ. The curve $y(t) = \psi \circ p(t)$ is a curve in $\psi(V)$ defined in some open neighborhood of t_0. Furthermore, $y(t) = \psi \circ \phi^{-1}(x(t))$, and

$$\frac{dy_i}{dt} = \sum_{j=1}^{n} \frac{\partial y_i}{\partial x_j}(x(t)) \frac{dx_j}{dt} = \sum_{j=1}^{n} \frac{\partial y_i}{\partial x_j}(x(t)) X_j(x(t)).$$

Because $(Y_1, \ldots, Y_n)$ and $(X_1, \ldots, X_n)$ are the coordinates of the same tangent vector $X(p)$, they are related through

$$Y_i(y) = \sum_{j=1}^{n} \frac{\partial y_i}{\partial x_j} X_j(x), \qquad i = 1, 2, \ldots, n.$$

Therefore, $y(t)$ is a solution curve of the system

$$\frac{dy_i}{dt}(t) = Y_i(y(t)), \qquad i = 1, 2, \ldots, n.$$

Let $\hat{y}(t)$ be the solution curve of this differential system in $\psi(V)$ that passes through $y^0 = \psi \circ p(t_0)$ at $t = t_0$, and denote $\hat{p}(t) = \psi^{-1} \circ \hat{y}(t)$. It then follows that the two integral curves $p(t)$ and $\hat{p}(t)$ on M agree at all values of t for which they are both defined.

Definition 3 We shall say that an integral curve $p(t)$ of X is the integral curve through $p_0 \in M$ if $p(0) = p_0$ and the domain $I \subset \mathbb{R}^1$ of $p(t)$ is maximal. That is, if $q(t)$ is any integral curve of X that satisfies $q(0) = p_0$, then its domain can be extended to I, so that $q(t) = p(t)$ for all t.

Definition 4 A vector field X is called *complete* if the integral curves through each point p in M are defined for all values of t in $\mathbb{R}^1$. In such a case, X is said to define a *flow* Φ on M. $\Phi : \mathbb{R}^1 \times M \to M$ is defined by $\Phi(t, p_0) = p(t)$, where $p(t)$ is the integral curve through p_0.

The following properties of Φ are basic:

1. $\Phi(0, p) = p$ for all $p \in M$.
2. $\Phi(t + s, p) = \Phi(t, \Phi(s, p))$ for all (s, t) in $\mathbb{R}^1$ and all $p \in M$.
3. $(\partial/\partial t)\Phi(t, p) = X \circ \Phi(t, p)$ for all (t, p) in $\mathbb{R}^1 \times M$.
4. The mapping Φ is smooth whenever X is smooth.

Properties 1, 2, and 3 follow directly from the uniqueness of integral curves, and property 4 is a consequence of the smooth dependence of solution curves on the initial data.

We shall call Φ the flow generated by X. It is clear from properties 1 and 2 that for each fixed t, the mapping $p \to \Phi(t, p)$ is invertible. It is also clear from 4 that such mappings are *smooth*. A mapping from a smooth manifold M to a smooth manifold N is said to be a *diffeomorphism* if both it and its inverse are smooth. For each t, the mapping $\Phi_t(p) = \Phi(t, p)$ is a diffeomorphism on M. The collection $\{\Phi_t : t \in \mathbb{R}^1\}$ forms a group under the composition of mappings. This group is called the one-parameter group of diffeomorphisms induced by X. We shall also use $\exp tX$ to denote the mapping Φ_t. Each notation is fairly standard, and each has different merits, depending on the context.

Each smooth flow on M, that is, any mapping Φ satisfying properties 1–4, is generated by a vector field X, which is called the infinitesimal generator of Φ. The relation between X and Φ is given by $X(p) = (\partial/\partial t)\Phi(t, p)|_{t=0}$. Therefore, there is a one-to-one correspondence between complete vector fields and flows.

Even though we shall deal mostly with systems of complete vector fields, elementary algebraic operations with vector fields that will be used later do not necessarily respect completeness, and for that reason it will be necessary to consider *local flows*.

In order to define the local flow of a vector field at p in M, it is first necessary to define the escape times of the integral curve of X through p. The positive escape time $e^+(p)$ is defined to be the supremum of t such that an integral curve passing through p can be defined at t. The negative escape time $e^-(p)$ is defined similarly. Let $\Delta = \{(t, p) : e^-(p) < t < e^+(p)\}$. Then Δ is an open subset of $\mathbb{R}^1 \times M$ and a neighborhood of $\{0\} \times M$. The *local flow* Φ of X is defined on Δ, where it satisfies the following:

1. $\Phi(0, p) = p$ for all $p \in \Delta$.
2. $\Phi(t + s, p) = \Phi(t, \Phi(s, p))$ whenever each of (s, p) and $(t, \Phi(s, p))$ is contained in Δ.
3. $(\partial \Phi / \partial t)(t, p) = X \circ \Phi(t, p)$ for all (p, t) in Δ.
4. Φ is C^{k-1} whenever X is C^k. In particular, Φ is smooth (analytic) whenever X is smooth (analytic).

Example 7 Let $M = \mathbb{R}^n$, and let X be a constant vector field (i.e., $X(x) = a$ for all $x \in \mathbb{R}^n$). In the "derivation notation," $X(x) = \sum_{i=1}^{n} a_i (\partial / \partial x_i)$, with $a_1, \ldots, a_n$ equal to the coordinates of a. The flow of X is $\Phi(x, t) = x + at$ for $x \in \mathbb{R}^n$, $t \in \mathbb{R}^1$. The integral curves of X are parallel lines, all in the direction of a. For each t, the mapping $x \to \Phi(t, x)$ is a translation of x by at. Hence, $\{\Phi_t : t \in \mathbb{R}^1\}$ is a one-parameter group of translations in $\mathbb{R}^n$.

Example 8 Let $M = \mathbb{R}^n$, and let A be an $n \times n$ matrix. Let X be the vector field with coordinates $X_1, \ldots, X_n$ defined by $X_i(x_1, \ldots, x_n) = \sum_{j=1}^{n} A_{ij} x_j$ for $i = 1, \ldots, n$. Each integral curve is of the form $(\exp tA)(x)$, where $\exp tA = \sum_{k=1}^{\infty} (t^k / k!) A^k$. Thus $\Phi(x, t) = (\exp tA)(x)$, and therefore $\{\Phi_t : t \in \mathbb{R}^1\}$ is a subgroup of all linear transformations. Here are some familiar cases:

$$n = 2, \qquad A = \begin{pmatrix} 0 & 1 \\ -1 & 0 \end{pmatrix}, \qquad \exp tA = \begin{pmatrix} \cos t & \sin t \\ -\sin t & \cos t \end{pmatrix} \qquad (6a)$$

$\{\Phi_t : t \in \mathbb{R}^1\}$ is equal to the group of rotations of the plane, and the integral curves are concentric circles centered at the origin (Figure 1.3a),

$$n = 2, \qquad A = \begin{pmatrix} 0 & 1 \\ 1 & 0 \end{pmatrix}, \qquad \exp tA = \begin{pmatrix} \cosh t & \sinh t \\ \sinh t & \cosh t \end{pmatrix}. \qquad (6b)$$

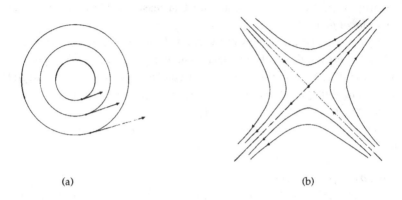

(a)   (b)

Fig. 1.3.

The one-parameter group in this case is a subgroup of $SL_2(\mathbb{R}^1)$. The integral curves are hyperbolas (Figure 1.3b).

$$n = 3, \qquad A = \begin{pmatrix} -1 & 1 & 0 \\ -1 & -1 & 0 \\ 0 & 0 & -2 \end{pmatrix},$$

$$\exp tA = \begin{pmatrix} e^{-t} \cos t & e^{-t} \sin t & 0 \\ -e^{-t} \sin t & e^{-t} \cos t & 0 \\ 0 & 0 & e^{-2t} \end{pmatrix}. \tag{6c}$$

The integral curves are spirals centered on the x_3 axis (Figure 1.4).

$$n = 4, \qquad A = \begin{pmatrix} 0 & a & 0 & 0 \\ -a & 0 & 0 & 0 \\ 0 & 0 & 0 & b \\ 0 & 0 & -b & 0 \end{pmatrix},$$

$$\exp tA = \begin{pmatrix} \cos at & \sin at & 0 & 0 \\ -\sin at & \cos at & 0 & 0 \\ 0 & 0 & \cos bt & \sin bt \\ 0 & 0 & -\sin bt & \cos bt \end{pmatrix}. \tag{6d}$$

The integral curves in this example satisfy the following relations:

$$x_1^2(t) + x_2^2(t) = x_1(0)^2 + x_2(0)^2 \quad \text{and} \quad x_3^2(t) + x_4^2(t) = x_3(0)^2 + x_4(0)^2.$$

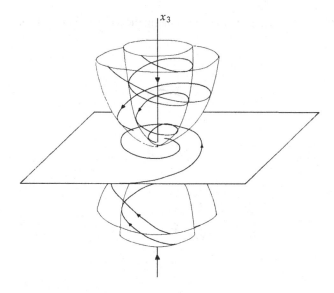

Fig. 1.4.

So if T_2 is the torus $\{(x_1, x_2, x_3, x_4) : x_1^2 + x_2^2 = R_1^2, x_3^2 + x_4^2 = R_2^2\}$, then integral curves of X that originate on T_2 remain on T_2 for all times t. Either each integral curve is periodic (when a/b is rational) or its graph is dense in T_2 (when a/b is irrational) (Figure 1.5).

Fig. 1.5.

Example 9 $M = \mathbb{R}^3$, $X(x) = Ax + a$, where

$$A = \begin{pmatrix} 0 & 1 & 0 \\ -1 & 0 & 0 \\ 0 & 0 & 0 \end{pmatrix}, \quad \text{and} \quad a = e_3.$$

Then $\Phi(t, x) = (\exp tA)(x) + ta$, with

$$\exp tA = \begin{pmatrix} \cos t & \sin t & 0 \\ -\sin t & \cos t & 0 \\ 0 & 0 & 1 \end{pmatrix}.$$

Integral curves are helices with centers along the x_3 axis (Figure 1.6).

Fig. 1.6.

Example 10 $M = \mathrm{GL}_2^+(R)$ is the group of 2×2 nonsingular matrices with positive determinant. For

$$A = \begin{pmatrix} 0 & 1 \\ 0 & 0 \end{pmatrix},$$

let X be the vector field on M defined by $x \to Ax$ for each $x \in M$. Then,

$$\Phi(t, x) = \begin{pmatrix} 1 & t \\ 0 & 1 \end{pmatrix} x, \quad x \in M.$$

$\Phi_t(x)$ is the matrix multiplication of x by $\exp t A$ from the left for each t.

3 Control systems

A control system can be viewed as a dynamical system whose dynamical laws are not entirely fixed as they are in the problems of classic physics, but depend on parameters, called controls, that can vary and with which one can control the behavior of the system; the basic challenge is to understand the effects of controls on the dynamics of the system.

It is natural to assume that the space of all possible configurations of the system is an n-dimensional manifold M and that the dynamics of the system are described by vector fields that (in contrast to the classic situation) depend on control parameters as well. The motion of the system at each point on M can follow a number of tangent directions, depending on the choice of the control. We shall be primarily interested in the following topics:

(a) the existence of a control function that can transfer the system from a given initial configuration to a prescribed terminal configuration,
(b) the length of time required to reach the terminal state, and
(c) the existence and properties of an optimal control.

The control systems that we shall consider will always be of the following form:

The state space M is an n-dimensional smooth manifold, or a real analytic manifold; the control set U is an arbitrary (usually closed) subset of $\mathbb{R}^m$, and the dynamics are described by a mapping $F : M \times U \to TM$ such that for each $u \in U$, $F_u : M \to TM$ defined by $F_u(x) = F(x, u)$ for x in M is a smooth vector field. Most often F will also be a smooth mapping on $M \times U$, although in the first part of this book it will not be necessary to make any use of the smoothness properties relative to the control parameters u.

The control functions can be of several types. A control u is called a feedback control, or a closed-loop control, if $u : M \longrightarrow U$. When u is a smooth map, and when F is smooth, the corresponding vector field $x \to F(x, u(x))$ is a smooth vector field. Any integral curve of this vector field is called a closed-loop trajectory.

A control u is called an open-loop control if $u : \mathbb{R}^1 \to U$. Open-loop controls are curves with values in U. Its trajectories are integral curves of the time-varying vector field $x \to F(x, u(t))$. Finally, a control can be a combination of both types, that is, a mapping $u : M \times \mathbb{R}^1 \to U$. The trajectories that correspond to such a choice of control are the solutions of the time-varying differential system

$$\frac{dx}{dt} = F(x, u(x, t)).$$

Each trajectory $x(t)$ of the preceding differential system corresponds to a trajectory generated by an open-loop control $v(t)$, defined through $v(t) = u(x(t), t)$. Therefore, all states that can be reached by feedback-type controls can also be reached by open-loop controls. For that reason we shall initially admit only open-loop controls.

For the first part of this exposition it will suffice to consider only piecewise-constant control functions with discontinuities occurring at discrete points of $\mathbb{R}^1$. This approach permits us to establish a geometric base without any theoretical considerations of measure, and it avoids technical issues concerning differential equations with measurable dependences.

The following examples involve important applications and provide context for further geometric and theoretical considerations.

Example 11 (Control of Liénard's equation) A general nonlinear oscillator with an external force u is described by the equation

$$\frac{d^2q}{dt^2} + f(q)\frac{dq}{dt} + g(q) = u\left(q, \frac{dq}{dt}, t\right),$$

which is known as Liénard's equation (Lefschetz, 1963).

As is customary in the literature, this equation can be expressed as a system in the phase plane by introducing coordinates

$$\begin{pmatrix} x_1 \\ x_2 \end{pmatrix},$$

defined by $x_1 = q$ and $x_2 = dq/dt$. Then $dx_1/dt = x_2$, and $dx_2/dt = d^2q/dt^2 = -f(x_1)x_2 - g(x_1) + u$. If X denotes the vector field with coordinates

$$\begin{pmatrix} x_2 \\ -f(x_1)x_2 - g(x_1) \end{pmatrix},$$

and Y denotes the vector field

$$\begin{pmatrix} 0 \\ 1 \end{pmatrix},$$

then the preceding system becomes

$$\frac{dx}{dt} = X(x) + u(x, t)Y(x), \tag{7}$$

with

$$x = \begin{pmatrix} x_1 \\ x_2 \end{pmatrix}.$$

We shall assume that the external force u plays the role of a control.

Case 1 No constraints on the control u. Because u is an arbitrary function, any differentiable curve

$$\sigma(t) = \begin{pmatrix} \sigma_1(t) \\ \sigma_2(t) \end{pmatrix}$$

is a trajectory of system (7) provided that $d\sigma_1/dt = \sigma_2$. The corresponding control is given by $u(t) = f(\sigma_1(t))\sigma_2(t) + g(\sigma_1(t)) - (d\sigma_2/dt)(t)$.

It follows that any initial state in the phase space can be transferred to any terminal state along a trajectory of the system, and the transfer can be accomplished along an arbitrary curve σ, with $\sigma_2 = d\sigma_1/dt$. The time of transfer can be made arbitrarily short by taking advantage of different parametrizations. ∎

The problem of minimizing an integral functional $\int_0^T f_0(x(t), u(t))\, dt$ over the trajectories of (7) that transfer a given initial state a to a given terminal b in T units of time becomes a standard problem in the calculus of variations, because all differentiable curves are trajectories of the system.

Case 2 Constrained controls. Assume now that the magnitude of the external force u is bounded by some constant c. This constraint dramatically reduces the space of trajectories of (7) and raises serious mathematical problems in dealing with the topics posed earlier. Even the simplest cases contain mathematical subtleties not encountered in the classic theory of the calculus of variations. For instance, the case of $f = 0$ and $g = 0$ reduces to $dx_1/dt = x_2, dx_2/dt = u$.

Constant controls produce trajectories $x_1 = \frac{1}{2}ut^2 + a_1 t + a_2$ and $x_2 = ut + a_1$ that trace a path along the parabolas $x_1 = (1/2u)(x_2 - a_1)^2 + (a_1/u)(x_2 - a_1) + a_2$. Piecewise-constant controls generate trajectories that consist of arcs of these parabolas. The reader may want to show that any points a and b in $\mathbb{R}^2$ can be connected to each other by a concatenation of two such arcs, no matter what the positive bound c is (recall that $|u(t)| \leq c$). Of course, the total time of transfer increases as c decreases. We shall show later that the time-optimal transfer is done along the arcs generated by the extremal values of the control $u = \pm c$.

The problem of minimizing $\frac{1}{2}\int_0^T x_1^2(t)\, dt$ over the trajectories of the system that satisfy the given boundary conditions shows an even more striking departure from the classic theory. We shall prove later that the optimal trajectories pass through the origin, where they remain for some time before exiting to the terminal state. But in order either to enter or to leave the origin optimally, the corresponding control is forced to oscillate infinitely often between the extreme values. This phenomenon, known as Fuller's phenomenon, occurs in many other situations (Borisov and Zelikin, 1993). ∎

Example 12 (Mechanical system with damping controls) Consider now the problem of controlling a mechanical system $d^2q/dt^2 + u(dq/dt) + g(q) = 0$ by a damping control function u. The equivalent first-order system is given by

$$\frac{dx_1}{dt} = x_2, \qquad \frac{dx_2}{dt} = -g(x) - ux_2. \tag{8}$$

For the sake of simplicity, assume that $g(x) = kx$ for some constant k. Then the foregoing system can be rewritten as

$$\frac{dx}{dt} = (A + uB)x \quad \text{for the matrices} \quad A = \begin{pmatrix} 0 & 1 \\ -k & 0 \end{pmatrix}, \qquad B = \begin{pmatrix} 0 & 0 \\ 0 & -1 \end{pmatrix},$$

with

$$x = \begin{pmatrix} x_1 \\ x_2 \end{pmatrix}.$$

For any constant control function u, the resulting system is linear. The characteristic polynomial of $A + uB$ is equal to

$$\lambda^2 + \lambda u + k = 0.$$

Case 3 $k > 0$. The integral curves that correspond to $u = 0$ are the concentric ellipses centered at the origin. Let u_1 be a number such that both eigenvalues are negative, and u_2 a number for which both eigenvalues are positive. The linear system $A + u_1 B$ is a sink, whereas $A + u_2 B$ is a source. Evidently any initial nonzero state a can be transferred to any nonzero state b by a piecewise-constant control that takes only the values 0, u_1, and u_2. ∎

Case 4 $k \le 0$. Let $x(t)$ be any trajectory of (8) generated by a control $u(t)$. Denote by $\phi(t)$ the product $\phi(t) = x_1(t)x_2(t)$. Then $(d/dt)\phi + u\phi = x_2^2 - kx_1^2$. Because $k \le 0$, the right-hand side is always non-negative. Therefore $\phi(t) \ge 0$ for all $t \ge 0$ whenever $\phi(0) \ge 0$, because

$$\phi(t) = e^{-\int_0^t u}\left(\phi(0) + \int_0^t e^{\int_0^s u}\left(x_2^2 - kx_1^2\right) ds\right).$$

Hence, the first quadrant $x_1 \ge 0$, $x_2 \ge 0$ is an invariant set no matter what control function u is chosen. This example shows that there are states that cannot be reached from an initial state a, even though there are no constraints on the size of controls. ∎

Although not classically regarded as a control problem, the next example, dealing with curves and their moving frames, fits nicely the theoretical formalism of this exposition and illustrates its interdisciplinary character.

Example 13 (Control of Serret-Frenet systems) A differentiable curve $\gamma(t)$ in the Euclidean plane E^2, parametrized by arc length, can be lifted to the group of motions of E^2 by means of a positively oriented orthonormal frame v_1, v_2 defined along the curve γ. v_1 is equal to the tangent vector $d\gamma/dt$. Because v_1 is a unit vector, its derivative dv/dt is perpendicular to v_1. Let v_2 denote the unit vector to v_1 positively oriented relative to v_1. Then there exists a function $k(t)$, called the curvature of γ, such that

$$\frac{d\gamma(t)}{dt} = v_1(t), \qquad \frac{dv_1}{dt} = k(t)v_2(t),$$

Expressing dv_2/dt as a linear combination of v_1 and v_2 leads to $dv_2/dt = -k(t)v_1$.

The frame v_1, v_2 is called the moving frame along γ. The moving frame can be expressed by a rotation matrix $R(t)$ whose columns consist of the coordinates of v_1 and v_2 relative to a fixed orthonormal frame e_1, e_2 in E^2. Omitting any notational distinctions between vectors and their coordinate vectors, then

$$R(t)e_i = v_i.$$

The curve $\gamma(t)$ along with its moving frame can be represented as an element $g(t)$ of the group of motions of E^2. We shall use E_2 to denote this group. Its elements can be viewed as 3×3 matrices of the form

$$g = \begin{pmatrix} 1 & 0 \\ \gamma & R \end{pmatrix},$$

with γ a column vector in $\mathbb{R}^2$, and R an element of the group of rotation $SO_2(R)$. Then the Serret-Frenet differential system $d\gamma/dt = v_1, dv_1/dt = kv_2$, $dv_2/dt = -kv_1$ becomes a matrix differential equation:

$$\frac{dg}{dt} = g(t) \begin{pmatrix} 0 & 0 & 0 \\ 1 & 0 & -k(t) \\ 0 & k(t) & 0 \end{pmatrix}. \tag{9}$$

Expressing an arbitrary element g as a point

$$\begin{pmatrix} x_1 \\ x_2 \\ x_3 \end{pmatrix}$$

in $\mathbb{R}^3$, with

$$g = \begin{pmatrix} 1 & 0 & 0 \\ x_1 & \cos x_3 & -\sin x_3 \\ x_2 & \sin x_3 & \cos x_3 \end{pmatrix},$$

realizes E_2 as a three-dimensional manifold and describes the preceding differential system in $\mathbb{R}^3$ as follows:

$$\frac{dx_1}{dt} = \cos x_3(t), \qquad \frac{dx_2}{dt} = \sin x_3(t), \qquad \frac{dx_3}{dt} = k(t). \qquad (10)$$

If the curvature function is regarded as a control function, then many classic variational problems in geometry become problems in optimal control. For example, the problem of finding a curve $\gamma(t)$ that will satisfy the given boundary conditions $\gamma(0) = a$, $(d\gamma/dt)(0) = \dot{a}$, will pass through the given terminal conditions $\gamma(T) = b$, $(d\gamma/dt)(T) = \dot{b}$, and will minimize $\int_0^T k^2(t)\, dt$ goes back to Euler, and its solutions are known as the *elastica* ($\dot{a}$ and $\dot{b}$ denote the tangent vectors at a and b). More generally, the problem of minimizing a functional of the form $\int_0^T f(k(t))\, dt$ is called the problem of J. Radon, according to Blaschke (1930). Much later, in 1957, Dubins considered and solved the problem of finding the curves of minimal length that would connect two given configurations $(a, \dot{a})$ and $(b, \dot{b})$ in the tangent bundle of E^2 and would satisfy an additional constraint that $|k(t)| \leq 1$. The problem of Dubins becomes a time-optimal control problem for system (9) in which the control $k(t)$ is restricted to take values in the interval $[-1, 1]$.

Curves in E^3 lift to curves in the group of motions of E^3, which is a six-dimensional manifold. The corresponding Serret-Frenet system is described by the curvature $k(t)$ and the torsion $\tau(t)$, which when regarded as control functions lead to a number of intriguing control problems. For example, Griffiths suggested the functional $\frac{1}{2}\int_0^T (k^2(t) + \tau^2(t))\, dt$ as the natural "elastic" energy of a curve and posed the question of finding all curves that would satisfy the given boundary conditions and minimize the elastic energy. We shall return to this class of problems in our chapter on variational problems on Lie groups.

3.1 Families of vector fields and control systems

Let $F : M \times U \to TM$ be a control system, as described in the preceding section. For each $u \in U$, F_u is the corresponding field. We shall use $\mathcal{F}$ to denote the family of vector fields $\mathcal{F} = \{F_u : u \in U\}$ generated by F. The first task is to relate the trajectories corresponding to the piecewise-constant controls with values in U to the integral curves of elements of $\mathcal{F}$.

Definition 5 A continuous curve $x(t)$ in M, defined on an interval $[0, T]$, is called an integral curve of $\mathcal{F}$ if there exist a partition $0 = t_0 < t_1 < \cdots < t_m = T$ and vector fields $X_1, \ldots, X_m$ in $\mathcal{F}$ such that the restriction of $x(t)$ to each open interval (t_{i-1}, t_i) is differentiable, and $dx(t)/dt = X_i(x(t))$ for $i = 1, \ldots, m$.

Because the elements of $\mathcal{F}$ are parametrized by controls, it follows that each X_i is equal to F_{u_i} for some u_i in U. Hence $x(t)$ is the solution curve of the time-varying vector field $F(x, u(t))$, with $u(t)$ equal to the piecewise-constant control, which takes constant value u_i in each interval $[t_{i-1}, t_i]$, and $x(t)$ can be visualized as a "broken" continuous curve consisting of pieces of integral curves of vector fields corresponding to different choices of control values (Figure 1.7).

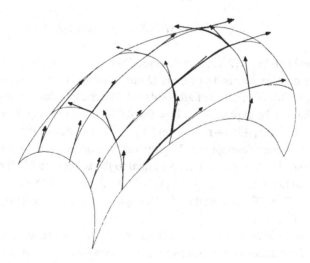

Fig. 1.7.

The reader should note that the foregoing definition applies to arbitrary families of vector fields, not just to those parametrized by the controls. This fact will be significant for some further geometric considerations encountered later in the text.

In any family of vector fields, our basic objects of interest will be their "reachable sets."

Definition 6

(a) For each $T > 0$, and each x_0 in M, the set of points reachable from x_0 at time T, denoted by $\mathcal{A}(x_0, T)$, is equal to the set of the terminal points $x(T)$ of integral curves of $\mathcal{F}$ that originate at x_0.

(b) The union of $\mathcal{A}(x_0, T)$, for $T \geq 0$, is called the set reachable from x_0.

On occasion it will also be necessary to consider the set of points reachable in T or fewer units of time. We shall use $\mathcal{A}(x_0, \leq T)$ to denote such a set. It is defined as the union of $\mathcal{A}(x_0, t)$, with $t \leq T$.

The reachable sets admit further geometric descriptions through the following formalism. Assuming that the elements of $\mathcal{F}$ are all complete vector fields, then each element X in $\mathcal{F}$ generates a one-parameter group of diffeomorphisms $\{\exp tX : t \in \mathbb{R}^1\}$. Let $G(\mathcal{F})$ denote the subgroup of the group of diffeomorphisms in M generated by the union of $\{\exp tX : t \in \mathbb{R}^1, X \in \mathcal{F}\}$. Each element Φ of $G(\mathcal{F})$ is a diffeomorphism of M of the form

$$\Phi = (\exp t_k X_k)(\exp t_{k-1} X_{k-1}) \cdots (\exp t_1 X_1)$$

for some real numbers $t_1, \ldots, t_k$ and vector fields $X_1, \ldots, X_k$ in $\mathcal{F}$. $G(\mathcal{F})$ acts on M in the obvious way and partitions M into its orbits. Then the set reachable through x_0 at time T consists of all points $\Phi(x_0)$ corresponding to elements Φ of $G(\mathcal{F})$ that can be expressed as $\Phi = (\exp t_k X_k) \cdots (\exp t_1 X_1)$, with $t_1 \geq 0$, $t_2 \geq 0, \ldots, t_k \geq 0, t_1 + \cdots + t_k = T$, and $X_1, \ldots, X_k$ in $\mathcal{F}$. The other reachable sets have analogous descriptions. In particular, $\mathcal{A}(x_0)$ is equal to the orbit of the semigroup $S_{\mathcal{F}}$ through x_0, with $S_{\mathcal{F}}$ equal to the semigroup of all elements Φ in $G(\mathcal{F})$ of the form $\Phi = (\exp t_k X_k) \cdots (\exp t_1 X_1)$, with $t_1 \geq 0, \ldots, t_k \geq 0$, and $X_1, \ldots, X_k$ in $\mathcal{F}$. The orbit of $S_{\mathcal{F}}$ through x_0, written as $S_{\mathcal{F}}(x_0)$, is equal to $\{\Phi(x_0) : \Phi \in S_{\mathcal{F}}\}$.

When some elements of $\mathcal{F}$ are not complete, then it becomes necessary to replace the corresponding groups of diffeomorphisms by local groups, and everything else remains the same. We now end this chapter with a few illustrative examples that will connect these ideas to the material in the next chapter.

Example 14 Let $M = \mathbb{R}^n$, and $\mathcal{F} = \{X, Y\}$, where X and Y are constant vector fields. Let $X(x) = a$, $Y(x) = b$ for all $x \in \mathbb{R}^n$. Then $(\exp tX)(x) = x + ta$ and $(\exp tY)(x) = x + tb$. Denote by G the subgroup of the group of diffeomorphisms generated by the exponentials in $\mathcal{F}$. It follows that G is equal to the group of translations of $\mathbb{R}^n$ by the elements in the vector space spanned by a and b.

For each $x \in \mathbb{R}^n$, the orbit $G(x)$ is equal to $x + V$, where V is the vector space generated by vectors a and b. $S_{\mathcal{F}}(x)$ is the positive quadrant in V given by $x + \alpha a + \beta b$ as α and β range through non-negative numbers. For each $T > 0$, the reachable set of $\mathcal{F}$ though x at time T is the line segment $\{x + \alpha a + (T - \alpha)b : 0 \leq \alpha \leq T\}$ (Figure 1.8).

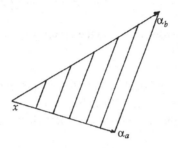

Fig. 1.8.

Example 15 Let $M = \mathbb{R}^n$, and $\mathcal{F} = \{X, Y\}$, where X is a linear field $X(x) = Ax$ for some matrix A, and Y is a constant vector field equal to a. Let G be the group generated by $\{\exp tX : t \in \mathbb{R}^1\} \cup \{\exp tY : t \in \mathbb{R}^1\}$. It follows that

$$((\exp tX) \circ (\exp sY) \circ (\exp - tX))(x) = e^{tA}(e^{-tA}x + sa) = x + s(e^{tA}(a)).$$

Thus G includes the group G_0 of the translations in the directions $se^{tA}a$ for each t and s in $\mathbb{R}^n$. Let C denote the vector space spanned by $\{e^{tA}a : t \in \mathbb{R}^n\}$. It will be shown subsequently that C is the linear span of $\{a, Aa, A^2a, \ldots, A^{n-1}a\}$. It follows that $G_0 = \{T : Tx = x + c, \ x \in \mathbb{R}^n, \ c \in C\}$. Every element of G is of the form $T \circ \exp tA$ for some $t \in \mathbb{R}^n$ and T in G_0. (It is easy to check that $\exp tA \circ T = T_t \circ \exp tA$, where $T_t \in G_0$.) The orbit of G through the origin is equal to C, and the orbit through an arbitrary point $x \in \mathbb{R}^n$ is the manifold $\{\exp tAx + c : t \in \mathbb{R}^1, \ c \in C\}$.

Example 16 Let $M = \mathbb{R}^3$, and $\mathcal{F} = \{X, Y\}$, where X and Y are linear fields given respectively by the matrices

$$A = \begin{pmatrix} 0 & 1 & 0 \\ -1 & 0 & 0 \\ 0 & 0 & 0 \end{pmatrix} \quad \text{and} \quad B = \begin{pmatrix} 0 & 0 & 0 \\ 0 & 0 & 1 \\ 0 & -1 & 0 \end{pmatrix}.$$

Then

$$\exp tX = e^{tA} = \begin{pmatrix} \cos t & \sin t & 0 \\ -\sin t & \cos t & 0 \\ 0 & 0 & 1 \end{pmatrix},$$

$$\exp tY = e^{tB} = \begin{pmatrix} 1 & 0 & \\ 0 & \cos t & \sin t \\ 0 & -\sin t & \cos t \end{pmatrix}.$$

We note that $S_{\mathcal{F}}$ in this case is equal to the group G generated by $\{\exp tX : t \in \mathbb{R}^1\} \cup \{\exp tY : t \in \mathbb{R}^1\}$.

As we shall see later, G is equal to $SO_3(R)$, the group of all orthogonal matrices with determinant equal to 1. Because such matrices preserve the lengths of vectors, it follows that the orbit of G through x is the sphere of radius $\|x\|$.

Example 17 Let

$$f(x) = \begin{cases} \exp(-\frac{1}{x^2}) & \text{for } x > 0, \\ 0 & \text{for } x \leq 0. \end{cases}$$

Then f is a C^∞ function on the real line. Let $X = \partial/\partial x_1$, and $Y(x_1, x_2) = (\partial/\partial x_1) + f(x_1)(\partial/\partial x_2)$, and consider $\mathcal{F} = \{\pm X, \pm Y\}$. In this case, $S_{\mathcal{F}}$ is the group generated by $\{\exp tX : t \in \mathbb{R}^1\} \cup \{\exp tY : t \in \mathbb{R}^1\}$. In contrast to the previous examples, $S_{\mathcal{F}}$ is infinite-dimensional.

It is easy to check that $(\exp tX)(x) = x + te_1$ and that $(\exp tY)(x_1, x_2) = (x_1 + t, \int_0^t f(x_1 + \tau)d\tau)$. The positive orbit of $S_{\mathcal{F}}$ through each point $x \in \mathbb{R}^2$ is equal to $\mathbb{R}^2$, as can be easily seen from Figure 1.9.

Problem 5 Consider the equation for the harmonic oscillator with external forces

$$\frac{dx_1}{dt} = x_2 \quad \text{and} \quad \frac{dx_2}{dt} = -x_1 + u, \quad \text{with } |u| \leq 1.$$

Show that for a constant u, the solutions are circles centered at $(u, 0)$. Also show that any two points a and b in $\mathbb{R}^2$ can be joined by a trajectory of the foregoing system. What can you say about the time-optimal trajectories?

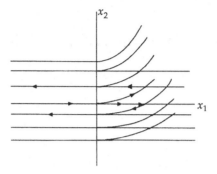

Fig. 1.9.

Problem 6 Consider the family $\mathcal{F}$ of linear vector fields generated by system (8), with $g(x) = kx$ in Example 1.2. Compute the group $G(\mathcal{F})$.

Problem 7 Calculate the flow of the vector field in equation (9) in Example 13 that corresponds to constant curvature k. Show that its projection on $\mathbb{R}^2$ traverses circles.

Problem 8 Show that any two points of the group of motions of the plane can be joined by a trajectory of (9) in Example 13.

Notes and sources

The reader not familiar with the basic theory of manifolds may want to consult other texts for additional information, such as Boothby (1975), Spivak (1970), or Sternberg (1964).

The reader may also want to have more information about the basic properties of Liénard's equation. The unforced Liénard equation was discussed in considerable detail by Lefschetz (1963). Guckenheimer and Holmes (1986) considered the behavior of its solutions under periodic inputs. Apart from those sources, there is not much literature on the control of this equation.

The geometric interpretation of a control system as a family of vector fields goes back to Hermann (1963). The merits of that point of view were further demonstrated by Lobry (1970).

2

Orbits of families of vector fields

A basic property of families of vector fields is that their orbits are manifolds. This fact, known as the "orbit theorem," marks a point of departure for geometric control theory, although from a more general mathematical perspective the theorem can also be seen as a fundamental result serving the needs of geometry, dynamical systems, mechanics, and control theory.

This chapter contains a proof of the orbit theorem, along with a self-contained treatment of the closely related integrability results, including the Frobenius integrability theorem and the Hermann-Nagano theorem concerning the orbits of families of real analytic fields. The latter theorem shows that the local structure of each orbit defined by a family of analytic vector fields is determined by the local properties of the vector fields in the family. This property of orbits, essential for geometric control theory, defines a distinguished class of families of vector fields, called "Lie-determined," large enough to include families of real analytic vector fields, whose orbits admit easy descriptions in terms of Lie theoretical and algebraic criteria.

The basic theory developed in the first part of this chapter is directed to Lie groups, homogeneous spaces, and families of vector fields subordinated to a group action, partly to illustrate its use in the classic theory of Lie groups, but more importantly to establish the conceptual framework required for subsequent analysis of differential systems on Lie groups. The chapter ends with the fundamental properties of zero-time orbits required for the study of reachable sets in the next chapter.

1 The orbit theorem

We shall state the orbit theorem in its most general form, for vector fields of class C^k, with $k \geq 1$, even though all the subsequent applications of this theorem deal with smooth or analytic differential systems. As will soon become clear, the more stringent condition on the differentiability class of vector fields does not complicate the proof, but rather elucidates its essential arguments based on the qualitative theory of ordinary differential equations.

To make matters precise, let $\mathcal{F}$ be an arbitrary family of C^k vector fields on a C^k manifold M, with $k \geq 1$. We shall continue with the notation from Chapter 1, in which $G(\mathcal{F})$, or simply G, denotes the group (pseudogroup) of diffeomorphisms (local) generated by $\{\exp tX : t \in \mathbb{R}^1, X \in \mathcal{F}\}$. Recall that the orbit of $\mathcal{F}$ through x is equal to $G(x)$, with $G(x) = \{\phi(x) : \phi \in G\}$.

Theorem 1 (the orbit theorem) *The orbit of $\mathcal{F}$ through each point x of M is a connected submanifold of M.*

The proof of this theorem will first require the development of several additional concepts. We begin with the appropriate definition of a submanifold.

1.1 Submanifolds

The notion of a submanifold lends itself to several reasonable interpretations, depending on the context and needs, and consequently it is a notion that is not firmly fixed in the existing literature. For our purposes, it will be essential to adopt the definitions that will make the orbit theorem true. The correct definition will be introduced through the notion of an immersion: A differentiable mapping f between two differentiable manifolds is called an *immersion* if the rank of the tangent map of f at each point is equal to the dimension of the domain manifold. Then our definition of a submanifold (also known as an immersed manifold) is as follows:

Definition 1 A subset N of the differentiable manifold M is called a submanifold of M if

(a) N is a differentiable manifold, and
(b) the inclusion map $i : N \to M$ is an immersion.

Because the inclusion map is continuous, it follows that the intersections of open sets of M (in its own topology) with N are open in the manifold topology of N. But it may happen that there will be open sets in N that will not be intersections of sets open in M. The densely winding line on the torus [equation (6d) in Chapter 1] illustrates this point well, and also justifies the present definition.

It will follow from the orbit theorem that when $\mathcal{F}$ consists of a single vector field X, then each orbit either is a single equilibrium point of X or is a one-dimensional submanifold of M. In the latter case, the orbit is equal to the trajectory through a nonequilibrium point of M.

It will be seen that for each $y = (\exp tX)(x)$, the set $U_\varepsilon = \{(\exp \tau X)(y) : |\tau| < \varepsilon\}$ is a coordinate neighborhood of y in the orbit through x, and $(\exp \tau X)(y) \to \tau$ is a local chart at y. When the orbit of X is dense on M, as in the torus example, the intersection of each orbit with an open set U in M consists of countably many line segments, as described earlier. Therefore, in this case the orbit topology is finer than the induced topology from the torus, because the sets U_ε are not intersections of open sets with the orbit in the ambient space.

Submanifolds whose manifold topology coincides with the relative topology induced by the ambient space are called *embedded* submanifolds (Figure 2.1).

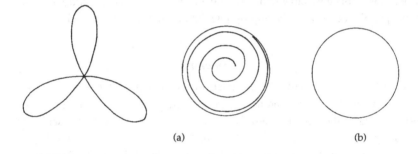

(a)   (b)

Fig. 2.1.

A general submanifold N of an ambient manifold M is locally embedded in M in the following sense: Each point p of N is contained in an open submanifold U_p of N that is an embedded submanifold of M. It follows from this property that the restrictions of differentiable functions in N to U_p are also the restrictions of functions differentiable in M. Therefore, the differentiable structure of N is determined by that of M.

The local property of N described earlier follows from this (stronger) version of the constant-rank theorem:

Theorem 2 (the constant-rank theorem) *Let $F : N \to M$ denote a differentiable map between manifolds N and M whose tangent map has constant rank k at all points of N. Then,*

(a) *for each y in the range of F, the inverse image $f^{-1}(y)$ is an embedded submanifold of N of dimension $n - k$, with $n = \dim N$;*

(b) *for each p in N there exists a neighborhood U_p of p in N such that $F(U_p)$ is an embedded submanifold of M of dimension k.*

Remark Statement (a) is a paraphrase of the version of the constant-rank the-orem stated in Chapter 1. It will be convenient for future reference to keep the same name for the stronger version stated here.

When N is a submanifold of M, and F is the inclusion map, part (b) of Theorem 2 coincides with the statement made earlier about the locally embed-ded nature of N. ∎

Proof of Theorem 2 Part (a) was essentially proved in Chapter 1. We shall use the chart constructed in that proof to prove part (b). Assume that p is a given point in N. Let (ϕ, U) be a chart at p in N, with coordinates $x_1, \ldots, x_n$. Let (ψ, V) be a chart at $F(p)$ in M, with coordinates $y_1, \ldots, y_m$. If necessary, refine U so that $F(U) \subset V$. Denote the coordinates of

$$\psi \circ F \circ \phi^{-1} : \phi(U) \subset \mathbb{R}^n \to \psi(V) \subset \mathbb{R}^m$$

by $y_i = F_i(x_1, \ldots, x_n), i = 1, \ldots, m$.

Following the construction in Chapter 1, we can assume that the coordinates $x_1, \ldots, x_n$ are so chosen that $x_i = F_i(x_1, \ldots, x_n)$ for $i = 1, \ldots, k$. It then follows from the rank assumption that the remaining functions F_i, for $i > k$, do not depend on the coordinates $x_{k+1}, \ldots, x_n$.

We shall note this fact explicitly by writing $y_i = F_i(x_1, \ldots, x_k)$, for $i > k$. Now we define new coordinates $z_1, \ldots, z_m$ at $F(p)$ through the transformation

$$z_i = x_i, \qquad i = 1, \ldots, k, \quad \text{and} \quad z_i = y_i - F_i(z_1, \ldots, z_k), \qquad i > k.$$

In terms of the new coordinates, $F(q) = (z_1, z_2, \ldots, z_k, 0, 0, \ldots, 0)$ for each $q \in U$. Therefore, $z_1, \ldots, z_k$ will serve as a system of coordinates for $F(U)$, and the proof is finished. ∎

1.2 Integral submanifolds

Definition 2

(a) A submanifold N of a manifold M is an integral manifold of a vector field X on M if $X(p)$ belongs to $T_p N$ for each p in N. We shall also say that X is tangent to N.

(b) A submanifold N of M is said to be an integral manifold for a family of vector fields $\mathcal{F}$ on M if each X in $\mathcal{F}$ is tangent to N.

The restriction of a vector field X that is tangent to N defines a vector field X_N on N, and the following holds:

Proposition 1 *Assume that* $k \geq 1$. *Then* X_N *is of class* C^k *whenever* X *is of class* C^k, *and for each* p *in* N,

$$(\exp t X)(p) = i((\exp t X_N)(p))$$

for each t *for which* $(\exp t X)(p)$ *is defined, where* i *denotes the inclusion mapping from* N *into* M.

Remark If N is an open submanifold of M, then every vector field X on M is tangent to N. The flow of X through an initial point of N may eventually leave N, and therefore X_N may not be complete even if X is complete.   ■

Exercise 1 Suppose that $X(x) = Ax$ is a linear vector field on $\mathbb{R}^n$, with A an antisymmetric matrix. Show that each sphere $\{x : \|x\|^2 = \text{constant}\}$ is an integral manifold of X.

Looking ahead toward the orbit theorem, it becomes obvious that if an orbit of $\mathcal{F}$ were a submanifold N of M, then $\mathcal{F}$ would be tangent to N, and the flow of each element X of $\mathcal{F}$ through any point of N would remain in N for all time. Therefore, it would follow that $G(\mathcal{F})(N) \subset N$.

1.3 Proof of the orbit theorem

For reasons of notational simplicity we shall not distinguish flows and local flows. The construction will be done through the mappings F of the form

$$(t_1, \ldots, t_p) \;\rightarrow\; (\exp t_p X_p) \circ \cdots \circ (\exp t_1 X_1)(x), \tag{1}$$

with $\{X_1, \ldots, X_p\} \subset \mathcal{F}$ and $x \in M$. When the vector fields are complete, the foregoing composition of diffeomorphisms is defined for each $(t_1, \ldots, t_p)$ in $\mathbb{R}^p$, and F is a C^k mapping from $\mathbb{R}^p$ into M. If some of the vector fields in (1) are not complete, then F is defined only for certain choices of $(t_1, \ldots, t_p)$ in $\mathbb{R}^p$. The basic facts required for the proof are the following:

(a) If F is defined for some $(t_1, \ldots, t_p)$ of $\mathbb{R}^p$, then F is defined in a neighbor-hood of $(t_1, \ldots, t_p)$ in $\mathbb{R}^p$. Hence $\text{dom}(F)$ is an open subset of $\mathbb{R}^p$.
(b) F is a C^k mapping from $\text{dom}(F)$ into M.
(c) For each fixed $t = (t_1, \ldots, t_p)$ in $\text{dom}(F)$, the mapping Φ_t, given $y \rightarrow (\exp t_p X_p) \circ \cdots \circ (\exp t_1 X_1)(y)$, is a local diffeomorphism from an open neighborhood of x onto a neighborhood of $F(t)$. The local inverse Φ_t^{-1} is equal to $(\exp -t_1 X_1) \circ \cdots \circ (\exp -t_p X_p)$.

It is important to note that the foregoing mappings depend on both the number and the choice of vector fields, as well as the choice of the initial point x in M.

Let N denote a fixed orbit of $\mathcal{F}$, and let x be a point in N. For each integer $p > 0$, and each choice of vector fields $X_1, \ldots, X_p$, the corresponding mapping F defined by (1) can be viewed as a mapping into N. We topologize N by the strongest topology under which each such mapping F is continuous as a mapping from $\mathbb{R}^p$ into N. We shall use Γ to denote this topology.

Because each mapping F is a continuous mapping into M, it follows that intersections of open sets in M with N are in Γ. Consequently, any distinct points of N can be separated by neighborhoods in Γ, and therefore (N, Γ) is a Hausdorff topological space. Being arcwise-connected, N is connected. The local manifold structure N will be constructed through the following procedure:

Let $\{X_1, \ldots, X_p\} \subset \mathcal{F}$ be such that the mapping F in (1) satisfies

(i) $F(\hat{t}) = x$ for some $\hat{t} \in \mathbb{R}^p$, and
(ii) the rank of the tangent map TF of F at $\hat{t}$ is maximal among the ranks of $TG(\bar{t})$, with G equal to any other mapping $G(t_1, \ldots, t_q) = (\exp t_q Y_q) \circ \cdots \circ (\exp t_1 Y_1)(x)$ of form (1) that satisfies $G(\bar{t}) = x$ for some value of $\bar{t}$ in the domain of G.

We shall use $k(x)$ to denote the rank of $TF(\hat{t})$. Because $F(0) = x$, there is at least one point $t \in \mathbb{R}^p$ that satisfies (i), and hence $k(x)$ is a well-defined integer such that $k(x) \leq \dim M$.

Our next task is to show that $k(x)$ is constant as x varies over N. Let y be any point of N, and let G denote the corresponding mapping that defines $k(y)$. That is, $\mathrm{dom}(G) \subset \mathbb{R}^q$ for some integer q,

$$G(s_1, \ldots, s_q) = (\exp s_q Y_q) \circ \cdots \circ (\exp s_1 Y_1)(y)$$

for some vector fields $Y_1, \ldots, Y_q$ in $\mathcal{F}$, $G(\hat{s}) = y$ for some $\hat{s}$ in $\mathrm{dom}(G)$, and the rank of its tangent map at $\hat{s}$ is equal to $k(y)$.

Because x and y are points in the same orbit, there exists an element Φ in $G(\mathcal{F})$ such that $y = \Phi(x)$. Let $\Phi = (\exp \hat{u}_r Z_r) \circ \cdots \circ (\exp \hat{u}_1 Z_1)$ for some numbers $\hat{u}_1, \hat{u}_2, \ldots, \hat{u}_r$ and vector fields $Z_1, \ldots, Z_r$ in $\mathcal{F}$. The composite mapping H defined by $((\exp -u_1 Z_1) \circ \cdots \circ (\exp -u_r Z_r))((\exp s_q Y_q) \circ \cdots \circ (\exp s_1 Y_1))((\exp v_r Z_r) \circ \cdots \circ (\exp v_1 Z_1)(x))$ as a mapping from $\mathbb{R}^r \times \mathbb{R}^q \times \mathbb{R}^r$ satisfies $H(\hat{u}, \hat{s}, \hat{u}) = x$. Therefore the rank of its tangent map at $(\hat{u}, \hat{s}, \hat{u})$ must be less than or equal to $k(x)$. Our next claim is that $k(x) \geq k(y)$.

The restriction of H to $u = \hat{u}$ and $v = \hat{u}$ is equal to $\Phi^{-1} \circ G$. Because Φ is a local diffeomorphism, the rank of the tangent map of $\Phi^{-1} \circ G$ at any point s

is equal to the rank of the tangent map of G at s. In particular, for $s = \hat{s}$,

$$k(y) = (\text{rank } TG)(\hat{s}) = (\text{rank } T(\Phi^{-1} \circ G))(\hat{s}) \leq (\text{rank } TH)(\hat{u}, \hat{s}, \hat{u}) \leq k(x).$$

Because x and y are arbitrary points of an orbit, the foregoing can hold only if $k(x) = k(y)$.

We shall now use k to denote the common value of $k(x)$. Note that the tangent map TF of any mapping F defined by (1) has rank less than or equal to k at any point t in the domain of F. The argument is analogous to the one used earlier and goes as follows:

If $y = (\exp t_p X_p) \circ \cdots \circ (\exp t_1 X_1)(x)$, then consider the map G on $\mathbb{R}^p \times \mathbb{R}^p$ defined by

$$G(u, v) = ((\exp -u_1 X_1) \circ \cdots \circ (\exp -u_p X_p))((\exp v_p X_p)$$
$$\circ \cdots \circ (\exp v_1 X_1)(x)).$$

G is a mapping of form (1) for which $G(-t, t) = x$; hence the rank of its tangent map is less than or equal to k at this point. The restriction of G to $u = t$ is equal to $\Phi \circ F$, with $\Phi = (\exp -t_1 X_1) \circ \cdots \circ (\exp -t_p X_p)$. The rank of the tangent map of $\Phi \circ F$ is equal to the rank of the tangent map of F at t, and the rank of $T(\Phi \circ F)$ is less than or equal to $TG(-t, t)$. Hence $TF(t) \leq k$.

The preceding observation implies that if the rank of the tangent map of some mapping F defined by (1) is k at some point $\hat{t}$ in the domain of F, then the rank must be k for all points t in some neighborhood of $\hat{t}$. We shall now apply the constant-rank theorem to such an F to conclude that $F(U)$ is an embedded submanifold of M for some neighborhood U of $\hat{t}$. Assuming that $F(\hat{t}) = x$, then $F(U)$ defines the local manifold structure of N at x.

It remains to show that the local manifold structure is compatible with the topology Γ defined earlier and that the local manifolds patch properly. Both of these properties follow from the fact that each local manifold $F(U)$ is an integral manifold of $\mathcal{F}$ of maximal dimension. The argument is as follows:

Let O denote an open subset of the local submanifold $F(U)$ constructed earlier. In order to show compatibility with Γ, we need to show that O is an element of Γ, that is, we need to show that the inverse image of O under any mapping of form (1) is open in $\mathbb{R}^p$. In order to keep the notation clear, we shall assume that $G(s_1, \ldots, s_q) = (\exp s_q Y_q) \circ \cdots \circ (\exp s_1 Y_1)(y)$ denotes an arbitrary mapping and that F is the particular mapping that defines a local manifold $F(U)$ through x. It follows by the maximality of the rank of F that the tangent map at $G(\hat{s})$ must be tangent to $F(U)$ at each point $G(\hat{s})$ contained in $F(U)$. Assume now that $G(\hat{s}) \in O$. For each index $i = 1, \ldots, q$, consider the curve $\sigma_i(t)$ defined by

$$\sigma_i(t) = G(\hat{s}_1, \ldots, \hat{u}_{i-1}, t, \hat{s}_{i+1}, \ldots, \hat{s}_q).$$

Then

$$\frac{d\sigma_i}{dt} = T((\exp \hat{s}_q Y_q) \circ \cdots \circ (\exp \hat{s}_{i+1} Y_{i+1})) Y_i ((\exp t Y_i)(\exp \hat{s}_{i-1} Y_{i-1})$$
$$\circ \cdots \circ (\exp \hat{s}_1 Y_1))(y)),$$

with $T((\exp \hat{s}_q Y_q) \circ \cdots \circ (\exp \hat{s}_{i+1} Y_{i+1}))$ denoting the tangent map of the local diffeomorphism $\Phi_i = (\exp \hat{s}_q Y_q) \circ \cdots \circ (\exp \hat{s}_{i+1} Y_{i+1})$. Therefore,

$$\frac{d\sigma_i}{dt} = (T\Phi_i)\big(Y_i \circ \Phi_i^{-1}(\sigma_i(t))\big) \quad \text{for all } t,$$

and consequently $\sigma_i(t)$ is an integral curve of the vector field $(T\Phi_i)(Y_i \circ \Phi_i^{-1})$. If $\sigma_i(s_i) \in O$, it then follows by the existence theory of ordinary differential equations that $\sigma_i(t + s_i) \in O$ for all t in some interval $[s_i - \varepsilon, s_i + \varepsilon]$.

Because

$$\sigma_i(\hat{s}_i) \in O, \qquad (\hat{s}_1, \ldots, \hat{s}_{i-1}, t + \hat{s}_i, \hat{s}_{i+1}, \ldots, \hat{s}_q) \in G^{-1}(O)$$

for all $t \in [\hat{s}_i - \varepsilon, \hat{s}_i + \varepsilon]$. A repetition of this argument applied to the integral curves of another vector field $(T\Phi_j)(Y_j \circ \Phi_j^{-1})$ that originate along the curve $\sigma_i(t)$ in O shows that for some $\varepsilon > 0$, $(\hat{s}_1, \ldots, \hat{s}_{j-1}, \hat{s}_j + t_j, \hat{s}_{j+1}, \ldots, \hat{s}_{i-1}, \hat{s}_i + t_i, \hat{s}_{i+1}, \ldots, \hat{s}_q)$ is contained in $G^{-1}(O)$ for all $t_i \in [\hat{s}_i - \varepsilon, \hat{s}_i + \varepsilon]$ and all t_j in $[\hat{s}_j - \varepsilon, \hat{s}_j + \varepsilon]$. This procedure repeated q times shows that the q-dimensional cube $\{s : |s_i - \hat{s}_i| < \varepsilon, i = 1, \ldots, q\}$ is contained in $G^{-1}(O)$, and therefore $G^{-1}(O)$ is open.

The same argument shows that the intersection of local manifolds $F_1(U_1) \cap F_2(U_2)$ is open in Γ and that the local manifolds patch properly to give global manifold structure on N.

The fact that each point p of N is contained in a coordinate neighborhood N_p that is an embedded submanifold of a manifold M implies that N can be covered by a countable collection of open charts, and therefore N is a submanifold of M. We shall not prove this fact. An interested reader may find its proof in the work of Sternberg (1964). Our proof of the orbit theorem is now finished. ∎

It follows from the preceding construction that an orbit of a single vector field X through a point x, such that $X(x) \neq 0$, is a one-dimensional submanifold of

M, with the local charts (ϕ, U) given by

$$U = \{(\exp(t + \tau)X)(x) : |\tau| < \varepsilon\}, \qquad \phi((\exp(t + \tau)X)(x)) = \tau,$$

at any point $y = (\exp tX)(x)$ of the orbit. The tangent space of the orbit at y is the set of all scalar multiples of $X(y)$.

However, when $\mathcal{F}$ consists of several non-commuting vector fields, then it becomes difficult to determine the structure of each tangent space in terms of the vector fields that define $\mathcal{F}$. The principal reason for this difficulty is that the manifold structure of an orbit is not, in general, determined by the local properties of the elements of $\mathcal{F}$. This phenomenon is easily seen through Example 17 in Chapter 1: Nonhorizontal tangent vectors at a point $x \in \mathbb{R}^2$, with $x_1 < 0$, are determined by the behavior of Y very far from x.

2 Lie brackets of vector fields and involutivity

In this section we shall concentrate on the conditions under which the local orbit structure is determined by the local properties of the family $\mathcal{F}$. In order to arrive at the appropriate theorems, it will be necessary to assume that M is a C^∞ manifold and that $\mathcal{F}$ is a family of smooth vector fields.

2.1 Lie brackets of vector fields

Smooth vector fields act as derivations on the space of smooth functions. If X denotes a smooth vector field, and f a smooth function on M, then Xf will denote the function $x \to X(x)(f)$.

For any smooth vector fields X and Y, their Lie bracket $[X, Y]$ is defined by $[X, Y]f = Y(Xf) - X(Yf)$. The reader may check that $[X, Y]$ is a smooth vector field. In terms of the local coordinates, $[X, Y]$ is given by the following relations:

Let $X(x) = \sum_{i=1}^{n} a_i(x_1, \ldots, x_n)(\partial/\partial x_i)$, $Y(x) = \sum_{i=1}^{n} b_i(x_1, \ldots, x_n)$ $(\partial/\partial x_i)$, and $[X, Y](x) = \sum_{i=1}^{n} c_i(x_1, \ldots, x_n)(\partial/\partial x_i)$. Then

$$c_i = \sum_{j=1}^{n} \frac{\partial a_i}{\partial x_j} b_j - \frac{\partial b_i}{\partial x_j} a_j, \qquad i = 1, 2, \ldots, n.$$

We single out some easy examples in $\mathbb{R}^n$:

Example 1 For constant vector fields X and Y in $\mathbb{R}^n$, the coefficients a_i and b_i are constant as x varies. Therefore, $[X, Y] = 0$.

Example 2 When X is a linear field in $\mathbb{R}^n$, then each coordinate $a_i(x_1, \ldots, x_n)$ is a linear function of $(x_1, \ldots, x_n)$ and can be written as $a_i(x_1, \ldots, x_n) = \sum_{j=1}^n a_{ij} x_j$ for some matrix $A = (a_{ij})$. Assuming that Y is a constant vector field $\sum_{i=1}^n b_i(\partial/\partial x_i)$, it follows that

$$c_i = \sum_{j=1}^n a_{ij} b_j.$$

Therefore $[X, Y]$ is a constant vector field, with $c = Ab$.

Example 3 If X and Y are linear fields, then their Lie bracket is also a linear field, with

$$c_i(x) = \sum_{j,k}^n a_{ij} b_{jk} x_k - b_{ij} a_{jk} x_k.$$

If $c_i(x_1, \ldots, x_n) = c_{ij} x_j$, then the matrix $C = (c_{ij})$ is equal to the commutator of $A = (a_{ij})$ and $B = (b_{ij})$, i.e., $C = AB - BA$.

X and Y are said to *commute* if $[X, Y] = 0$. The following proposition explains the role of commutativity and is quite useful for further developments.

Proposition 2 *Vector fields X and Y commute if and only if* $(\exp tX) \circ (\exp sY) = (\exp sY) \circ (\exp tX)$ *for all s and t. That is, X and Y commute if and only if their corresponding flows commute.*

Proof For each fixed t, $\{(\exp tX) \circ (\exp sY) \circ (\exp -tX) : s \in \mathbb{R}^1\}$ is a one-parameter group of diffeomorphisms on M. Recall that every one-parameter group is generated by a vector field, called its infinitesimal generator. Denoting the infinitesimal generator of the foregoing flow by Z_t, it follows that

$$(\exp tX) \circ (\exp sY) \circ (\exp -tX) = (\exp sZ_t).$$

Using $Z_t = (\partial/\partial s)(\exp sZ_t)|_{s=0}$ leads to

$$Z_t = (\exp tX)_*(Y) \circ (\exp -tX),$$

with $(\exp tX)_*$ denoting the tangent map of $\exp tX$. Therefore, for each function f,

$$Z_t f = (\exp tX)_*(Y)(\exp -tX)(f),$$

and therefore

$$\frac{d}{dt} Z_t f = (\exp tX)_*[X, Y](\exp tX)(f).$$

So if $[X, Y] = 0$, then Z_t is constant. Because $Z_t|_{t=0} = Y$, it follows that $Z_t = Y$ for all t. Therefore $(\exp tX) \circ (\exp sY) \circ (\exp -tX) = (\exp sY)$, or $(\exp tX) \circ (\exp sY) = (\exp sY) \circ (\exp tX)$. The converse follows from reversal of the preceding steps.   ∎

Remark Note that this proof is valid even if the fields are only C^1.   ∎

2.2 *Lie algebras*

Let $F^\infty(M)$ denote the space of all smooth vector fields on M. $F^\infty(M)$ is a real vector space under the pointwise addition of vectors

$$(\alpha X + \beta Y)(x) = \alpha X(x) + \beta Y(x) \quad \text{for all} \quad x \in M \tag{2}$$

for each set of real numbers α and β and vector fields X and Y.

We shall regard $F^\infty(M)$ as an algebra, with the addition given by (1) and with the product given by the Lie bracket.

It is easy to check that the Lie bracket satisfies

(a) $[X, Y] = -[Y, X]$ (antisymmetry rule),
(b) $[X, \alpha Y + \beta Z] = \alpha[X, Y] + \beta[X, Z]$ (bilinearity rule), and
(c) $[X, [Y, Z]] + [Z, [X, Y]] + [Y, [Z, X]] = 0$ (the Jacobi identity).

Any algebra that satisfies (a), (b), and (c) is called a Lie algebra. In addition to its Lie-algebra structure, $F^\infty(M)$ admits another algebraic operation: For any $f \in C^\infty(M)$ and any $X \in F^\infty(M)$, fX is a vector field defined by $fX(x) = f(x)X(x)$. (Do not confuse fX and Xf – the latter is a function.) With this operation, $F^\infty(M)$ becomes a module over the ring of smooth functions on M. We shall mostly rely on the Lie-algebra structure of $F^\infty(M)$, although in several instances it will be useful to consider the module structure as well.

For any family of vector fields $\mathcal{F}$, $\text{Lie}(\mathcal{F})$ will denote the Lie algebra of vector fields generated by $\mathcal{F}$. It follows that $\text{Lie}(\mathcal{F})$ is equal to the smallest vector subspace S of $F^\infty(M)$ that also satisfies $[X, S] \subset S$ for any X in $\mathcal{F}$. $\text{Lie}(\mathcal{F})$ can also be described in terms of the following notation: For each X in $F^\infty(M)$, let $\text{ad} X : F^\infty(M) \to F^\infty(M)$ denote the mapping $\text{ad} X(Y) = [X, Y]$ for Y in $F^\infty(M)$. Then $\text{Lie}(\mathcal{F})$ is equal to the smallest vector subspace S of $F^\infty(M)$ for which $\text{ad} X_p \circ \text{ad} X_{p-1} \circ \cdots \circ \text{ad} X_1(X_0) \in S$ for any finite set of vector fields $X_0, \ldots, X_p$ in $\mathcal{F}$. In general, $\text{Lie}(\mathcal{F})$ is an infinite-dimensional subspace of $F^\infty(M)$.

For any point x in M, $\text{Lie}_x(\mathcal{F})$ denotes the set of all tangent vectors $V(x)$ with V in $\text{Lie}(\mathcal{F})$. It follows that $\text{Lie}_x(\mathcal{F})$ is a linear subspace of $T_x M$ and hence is finite-dimensional.

Example 4 Let X be a linear field in $\mathbb{R}^n$ given by $x \to Ax$, with A an $n \times n$ matrix, and let Y be a constant field equal to b, $b \in \mathbb{R}^n$. Then $\text{Lie}(\{X, Y\})$ is equal to the vector space spanned by X and $(\text{ad})^k X(Y)$. Each of the fields $(\text{ad})^k X(Y)$ is constant and is equal to $A^k b$. The Cayley-Hamilton theorem says that A^n is linearly dependent on $A^0, A^1, \ldots, A^{n-1}$, and therefore each $(\text{ad})^k X(Y)$, for $k \geq n$, is linearly dependent on the previous powers. Therefore, $\text{Lie}(\{X, Y\})$ is a finite-dimensional Lie algebra. In particular, $\text{Lie}_x\{X, Y\}$ is equal to the linear span of $\{Ax, b, Ab, \ldots, A^{n-1}b\}$. So when $x = 0$, $\text{Lie}_x(\{X, Y\})$ is the linear span of $\{b, Ab, \ldots, A^{n-1}b\}$.

Example 5 Consider now $\mathcal{F} = \{X, Y\}$, with X and Y linear fields in $\mathbb{R}^n$ defined by the matrices A and B. $\text{Lie}(\mathcal{F})$ is contained in the n^2-dimensional space of all linear fields and is therefore finite-dimensional. The set of matrices corresponding to vector fields in $\text{Lie}(\mathcal{F})$ coincides with the smallest vector space of matrices containing A and B that is closed under the commutation. The evaluation of $\text{Lie}(\mathcal{F})$ at a point x may depend on x. In particular, $\text{Lie}(\mathcal{F})$ evaluated at the origin consists of zero alone. At other points, its dimension may be larger. For instance, if

$$A = \begin{pmatrix} 0 & 1 \\ 1 & 0 \end{pmatrix} \quad \text{and} \quad B = \begin{pmatrix} 1 & 0 \\ 0 & -1 \end{pmatrix},$$

then the space of matrices containing A and B that is closed under commutation is equal to all 2×2 matrices with trace equal to zero. It then follows that $\text{Lie}_x(\mathcal{F}) = \mathbb{R}^2$ if $x \neq 0$.

Example 6 Let X and Y denote vector fields in $\mathbb{R}^2$, with $X(x) = \partial/\partial x_1$, and $Y(x) = (\partial/\partial x_1) + f(x_1)(\partial/\partial x_2)$, with $f \in C^\infty(\mathbb{R}^1)$. Then $\text{Lie}(\{X, Y\})$ is spanned by X and $f^{(k)}(x_1)(\partial/\partial x_2)$ for $k = 0, 1, 2, \ldots$. Thus $\text{Lie}(\mathcal{F})$ is infinite-dimensional whenever the derivatives of f span an infinite-dimensional space of functions.

2.3 Involutivity and integral manifolds

As we have already seen, for any family of smooth vector fields $\mathcal{F}$ and for any orbit N of $\mathcal{F}$, each vector field X in $\mathcal{F}$ is tangent to N. In general, there may be other vector fields that are also tangent to N. The following proposition will be used as a stepping stone to more general tangency conditions.

Proposition 3 *Suppose that X and Y are any vector fields on M. Then*

(a) $(d/dt)(\exp t\alpha X) \circ (\exp t\beta Y)(x)|_{t=0} = \alpha X(x) + \beta Y(x)$ *for any $x \in M$,*
and

(b) $(d/dt)((\exp -\sqrt{t}Y) \circ (\exp -\sqrt{t}X) \circ (\exp \sqrt{t}Y) \circ (\exp \sqrt{t}X(x))|_{t=0} = [X, Y](x)$.

We leave the proof of this proposition to the reader. Instead, we return to our main theme, the tangency of vector fields to integral submanifolds, with the following:

Proposition 4 *If X and Y are vector fields that are tangent to a submanifold N, then their Lie bracket is also tangent to N, and so is any linear combination $\alpha X + \beta Y$.*

Proof Assume that X and Y are tangent to N, and let x denote an arbitrary point of N. Because X and Y are tangent to N, it follows that for small t, both $(\exp tX)(y)$ and $(\exp tY)(y)$ remain in N for any point y in some neighborhood of x in N (Proposition 1).

The composite curve in part (b) of Proposition 3 remains in N for small t, and therefore its derivative at $t = 0$ is tangent to N at x. Therefore, $[X, Y](x) \in T_x N$ for all x in N.

The statement about linear combinations of X and Y follows from part (a) in Proposition 3, and our proof is finished.   ∎

It follows by an inductive use of Proposition 4 that if $\mathcal{F}$ is tangent to N, so is Lie($\mathcal{F}$). This observation immediately leads to the following theorem.

Theorem 3 *Suppose that $\mathcal{F}$ is such that $\mathrm{Lie}_x(\mathcal{F}) = T_x M$ for some x in M. Then the orbit $G(x)$ of $\mathcal{F}$ through x is open. If, in addition, $\mathrm{Lie}_y(\mathcal{F}) = T_y M$ for each y in M, and if M is connected, then there is only one orbit of $\mathcal{F}$ equal to M.*

Proof The orbit of $\mathcal{F}$ through each point x in M is a submanifold of M, and therefore its tangent space is of constant dimension at each of its points. In particular, if $\mathrm{Lie}_x(\mathcal{F}) = T_x M$, then the tangent space of $G(x)$ at x must be equal to $T_x M$, since $\mathrm{Lie}_x(\mathcal{F})$ is tangent to $G(x)$ at x. Therefore, $\dim G(x) = \dim M$, and hence $G(x)$ is open in M. If each orbit is open, then each orbit must also be closed. Because M is connected, each orbit must be equal to M, and our proof is finished.   ∎

Definition 3 A family $\mathcal{F}$ is said to be involutive if for any vector fields X and Y in $\mathcal{F}$, $[X, Y](x)$ is contained in the linear span of the elements in $\mathcal{F}(x)$ for each x in M. $\mathcal{F}(x)$ is the evaluation of $\mathcal{F}$ at x equal to $\{V(x) : V \in \mathcal{F}\}$.

Theorem 4 (Frobenius) *Let $\mathcal{F}$ be an involutive family of smooth vector fields for which the dimension of the linear span of $\mathcal{F}(x)$ is constant for all x in M. Then the tangent space at a point x of an orbit of $\mathcal{F}$ is equal to the linear span of $\mathcal{F}(x)$.*

This theorem is usually stated in terms of the distribution spanned by $\mathcal{F}$ and the existence of the maximal integral manifolds of this distribution. The passage to this terminology is easy, because the orbits of $\mathcal{F}$ are the same as the maximal integral manifolds of the distribution defined by $\mathcal{F}$. Because our proof of the theorem of Frobenius begins with the orbit theorem, it will first be necessary to further elaborate the tangency properties of the orbit manifolds.

Recall that $G(\mathcal{F})$ denotes the group (pseudogroup) of diffeomorphisms (local) generated by $\{\exp tX : X \in \mathcal{F}, t \in \mathbb{R}^1\}$. For each $\Phi \in G(\mathcal{F})$, and each $X \in \mathcal{F}$, we shall consider the vector field $\Phi_* \circ X \circ \Phi^{-1}$. (In this notation, Φ_* denotes the tangent map of Φ.)

It is transparent from the proof of the orbit theorem not only that each X in $\mathcal{F}$ is tangent to the orbits of $\mathcal{F}$ but also that $\Phi_* X \circ \Phi^{-1}$ must be tangent for each $\Phi \in G(\mathcal{F})$. In general, the latter family is larger than $\mathrm{Lie}(\mathcal{F})$ (as in Example 17, Chapter 1).

In fact, the following result clarifies the connection between the distribution spanned by $\{\Phi_* \circ X \circ \Phi^{-1} : \Phi \in G(\mathcal{F}), X \in \mathcal{F}\}$ and that of $\mathrm{Lie}(\mathcal{F})$.

Theorem 5 *The linear span of $\{\Phi_* \circ X \circ \Phi^{-1}(x) : X \in \mathcal{F}, \Phi \in G(\mathcal{F})\}$ is equal to $\mathrm{Lie}_x(\mathcal{F})$ for all $x \in M$ if and only if $(\exp tX)_* \circ (Y) \circ (\exp -tX)(x) \in \mathrm{Lie}_x(\mathcal{F})$ for each X in $\mathcal{F}$, each Y in $\mathrm{Lie}(\mathcal{F})$, and all $t \in \mathbb{R}^1$.*

Proof Let $P(x)$ denote the linear span of $\{(\Phi_* \circ X \circ \Phi^{-1})(x) : \Phi \in G(\mathcal{F}), X \in \mathcal{F}\}$. It follows from the construction of the orbit manifold that $P(x)$ is equal to the tangent space of the orbit of $\mathcal{F}$ through x. Because $\mathrm{Lie}(\mathcal{F})$ is tangent to each orbit of $\mathcal{F}$, $\mathrm{Lie}_x(\mathcal{F}) \subset P(x)$.

Let Y be any element of $\mathrm{Lie}(\mathcal{F})$. Then for any X in $\mathcal{F}$, the curve $\sigma_t(s) = (\exp tX)(\exp sY)(\exp -tX)(x)$ is contained in the orbit of $\mathcal{F}$ through x for small s, because Y is tangent to each orbit of $\mathcal{F}$. Therefore, $(d\sigma_t/ds)|_{s=0}$ is contained in the tangent space of the orbit of $\mathcal{F}$ through x. But

$$\left. \frac{d\sigma_t}{ds} \right|_{s=0} = (\exp tX)_* \circ (Y) \circ (\exp -tX)(x).$$

So if $\mathrm{Lie}_x(\mathcal{F}) = P(x)$, it follows that $(\exp tX)_* \circ (Y) \circ (\exp -tX)(x) \in \mathrm{Lie}_x(\mathcal{F})$.

Conversely, if $(\exp tX)_* \circ (Y) \circ (\exp -tX)(x) \in \mathrm{Lie}_x(\mathcal{F})$ for all x in M and all vector fields X in $\mathcal{F}$ and Y in $\mathrm{Lie}(\mathcal{F})$, then the same conclusion applies to arbitrary elements Φ in $G(\mathcal{F})$, because each Φ in $G(\mathcal{F})$ is of the form $\Phi = (\exp t_p X_p) \circ \cdots \circ (\exp t_1 X_1)$, and $(\Phi_* \circ X \circ \Phi^{-1})(x) = (\exp t_p X_p)_* \circ (\exp t_{p-1} X_{p-1})_* \circ \cdots \circ (\exp t_1 X_1)_* \circ (X) \circ (\exp -t_1 X_1) \circ \cdots \circ (\exp -t_p X_p)(x)$ for some vector fields $X_1, \ldots, X_p$ in $\mathcal{F}$. Evidently $(\Phi_* \circ X \circ \Phi^{-1})(x) \in \mathrm{Lie}_x(\mathcal{F})$ when each $(\exp t X_i)_* \circ (Y) \circ (\exp -t X_i)(x)$ belongs to $\mathrm{Lie}_x(\mathcal{F})$ for each $Y \in \mathrm{Lie}(\mathcal{F})$. Hence $P(x) = \mathrm{Lie}_x(\mathcal{F})$, and the proof is now finished.　■

We return now to the proof of Theorem 4:

Proof of Theorem 4 The fact that $\mathcal{F}$ is involutive implies that $\mathrm{Lie}_x(\mathcal{F})$ is equal to the linear span of $\mathcal{F}(x)$. It follows from Theorem 5 that it suffices to show that for each X and Y in $\mathcal{F}$, $(\exp t X)_* \circ (Y) \circ (\exp -t X)(x)$ is contained in the linear span of $\mathcal{F}(x)$ for all t and each x.

Let $X_1, \ldots, X_n$ be any elements of $\mathcal{F}$ such that $X_1(x), \ldots, X_n(x)$ form a maximal number of linearly independent vectors in $\mathcal{F}(x)$. Then $X_1(y), \ldots, X_n(y)$ are also linearly independent in some neighborhood of x. It follows from the constancy of the rank assumption that n is the maximal number of linearly independent vectors in each fiber $\mathcal{F}(y)$. Thus $X_1(y), \ldots, X_n(y)$ form a basis for the linear span of $\mathcal{F}(y)$ for each y in some neighborhood U of x.

For each X in $\mathcal{F}$, $[X, X_j](y)$ is a linear combination of $X_1(y), \ldots, X_n(y)$ for all y in U. Let $f_j^i(y)$ be the functions on U such that

$$[X, X_j](y) = \sum_{i=1}^{n} f_j^i(y) X_i(y) \quad \text{for all } y \text{ in } U.$$

Consider now the curves $\sigma_j(t) = (\exp t X)_* \circ (X_j) \circ (\exp -t X)(x)$. It follows that

$$\frac{d\sigma_j}{dt} = (\exp t X)_* \circ [X, X_j] \circ (\exp -t X)(x).$$

For small t, $(\exp -t X)(x) \in U$, and therefore

$$[X, X_j](\exp -t X)(x) = \sum_{i=1}^{n} f_j^i((\exp -t X)(x)) X_i((\exp -t X)(x)).$$

But then,

$$\frac{d\sigma_j}{dt} = (\exp t X)_* \left(\sum_{i=1}^{n} f_j^i((\exp -t X)(x)) X_i(\exp -t X(x)) \right)$$

$$= \sum_{i=1}^{n} f_j^i((\exp -t X)(x)) \sigma_i(t).$$

Consider now the curve $\gamma(t) = (\gamma_1(t), \ldots, \gamma_n(t))$ given by $\gamma_j(t) = \alpha(\sigma_j(t))$, $i = 1, \ldots, n$, for an element α in $T_x^* M$.

$$\frac{d\gamma_j}{dt} = \alpha\left(\frac{d\sigma_j}{dt}\right) = \sum_{i=1}^{n} f_j^i((\exp -tX)(x))\alpha(\sigma_i(t))$$

$$= \sum_{i=1}^{n} f_j^i((\exp -tX)(x))\gamma_i(t).$$

Therefore $\gamma(t)$ is a solution of a linear system of differential equations in $\mathbb{R}^n$. It follows that $\gamma(0) = 0$ implies that $\gamma(t) = 0$ for all t.

Assuming that α annihilates $\mathcal{F}(x)$, $\gamma(0) = 0$, and therefore α annihilates each vector $\sigma_1(t), \ldots, \sigma_n(t)$. Hence $\sigma_1(t), \ldots, \sigma_n(t)$ remain in $\mathcal{F}(x)$ for all t, for which these curves are defined.

We have proved that for each x in M, there exists an $\varepsilon > 0$ such that $((\exp tX)_* Y(\exp -tX)(x))$ is contained in the linear span of $\mathcal{F}(x)$ for $|t| \leq \varepsilon$. But then it follows that the same must be true for all t, because the set of all t for which the foregoing holds is both open and closed (for a fixed x). Our proof is now finished.   ■

Corollary *Let $\mathcal{F}$ be any family of smooth vector fields such that the dimension of each vector space $\mathrm{Lie}_x(\mathcal{F})$ is constant as x varies over M. Let k denote the dimension of $\mathrm{Lie}_x(\mathcal{F})$. Then for each x in M, the tangent space at x of the orbit of $\mathcal{F}$ through x coincides with $\mathrm{Lie}_x(\mathcal{F})$. Consequently, each orbit of $\mathcal{F}$ is a k-dimensional submanifold of M.*

Proof $\mathcal{F}$ and $\mathrm{Lie}(\mathcal{F})$ have the same orbits, but $\mathrm{Lie}(\mathcal{F})$ satisfies the conditions of Theorem 4. Our statements follow from Theorem 4 applied to $\mathrm{Lie}(\mathcal{F})$.   ■

The class of differential systems for which the rank of the Lie algebra determines the dimension of the orbits seems to combine the local properties of $\mathcal{F}$, expressed through the Lie derivatives of its elements at a given point, with the global properties of the orbit of $\mathcal{F}$ and, as such, does not lend itself to easy characterizations.

R. Hermann was the first to extend the Frobenius class to a more general class having the foregoing property, which he called "locally finitely generated." It was later shown by Lobry that the locally finitely generated class includes the class of real analytic systems, and subsequent studies of the latter gave impetus to geometric control theory. In this respect, the work of Hermann played an important part in the development of the subject. The interested reader may want to go to the original sources listed in the References for further information concerning various attempts to characterize the foregoing class. We shall simply use Example 6 to illustrate the difficulties, and show that the constancy of rank is essential for the conclusions of Theorem 4 to hold in the class of smooth systems.

Example 7 Let $M = \mathbb{R}^2$, and let $\mathcal{F} = \{X, Y\}$, with $X(x_1, x_2) = \partial/\partial x_1$, and $Y(x_1, x_2) = (\partial/\partial x_1) + f(x_1)(\partial/\partial x_2)$. f is a C^∞ function such that $f(x) = 0$ if $x \le 0$, and $f(x) > 0$ if $x > 0$.

It is evident that $((\exp tX)_* \circ Y \circ (\exp -tX))(x)$ is not contained in $\mathrm{Lie}_x(\mathcal{F})$, for $(\exp tX)_*$ is the identity map, and so $((\exp tX)_* \circ Y \circ (\exp -tX))(x) = (\partial/\partial x_1) + f(x_1 - t)(\partial/\partial x_2)$. If $x_1 < 0$, then $\mathrm{Lie}_{(x_1, x_2)}(\mathcal{F})$ is equal to the vector space spanned by $\partial/\partial x_1$. However, if t is such that $x_1 - t_1 > 0$, then $((\exp tX)_* \circ Y \circ (\exp -tX))(x) \notin \mathrm{Lie}_x(\mathcal{F})$.

In Example 6 we showed that $\mathrm{Lie}(\mathcal{F})$ is equal to the linear span of $\partial/\partial x_1$ and $f^{(k)}(x_1)(\partial/\partial x_2)$ for $k = 0, 1, 2, \dots$. $\mathrm{Lie}(\mathcal{F})$ is an involutive family for which $\dim \mathrm{Lie}_x(\mathcal{F})$ is not constant, and therefore it provides a counterexample to the Frobenius theorem when the constancy of rank is dropped.

3 Analytic vector fields and their orbit properties

Rather than proving that the class of real analytic vector fields is locally finitely generated and proceeding along the lines of Lobry and Hermann, it will be easier to proceed directly. We shall begin with the fact that when X and Y are real analytic vector fields, then the curve $\sigma(t) = ((\exp tX)_* \circ Y \circ (\exp -tX))(x)$ is an analytic curve on M for each x in M. σ is therefore represented by its Taylor series $\sum_{k=0}^\infty (t^k/k!)a_k$.

It follows that $d\sigma/dt = ((\exp tX)_*[X, Y](\exp -tX))(x)$, and therefore

$$\frac{d^k \sigma}{dt^k} = (\exp tX)_* \circ ((\mathrm{ad}X)^k(Y)) \circ (\exp -tX)(x).$$

In particular, $(d^k\sigma/dt^k)|_{t=0} = ((\mathrm{ad}X)^k(Y))(x) = a_k$, and hence for small t, $\sigma(t) = \sum_{k=0}^\infty (t^k/k!)((\mathrm{ad}X)^k(Y))(x)$. This formula leads to the following:

Theorem 6 (Hermann–Nagano) *Let M be an analytic manifold, and $\mathcal{F}$ a family of analytic vector fields on M. Then*

(a) *each orbit of $\mathcal{F}$ is an analytic submanifold of M, and*
(b) *if N is an orbit of $\mathcal{F}$, then the tangent space of N at x is given by $\mathrm{Lie}_x(\mathcal{F})$. In particular, the dimension of $\mathrm{Lie}_x(\mathcal{F})$ is constant as x varies over N.*

Proof We shall use the results from Theorem 5, for when X is an element of $\mathcal{F}$, and Y an element of $\mathrm{Lie}(\mathcal{F})$, then each $(\mathrm{ad}X)^k(Y)$ is an element of $\mathrm{Lie}(\mathcal{F})$. Because $\mathrm{Lie}_x(\mathcal{F})$ is a finite-dimensional vector space, it is closed, and therefore $\sum_{k=0}^{\infty}(t^k/k!)((\mathrm{ad}X)^k(Y))(x)$ belongs to $\mathrm{Lie}_x(\mathcal{F})$. Therefore, $((\exp tX)_* \circ Y \circ (\exp -tX))(x)$ belongs to $\mathrm{Lie}_x(\mathcal{F})$ for small t. But then the same must be true for all t, and therefore Theorem 5 is applicable. The proof is now finished. $\blacksquare$

The preceding theorem can also be stated in terms of integrable distributions, as follows: A family $\mathcal{F}$ of vector fields defines a *distribution* on M if $\mathcal{F}(x)$ is a linear subspace of $T_x M$ for each x in M. A distribution $\mathcal{F}$ is said to have the *integral-manifold property* if for each $x \in M$ there exists a submanifold N_x of M that contains x and such that the tangent space of N_x at each point y is equal to $\mathcal{F}(y)$. In this language, Theorem 6 admits the following phrasing:

Let $\mathcal{F}$ be any family of analytic vector fields on M. Then $\mathrm{Lie}(\mathcal{F})$ has the integral-manifold property. In fact, the orbit of $\mathcal{F}$ through x is the maximal integral manifold of $\mathrm{Lie}(\mathcal{F})$ through x.

Example 8 Let $M = \mathbb{R}^n$, and let $\mathcal{F} = \{X, Y\}$, with X a linear field and Y a constant field. Both fields are analytic. If $X(x) = Ax$ and if $Y(x) = a$ for all x in $\mathbb{R}^n$, the orbit of $\mathcal{F}$ through an arbitrary point x in $\mathbb{R}^n$ is given by

$$\{(\exp tA)(x) + c : t \in \mathbb{R}^1 \quad \text{and} \quad c \in C\},$$

with C equal to the linear span of $\{a, Aa, A^2a, \ldots, A^{n-1}a\}$.

Evidently the tangent space at each point x of an orbit of $\mathcal{F}$ is equal to the linear span of Ax and C. Because each $A^j a$ is equal to $(\mathrm{ad})^j X(Y)$, it follows that $\mathrm{Lie}_x(\mathcal{F})$ is equal to the tangent space of the orbit at x. This finding coincides with the conclusions of Theorem 6, because both X and Y are analytic.

The orbit of $\mathcal{F}$ through any point of C is equal to C, because $(\exp tA)(C) \subseteq C$. Therefore, the dimension of C is minimal among the orbits of $\mathcal{F}$. It may happen that the orbits of $\mathcal{F}$ are not of the same dimension when C is a proper subspace of $\mathbb{R}^n$, thus violating the conditions of the Frobenius theorem.

Example 9 Let $M = \mathbb{R}^3$, and $\mathcal{F} = \{X, Y\}$, where X and Y are linear fields given respectively by the matrices

$$A = \begin{pmatrix} 0 & 1 & 0 \\ -1 & 0 & 0 \\ 0 & 0 & 0 \end{pmatrix} \quad \text{and} \quad B = \begin{pmatrix} 0 & 0 & 0 \\ 0 & 0 & 1 \\ 0 & -1 & 0 \end{pmatrix}.$$

Then, as we have already observed, $[X, Y]$ is a linear field and is given by

$$\begin{pmatrix} 0 & 0 & 1 \\ 0 & 0 & 0 \\ -1 & 0 & 0 \end{pmatrix}.$$

In this case, $\text{Lie}(\mathcal{F})$ consists of all linear fields given by 3×3 skew-symmetric matrices.

The orbit of $\mathcal{F}$ through each point x of $\mathbb{R}^3$ is given by the action of $SO_3(R)$ on the point x. It follows that the orbit of $\mathcal{F}$ through x is the sphere centered at the origin with radius equal to $\|x\|$. The sphere reduces to a single point when $x = 0$.

3.1 Lie groups

Lie groups form an important class of analytic manifolds. Their prototype is any finite-dimensional group of linear transformations on a vector space V. By way of introduction, we shall consider such a situation first.

Let V be an n-dimensional vector space. It follows from Example 3 in Chapter 1 that $GL(V)$, the group of all linear transformations on V, is an n^2-dimensional real analytic manifold. Any choice of basis in V induces a linear isomorphism Φ from $GL(V)$ onto the space of $n \times n$ nonsingular matrices. Any such matrix can be identified with a point in $\mathbb{R}^{n^2}$ by, for instance, enumerating the columns of the matrix in some order. Let Ψ denote any such correspondence. Then $\Psi \circ \Phi$ defines a global coordinate chart of $GL(V)$. The coordinates of any composition ST of elements in $GL(V)$ are polynomial expressions of the coordinates of S and T, and the coordinates of S^{-1} are rational functions of the coordinates of S. It therefore follows that both group operations $(S, T) \to ST$ and $S \to S^{-1}$ are analytic mapping from $GL(V) \times GL(V)$ and $GL(V)$, respectively, onto $GL(V)$.

Definition 4 Any group G that is an analytic manifold and that in addition satisfies the property that the operations

$$(g, h) \to g \cdot h \quad \text{and} \quad g \to g^{-1}$$

are analytic is called a Lie group. (G is a complex Lie group if the underlying manifold is complex.)

In particular, $GL(V)$ is a Lie group, called the *general linear group of V*. We shall follow the usual terminology and write $GL_n(\mathbb{R})$ for $GL(\mathbb{R}^n)$.

For any Lie group G, the tangent space at the group identity e plays a special role and is called the Lie algebra of G. The reason for this terminology is that

T_eG, in addition to being a vector space, is a Lie algebra in its own way. The Lie-algebra structure is defined in terms of invariant vector fields, as described in the next section.

3.2 Group translations and invariant vector fields

For any $g \in G$, denote by R_g the right-translation on G by g; that is, $R_g(x) = xg$ for all x in G. L_g will denote the left-translation by g; that is, $L_g(x) = gx$. For each $g \in G$, both R_g and L_g are analytic transformations on G. Because they both take the group identity e into g, it follows that their tangent maps take T_eG onto T_gG.

Definition 5 A vector field X on G is called right-invariant (respectively, left-invariant) if $(R_g)_*X(e) = X(g)$ (respectively, if $(L_g)_*X(e) = X(g)$).

It follows that a vector field that is either right- or left-invariant is determined by its value at the identity. Because both R_g and L_g are analytic, it follows that invariant vector fields are necessarily analytic.

Proposition 5 *Let X and Y be any right-invariant (respectively, left-invariant) vector fields. Then $[X, Y]$ is a right-invariant (respectively left-invariant) vector field.*

Proof For any diffeomorphism Φ on G and any vector field X on G, Φ and X are said to be related if $\Phi_*X(x) = X(\Phi(x))$. It is easy to check that if X and Y are both related to Φ, then so is their Lie bracket $[X, Y]$. Because, in particular, right-invariant (respectively, left-invariant) fields are related by their right (respectively, left) translations, the statement of the proposition follows. ∎

On the basis of this proposition, each of the right- and left-invariant vector fields on G forms a Lie algebra on G. Each of these algebras must be of the same dimension as the tangent space of G at e. In particular, each algebra is finite-dimensional.

There is a natural isomorphism between the right- and the left-invariant fields. It is equal to the tangent map of $\Phi(x) = x^{-1}$. The argument is as follows:

The identity e is a fixed point of Φ, and therefore $\Phi_* : T_eG \to T_eG$. If X_r is a right-invariant field, let $X_\ell = \Phi_*X_r \circ \Phi^{-1}$. The fact that $\Phi R_{g^{-1}}\Phi^{-1} = L_g$ implies that $\Phi_*(R_{g^{-1}})_* = (L_g)_*\Phi_*$. Then

$$(L_g)_*X_\ell(e) = (L_g)_*\Phi_*X_r\Phi^{-1}(e) = \Phi_*(R_{g^{-1}})_*X_r(e)$$
$$= \Phi_*X_r(\Phi^{-1}(g)) = X_\ell(g),$$

and so X_ℓ is a left-invariant field. The correspondence $X_r \to X_\ell$ yields the desired isomorphism.   ∎

Definition 6 For any A and B in $T_e G$, we define their Lie product $[A, B]$ by the following rule: Let X and Y be right-invariant vector fields such that $X(e) = A$ and $Y(e) = B$. Then $[A, B] = [X, Y](e)$. With this product rule, $T_e G$ becomes a Lie algebra.

Remark We could easily have defined $[A, B]$ in terms of left-invariant vector fields, but the corresponding algebra induced on $T_e G$ would then be isomorphic with the one given in terms of right-invariant fields. The Lie product induced by the left-invariant fields is equal to the negative of the Lie product induced by the right-invariant fields.   ∎

Definition 7 $T_e G$ endowed with the foregoing algebra will be called the *Lie algebra of G* and will be denoted by $L(G)$.

It is clear from the definition that $L(G)$ is isomorphic with the algebra of right- and left-invariant vector fields on G.

3.3 Orbits of invariant vector fields

We shall now consider the basic properties of flows generated by right-invariant vector fields on a Lie group G. Left-invariant fields have completely analogous properties.

Let Φ be the flow corresponding to a right-invariant vector field X. Let $e(t) = \Phi(t, e)$ denote the integral curve through the group identity of G. For any g in G, the curve $g(t)$, defined by $g(t) = e(t) \cdot g = R_g(e(t))$, satisfies

$$\frac{d}{dt} g(t) = (R_g)_* \frac{d}{dt} e(t) = (R_g)_* X \circ e(t) = X(e(t) \cdot g) = X(g(t)).$$

So $g(t)$ is the integral curve of X through g, which furthermore satisfies $\Phi(t, g) = \Phi(t, e) \cdot g$ for each $g \in G$ and each $t \in \mathbb{R}^1$. The foregoing equality has several basic implications:

1. The integral curve $e(t)$ is defined for all t in $\mathbb{R}^1$. $H = \{e(t) : t \in \mathbb{R}^1\}$ is an abelian subgroup of G. The proof is very simple:

$$\Phi(t + s, e) = \Phi(t, \Phi(s, e)) = \Phi(s, \Phi(t, e)).$$

The preceding flow relations can also be expressed as $e(t+s) = e(t) \cdot e(s) = e(s)e(t)$, from which it follows that H is an abelian group. That $e(t)$ is defined for all t also follows from the preceding equality: if the curve e is defined for a particular value of t, then e must be defined for $t + \epsilon$, where ϵ is independent of t, since $e(t + \varepsilon) = e(\varepsilon)e(t)$. Therefore $e(t)$ is defined for all t.

2. X is a complete vector field, because $\Phi(t, g) = e(t) \cdot g$, and therefore $\Phi(t, g)$ is defined for all t in $\mathbb{R}^1$.

3. If X were left-invariant, then its flow Φ would satisfy $\Phi(t, g) = g \cdot \Phi(t, e)$. It therefore follows that implications 1 and 2 are also true for the left-invariant vector fields.

4. In particular, if a right-invariant vector field and a left-invariant vector field are equal to each other at the identity, then their orbits through the identity are the same.

Definition 8 For any right-invariant vector field X on G, e^{tA} will denote the integral curve of X through the identity with $A = X(e)$. In terms of this notation, $(\exp t X)(g) = e^{tA} \cdot g$.

The orbit theorem applied to families of right-invariant vector fields on a Lie group G becomes an effective tool for understanding some aspects of the classic theory of Lie groups. By way of illustration, let us note the following facts:

Let $\mathcal{F}$ be any family of right-invariant vector fields on a Lie group G, and let $\Gamma = \{X(e) : X \in \mathcal{F}\}$. Then $\Gamma \subset L(G)$. Let H denote the orbit of $\mathcal{F}$ through the identity. We know from the orbit theorem that H is an analytic submanifold of G. But this orbit is also a subgroup of G generated by $\{e^{tA} : t \in \mathbb{R}^1 : A \in \Gamma\}$.

So H is a connected Lie group, and its Lie algebra is equal to the Lie subalgebra of $L(G)$ generated by Γ. The foregoing reasoning outlines the proof of a classic theorem in Lie groups saying that to any subalgebra $\mathcal{L}$ of $L(G)$ there corresponds exactly one connected Lie subgroup of G. In that context, Lie(Γ) is the given algebra $\mathcal{L}$, and H is its corresponding Lie subgroup of G. But the orbit theorem also shows that given any set Γ such that Lie$(\Gamma) = \mathcal{L}$, then every element of H can be written as a group product of the form

$$e^{t_1 A_1} e^{t_2 A_2} \cdots e^{t_p A_p}$$

for some elements $A_1, , \ldots, A_p$ in Γ.

Our discussion brings us close to another famous theorem in Lie groups: A closed subgroup H of a Lie group G is itself a Lie group. Its proof is along the following lines. Let

$$\Gamma = \{A : e^{tA} \in H \quad \text{for all} \quad t\} .$$

It is known that Γ is a Lie subalgebra of $L(G)$. If we define $\mathcal{F}$ as the family of right-invariant vector fields whose values at the identity are in Γ, then the orbit of $\mathcal{F}$ through the identity is a Lie group. It is equal to H, whenever H is connected.

3.4 $GL_n(R)$ and its subgroups

When G is a Lie subgroup of $GL_n(R)$, the Lie algebraic operations become operations with matrices, and much of the theory of such groups can be explained by elementary means. In order to get a better grasp on the basic theory, and also to set the stage for future applications, assume that G is such a group.

For each matrix A in $M_n(R)$, $\exp tA$ is a curve in $GL_n(R)$ whose tangent vector at the identity is equal to A. Thus $M_n(R)$ is contained in the tangent space at the identity of $GL_n(R)$, and because $GL_n(R) \subset M_n(R)$, they must be equal. The right-translations (respectively, left-translations) in $GL_n(R)$ are matrix multiplications from the right (respectively, left). Hence, right-invariant vector fields are of the form $X(g) = Ag$ for A in $M_n(R)$ and $g \in GL_n(R)$. A is the value of X at the identity. Left-invariant vector fields in $GL_n(R)$ are of the form $X(g) = gA$. Considering the matrix entries (g_{ij}) of g as the coordinates at g, each right-invariant vector field can also be written as

$$X(g) = \sum_{i,j=1}^{n} X_{ij}(g) \frac{\partial}{\partial g_{ij}} \quad \text{and} \quad X_{ij}(g) = \sum_{k=1}^{n} a_{ik}g_{kj} .$$

If $Y(g) = \sum_{i,j=1}^{n} Y_{ij}(g)(\partial/\partial g_{ij})$ is another right-invariant vector field, then $Y_{ij}(g) = \sum_{k=1}^{n} b_{ik}g_{kj}$ for some matrix $B = (b_{ij})$ in $M_n(R)$, and the reader can easily verify that

$$[X, Y](g) = (AB - BA)g = [A, B]g \quad \text{for all } g \text{ in } GL_n(R).$$

Therefore the Lie algebra of $GL_n(R)$ is equal to $M_n(R)$ equipped with the matrix commutator as its Lie bracket.

A Lie subgroup G of $GL_n(R)$ is called a linear group. The Lie algebra of any such group is a subalgebra of $M_n(R)$. For instance, the set of all skew-symmetric matrices is the Lie algebra of the orthogonal group $O_n(R)$. The set of all matrices with traces equal to zero is the Lie algebra of the special linear group $SL_n(R)$.

Any collection of matrices Γ determines a family of right- or left-invariant vector fields $\mathcal{F}_\Gamma$. The orbit of $\mathcal{F}_\Gamma$ through the identity of $GL_n(R)$ is a linear subgroup G consisting of all matrix products of the form

$$e^{t_1 A_1} \cdots e^{t_p A_p}$$

for some matrices $A_1, \ldots, A_p$ in Γ and real numbers $t_1, \ldots, t_p$. The orbit of $\mathcal{F}_\Gamma$ through any other point g in $\mathrm{GL}_n(R)$ is the right or left coset of G (depending on whether $\mathcal{F}_\Gamma$ is right- or left-invariant). $G = \mathrm{GL}_n^+(R)$ if and only if $\mathrm{Lie}(\Gamma) = M_n(R)$.

3.5 Homogeneous spaces

An important class of analytic manifolds with natural analytic vector fields is furnished by the following situation:

Definition 9 Let G be a Lie group, and M an analytic manifold. G is said to act on M if there exists an analytic mapping $\theta : G \times M \to M$ that satisfies

(1) $\theta(g_1 g_2, x) = \theta(g_1, \theta(g_2, x))$ for any g_1, g_2 in G and any x in M, and
(2) $\theta(e, x) = x$ for each x in M.

For each $g \in G$, let $\theta_g : M \to M$ be given by $\theta_g(x) = \theta(g, x)$. The mapping $g \to \theta_g$ is called an action of G on M. Any such action is a group homomorphism from G into the group of analytic diffeomorphisms on M. In particular, if A is an element of $L(G)$, then $\{\theta_{\exp tA}\}$ is a one-parameter group of diffeomorphisms on M. Let X_A denote its infinitesimal generator. The correspondence $A \to X_A$ is a Lie-algebra homomorphism from $L(G)$ into the Lie algebra of analytic vector fields on M. Therefore the family $\mathcal{F} = \{X_A : A \in L(G)\}$ is a finite-dimensional Lie algebra of analytic vector fields on M. We shall refer to its elements as the vector fields subordinated to G. The elements of $\mathcal{F}$ are necessarily complete vector fields on M.

G is said to act *transitively* on M if each orbit $\{\theta(g, x) : g \in G\}$ is equal to M. If G acts transitively on M, then we have the following version of the orbit theorem:

Theorem 7 *Let Γ be an arbitrary subset of $L(G)$, and let $\mathcal{F} = \{X_A : A \in \Gamma\}$. Denote by H the subgroup of G generated by $\{\exp tA : A \in \Gamma, \ t \in \mathbb{R}^1\}$. Then the orbit of $\mathcal{F}$ through each point x in M is given by the action of H on x; that is, $G(\mathcal{F})(x) = \{\theta_h(x) : h \in H\}$. In particular, if Γ generates the Lie algebra of G, then $G(\mathcal{F})(x) = M$ for each $x \in M$.*

Manifolds that admit transitive actions of Lie groups are called *homogeneous spaces*. They are precisely equal to the class of analytic manifolds that can be realized as the quotients of Lie groups. In fact, if G acts transitively on M, then the isotropy group H at a given point x is defined by $H = \{g \in G : \theta_g(x) = x\}$. The mapping $gH \to \theta_g(x)$ realizes M as G/H.

The geometry of homogeneous spaces is intimately linked with the Lie groups that act on them and plays a crucial role in understanding the structure of differential systems defined on these spaces. We shall encounter several situations in which we can take advantage of this formalism.

4 Zero-time orbits of families of vector fields

Orbits of families of vector fields are foliated by the "leaves," called zero-time orbits, that contain important information concerning the time propagations of the reachable sets of control systems, as will be explained in the next chapter. The material in this section will establish the basic properties of zero-time orbits required for the subsequent applications to control systems.

Definition 10 Let $\mathcal{F}$ be any family of C^k ($k \geq 1$) vector fields on a C^k manifold M. The zero-time orbit of $\mathcal{F}$ through x consists of all points y in M such that

$$y = (\exp t_p X_p) \circ \cdots \circ (\exp t_1 X_1)(x)$$

for some integer p, a subset $\{X_1, \ldots, X_p\}$ of $\mathcal{F}$, and some $(t_1, \ldots, t_p)$ in $\mathbb{R}^p$ for which $t_1 + t_2 + \cdots + t_p = 0$.

When each vector field in $\mathcal{F}$ is complete, then the zero-time orbits of $\mathcal{F}$ admit an easy algebraic description, for then the set of all diffeomorphisms in $G(\mathcal{F})$,

$$(\exp t_1 X_1) \circ \cdots \circ (\exp t_p X_p),$$

that satisfy $t_1 + \cdots + t_p = 0$ will be a normal subgroup of $G(\mathcal{F})$. As a subgroup of diff(M), this group acts on points of M, and its orbit through a point x in M is equal to the zero-time orbit of $\mathcal{F}$ through x.

In general, the zero-time orbits induce an equivalence relation on M in which equivalent points belong to the same zero-time orbit. Hence, zero-time orbits partition M, as well as each orbit of $\mathcal{F}$, into equivalence classes.

Example 10 Let $X(x) = a$ and $Y(x) = b$ be two constant, non-collinear vector fields in $\mathbb{R}^n$. Then the zero-time orbit of $\mathcal{F} = \{X, Y\}$ through x will consist of all y in $\mathbb{R}^n$ such that $y = x + t(a - b)$ for some t in $\mathbb{R}^1$.

Each orbit of $\mathcal{F}$ is of the form $x + P$, with P equal to the plane spanned by a and b. The zero-time orbits of $\mathcal{F}$ are the affine translates of the line $\ell = \{t(b - a) : t \in \mathbb{R}^1\}$. It then follows that each orbit of $\mathcal{F}$ is a disjoint union of its zero-time orbits.

Example 11 Suppose that $M = \mathbb{R}^2 - \{(x_1, 0) : x_1 \geq 0\}$. Consider $\mathcal{F} = \{X, Y\}$ on M, with $X(x) = e_1$ and $Y(x) = e_2$ for all x in M. Neither X nor Y is a complete vector field on M. The zero-time orbit through a point x in M is the line $y = x + t(e_1 - e_2)$, provided that $y_1 < 0$ when $y_2 = 0$. But $y_2 = 0$ when $t = -x_2$, and then $y_1 = x_1 + x_2$. So when the initial point x satisfies $x_1 + x_2 \geq 0$, $x + t(e_1 - e_2)$ is not in M for $t = -x_2$. In this case, the zero-time orbit consists of two open and disjoint line segments (Figure 2.2a).

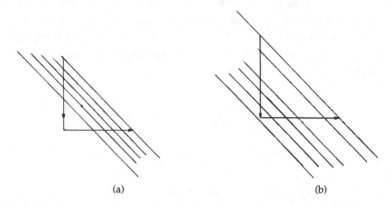

(a)   (b)

Fig. 2.2.

Remark It is easy to modify the preceding example so that a zero-time orbit will consist of infinitely many connected disjoint and open line segments (Figure 2.2b).   ∎

Theorem 8 *Assume that $\mathcal{F}$ is any C^k family of vector fields on a C^k manifold M ($k \geq 1$) that contains at least one complete vector field. Then we have the following:*

(a) *Each zero-time orbit of $\mathcal{F}$ is a connected submanifold of M.*
(b) *The dimension of each zero-time orbit through the points of a fixed orbit N of $\mathcal{F}$ is constant. It is equal to either $(\dim N) - 1$ or $\dim N$.*
(c) *If the dimension of some zero-time orbit is the same as the dimension of the orbit of $\mathcal{F}$ through the same point, then the $\mathcal{F}$ orbit and the zero-time orbit coincide.*

Proof For each integer p, E_p will denote the $(p-1)$-dimensional submanifold $\{t \in \mathbb{R}^p,\ t_1 + \cdots + t_p = 0\}$. We shall consider mappings from E_p into M of the form $y = (\exp t_p X_p) \circ \cdots \circ (\exp t_1 X_1)(x)$ for some elements $X_1, \ldots, X_p$ of $\mathcal{F}$ and $t = (t_1, \ldots, t_p)$ in E_p. Because all the properties of the flows (local flows) required for proof of the orbit theorem apply to the restrictions to manifolds

E_p, the proof of part (a) is quite analogous to the proof of the orbit theorem, and its steps will not be repeated here. So we shall assume that each zero-time orbit N_0 is a manifold.

To show that N_0 is connected, we have to use the completeness assumption. Let X denote a complete vector field in $\mathcal{F}$. Let y and x be any points of N_0 such that $y = (\exp t_p X_p) \circ \cdots \circ (\exp t_1 X_1)(x)$, with $t_1 + \cdots + t_p = 0$.

For each integer k ($1 \leq k < p$), let $\sigma_{k+1}(t) = (\exp -(t + \sum_{i=1}^{k} t_i))X$ $(\exp t X_{k+1})(y_k)$, with $y_k = (\exp t_k X_k) \circ \cdots \circ (\exp t_1 X_1)(x)$. Because X is complete, each curve σ_k is defined on the entire interval $[0, t_{k+1}]$. Let $\sigma_1(t) = (\exp -tX)(\exp t X_1)(x)$. Then $\sigma_1(0) = x$, and $\sigma_1(t_1) = (\exp -t_1 X)(y_1)$. Because $\sigma_1(t)$ belongs to N_0 for all t, $\sigma_1(t_1)$ also belongs to N_0. $\sigma_2(0) = \sigma_1(t_1)$, and therefore $\sigma_2(t_2) \in N_0$. In general,

$$\sigma_k(0) = \left(\exp\left(-\sum_{i=1}^{k} t_i\right)X\right)(y_k) = \sigma_{k-1}(t_k),$$

and $\sigma_{p-1}(t_p) = (\exp(-\sum_{i=1}^{p} t_i)X(\exp t_p X_p))(y_{p-1}) = y$. Therefore y is path-connected to x, and hence N_0 is connected.

To prove (b) and (c), let N_0 denote a fixed zero-time orbit of $\mathcal{F}$ contained in an orbit N of $\mathcal{F}$. If $\dim N_0 = \dim N$, then N_0 is an open submanifold of N. Hence every element X of $\mathcal{F}$ must be tangent to N_0. But then the orbit of $\mathcal{F}$ through a point x of N_0 is contained in N_0. Hence $N = N_0$. This argument proves part (c).

In the remaining case, $\dim N_0 < \dim N$. This condition implies that there are no vector fields of $\mathcal{F}$ that are tangent to N_0, for if one vector field X in $\mathcal{F}$ were tangent to N_0, then every vector field Y in $\mathcal{F}$ would be tangent to N_0, since their difference $X - Y$ is tangent to N_0 (Proposition 3, part (a)). It remains to show that $\dim N_0 = (\dim N) - 1$.

Let $x \in N_0$, and assume that $X_1, \ldots, X_p$ are some elements of $\mathcal{F}$ such that the mapping $F(t_1, \ldots, t_p) = (\exp t_p X_p) \circ \cdots \circ (\exp t_1 X_1)(x)$ satisfies $F(\hat{t}) = x$ for some $\hat{t}$ in $\mathbb{R}^p$, and the rank of the tangent map $TF(\hat{t})$ is equal to the dimension of N.

Let $\alpha = \sum_{i=1}^{p} \hat{t}_i$, and $E = \{t \in \mathbb{R}^p : \sum_{i=1}^{p} t_i = \alpha\}$. E is a hypersurface in $\mathbb{R}^p$. Use F_E to denote the restriction of F to E. The assumption that $\dim N_0 < \dim N$ implies that $TF_E(\hat{t})$ has rank equal to $(\dim N) - 1$, for the rank of $TF_E(\hat{t})$ is equal either to the rank of $TF(\hat{t})$ or to the rank of $TF(\hat{t}) - 1$, since E is of codimension 1 in $\mathbb{R}^p$.

Let X be any complete vector field, and consider the mapping H from E into N_0 defined by $H(t_1, \ldots, t_p) = (\exp -\alpha X) \circ F_E(t_1, \ldots, t_p)$. It follows that $TH(t) = (\exp -\alpha X)_* TF_E(t)$. Because $\exp -\alpha X$ is a diffeomorphism, the rank of $TH(\hat{t})$ is equal to the rank of $TF_E(\hat{t})$. Hence dim $N_0 = (\dim N) - 1$. This argument proves part (b). Our proof is finished, since part (c) has already been proved. ∎

4.1 Zero-time orbits of analytic vector fields

When $\mathcal{F}$ is a family of real analytic vector fields on an analytic manifold M, then each zero-time orbit is an analytic submanifold of M. The tangent space of each orbit is determined by the evaluations of $\text{Lie}(\mathcal{F})$. It turns out that there is a subalgebra $\mathcal{I}(\mathcal{F})$ that determines the tangent spaces of the zero-time orbits, analogous to the way in which $\text{Lie}(\mathcal{F})$ determines the tangent spaces of the orbits of $\mathcal{F}$.

Definition 11 For any family $\mathcal{F}$, let $\mathcal{D}(\mathcal{F})$ denote its derived algebra. That is, every element of $\mathcal{D}(\mathcal{F})$ is a linear combination of iterated Lie brackets of elements of $\mathcal{F}$.

Because for any X in $\text{Lie}(\mathcal{F})$, and any Y in $\mathcal{D}(\mathcal{F})$, $[X, Y] \in \mathcal{D}(\mathcal{F})$, $\mathcal{D}(\mathcal{F})$ is an ideal of $\text{Lie}(\mathcal{F})$.

Definition 12 The zero-time ideal of $\mathcal{F}$ denoted by $\mathcal{I}(\mathcal{F})$ is equal to the linear span of elements in $\mathcal{D}(\mathcal{F})$ and differences $X - Y$ of elements in $\mathcal{F}$. That is, elements of $\mathcal{I}(\mathcal{F})$ are expressions of the form $\alpha_1 D_1 + \cdots + \alpha_n D_n + D$, where $\alpha_1, \ldots, \alpha_n$ are real numbers, where each $D_i = X_i - Y_i$ for some vector fields X_i and Y_i in $\mathcal{F}$, and where D is an element of $\mathcal{D}(\mathcal{F})$.

Evidently $\mathcal{I}(\mathcal{F})$ is a Lie-ideal of $\text{Lie}(\mathcal{F})$, and its relation to $\text{Lie}(\mathcal{F})$ is as described by the following proposition.

Proposition 6 *Let X be any element of $\mathcal{F}$, and let $RX = \{\lambda X : \lambda \in \mathbb{R}^1\}$. Then*

$$\text{Lie}(\mathcal{F}) = RX + \mathcal{I}(\mathcal{F}).$$

Proof Let span($\mathcal{F}$) denote the linear span of $\mathcal{F}$. Then $\text{Lie}(\mathcal{F}) = \text{span}(\mathcal{F}) + \mathcal{D}(\mathcal{F})$. We need to show only that if V is in span($\mathcal{F}$), then $V = \lambda X + W$ for some W in $\mathcal{I}(\mathcal{F})$. But

$$\alpha_1 X_1 + \cdots + \alpha_k X_k = \sum_{i=1}^{k} \alpha_i X + \alpha_1(X_1 - X) + \cdots + \alpha_k(X_k - X),$$

and therefore the proof is complete. ∎

The precise relation of $\mathcal{I}(\mathcal{F})$ to the zero-time orbits of $\mathcal{F}$ is given by the following theorem.

Theorem 9 *Let $\mathcal{F}$ be any family of analytic vector fields on an analytic manifold M. Let N denote an orbit of $\mathcal{F}$, and N_0 a zero-time orbit of $\mathcal{F}$ contained in N. Then we have the following:*

(a) *Each connected component of N_0 is an orbit of $\mathcal{I}(\mathcal{F})$.*

(b) *For each $x \in N_0$, the tangent space of N_0 at x is equal to the evaluation of $\mathcal{I}(\mathcal{F})$ at x.*

(c) *The dimension of $\mathcal{I}_x(\mathcal{F})$ is constant as x varies over N. It is equal either to $(\dim(\mathrm{Lie}_x(\mathcal{F}))) - 1$ or to $\dim(\mathrm{Lie}_x(\mathcal{F}))$.*

(d) *$\dim(\mathrm{Lie}_x(\mathcal{F})) = \dim(\mathcal{I}_x(\mathcal{F}))$ if and only if $X(x)$ belongs to $\mathcal{I}_x(\mathcal{F})$ for some X in $\mathcal{F}$.*

Proof For any vector fields X and Y in $\mathcal{F}$ and any x in N_0, the curve $\sigma(t) = (\exp tX)(\exp -tY)(x)$ is contained in N_0, and its tangent at $t = 0$ is equal to $(X - Y)(x)$. Hence each difference $X - Y$ is tangent to N_0.

We can use Proposition 3, part (b), to show that each Lie bracket $[X, Y]$ is also tangent to N_0. An inductive use of this argument based on the length of iterated Lie brackets shows that every element of $\mathcal{I}(\mathcal{F})$ is also tangent to N_0. Hence, $\mathcal{I}(\mathcal{F})$ is tangent to N_0, and therefore the orbit of $\mathcal{I}(\mathcal{F})$ through each point of N_0 is contained in the connected component of N_0 through this point.

In order to show the equality between these manifolds, we shall denote by E_p the manifold $\{(t_1, \ldots, t_p) : \sum_{i=1}^{p} t_i = 0\}$ and consider the mappings $F : E_p \to N_0$ given by

$$(t_1, \ldots, t_p) \to (\exp t_p X_p) \cdots (\exp t_1 X_1)(x_0)$$

for various choices of p and for vector fields $X_1, \ldots, X_p$ in $\mathcal{F}$. If $T_t E_p$ denotes the tangent space of E_p at $t = (t_1, \ldots, t_p)$, then $TF(T_t E_p) \subset T_{F(t)} N_0$ for each t in E_p. Moreover, for some mapping F of the preceding form,

$$TF(T_t E_p) = T_{F(t)} N_0.$$

But, $TF(T_t E_p)$ is spanned by the vectors of the form

$$-X_p(F(t)) + (\exp t_p X_p)_* \circ \cdots \circ (\exp t_k X_k)_*(X_{k-1}) \circ (\exp -t_k X_k)$$
$$\circ \cdots \circ (\exp -t_p X_p)(F(t)),$$

for $k = 2, \ldots, p$.

Let $W_k = (\exp t_p X_p)_* \circ \cdots \circ (\exp t_k X_k)_*(X_{k-1}) \circ (\exp -t_k X_k) \circ \cdots \circ (\exp -t_p X_p)(F(t))$

An argument based on the formula $(\exp tX)_* Y (\exp -tX) = \sum_{m=0}^{\infty} (t^m/m!)$ $(\mathrm{ad}X)^m(Y)$ shows that $W_k = X_{k-1} + Z$ for some element Z in $\mathcal{D}(\mathcal{F})$. Therefore,

$$-X_p(F(t)) + W_k(F(t)) = (X_{k-1} - X_p)(F(t)) + Z(F(t)).$$

Thus the tangent space of N_0 at $F(t)$ is spanned by $\mathcal{I}_{F(t)}(\mathcal{F})$.

The preceding fact implies that each orbit of $\mathcal{I}(\mathcal{F})$ through a point of N_0 is a connected, open submanifold of N_0. Because orbits of families of vector fields are maximal integral manifolds of the family, it follows that the orbits of $\mathcal{I}(\mathcal{F})$ are equal to the connected components of N_0. We have now proved both (a) and (b).

Part (c) is an immediate consequence of Theorem 8(b) and parts (a) and (b) herein, and part (d) follows from Proposition 6. The proof of this theorem is now finished. ∎

Example 11 provides a simple illustration for the foregoing theorem: Because X and Y commute $\mathcal{I}(\mathcal{F})$ is the vector space spanned by the difference $X - Y$, the integral manifolds of $\mathcal{I}(\mathcal{F})$ are the maximal line segments with slope equal to -1 contained in M.

In the case in which zero-time orbits are connected, then each zero-time orbit is an integral manifold of $\mathcal{I}(\mathcal{F})$, and each orbit of $\mathcal{F}$ is foliated by the integral manifolds of $\mathcal{I}(\mathcal{F})$.

Problem 1 Let $X(x) = Ax$ be the linear vector field in $\mathbb{R}^3$ defined by the matrix

$$A = \begin{pmatrix} 0 & -1 & 0 \\ 1 & 0 & 0 \\ 0 & 0 & 0 \end{pmatrix},$$

and let Y be the constant vector field

$$\begin{pmatrix} 1 \\ 1 \\ 0 \end{pmatrix}.$$

Describe the orbits of $\mathcal{F} = \{X, Y\}$.

Problem 2 Change Y in Problem 1 to the vector field $Y(x) = Bx$, with

$$B = \begin{pmatrix} 1 & 0 & 0 \\ 0 & -1 & 0 \\ 0 & 0 & 1 \end{pmatrix}.$$

Describe all orbits of $\mathcal{F} = \{X, Y\}$.

Problem 3 Show that the zero-time orbits coincide with the orbits for each of the preceding problems.

Problem 4 Let $X(g) = gA$ and $Y(g) = gB$ be left-invariant vector fields on the group of motions of the plane E_2 such that their values at the identity are given by the matrices

$$A = \begin{pmatrix} 0 & 0 & 0 \\ 1 & 0 & 0 \\ 0 & 0 & 0 \end{pmatrix} \quad \text{and} \quad B = \begin{pmatrix} 0 & 0 & 0 \\ 0 & 0 & -1 \\ 0 & 1 & 0 \end{pmatrix}.$$

Calculate the Lie algebra generated by $\mathcal{F} = \{X, Y\}$, and also calculate its zero-time ideal $\mathcal{I}(\mathcal{F})$.

What conclusions can you draw about the reachable sets of the Serret-Frenet system

$$\frac{dg}{dt} = X(g) + k(t)Y(g)$$

by regarding the curvature k as the control function?

Problem 5 Consider the Serret-Frenet system of curves in $\mathbb{R}^3$ as a control system in the group of motions of $\mathbb{R}^3$:

$$\frac{dg}{dt} = X_0(g) + kX_1(g) + \tau X_2(g),$$

with

$$X_0(e) = \begin{pmatrix} 0 & 0 & 0 & 0 \\ 1 & 0 & 0 & 0 \\ 0 & 0 & 0 & 0 \\ 0 & 0 & 0 & 0 \end{pmatrix}, \quad X_1(e) = \begin{pmatrix} 0 & 0 & 0 & 0 \\ 0 & 0 & -1 & 0 \\ 0 & 1 & 0 & 0 \\ 0 & 0 & 0 & 0 \end{pmatrix},$$

$$X_2(e) = \begin{pmatrix} 0 & 0 & 0 & 0 \\ 0 & 0 & 0 & 0 \\ 0 & 0 & 0 & -1 \\ 0 & 0 & 1 & 0 \end{pmatrix}.$$

Calculate the orbits of the induced family

$$\mathcal{F} = \{X_0 + kX_1 + \tau X_2 : k \ge 0, \ \tau \in \mathbb{R}^1\}.$$

Notes and sources

The presentation of the orbit theorem is essentially a paraphrase of the original paper by Sussmann (1973). An independent proof of the orbit theorem can be found in the paper by Stefan (1974). Both of those papers were fundamentally influenced by the accessibility problem of control, and they marked a conceptual shift in approach to investigations of the existence of integral manifolds for distributions. The results of Sussmann and Stefan can be viewed as generalizations of an important paper by Hermann (1962b) dealing with singular foliations, which he called "locally finitely generated." Except for a minor error, later discovered by Stefan (1974), Lobry (1970) showed that distributions spanned by analytic vector fields are locally finitely generated and therefore are integrable, whenever involutive. The same result was obtained by Nagano (1966), independently of developments in control theory.

The notion of a zero-time orbit is from control theory and was largely inspired by a paper by Sussmann and Jurdjevic (1972). The origins of the zero-time ideal can be traced further back, to Kučera (1966).

3

Reachable sets of Lie-determined systems

Because control theory is fundamentally concerned with the forward time evolution of differential systems, reachable sets become the basic objects of mathematical interest associated with any family of vector fields. In contrast to the reachable sets of arbitrary families of vector fields, which show little mathematical structure, the reachable sets of Lie-determined systems have remarkable regularity properties that serve as a basis for geometric control theory. The main purpose of this chapter is to describe these properties. We shall begin with the definition of Lie-determined systems.

Definition 1 A smooth family of vector fields $\mathcal{F}$ is Lie-determined if the tangent space of each point x in an orbit of $\mathcal{F}$ coincides with the evaluation at x of the Lie algebra generated by $\mathcal{F}$.

The essential property of a Lie-determined system $\mathcal{F}$, which accounts for its name, is that its orbit structure is determined by the local properties of the elements of $\mathcal{F}$ and their Lie derivatives. As we saw in Chapter 2, all real analytic systems are Lie-determined, and so are all smooth families of vector fields whose Lie algebra evaluated at each point of the ambient manifold M spans the tangent space of M at that point.

The significance of Lie-determined systems becomes clear when discussing the topological properties of reachable sets. For such systems, each reachable set $\mathcal{A}(x, \leq T)$ has a non-void interior in the topology of the orbit manifold, and the set of interior points grows regularly with T (Theorem 2). Each reachable set $\mathcal{A}(x, T)$ is contained in some zero-time orbit and has a nonempty interior in its topology, provided that the zero-time ideal of $\mathcal{F}$ is Lie-determined (Theorem 3).

There are examples of differential systems for which each reachable set $\mathcal{A}(x, T)$ has an empty interior, but that are nevertheless controllable on M, in the sense that $\mathcal{A}(x) = M$ for each x in M. Such examples can occur on cylinders or toruses, but not on $\mathbb{R}^n$ nor on any manifold M whose fundamental group contains no elements of infinite order (Theorem 4).

The fact that reachable sets need not be closed causes serious theoretical difficulties that persist even after one enlarges the class of allowable control functions. The closure of reachable sets is an issue that will confront us frequently in the development of the subject, particularly in the latter parts of this book dealing with optimality. In the context of this chapter, we shall take a general point of view and consider equivalence classes of systems whose reachable sets have the same closure, with the ultimate objective to describe the invariance properties of each equivalence class. For Lie-determined systems, each equivalence class has a canonical representative that is called the Lie saturate, or the strong Lie saturate, depending on whether $A(x)$ or $A(x, \leq T)$ is used as the criterion for equivalence of systems. In the sense of set inclusion, the saturate is the largest Lie-determined system that is equivalent to a given system. Its basic properties are assembled in Theorem 11. That theorem will be a useful reference for the subsequent proofs of controllability and optimality. Finally, Theorem 12 describes the essential property of the Lie saturates as the criterion of controllability for the original system.

1 Topological properties of reachable sets

1.1 Reachable sets of the form $\mathcal{A}_{\mathcal{F}}(x, \leq T)$

Figures 3.1 and 3.2 illustrate the discontinuities encountered in the propagation of reachable sets $A(x, \leq T)$ that, as will be seen presently, cannot occur in Lie-determined systems.

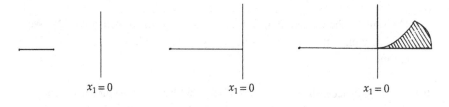

$$x_1 = 0 \qquad\qquad x_1 = 0 \qquad\qquad x_1 = 0$$

Fig. 3.1.

Figure 3.1 depicts the reachable sets of a family $\mathcal{F}$ defined by two vector fields in a plane with $X_1 = \partial/\partial x_1$ and $X_2 = f(x_1, x_2)(\partial/\partial x_2)$, where f is any smooth function equal to zero in the left-hand plane $x_1 \leq 0$, and positive otherwise. The reachable set $\mathcal{A}(x, \leq T)$ is a line segment, provided that $x_1 < 0$. This line segment grows until $T = -x_1$, after which the effect of X_2 combines with that of X_1 to produce a closed set with a nonempty interior.

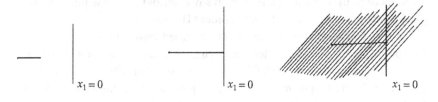

Fig. 3.2.

Figure 3.2 shows more dramatic discontinuities, with the reachable set grow-
ing as a line segment up to some finite instant of time, after which it becomes
a dense subset of the plane, omitting only the open half-line to the left of the
initial point. This pattern can be realized by adjoining to the foregoing family
all scalar multiples of vector fields X_3 and X_4, with $X_3 = g(x)(\partial/\partial x_2)$ and
$X_4 = g(x)(\partial/\partial x_1)$ for a smooth function $g(x)$ that vanishes along $x_2 = 0$, and
is otherwise positive.

We leave it to the reader to construct further examples of reachable sets
exhibiting complicated and discontinuous behaviors. Instead, we return to Lie-
determined systems with theorems that will rule out such patterns and justify
the claim made at the beginning of this chapter. It will be convenient to begin
with the following notion:

Definition 2 A point y in M is *normally accessible* from a point x in M by
$\mathcal{F}$ if there exist elements $X_1, \ldots, X_p$ in $\mathcal{F}$ and $\hat{t} \in \mathbb{R}^p$ with positive coordinates $\hat{t}_1, \ldots, \hat{t}_p$ such that the mapping $F(t_1, \ldots, t_p) = (\exp t_p X_p) \circ \cdots \circ$
$(\exp t_1 X_1)(x)$ as a mapping from $\mathbb{R}^p$ into M satisfies the following:

(a) $F(\hat{t}) = y$.
(b) The rank of the tangent map TF at $\hat{t}$ is equal to the dimension of M. We
    shall conveniently say that y is normally accessible from x by $X_1, \ldots, X_p$
    at time $\sum_{i=1}^{p} \hat{t}_i$.

If y is normally accessible from x, then it follows by the constant-rank theorem
that there exists an open neighborhood U of $\hat{t}$ in $\mathbb{R}^p$ such that $F(U)$ is an open
neighborhood of y. Therefore, any point that is normally accessible from x at
a time less than T must be an interior point of the reachable set $\mathcal{A}_{\mathcal{F}}(x, \leq T)$.
Our first theorem deals with the existence of normally accessible points.

Theorem 1 *Suppose that $\mathcal{F}$ is any family of smooth vector fields on a manifold
M such that* $\text{Lie}_x(\mathcal{F}) = T_x M$ *for some x in M. Then for any neighborhood U
of x and any $T > 0$ there exists a point y in U that is normally accessible from*

x by $\mathcal{F}$ at a time less than T. Consequently, $\mathcal{A}_{\mathcal{F}}(x, \leq T) \cap U$ contains an open set.

Proof Suppose that U is a neighborhood of a point x in M for which $\mathrm{Lie}_x(\mathcal{F}) = T_x M$. Let X_1 be any vector field in $\mathcal{F}$ such that $X_1(x) \neq 0$. If each X in $\mathcal{F}$ were to be equal to zero at p, $\mathrm{Lie}_x(\mathcal{F})$ would be zero-dimensional, which in turn would imply that $M = \{p\}$, and there would be nothing to prove. Let $\varepsilon_1 > 0$ be such that $X_1 \circ (\exp t X_1)(x) \neq 0$ for all $t \in (0, \varepsilon_1)$. For sufficiently small ε_1, $\varepsilon_1 < T$, and $M_1 = \{(\exp t X_1)(x) : t \in (0, \varepsilon_1)\}$ is contained in U. By the constant-rank theorem, M_1 is a one-dimensional submanifold of M, with $t \to (\exp t X_1)(x)$ its coordinate map. If $\dim M = \dim M_1$, the proof is finished, because points in M_1 are normally accessible by the preceding coordinate map. Otherwise, there exists a vector field X_2 in $\mathcal{F}$ such that X_2 is not tangent to M_1.

Let x_1 be a point in M_1 such that $X_2(x_1)$ is not collinear with $X_1(x_1)$. Denote by $\hat{t}_1$ the point in $(0, \varepsilon_1)$ such that $(\exp \hat{t}_1 X_1)(x) = x_1$. Then X_1 and X_2 are not collinear for all points in a neighborhood x_1. Therefore there exist a neighborhood I_1 of $\hat{t}_1$ in $(0, \varepsilon_1)$ and $\varepsilon_2 > 0$ such that the mapping F given by $(\exp t_2 X_2) \circ (\exp t_1 X_1)(x)$ from $I_1 \times (0, \varepsilon_2)$ into M satisfies the following:

(a) $\varepsilon_1 + \varepsilon_2 < T$.
(b) $F(I_1 \times (0, \varepsilon_2)) \subset U$.
(c) The tangent map of F has rank 2 at each point of $I_1 \times (0, \varepsilon_2)$.

Continuing in this way, we obtain a sequence $X_1, X_2, \ldots, X_k$ of vector fields in $\mathcal{F}$ with the property that the tangent map of the mapping $F_k(t_1, \ldots, t_k) = (\exp t_k X_k) \circ \cdots \circ (\exp t_1 X_1)(x)$ has rank k at a point $\hat{t}_1 = (\hat{t}_1, \ldots, \hat{t}_k)$, which further satisfies $\hat{t}_1 > 0, \ldots, \hat{t}_k > 0$, $\hat{t}_1 + \cdots + \hat{t}_k < T$, and $F(\hat{t}_1, \ldots, \hat{t}_k) \in U$. Again by the constant-rank-theorem, the image of a neighborhood of $\hat{t}_1, \ldots, \hat{t}_k$ under the mapping F_k is a k-dimensional submanifold M_k of $M \cap U$.

This procedure stops precisely when each element of $\mathcal{F}$ is tangent to M_k. But then M_k is necessarily open, because $\dim M_k \geq \dim(\mathrm{Lie}_x(\mathcal{F})) = \dim M$, and each point y in M_k is normally accessible from x. The proof is now finished. ∎

Corollary *Suppose that U is an open set in an orbit of a Lie-determined family of vector fields $\mathcal{F}$. Then for any x in U and any $T > 0$, $\mathcal{A}_{\mathcal{F}}(x, \leq T) \cap U$ contains an open set in the orbit topology.*

Proof Let N denote an orbit of $\mathcal{F}$. Because $\mathcal{F}$ is Lie-determined, $T_x N = \mathrm{Lie}_x(\mathcal{F})$ for each x in N. Consider the restrictions of elements of $\mathcal{F}$ to N, and apply Theorem 1. ∎

To describe the topological properties of reachable sets, we shall use int($\mathcal{A}$) to denote the topological interior of a set $\mathcal{A}$, and we shall use cl($\mathcal{A}$) to denote the topological closure.

Theorem 2 *Suppose that $\mathcal{F}$ is a smooth family of vector fields on M such that* $\text{Lie}_x(\mathcal{F}) = T_x M$ *for some $x \in M$. Then, for each $T > 0$ and each $\varepsilon > 0$,*

(a) $\mathcal{A}_{\mathcal{F}}(x, \leq T) \subset \text{cl}(\text{int}(\mathcal{A}_{\mathcal{F}}(x, \leq T)))$,
(b) $\text{int}(\text{cl}(\mathcal{A}_{\mathcal{F}}(x, \leq T))) \subset \text{int}(\mathcal{A}_{\mathcal{F}}(x, \leq T + \varepsilon))$, *and therefore*
(c) $\text{int}(\text{cl}(\mathcal{A}_{\mathcal{F}}(x))) = \text{int}(\mathcal{A}_{\mathcal{F}}(x))$.

Proof In order to show the first inclusion, assume that $y \in \mathcal{A}_{\mathcal{F}}(x, \leq T)$. It suffices to show that for any neighborhood U of y there exists a point z in $\text{int}(\mathcal{A}_{\mathcal{F}}(x, \leq T)) \cap U$. Let $X_1, \ldots, X_p$ be elements in $\mathcal{F}$, such that $y = (\exp \hat{t}_p X_p) \circ \cdots \circ (\exp \hat{t}_1 X_1)(x)$, for non-negative numbers $\hat{t}_1, \ldots, \hat{t}_p$, with $\sum_{i=1}^{p} \hat{t}_i \leq T$.

Then there exist $t \in \mathbb{R}^p$, $\sum_{i=1}^{p} t_i < T$, and an open neighborhood V of x such that $(\exp t_p X_p) \circ \cdots \circ (\exp t_1 X_1)(V) \subset U$. Let $\varepsilon > 0$ be any number such that $\sum_{i=1}^{p} t_i + \varepsilon < T$. By Theorem 1, $\text{int}(\mathcal{A}_{\mathcal{F}}(x, \leq \varepsilon)) \cap V$ is nonempty. If $u \in \text{int}(\mathcal{A}_{\mathcal{F}}(x, \leq \varepsilon)) \cap V$, then $z = (\exp t_p X_p) \circ \cdots \circ (\exp t_1 X_1)(u)$ is an element of $\text{int}(\mathcal{A}_{\mathcal{F}}(x, \leq T)) \cap U$. Therefore, part (a) is proved.

To prove the second inclusion, assume that $y \in \text{int}(\text{cl}(\mathcal{A}_{\mathcal{F}}(x, \leq T)))$ for some $T > 0$, and assume that $\varepsilon > 0$ is given. Let U be an open neighborhood of y contained in $\text{int}(\text{cl}(\mathcal{A}_{\mathcal{F}}(x, \leq T)))$. Applying Theorem 1 to the family $-\mathcal{F}$ shows that $\text{int}(\mathcal{A}_{-\mathcal{F}}(y, \leq \varepsilon)) \cap U$ is not empty. It follows by inclusion (a) that $\text{int}(\mathcal{A}_{\mathcal{F}}(x, \leq T))$ is dense in $\text{cl}(\mathcal{A}_{\mathcal{F}}(x, \leq T))$. Hence there exists a point z contained in $\text{int}(\mathcal{A}_{-\mathcal{F}}(y, \leq \varepsilon)) \cap \text{int}(\mathcal{A}_{\mathcal{F}}(x, \leq T))$.

Let $z = (\exp -t_p X_p) \circ \cdots \circ (\exp -t_1 X_1)(y)$ for some non-negative numbers $t_1, \ldots, t_p$, with $\sum_{i=1}^{p} t_i \leq \varepsilon$, and elements $X_1, \ldots, X_p$ in $\mathcal{F}$. If V denotes an open neighborhood of z contained in $\text{int}(\mathcal{A}_{-\mathcal{F}}(y, \leq \varepsilon)) \cap \text{int}(\mathcal{A}_{\mathcal{F}}(x, \leq T))$, then $(\exp t_1 X_1) \circ \cdots \circ (\exp t_p X_p)(V)$ is a neighborhood of y that is contained in $\text{int}(\mathcal{A}_{\mathcal{F}}(x, \leq T + \varepsilon))$. The proof is now finished. ∎

This theorem, too, has an obvious corollary:

Corollary 1 *Let N denote an orbit of a Lie-determined family of vector fields $\mathcal{F}$. Then for each x in N, for each $T > 0$, and for $\varepsilon > 0$,*

(a) $\mathcal{A}_{\mathcal{F}}(x, \leq T) \subset \text{cl}(\text{int}(\mathcal{A}_{\mathcal{F}}(x, \leq T)))$, *and*
(b) $\text{int}(\text{cl}(\mathcal{A}_{\mathcal{F}}(x, \leq T))) \subset \text{int}(\mathcal{A}_{\mathcal{F}}(x, \leq T + \varepsilon))$, *with the interior and the closure taken in the orbit topology of N.*

Figure 3.3 depicts the growth of the reachable sets as described by Theorem 2. Part (*b*) of Theorem 2 may be false if ε is taken to be equal to zero, as can be seen from the next example.

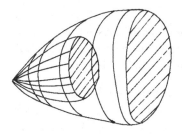 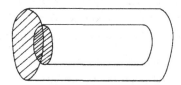

Fig. 3.3.

Example 1 Let M be the unit circle S^1, and X the rotation vector field $X(x_1, x_2) = (-x_2, x_1)$. Let f denote any monotonically increasing function that satisfies $\lim_{u \to -\infty} f(u) = -\frac{1}{2}$ and $\lim_{u \to \infty} f(u) = \frac{1}{2}$. Consider $\mathcal{F} = \{f(u)X : u \in \mathbb{R}^1\}$. Then

$$\mathcal{A}_{\mathcal{F}}(x, \leq 1) = S^1 - \{-x\} \quad \text{for any } x \in S^1.$$

Hence $\text{int}(\text{cl}(\mathcal{A}_{\mathcal{F}}(x, \leq 1))) = S^1$, which is not contained in $\text{int}(\mathcal{A}_{\mathcal{F}}(x, \leq 1))$. The growth of the reachable sets is shown in Figure 3.4.

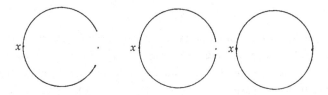

Fig. 3.4.

Corollary 2 *For Lie-determined systems, the reachable sets $\mathcal{A}_{\mathcal{F}}(x)$ cannot be dense in an orbit of $\mathcal{F}$ without being equal to the entire orbit.*

Proof If $\mathcal{A}_{\mathcal{F}}(x)$ is dense in an orbit N, so is its interior. But Corollary 1, part (b), implies that $N \subset \mathcal{A}_{\mathcal{F}}(x)$, because $N = \text{int}(\text{cl}(\mathcal{A}_{\mathcal{F}}(x)))$ is contained in $\text{int}(\mathcal{A}_{\mathcal{F}}(x))$. ∎

1.2 Reachable sets of the form $\mathcal{A}_{\mathcal{F}}\,(x,T)$

The topological properties of sets reachable in exactly T units of time are best analyzed in terms of the topology of zero-time orbits, as described by the following example.

Example 2 Suppose that $M = \mathbb{R}^2$ and that X and Y are the constant vector fields $X(x, y) = e_1$ and $Y(x, y) = e_2$. Because X and Y commute, the zero-time ideal $\mathcal{I}(\mathcal{F})$ of $\mathcal{F} = \{X, Y\}$ consists of the scalar multiples of the difference $X - Y$.

Each zero-time orbit is equal to the line with slope equal to -1. For each initial point (x, y), the reachable set at time T is the line segment $\{(x + t_1, y + t_2) : t_1 \geq 0,\ t_2 \geq 0,$ and $t_1 + t_2 = T\}$. Because the slope of this line segment is -1, $\mathcal{A}_{\mathcal{F}}((x, y), T)$ has a nonempty interior in the zero-time orbit of $\mathcal{F}$ that passes through a point of $\mathcal{A}_{\mathcal{F}}((x, y), T)$. Moreover, the entire reachable set is contained in the same zero-time orbit (Figure 3.5)

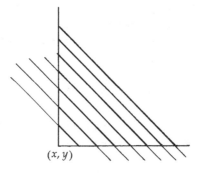

(x, y)

Fig. 3.5.

In many ways, the situation described by the previous example extends to arbitrary families of vector fields. To begin with, if $\mathcal{A}_{\mathcal{F}}(x, T)$ intersects a zero-time orbit N of $\mathcal{F}$, then it must be contained in it. The argument is simple:

Suppose that $y = (\exp t_p X_p) \circ \cdots \circ (\exp t_1 X_1)(x)$ is a point that belongs to some zero-time orbit N of $\mathcal{F}$. If z is any point in $\mathcal{A}_{\mathcal{F}}(x, T)$, then $z = (\exp s_m Y_m) \circ \cdots \circ (\exp s_1 Y_1)(x)$ for some elements $Y_1, \ldots, Y_m$ in $\mathcal{F}$ and nonnegative numbers $s_1, \ldots, s_m$, with $\sum_{i=1}^{m} s_i = T$. But then

$$z = (\exp s_m Y_m) \circ \cdots \circ (\exp s_1 Y_1) \circ (\exp -t_1 X_1) \circ \cdots \circ (\exp -t_p X_p)(y).$$

Hence z is an element of the zero-time orbit of $\mathcal{F}$ through y. Thus, $\mathcal{A}_{\mathcal{F}}(x, T) \subset N$.

Definition 3 Suppose that $\mathcal{F}$ contains at least one complete vector field X. If N is a zero-time orbit of $\mathcal{F}$, then $N_t = (\exp t X)(N)$ is called the t-translate of N. It follows that N_t is a zero-time orbit of $\mathcal{F}$ independent of the field X by which N is translated.

Evidently each reachable set $\mathcal{A}_\mathcal{F}(x, T)$ is contained in the T-translate of the zero-time orbit of $\mathcal{F}$ through x.

Theorem 3 *Suppose that $\mathcal{F}$ is a family of vector fields such that both $\mathcal{F}$ and its zero-time ideal $\mathcal{I}(\mathcal{F})$ are Lie-determined. In addition, assume that $\mathcal{F}$ contains at least one complete vector field. Then*

(a) *for each $x \in M$ and $T > 0$, $\mathcal{A}_\mathcal{F}(x, T)$ is a connected subset of the T-translate N_T of the zero-time orbit of $\mathcal{F}$ through x, and*

(b) *$\mathcal{A}_\mathcal{F}(x, T)$ has a nonempty interior in the manifold topology of N_T. Moreover, the set of interior points is dense in $\mathcal{A}_\mathcal{F}(x, T)$.*

Proof Let X denote a complete vector field in $\mathcal{F}$. Part (a) will be proved by constructing a continuous curve $\sigma(t)$ in the interval $[0, T]$ that lies entirely in $\mathcal{A}_\mathcal{F}(x, T)$ for all t and that connects $(\exp TX)(x)$ to an arbitrary point y of $\mathcal{A}_\mathcal{F}(x, T)$.

Let $y = (\exp t_p X_p) \circ \cdots \circ (\exp t_1 X_1)(x)$ for some positive numbers $t_1, \ldots, t_p$, with $\sum_{i=1}^p t_i = T$, and some elements $X_1, \ldots, X_p$ in $\mathcal{F}$. The curve $\sigma(t) = (\exp(T - t))(X) \circ (\exp tX_1)(x)$ is contained in $\mathcal{A}_\mathcal{F}(x, T)$, for each t in the interval $[0, t_1]$, and connects $(\exp TX)(x)$ to $(\exp(T - t_1))(X) \circ (\exp t_1 X_1)(x)$. Extend σ inductively to the interval $[\sum_{i=1}^{k-1} t_i, \sum_{i=1}^k t_i]$ by the formula

$$\sigma(t) = (\exp(T - t))(X) \circ \left(\exp\left(t - \sum_{i=1}^{k-1} t_i\right)\right)(X_k)$$

$$\circ (\exp t_{k-1} X_{k-1}) \circ \cdots \circ (\exp t_1 X_1)(x).$$

$\sigma(t)$ is contained in $\mathcal{A}_\mathcal{F}(x, T)$ for each t, and it connects

$$\left(\exp\left(T - \sum_{i=1}^{k-1} t_i\right)\right)(X) \circ (\exp t_{k-1} X_{k-1}) \circ \cdots \circ (\exp t_1 X_1)(x)$$

to

$$\left(\exp\left(T - \sum_{i=1}^k t_i\right)\right)(X) \circ (\exp t_k X_k) \circ \cdots \circ (\exp t_1 X_1)(x).$$

When $k = p$, $T - \sum_{i=1}^{p} t_i = 0$, and therefore $\sigma(\sum_{i=1}^{p} t_i) = y$. Thus $\mathcal{A}_{\mathcal{F}}(x, T)$ is connected.

Part (b) will be proved by using the following intermediate result: Let M_x denote the orbit of $\mathcal{F}$ through x. For any neighborhood U of x, in the orbit topology of M_x, and for any $T > 0$, there exists $\hat{T} > 0$, $\hat{T} < T$ such that $\mathcal{A}_{\mathcal{F}}(x, \hat{T})$ has a nonempty interior in $N_{\hat{T}} \cap U$. The proof of this intermediate result splits into two separate arguments, depending on whether $\dim N_T = \dim(\text{Lie}_x(\mathcal{F}))$ or $\dim N_T = (\dim(\text{Lie}_x(\mathcal{F}))) - 1$.

Let $k = \dim(\text{Lie}_x(\mathcal{F}))$, and assume first that $\dim N_T = k - 1$. Then, according to Theorem 1 and its corollary, for any $T > 0$ and any neighborhood U of x, there exist vector fields $X_1, \ldots, X_k$ in $\mathcal{F}$, and positive numbers $\hat{t}_1, \ldots, \hat{t}_k$, with $\sum_{i=1}^{k} \hat{t}_i < T$, such that the mapping $F(t_1, \ldots, t_k) = (\exp t_k X_k) \circ \cdots \circ (\exp t_1 X_1)(x)$ has rank k at $\hat{t}$, and $F(\hat{t})$ belongs to U. Therefore F is a diffeomorphism from an open neighborhood of $\hat{t}$ in $\mathbb{R}^k$ onto an open neighborhood of $F(\hat{t})$ in $M_x \cap U$ in the orbit topology of M_x.

Let $\hat{T} = \sum_{i=1}^{k} \hat{t}_i$, and $E^k = \{t \in \mathbb{R}^k : \sum_{i=1}^{k} t_i = \hat{T}\}$. Then E^k is a submanifold of $\mathbb{R}^k$ of codimension 1, and F restricted to E^k has rank $k - 1$ in a neighborhood of $t = \hat{t}$. Therefore there is a neighborhood of $\hat{t}$ in E^k whose image under F is a $(k - 1)$-dimensional submanifold of $N_{\hat{T}} \cap U$. If this neighborhood is further restricted to points of E^k with positive coordinates, then its image under F is contained in $\mathcal{A}_{\mathcal{F}}(x, \hat{T})$. Therefore, $\mathcal{A}_{\mathcal{F}}(x, \hat{T})$ has a nonempty interior in $N_{\hat{T}} \cap U$.

In the remaining case, $\dim N_T = k$. Then every zero-time orbit in M_x coincides with M_x. Therefore, $X(x) \in \mathcal{I}_x(\mathcal{F})$ for each x in M_x and for any vector field X in $\mathcal{F}$, because $\mathcal{I}(\mathcal{F})$ is Lie-determined.

It will be convenient to consider momentarily the extended vector fields $\tilde{X}(x, \tau) = X(x) \oplus \partial/\partial\tau$ in the time-extended manifold $\tilde{M} = M \times \mathbb{R}^1$. Each vector field X in $\mathcal{F}$ extends to $\tilde{X}$ and defines the extended family $\tilde{\mathcal{F}}$. We shall presently show that $\dim(\text{Lie}_{(x,0)}(\tilde{\mathcal{F}})) = (\dim(\text{Lie}_x(\mathcal{F}))) + 1$.

It follows that $[V_1 \oplus \partial/\partial\tau, V_2 \oplus \partial/\partial\tau] = [V_1, V_2] \oplus 0$ for any vector fields V_1 and V_2 on M. Therefore, for any Y in the derived algebra $\mathcal{D}(\mathcal{F})$, $Y \oplus 0$ belongs to $\text{Lie}(\tilde{\mathcal{F}})$. Because each element of the zero-time ideal $\mathcal{I}(\mathcal{F})$ is of the form $\sum_{i=1}^{p} \alpha_i X_i + Y$ for some numbers $\alpha_1, \ldots, \alpha_p$ (with $\sum_{i=1}^{p} \alpha_i = 0$), for some elements $X_1, \ldots, X_p$ in $\mathcal{F}$, and for an element Y in the derived algebra $\mathcal{D}(\mathcal{F})$, it follows that $X(x) = \sum_{i=1}^{p} \alpha_i X_i(x) + Y(x)$. Then

$$\sum_{i=1}^{p} \alpha_i \tilde{X}_i(x, 0) + (Y \oplus 0)(x, 0) = X(x) \oplus 0.$$

Hence $X(x) \oplus 0 \in \text{Lie}_{(x,0)}(\tilde{\mathcal{F}})$ for any X in $\mathcal{F}$. But then $0 \oplus \partial/\partial t = X(x) \oplus (\partial/\partial t) - X \oplus 0$ belongs to $\text{Lie}_{(x,0)}(\tilde{\mathcal{F}})$. So $\text{Lie}_{(x,0)}(\tilde{\mathcal{F}})$ is spanned by $\text{Lie}_x(\mathcal{F}) \oplus 0$ and $0 \oplus \partial/\partial t$, and therefore $\dim(\text{Lie}_{(x,0)}(\tilde{\mathcal{F}})) = (\dim(\text{Lie}_x(\mathcal{F}))) + 1$.

According to Theorem 1, there exist $k+1$ vector fields $\tilde{X}_1, \ldots, \tilde{X}_{k+1}$ in $\tilde{\mathcal{F}}$, and positive numbers $\hat{t}_1, \ldots, \hat{t}_{k+1}$, with $\sum_{i=1}^{k+1} \hat{t}_i < T$ ($T > 0$ is an a priori given number), such that the mapping F that takes $(t_1, \ldots, t_{k+1})$ into $(\exp t_{k+1} \tilde{X}_{k+1}) \circ \cdots \circ (\exp t_1 \tilde{X}_1)(x, 0)$ has rank $k+1$ at $\hat{t} = (\hat{t}_1, \ldots, \hat{t}_{k+1})$. If F denotes the foregoing mapping, then $F(t) = ((\exp t_{k+1} X_{k+1}) \circ \cdots \circ (\exp t_1 X_1)(x), \sum_{i=1}^{k+1} t_i)$.

Let V be a neighborhood of $\hat{t}$ in $\mathbb{R}^{k+1}$ such that $F(V)$ is an open set in $\tilde{M}$. But then $\{(\exp t_{k+1} X_{k+1}) \circ \cdots \circ (\exp t_1 X_1)(x) : \sum_{i=1}^{k+1} t_i = \sum_{i=1}^{k+1} \hat{t}_i = \hat{T}\}$ is an open subset of M, and the proof of our intermediate result is finished.

The proof of (b) goes as follows: Let $T > 0$ be given. If $\hat{T} < T$ is such that $\mathcal{A}_{\mathcal{F}}(x, \hat{T})$ has an open set V in $N_{\hat{T}}$, then $((\exp(T - \hat{T}))X)(V)$ is an open subset of $\mathcal{A}_{\mathcal{F}}(x, T)$ in N_T. Therefore, $\mathcal{A}_{\mathcal{F}}(x, T)$ has a nonempty interior in N_T for any $T > 0$.

It remains to show that for any y in $\mathcal{A}_{\mathcal{F}}(x, T)$ and any neighborhood V of y in N_T there exists a point z in the interior of $\mathcal{A}_{\mathcal{F}}(x, T)$ contained in V. If $\dim(\text{Lie}_x(\mathcal{F})) = \dim N_T$, then V is an open subset of y in M_x. Let $y = (\exp t_p X_p) \circ \cdots \circ (\exp t_1 X_1)(x)$, with $\sum_{i=1}^{p} t_i = T$, and denote $\Phi_\varepsilon = (\exp t_p X_p) \circ \cdots \circ (\exp(t_1 - \varepsilon)X_1)$. For sufficiently small $\varepsilon > 0$, $\Phi_\varepsilon(x) \in V$, and $\Phi_\varepsilon^{-1}(V)$ is an open neighborhood of x. Choose a sufficiently small neighborhood U of x such that $(\exp t X_1)(U) \subset \Phi_\varepsilon^{-1}(V)$ for all t, $0 \leq t \leq \varepsilon$.

According to the intermediate result, there exists ε_1 ($\varepsilon_1 < \varepsilon$) such that $\mathcal{A}_{\mathcal{F}}(x, \varepsilon_1)$ has a nonempty interior in U. If U_0 denotes an open subset of $\text{int}(\mathcal{A}_{\mathcal{F}}(x, \varepsilon_1)) \cap U$, then $((\exp(\varepsilon - \varepsilon_1))(X_1))(U_0)$ is an open subset of $\mathcal{A}_{\mathcal{F}}(x, \varepsilon) \cap \Phi_\varepsilon^{-1}(V)$. Hence $\Phi_\varepsilon \circ ((\exp(\varepsilon - \varepsilon_1))(X_1))(U_0)$ is an open subset of $\mathcal{A}_{\mathcal{F}}(x, T) \cap V$, and therefore the interior is dense in $\mathcal{A}_{\mathcal{F}}(x, T)$.

Finally, assume that $\dim N_T = k - 1$. Then $V_0 = (\exp -t_1 X_1) \circ \cdots \circ (\exp -t_p X_p)(V)$ is a neighborhood of x in the zero-time orbit N_0 of $\mathcal{F}$ through x. Because each vector field X is transversal to N_0, it follows that $U = \{(\exp \varepsilon X_1)(z) : |\varepsilon| < \varepsilon_0, z \in V_0\}$ is a neighborhood of x for some $\varepsilon_0 > 0$. By using the intermediate result again, we get an $\varepsilon_1 (\varepsilon_1 < \varepsilon_0)$ such that $\mathcal{A}_{\mathcal{F}}(x, \varepsilon_1)$ has a nonempty interior in $N_{\varepsilon_1} \cap U$. Let U_0 be an open set in $N_{\varepsilon_1} \cap U$ contained in $\mathcal{A}_{\mathcal{F}}(x, \varepsilon_1)$. Then $(\exp t_p X_p) \circ \cdots \circ ((\exp(t_1 - \varepsilon_1))(X_1))(U_0)$ is an open set in $V \cap \mathcal{A}_{\mathcal{F}}(x, T)$. The proof is now finished. ∎

Example 3 Suppose that $M = \mathbb{R}^2$, $X = e_1$, and $Y = e_2$ are constant vector fields, and $\mathcal{F} = \{X, -X, Y\}$. Because $2X = X - (-X)$, it follows that X belongs to the zero-time ideal $\mathcal{I}(\mathcal{F})$. Hence, $\text{Lie}_x(\mathcal{F}) = \mathcal{I}_x(\mathcal{F})$ at each point x

in $\mathbb{R}^2$. According to the preceding theory, each zero-time orbit is equal to $\mathbb{R}^2$, and $\mathcal{A}_{\mathcal{F}}(x, T)$ has a nonempty interior in $\mathbb{R}^2$ for each $x \in \mathbb{R}^2$ and $T > 0$.

Indeed, $\mathcal{A}_{\mathcal{F}}(x, T)$ consists of all points in $\mathbb{R}^2$ of the form $x(T) = x + (t_1 - t_2)e_1 + t_3 e_2$, with $t_1 \geq 0$, $t_2 \geq 0$, $t_3 \geq 0$, and $t_1 + t_2 + t_3 = T$. This set is the closed triangle, as shown in Figure 3.6.

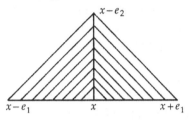

Fig. 3.6.

The preceding example is a special case of a situation in which the zero vector field belongs to the convex hull $\mathrm{co}(\mathcal{F})$ of $\mathcal{F}$. In such situations, $\mathcal{A}_{\mathrm{co}(\mathcal{F})}(x, \leq T) = \mathcal{A}_{\mathrm{co}(\mathcal{F})}(x, T)$, since the constant trajectory $x(t) = x$ is a trajectory of $\mathrm{co}(\mathcal{F})$. It will be shown in the next section that $\mathrm{cl}(\mathcal{A}_{\mathcal{F}}(x, T)) = \mathrm{cl}(\mathcal{A}_{\mathrm{co}(\mathcal{F})}(x, T))$, and therefore $\mathrm{cl}(\mathcal{A}_{\mathcal{F}}(x, \leq T)) = \mathrm{cl}(\mathcal{A}_{\mathcal{F}}(x, T))$. Of course, in this situation, the zero-time ideal $\mathcal{I}(\mathcal{F})$ is equal to $\mathrm{Lie}(\mathcal{F})$, because every element of $\mathcal{F}$ can be written as a linear combination of the differences of elements of $\mathcal{F}$.

It is interesting to note that $\mathrm{cl}(\mathcal{A}_{\mathcal{F}}(x, T)) = \mathrm{cl}(\mathcal{A}_{\mathcal{F}}(x, \leq T))$ may hold even when the convex hull of $\mathcal{F}$ does not contain any elements for which x is an equilibrium point, as can be seen from the following example.

Example 4 Let $\mathcal{F}$ consist of two linear vector fields X and Y in $\mathbb{R}^2$, described respectively by the matrices

$$A = \begin{pmatrix} 0 & 1 \\ 1 & 0 \end{pmatrix} \quad \text{and} \quad B = \begin{pmatrix} -1 & 0 \\ 0 & -2 \end{pmatrix}.$$

The integral curves of X are the hyperbolas $x_2^2 - x_1^2 = $ constant, and the integral curves of Y are the parabolas $x_2 = C x_1^2$.

These two families of curves are tangent to each other along the lines $x_1 = \pm\sqrt{2}x_2$. If x is any point on these lines, then $\mathrm{cl}(\mathcal{A}_{\mathcal{F}}(x, T)) = \mathrm{cl}(\mathcal{A}_{\mathcal{F}}(x, \leq T))$ for any $T > 0$, since there are closed trajectories of arbitrarily short time duration emanating from x, as shown in Figure 3.7.

The closed trajectories depicted in Figure 3.7 can be called cycles. In the foregoing example they can be of arbitrary duration. Their use suggests a natural question:

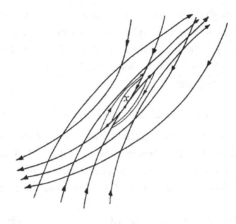

Fig. 3.7.

Problem 1 Is the existence of cycles necessary for $\mathrm{cl}(\mathcal{A}_{\mathcal{F}}(x, \leq T)) = \mathrm{cl}(\mathcal{A}_{\mathcal{F}}(x, T))$ to hold?

2 The closure of the reachable sets and its invariants

We shall now shift attention to the closure of reachable sets, in order to describe prolongations of a given family of vector fields that will leave the closure of its reachable sets invariant. Ultimately we shall be interested in the notion of a "completed system," and the results from this section will set the stage for its discussion in the next section.

Looking ahead toward further applications of this formalism, it seems appropriate to make additional comments that may help the reader gain a suitable perspective on the present point of view. The reader for whom the example of a single vector field with dense orbits casts serious doubt on the feasibility of using the closure of reachable sets as a criterion for classification of systems may be somewhat reassured to know that our subsequent investigations will be restricted to Lie-determined systems and that the classifications will be restricted to systems having the same Lie algebra. Apart from avoiding the obvious difficulties in the example with dense orbits, we shall further demonstrate the merits of this approach with several substantial theorems concerning controllability and optimality.

It may also be appropriate to remark that the system prolongations based on the invariance properties of the closure fall outside of the usual philosophical framework in which the system is regarded as a black box into which control functions are fed as inputs. From the black-box point of view, enlarging the class of admissible controls is the only means for obtaining the closure of reachable sets, although the formalism presented here allows for the possibility of changing the black box as well. This extra degree of freedom leads to new results by allowing systematic exploitation of the effects of large controls that act over small intervals. But all these details will become clear later on, and we are getting ahead of ourselves.

2.1 Closure and convexification of families of vector fields

Having taken the closure of the reachable sets as the basic invariant object, it becomes natural to pass to topologically closed families of vector fields. Keeping further applications in mind, it will be sufficient to stay in the class of smooth vector fields, although some of the theorems that will follow will be valid for more general classes.

Recall that $F^\infty(M)$ denotes the space of all C^∞ vector fields on M. $F^\infty(M)$ will be regarded as a topological vector space topologized by the C^∞ convergence on compact subsets of M. This means that a sequence of vector fields $\{X^{(m)}\}$ converges to X if for any chart (ϕ, U) and any compact set $K \subset U$, $\partial^k X_i^{(m)}/(\partial x_1^{i_1} \cdots \partial x_n^{i_n})$ converges uniformly to $\partial^k X_i/(\partial x_1^{i_1} \cdots \partial x_n^{i_n})$ on $\phi(K)$ in $\mathbb{R}^n$ for each integer k and for any multi-index $i_1 + i_2 + \cdots + i_n = k$. In this notation, $(X_1^{(m)}, \ldots, X_n^{(m)})$ are the coordinates of each vector field $X^{(m)}$, and $(X_1, \ldots, X_n)$ are the coordinates of the limiting vector field X.

The prolongation to topologically closed families of vector fields will be based on the following theorem.

Theorem 4 *Suppose that $\{X_m\}$ is a convergent sequence of smooth vector fields on M converging to a vector field X. Then we have the following:*

(a) *X is a smooth vector field.*
(b) *Let $x(t)$ denote an integral curve of X defined on an interval $[0, T]$. There exists a neighborhood U of $x(0)$ such that every integral curve of X that originates in U is defined for all t in $[0, T]$.*
(c) *Furthermore, for any sequence $\{y_m\}$ that converges to $x(0)$, the integral curves x_m of X_m, with $x_m(0) = y_m$, are defined on $[0, T]$ for sufficiently large m, and they converge uniformly to $x(t)$ on $[0, T]$ as $m \to \infty$.*

Because Theorem 4 is a paraphrase of well-known facts from the theory of ordinary differential equations, its proof will be omitted.

Example 5 Let $M = \mathbb{R}^n$, and let X_m be a sequence of linear vector fields on $\mathbb{R}^n$. If $X_m(x) = A_m x$ for an $n \times n$ matrix A_m, and if $X(x) = Ax$, then $X_m \to X$ if and only if the matrices A_m converge to the matrix A. It follows from Theorem 4 that the curves $e^{A_m t} y_m$ converge uniformly to $e^{At} x$ on each interval $[0, T]$ as y_m converge to x.

The next example shows that a limit of complete vector fields need not be complete.

Example 6 Let $M = \mathbb{R}^1$, and $X(x) = x^2$. X is not a complete vector field. Its local flow is given by

$$\Phi(x, t) = \frac{x}{1 - xt}.$$

Let f_n be a C^∞ function that is equal to 1 in the interval $[-n, n]$, and 0 outside of the interval $[-(n+1), n+1]$. Let $X_n = f_n(x)x^2$. Then each X_n is a complete vector field, and $\lim_{n \to \infty} X_n = X$.

Theorem 5 *Let* $\mathrm{cl}(\mathcal{F})$ *denote the topological closure of a smooth family of vector fields* $\mathcal{F}$. *Then*

$$\mathcal{A}_{\mathrm{cl}(\mathcal{F})}(x, T) \subseteq \mathrm{cl}(\mathcal{A}_{\mathcal{F}}(x, T))$$

for each x in M and each $T > 0$.

Proof Let $y = (\exp t_p X_p) \circ \cdots \circ (\exp t_1 X_1)(x)$ for $\{X_1, \ldots, X_p\} \subset \mathrm{cl}(\mathcal{F})$ and some non-negative numbers $t_1, \ldots, t_p$ that satisfy $t_1 + t_2 + \cdots + t_p = T$. It is required to show that $y \in \mathrm{cl}(\mathcal{A}_{\mathcal{F}}(x, T))$.

For each index i, $1 \le i \le p$, let $\{Y_m^{(i)}\} \subset \mathcal{F}$ be any sequence such that $\lim_{m \to \infty} Y_m^{(i)} = X_i$. Then, according to Theorem 4, $(\exp t X_1)(x) = \lim_{m \to \infty} ((\exp t Y_m^{(1)})(x))$ uniformly in the interval $[0, t_1]$. Hence $(\exp t X_1)(x)$ belongs to $\mathrm{cl}(\mathcal{A}_{\mathcal{F}}(x, t_1))$. A repetition of this argument shows that $\lim_{m \to \infty} ((\exp t Y_m^{(2)}) \circ (\exp t_1 Y_m^{(1)})(x))$ converges uniformly to $(\exp t X_2) \circ (\exp t_1 X_1)(x)$ on the interval $[0, t_2]$. Thus, $(\exp t_2 X_2) \circ (\exp t_1 X_1)(x)$ belongs to $\mathrm{cl}(\mathcal{A}_{\mathcal{F}}(x, t_1 + t_2))$. Further repetitions of the same argument will show that $y \in \mathrm{cl}(\mathcal{A}_{\mathcal{F}}(x, t_1 + t_2 + \cdots + t_p))$. End of the proof. ∎

Example 7 Let X be a complete vector field on $\mathbb{R}^n$ such that $X(0) = 0$. Assume that the origin is a local attractor for X; that is, assume that $\lim_{t \to \infty} (\exp t X(x)) = 0$ for all x in some neighborhood of the origin. Let $\mathcal{F} = \{\lambda X : \lambda \ge 0\}$. Then $\mathcal{F}$ is closed, and its reachable set from x at each time T is equal to the positive trajectory of X through x. The origin is its limit point, which is not in the set; hence $\mathcal{A}_{\mathcal{F}}(x, T)$ need not be closed, even when $\mathcal{F}$ is closed.

Definition 4 For any family of vector fields $\mathcal{F}$, $\mathrm{co}(\mathcal{F})$ will denote the convex hull of $\mathcal{F}$. The elements of $\mathrm{co}(\mathcal{F})$ are all elements of the form $\sum_{i=1}^m \lambda_i X_i$, with $\lambda_1, \ldots, \lambda_n$ positive numbers such that $\sum_{i=1}^{n} \lambda_i = 1$, and each $X_1, \ldots, X_n$ belonging to $\mathcal{F}$.

Theorem 6 *For any family of smooth vector fields $\mathcal{F}$,*

$$\mathrm{cl}(\mathcal{A}_{\mathcal{F}}(x, T)) = \mathrm{cl}(\mathcal{A}_{\mathrm{co}(\mathcal{F})}(x, T))$$

for any x in M and any $T > 0$.

This theorem is a consequence of the following:

Theorem 7 *Suppose that X and Y are any smooth (or even C^1) vector fields on M, and let $\mathcal{F} = \{X, Y\}$. For any number λ, $0 \leq \lambda \leq 1$, let $x(t)$ denote an integral curve of $\lambda X + (1 - \lambda)Y$ defined on an interval $[0, T]$. Then there exists a sequence $\{x^{(n)}\}$ of trajectories of $\mathcal{F}$, each defined on $[0, T]$, such that $x^{(n)}(0) = x(0)$, and $\{x^{(n)}(t)\}$ converges uniformly to $x(t)$ on $[0, T]$. Consequently, $(\exp(\lambda X + (1 - \lambda)Y)T)(x) \in \mathrm{cl}(\mathcal{A}_{\mathcal{F}}(x, T))$ for each $x \in M$ and each $T > 0$ for which $(\exp(\lambda X + (1 - \lambda)Y)T)(x)$ is defined.*

The proof of this theorem will be based on the following lemma:

Lemma *Let $0 = T_0 < T_1 < \cdots < T_{2n} = T$ be a partition of an interval $[0, T]$, with*

$$T_k - T_{k-1} = \frac{T}{2n} \quad \text{for each} \quad k = 1, 2, \ldots, 2n.$$

Let I_k be the interval $[T_k, T_{k-1}]$, and let $\phi_n(t)$ be the characteristic function of $I_2 \cup I_4 \cup \cdots \cup I_{2n}$. Then for any bounded and measurable function f on the interval $[0, T]$,

$$\lim_{n \to \infty} \int_0^T \phi_n(t) f(t) \, dt = \frac{1}{2} \int_0^T f(t) \, dt.$$

Proof Any function f in $L^1[0, T]$ is a pointwise limit of step functions $\{f_n\}$ for which $\int_0^T f(t) \, dt = \lim \int_0^T f_n(t) \, dt$. Therefore it suffices to prove the lemma for the case where f is a characteristic function of an interval $I \subset [0, T]$.

Let a and b $(a < b)$ be the end points of I. For each $\varepsilon > 0$, let N be such that $\frac{1}{2}(T/N) < \varepsilon$. Then for any integer n $(n > N)$,

$$\left| \int_0^T \phi_n(t) f(t) \, dt - \frac{1}{2}(b - a) \right| < \varepsilon.$$

End of the proof.   ∎

In the course of the next proof it will be necessary to make use of the following inequality, known in the theory of differential equations as *Gronwall's inequality* (Coddington and Levinson, 1955): Let α and ϕ be any integrable functions on an interval $[t_0, t_1]$ such that $\alpha \geq 0$ and

$$\phi(t) \leq \varepsilon + \int_{t_0}^{t} \alpha(s)\phi(s)\,ds.$$

Then $\phi(t) \leq \varepsilon e^{K}$ for all t in $[t_0, t_1]$, where $K = \int_{t_0}^{t_1} \alpha(t)\,dt$.

Proof of Theorem 7 Because the interval $[0, T]$ is compact, it suffices to prove the result for small T. If $x(t)$ is an integral curve of $\lambda X + (1 - \lambda)Y$, then for sufficiently small T, the curve segment $x(t)$, with $0 \leq t \leq T$, remains in a fixed chart.

Let $(x_1(t), \ldots, x_n(t))$, $(X_1, \ldots, X_n)$, and $(Y_1, \ldots, Y_n)$ be the coordinates of $x(t)$, X, and Y, respectively, valid in some open set U in $\mathbb{R}^n$. There is no loss of generality in assuming that U is a sufficiently small neighborhood of the curve segment $(x(t), \ldots, x_n(t))$, with $0 \leq t \leq T$, such that the following estimates will hold: For each $i = 1, 2, \ldots, n$ there exist numbers M and K such that

$$|X_i(x)| \leq M, \qquad |Y_i(x)| \leq M, \qquad |X_i(x) - X_i(y)| \leq K\|x - y\|,$$

and

$$|Y_i(x) - Y_i(y)| \leq K\|x - y\|$$

for all x and y in U. $\|x\|$ is the Euclidean distance $\sqrt{x_1^2 + \cdots + x_n^2}$.

Let $\phi_k(t)$ be the characteristic function from the preceding lemma. For each integer k, denote by $Z^{(k)}$ the time-varying vector field on U whose coordinates are defined by

$$Z_i^{(k)}(t, x) = 2\lambda\phi_k(t)X_i(x) + 2(1 - \lambda)(1 - \phi_k(t))Y_i(x).$$

Let $y^k(t)$ be the solution curve of the differential system $dy_i/dt = Z_i^{(k)}(t, y(t))$, with $i = 1, 2, \ldots, n$, that satisfies $y_i^k(0) = x_i(0)$, with $i = 1, 2, \ldots, n$.

Because $|Z_i^{(k)}(t, x)| \leq 4M$, there exists a common interval $[0, T_0]$ on which each solution curve $y^{(k)}$ is defined. There is no loss of generality if T_0 is taken to be equal to T.

Each solution curve $y^{(k)}(t)$ is a trajectory of $\mathcal{F}$ that in addition satisfies $y^{(k)}(T) \in \mathcal{A}_{\mathcal{F}}(x, T)$; for in the interval $[0, T/2k]$, $Z_i^{(k)}(t, x) = 2(1 - \lambda)Y_i(x)$, and therefore $y^{(k)}(t)$ is a solution curve of Y. In the next interval $[T/2k,$ $2T/(2(k+1))]$, $Z_i^{(k)}(t,x) = 2\lambda X_i(x)$, and therefore in this interval, $y^{(k)}(t)$ is a solution curve of X. In each subsequent interval, $y^k(t)$ is either an integral curve of X or an integral curve of Y. In the interval $[0, T]$, the total time that corresponds to X is equal to $2\lambda(T/2)$, and the total time that corresponds to Y is equal to $2(1-\lambda)(T/2)$. The sum of these two numbers is equal to T, and hence $y^{(k)}(T) \in \mathcal{A}_{\mathcal{F}}(x, T)$.

It remains to show that $y^{(k)}$ converges uniformly to x on $[0, T]$. Let $\varepsilon > 0$ be given. Choose k sufficiently large that both $|\int_0^T (\phi_k(t)X_i(x(t)) - \frac{1}{2}X_i(x(t)))\,dt|$ $\leq \varepsilon$ and $|\int_0^T (\phi_k Y_i(x(t)) - \frac{1}{2}Y_i(x(t)))\,dt| \leq \varepsilon$. Then

$$\left|y_i^{(k)}(t) - x_i(t)\right| \leq \left|\int_0^T 2\lambda\phi_k(t)X_i(y^{(k)}(t)) - \lambda X_i(x(t))\right.$$
$$\left. + 2(1-\lambda)(1-\phi_k(t))Y_i(y^{(k)}(t)) - (1-\lambda)Y_i(x(t))\,dt\right|.$$

On using the preceding estimates,

$$\left|y_i^{(k)}(t) - x_i(t)\right| \leq \int_0^T 2\lambda\phi_k(t)|(X_i(y^{(k)}(t)) - X_i(x(t))|$$
$$+ 2(1-\lambda)(1-\phi_k(t))|Y_i(y^{(k)}(t)) - Y_i(x(t))|\,dt + 2\varepsilon.$$

But then $|X_i(y^{(k)}(t)) - X_i(x(t))| \leq K\|y^{(k)}(t) - x(t)\|$, and $|Y_i(y^{(k)}(t)) - Y_i(x(t))| \leq K\|y^{(k)}(t) - x(t)\|$, and therefore

$$\left|y_i^{(k)}(t) - x_i(t)\right| \leq 2\int_0^T K\|y^{(k)}(t) - x(t)\|\,dt + 2\varepsilon \quad \text{for each } i = 1, 2, \ldots, n,$$

or

$$\|y^{(k)}(t) - x(t)\| \leq 2K\sqrt{n}\int_0^T \|y^{(k)}(t) - x(t)\|\,dt + \varepsilon\sqrt{n}.$$

Gronwall's inequality then yields $\|y^{(k)}(t) - x(t)\| \leq \varepsilon\sqrt{n}e^{2K\sqrt{n}T}$. Hence $y^{(k)}$ converges uniformly to x, and the proof is complete. ∎

Remark The statement of Theorem 7 and its proof constitute a special case of the following situation: If $dx/dt = F(x, u)$ is any control system convex in u, then any sequence of trajectories emanating from a common initial point that are produced by a weakly convergent sequence of controls is uniformly convergent. We shall see this argument in more detail in the next chapter. In the case of Theorem 7,

$$\frac{dx}{dt} = u(2\lambda X(x)) + (1 - u)2(1 - \lambda)Y(x), \qquad 0 \le u \le 1.$$

The sequence of controls ϕ_n constructed in the lemma converges weakly to $\frac{1}{2}$.

∎

Example 8 Let $M = \mathbb{R}^2$, and let $X = \partial/\partial x$, $Y = \partial/\partial y$. Each line segment in the plane with slope 1 is a uniform limit of "staircase" functions, as shown in Figure 3.8.

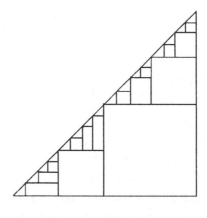

Fig. 3.8.

Example 9 Let A and B be any $n \times n$ matrices. Then the construction used in the proof of Theorem 7 can be used to show the following limit:

$$\exp\left(\frac{1}{2}(A + B)\right) = \lim_{n \to \infty} \left(\exp\frac{A}{2n}\right)\left(\exp\frac{B}{2n}\right)\left(\exp\frac{A}{2n}\right)\cdots\left(\exp\frac{B}{2n}\right).$$

Identify A and B with right-invariant (or left-invariant) vector fields X and Y on $GL_n(R)$: $X(g) = Ag$ and $Y(g) = Bg$ for all g in $GL_n(R)$. Then $\exp(\frac{1}{2}(A + B)) = (\exp\frac{1}{2}(X + Y))(e)$.

The same limit holds on any Lie group, with A and B arbitrary elements of its Lie algebra. The foregoing formula is often used to show that the set of vectors tangent to a closed subgroup of a Lie group forms a vector space.

Example 10 A convex combination of complete vector fields need not be complete. This fact has been known for some time, but the simplest example that comes to mind is due to Claude Lobry, and it goes as follows:

Let $M = \mathbb{R}^2$, $X(x, y) = 2y^2(\partial/\partial x)$, and $Y(x, y) = 2x^2(\partial/\partial y)$. Then $(\exp tX)(x, y) = (x + 2y^2t, y)$, and $(\exp tY)(x, y) = (x, y + 2x^2t)$. Consequently, both X and Y are complete, but $\frac{1}{2}(X + Y)$ is not complete, for if the initial point (x, y) is on the line $x = y$, then the corresponding integral curve of $\frac{1}{2}(X + Y)$ is given by $(1/(1 - tx), \ 1/(1 - tx))$.

The proof of Theorem 6 follows from an inductive use of Theorem 7 applied to the number of vector fields of $\mathcal{F}$ that make up the convex combination, as suggested by Figure 3.9.

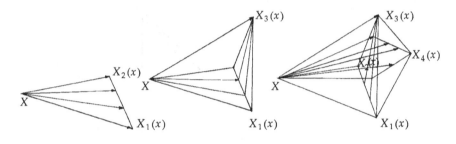

Fig. 3.9.

2.2 Time scaling and normalizers

For a single vector field X on a manifold M, it is quite evident that the trajectories of X and λX, for $\lambda \in \mathbb{R}^1$, are the same. Their corresponding integral curves agree up to a reparametrization. This simple idea naturally extends to families of vector fields, to yield the following:

Theorem 8 *Let $\mathcal{F}$ be a family of vector fields on M.*

(a) *Denote by $\mathcal{F}_1$ the set of all vector fields of the form $\sum_{i=1}^{n} \lambda_i X_i$, with each λ_i non-negative, $\sum_{i=1}^{n} \lambda_i \leq 1$, and each X_i in $\mathcal{F}$. Then, for each $x \in M$ and for $T > 0$, $\mathcal{A}_{\mathcal{F}_1}(x, \leq T) \subset \mathrm{cl}(\mathcal{A}_{\mathcal{F}}(x, \leq T))$.*

(b) *If $\mathcal{F}_1$ is replaced by the positive convex cone $\sum_{i=1}^{n} \lambda_i X_i$, with $\lambda_i \geq 0$ and X_i in $\mathcal{F}$, then*

$$\mathcal{A}_{\mathcal{F}_1}(x) \subset \mathrm{cl}(\mathcal{A}_{\mathcal{F}}(x)) \quad \text{for each } x \in M.$$

The reader may note that the family $\mathcal{F}_1$ in part (a) can also be described as the convex hull of $\mathcal{F}$ and the zero vector field. This observation makes the proof of part (a) transparent. In either case, we leave the proof to the reader.

Definition 5 Denote by diff$^\infty(M)$ the group of smooth diffeomorphisms on M.

(a) We shall say that Φ in diff$^\infty(M)$ is a strong *normalizer* for a family of vector fields $\mathcal{F}$ if $\Phi(A_\mathcal{F}(\Phi^{-1}(x), \leq T)) \subset \mathrm{cl}(A_\mathcal{F}(x, \leq T))$ for each $x \in M$ and $T \geq 0$.
(b) Φ will be said to be a normalizer for $\mathcal{F}$ if $\Phi(A_\mathcal{F}(\Phi^{-1}(x))) \subset \mathrm{cl}(A_\mathcal{F}(x))$ for each $x \in M$.

Remark It is clear that the set of normalizers is equal to diff$^\infty(M)$ whenever $\mathrm{cl}(A_\mathcal{F}(x, T)) = M$ for each $x \in M$ and each $T > 0$. ∎

The following lemma will be useful.

Lemma *A diffeomorphism Φ is a strong normalizer for $\mathcal{F}$ whenever both $\Phi(x)$ and $\Phi^{-1}(x)$ belong to $\mathrm{cl}(A_\mathcal{F}(x, \leq T))$ for each $x \in M$ and $T > 0$. Analogously, Φ is a normalizer for $\mathcal{F}$ whenever both $\Phi(x)$ and $\Phi^{-1}(x)$ belong to $\mathrm{cl}(A_\mathcal{F}(x))$ for each $x \in M$.*

Proof For a subset $\mathcal{S}$ of M, denote $A_\mathcal{F}(\mathcal{S}, \leq T) = \cup_{x \in \mathcal{S}} A_\mathcal{F}(x, \leq T)$. Then, for each $x \in M$, $s > 0$, $t > 0$, and $A_\mathcal{F}(\mathrm{cl}(A_\mathcal{F}(x, \leq t), \leq s)) \subset \mathrm{cl}(A_\mathcal{F}(x, \leq s+t))$. Because both $\Phi(x)$ and $\Phi^{-1}(x)$ belong to $\mathrm{cl}(A_\mathcal{F}(x, \leq\varepsilon))$ for all $x \in M$ and each $\varepsilon > 0$, it follows that

$$\mathrm{cl}(A_\mathcal{F}(\Phi^{-1}(x), \leq T)) \subset \mathrm{cl}(A_\mathcal{F}(\mathrm{cl}(A_\mathcal{F}(x, \leq\varepsilon), \leq T))) \subset \mathrm{cl}(A_\mathcal{F}(x, \leq T+\varepsilon)).$$

For each y in $\mathrm{cl}(A_\mathcal{F}(\Phi^{-1}(x), \leq T))$, $\Phi(y)$ belongs to $\mathrm{cl}(A_\mathcal{F}(x, \leq T + 2\varepsilon))$, because

$$\mathrm{cl}(A_\mathcal{F}(\mathrm{cl}(A_\mathcal{F}(\Phi^{-1}(x), \leq T), \leq\varepsilon))) \subset \mathrm{cl}(A_\mathcal{F}(\mathrm{cl}(A_\mathcal{F}(x, T + \varepsilon), \leq\varepsilon)))$$
$$\subseteq \mathrm{cl}(A_\mathcal{F}(x, T + 2\varepsilon)).$$

Because ε is arbitrary, $\Phi(\mathrm{cl}(A_\mathcal{F}(\Phi^{-1}(x), \leq T))) \subset \mathrm{cl}(A_\mathcal{F}(x, \leq T))$, and Φ is a strong normalizer. The proof concerning the normalizers of $\mathcal{F}$ is completely analogous and will be omitted. ∎

Definition 6 For each vector field X and each diffeomorphism Φ, we shall use $\Phi_\#(X)$ to denote the vector field $(\Phi_* \circ X) \circ \Phi^{-1}$. It follows that $\exp t\Phi_\#(X) = \Phi \circ (\exp tX) \circ \Phi^{-1}$. In particular, $\Phi_\#(X)$ is a complete vector field whenever X is complete.

Definition 7 We shall use $\mathcal{N}_S(\mathcal{F})$ to denote the set of strong normalizers of a family of vector fields $\mathcal{F}$. $\mathcal{N}(\mathcal{F})$ will denote the set of normalizers of $\mathcal{F}$.

Theorem 9 *Let $\mathcal{F}$ be a family of vector fields, and let $\mathcal{F}_1 = \{\Phi_\#(X) : \Phi \in \mathcal{N}_S(\mathcal{F}), X \in \mathcal{F}\}$. Then $\mathcal{A}_{\mathcal{F}_1}(x, \leq T) \subset \text{cl}(\mathcal{A}_{\mathcal{F}}(x, \leq T))$ for each x in M and all $T > 0$. Furthermore,*

$$\mathcal{A}_{\mathcal{F}_1}(x) \subset \text{cl}(\mathcal{A}_{\mathcal{F}}(x)) \quad \text{for each} \quad x \in M$$

whenever $\mathcal{F}_1$ is replaced by $\{\Phi_\#(X) : \Phi \in \mathcal{N}(\mathcal{F}), X \in \mathcal{F}\}$.

Proof We shall show that for each $X \in \mathcal{F}$, $x \in M$, $T > 0$, and $\Phi \in \mathcal{N}_S(\mathcal{F})$, $(\exp t\Phi_\#(X))(x) \in \text{cl}(\mathcal{A}_{\mathcal{F}}(x, \leq T))$ for all t, with $t \leq T$. Because $(\exp t\Phi_\# (X))(x) = \Phi((\exp tX)(\Phi^{-1}(x)))$, and $\Phi((\exp tX)(\Phi^{-1}(x))) \in \text{cl}(\mathcal{A}_{\mathcal{F}}(x, \leq T))$, it follows that $((\exp t\Phi_\#(X))(x) \in \text{cl}(\mathcal{A}_{\mathcal{F}}(x, \leq T))$. The second statement follows along analogous lines. ∎

The system invariants described in the two preceding sections can be used to generate an effective prolongation procedure leading up to the maximal system that contains remarkable information about the original system. Our next example, presented in the form of a theorem, illustrates the prolongation procedure and at the same time provides a beautiful proof of the well-known controllability result for linear systems.

Theorem 10 *Let M be a vector space, and let A be a linear vector field on M. Suppose further that b is any constant vector field on M, and consider the following control system on M:*

$$\frac{dx}{dt} = Ax + u(t)b,$$

with $u(t)$ any arbitrary piecewise-constant control function. Then $\mathcal{A}(x, \leq T) = M$ for any x in M and any $T > 0$ if and only if M is equal to the linear span of $b, Ab, \ldots, A^{n-1}b$.

Remark The fact that the control functions are restricted to be piecewise-constant is a matter of convenience. The theorem is valid for any other class of functions in which the piecewise-constant functions are dense. ∎

Proof Let $\mathcal{F}$ denote the equivalent family $\{A + ub : u \in \mathbb{R}^1\}$. Because $\mathcal{F}$ consists of real analytic vector fields, $\mathcal{F}$ is Lie-determined. The first prolongation of $\mathcal{F}$ consists of the union of $\mathcal{F}$ with the one-dimensional vector subspace containing the constant vector field b. We shall use $\mathcal{F}_1$ to denote this prolongation. It follows from Theorems 5 and 8 that $\mathrm{cl}(\mathcal{A}_{\mathcal{F}_1}(x, \leq T)) = \mathrm{cl}(\mathcal{A}_{\mathcal{F}}(x, \leq T))$ for any x in M and any $T > 0$. The argument is simple:

$$ub = \lim_{\lambda \to \infty} \frac{1}{\lambda}(A + (u\lambda)b) \quad \text{for any real number } u.$$

The second prolongation of $\mathcal{F}$ will be denoted by $\mathcal{F}_2$ and will consist of $\mathcal{F}$ along with the vector space of constant vector fields spanned by b and Ab, and so it becomes necessary to prove $\mathrm{cl}(\mathcal{A}_{\mathcal{F}_2}(x, \leq T)) = \mathrm{cl}(\mathcal{A}_{\mathcal{F}}(x, \leq T))$ for all x in M and $T > 0$.

The exponential map Φ_u defined by the constant vector field ub is the translation $\Phi_u(x) = x + ub$. It follows that $(\Phi_u)^{-1} = \Phi_{-u}$, and therefore each Φ_u belongs to the set of strong normalizers. According to Theorem 9, adjoining $(\Phi_u)_\#(A)$ to $\mathcal{F}$ does not change the closure of the reachable sets. But $(\Phi_u)_\#(A) = A - u(Ab)$, and therefore

$$\lim_{\lambda \to \infty} \frac{1}{\lambda} \Phi_{\lambda u}(A) = \lim_{\lambda \to \infty} \frac{1}{\lambda}(A - (u\lambda)(Ab)) = uAb$$

for each real number u.

Theorems 5, 8, and 6 show that the closure of the reachable sets of $\mathcal{F}$ is not changed when $\mathcal{F}$ is prolonged to $\mathcal{F}_2$.

Proceeding inductively, the kth prolongation $\mathcal{F}_k$ is the union of $\mathcal{F}$ and the vector space spanned by $b, Ab, \ldots, A^{k-1}b$. A repetition of the argument used earlier, when applied to the limit

$$uA^kb = \lim_{\lambda \to \infty} \frac{1}{\lambda}(A + (u\lambda)A^{k-1}b),$$

shows that $\mathrm{cl}(\mathcal{A}_{\mathcal{F}_k}(x, \leq T)) = \mathrm{cl}(\mathcal{A}_{\mathcal{F}}(x, \leq T))$ for all x in M and all $T > 0$.

Assuming that the linear span of $b, Ab, \ldots, A^{n-1}b$ is M implies that $\mathcal{A}_{\mathcal{F}_n}(x, \leq T) = M$, and therefore, by Corollary 2 of Theorem 2, $\mathcal{A}_{\mathcal{F}}(x, \leq T) = M$, because $\mathrm{cl}(\mathcal{A}_{\mathcal{F}}(x, \leq T)) = \mathrm{cl}(\mathcal{A}_{\mathcal{F}_n}(x, \leq T))$.

The converse statement follows from the fact that $\mathcal{A}_{\mathcal{F}}(0, \leq T)$ is contained in the orbit of $\mathcal{F}$ through the origin that is equal to the linear span of $b, Ab, \ldots, A^{n-1}b$. End of the proof. ∎

Figure 3.10 depicts the geometric procedure used in the preceding proof.

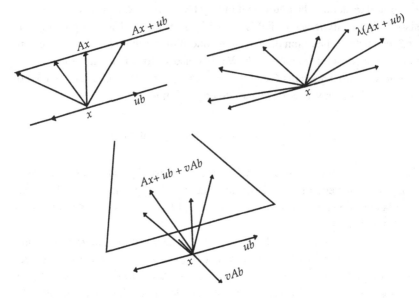

Fig. 3.10.

3 The Lie saturate and controllability

Each of the reachable sets $\mathcal{A}_{\mathcal{F}}(x, T)$, $\mathcal{A}_{\mathcal{F}}(x, \leq T)$, and $\mathcal{A}_{\mathcal{F}}(x)$, or their closures, can be used to define an equivalence relation among the families of vector fields in which all families having the same reachable sets (or their closures) are equivalent. Regardless of which reachable set is used as the criterion, it is always true that the set union of equivalent families is in the equivalence class of each of its representatives. Therefore, each equivalence class of families of vector fields contains the largest element, in the sense of set inclusion; it is equal to the union of all members of the equivalence class.

In order to benefit from the theorems developed in the first part of this chapter, we shall use the closure of the reachable sets as the criterion of equivalence. For many applications, the interplay of time scaling and large controls will produce new vector fields in the equivalence class of the original system. Looking ahead to those applications, we shall concentrate on the equivalence classes defined by the reachable sets $\mathcal{A}(x, \leq T)$ and $\mathcal{A}(x)$. The remaining sets $\mathcal{A}(x, T)$ do not allow for time scaling, because T is fixed.

Each prolongation of the system contains new information about the closure of its reachable sets, and somewhat remarkably for Lie-determined systems, that information is passed down to the reachable set itself.

These general remarks set the stage for the following:

Definition 8 Let $\mathcal{F}$ be a Lie-determined family of vector fields.

(a) The strong Lie saturate of $\mathcal{F}$ is the largest subset $\hat{\mathcal{F}}$ of Lie($\mathcal{F}$) with the property that cl($A_{\hat{\mathcal{F}}}(x, \leq T)$) = cl($A_{\mathcal{F}}(x, \leq T)$) for each $x \in M$ and each $T > 0$. It will be denoted by $\mathcal{LS}_s(\mathcal{F})$.
(b) The Lie saturate of $\mathcal{F}$, denoted by $\mathcal{LS}(\mathcal{F})$, is the largest subset $\hat{\mathcal{F}}$ of Lie($\mathcal{F}$) such that cl($A_{\hat{\mathcal{F}}}(x)$) = cl($A_{\mathcal{F}}(x)$) for each $x \in M$.

The following example, interesting in its own right, illustrates the difference between the two notions in the foregoing definition.

Example 11 Consider the Serret-Frenet control system

$$\frac{dg}{dt} = \vec{A}(g) + k(t)\vec{B}(g)$$

in the group of motions of the plane E_2. Recall that $\vec{A}$ and $\vec{B}$ are left-invariant vector fields whose values at the identity are given by the matrices

$$A = \begin{pmatrix} 0 & 0 & 0 \\ 1 & 0 & 0 \\ 0 & 0 & 0 \end{pmatrix} \quad \text{and} \quad B = \begin{pmatrix} 0 & 0 & 0 \\ 0 & 0 & -1 \\ 0 & 1 & 0 \end{pmatrix}.$$

The equivalent family of vector fields $\mathcal{F}$ is given by $\{\vec{A} + k\vec{B} : k \in \mathbb{R}^1\}$. (The reader who wishes to consider signed curvature should restrict k to non-negative numbers.)

According to the earlier theorems on closure and time scaling, the vector space containing $\vec{B}$ is contained in the strong Lie saturate of $\mathcal{F}$. The preceding statement is valid even if k is restricted to positive numbers, because the flow exp($(k\vec{B})t$) is a rotation

$$\begin{pmatrix} 1 & 0 & 0 \\ 0 & \cos kt & -\sin kt \\ 0 & \sin kt & \cos kt \end{pmatrix}.$$

Then $(\exp k\vec{B})_\#(\vec{A})$ is a left-invariant vector field continued in $\mathcal{LS}_s(\mathcal{F})$ for each k. At the identity, $(\exp k\vec{B})_\#(\vec{A})$ is given by the matrix

$$C(k) = \begin{pmatrix} 0 & 0 & 0 \\ \cos k & 0 & 0 \\ \sin k & 0 & 0 \end{pmatrix}.$$

The positive convex cone generated by $C(k)$ for different values of k is equal to the vector space spanned by

$$\begin{pmatrix} 0 & 0 & 0 \\ 1 & 0 & 0 \\ 0 & 0 & 0 \end{pmatrix} \quad \text{and} \quad \begin{pmatrix} 0 & 0 & 0 \\ 0 & 0 & 0 \\ 1 & 0 & 0 \end{pmatrix},$$

and therefore $\mathcal{LS}(\mathcal{F}) = \text{Lie}(\mathcal{F})$. However, the strong Lie saturate cannot be equal to $\text{Lie}(\mathcal{F})$, for that would imply that any two points in the plane could be joined to each other by an arc of arbitrarily short length.

Theorems 5–9 in this section apply to the Lie saturate and together with its maximality property give the following:

Theorem 11 *Suppose that* $\text{Lie}(\mathcal{F})$ *is a topologically closed set. Then,*

(a) *the strong Lie saturate is a closed convex body invariant under its strong normalizer, and*
(b) *the Lie saturate is a closed affine hull invariant under its normalizer.*

Remark A set of vector fields $\mathcal{F}$ on M is said to be *invariant* under a group of diffeomorphisms G of M if

$$\{\Phi_\#(X) : \ X \in \mathcal{F}, \ \Phi \in G\} = \mathcal{F}.$$

The reader may also recall that the affine hull of a set A consists of all linear combinations $\Sigma \lambda_i a_i$ of elements in A, with all coefficients λ_i positive. For the convex body, the coefficients must also satisfy $\Sigma \lambda_i \leq 1$.   ∎

Definition 9

(a) A family $\mathcal{F}$ is *strongly controllable* if for any $T > 0$ any point of M is reachable from any other point by $\mathcal{F}$ in T or fewer units of time.
(b) $\mathcal{F}$ is *controllable* if any point of M is reachable from any other point of M.

A family consisting of a single vector field X is never strongly controllable in its orbits. It is controllable on an orbit if and only if the orbit is compact, or, equivalently, if the integral curve of X that defines the orbit is closed. A family consisting of all scalar multiples of X is always strongly controllable in each orbit of X.

In general, when $\mathcal{F}$ consists of many vector fields, it is difficult to determine conditions on $\mathcal{F}$ that will ensure controllability. We have already proved that the linear system of Theorem 10 is strongly controllable whenever the linear span of $b, Ab, \ldots, A^{n-1}b$ is equal to M. Our arguments used in Example 11 show that the Serret-Frenet system is controllable, but not strongly controllable in the group of motions of the plane.

The next theorem provides an abstract criterion of controllability.

Theorem 12 *Suppose that $\mathcal{F}$ is a Lie-determined family of vector fields. Then $\mathcal{F}$ is strongly controllable if and only if the strong Lie saturate is equal to* $\mathrm{Lie}(\mathcal{F})$ *and* $\mathrm{Lie}_x(\mathcal{F}) = T_x M$ *for each $x \in M$. $\mathcal{F}$ is controllable if and only if the Lie saturate is equal to* $\mathrm{Lie}(\mathcal{F})$ *and* $\mathrm{Lie}_x(\mathcal{F}) = T_x M$.

Proof Evidently, if $\mathcal{F}$ is (strongly) controllable, then the (strong) Lie saturate is equal to $\mathrm{Lie}(\mathcal{F})$. Conversely, if the strong Lie saturate is equal to $\mathrm{Lie}(\mathcal{F})$ and if $\mathrm{Lie}_x(\mathcal{F}) = T_x M$, then $\mathrm{cl}(\mathcal{A}_{\mathcal{F}}(x, \leq T)) = M$ for each $x \in M$ and $T > 0$. But then, according to Theorem 2(b),

$$M = \mathrm{int}(\mathrm{cl}(\mathcal{A}_{\mathcal{F}}(x, \leq T))) \subset \mathcal{A}_{\mathcal{F}}(x, \leq T + \varepsilon).$$

Because $\varepsilon > 0$ is arbitrary, $\mathcal{A}_{\mathcal{F}}(x, \leq T) = M$ for each $T > 0$ and each $x \in M$. The argument is analogous for the case when $\mathrm{cl}(\mathcal{A}_{\mathcal{F}}(x)) = M$, that is, when the Lie saturate of $\mathcal{F}$ is equal to $\mathrm{Lie}(\mathcal{F})$ and $\mathrm{Lie}_x(\mathcal{F}) = T_x M$ for all x. End of the proof. ∎

As elegant as the foregoing criterion is, its applicability is nevertheless restricted to situations in which there are further symmetries that allow for explicit calculations of the Lie saturate. For that reason we are obliged to defer further uses of the Lie saturate to later chapters dealing with systems having additional mathematical structure.

4 Exact time controllability

Except for the convexification property, the closures of the exact-time reachable sets do not exhibit any other obvious invariants, and for that reason their role in the preceding theory was minimal. Yet in some situations there are simple relations between the exact-time reachable sets and other reachable sets. For instance, the strongly controllable linear systems of Theorem 10 are exactly time-controllable, because the origin is a critical point of the drift vector field. Every controllable family of vector fields on the two-dimensional sphere S^2 is exactly time-controllable for some T, because each vector field on S^2 has an equilibrium point. For the same reason, strongly controllable families on S^2 must be exact-time-controllable for every T.

Our next theorem reveals even more remarkable consequences of controllability and strong controllability on the exact-time reachable sets.

Theorem 13 *Let $\mathcal{F}$ be a Lie-determined family of vector fields on a manifold M, and suppose that, in addition, the zero-time ideal $\mathcal{I}(\mathcal{F})$ is Lie-determined.*

(a) *If $\mathcal{F}$ is controllable, and the fundamental group of M contains no elements of infinite order, then the zero-time ideal $\mathcal{I}(\mathcal{F})$ satisfies $\mathcal{I}_x(\mathcal{F}) = \mathrm{Lie}_x(\mathcal{F})$ for each $x \in M$. In particular, then $\mathcal{A}_{\mathcal{F}}(x, T)$ has a nonempty interior for each x in M and $T > 0$.*

(b) *If $\mathcal{F}$ is strongly controllable, then $\mathcal{A}_{\mathcal{F}}(x, T) = M$ for each x in M and $T > 0$, provided that $\mathcal{I}_x(\mathcal{F}) = \mathrm{Lie}_x(\mathcal{F})$ for each $x \in M$.*

Proof Suppose that $\mathcal{F}$ is a given family and that the zero-ideal $\mathcal{I}(\mathcal{F})$ is such that $\dim(\mathcal{I}_x(\mathcal{F})) = (\dim(\mathrm{Lie}_x(\mathcal{F}))) - 1$ for each $x \in M$. There exists a local chart (ϕ, U) at each $x \in M$ such that the coordinates of each vector field X in this chart are of the form $(\partial/\partial x_1) + \sum_{i=2}^{n} a_i(x)(\partial/\partial x_i)$. The argument is as follows:

The zero-time orbits of $\mathcal{F}$ are the integral manifolds of the distribution $\mathcal{I}(\mathcal{F})$ and have codimension 1, and each vector field X in $\mathcal{F}$ is transversal to each zero-time orbit of $\mathcal{F}$ (Theorem 9, Chapter 2). Let N_x be the zero-time orbit through a point x, and let $X \in \mathcal{F}$. The mapping $(t, p) \to (\exp tX)(p)$ is a local diffeomorphism from a neighborhood of $(0, x)$ in $\mathbb{R}^1 \times N_x$ onto a neighborhood of x in M. Let (ϕ_0, U_0) be a local chart in N_x at x, with coordinates $x_2, \ldots, x_n$. Let $x_1 = t$. Then $x_1, \ldots, x_n$ is the desired system of coordinates on M.

Suppose now that $\omega : [0, 1] \to M$ is any continuous path, and let (ϕ_i, U_i), with $i = 1, \ldots, r$, be a finite set of local charts, as described earlier, such that the union of $\{U_i\}$ covers ω. In each such chart the zero-time orbit of $\mathcal{F}$ through a point of U_i is given locally by $x_1 = \mathrm{constant}$. Because every path in M can be uniformly approximated by a trajectory of the symmetric family $\mathcal{F} \cup -\mathcal{F}$, it follows that there exist trajectories γ of $\mathcal{F} \cup -\mathcal{F}$ such that $\gamma(0) = \omega(0)$ and $\gamma(T) = \omega(1)$ and such that $\gamma[0, T]$ is contained in the union of all U_i.

Let $\gamma_1 : [0, T_1] \to M$ and $\gamma_2 : [0, T_2] \to M$ denote any two such trajectories. For each curve γ_i, denote by T_i^+ the total length of all time intervals in $[0, T_i]$ during which γ_i is an integral curve of $\mathcal{F}$. Analogously, T_i^- denotes the total length of all time intervals during which γ_i is an integral curve of $-\mathcal{F}$. We shall now prove that $T_1^+ - T_1^- = T_2^+ - T_2^-$. There exists a partition $0 = t_0 < t_1 < t_2 < \cdots < t_k = T_1$ of the interval $[0, T_1]$ such that $\gamma_1[t_i, t_{i+1}] \subset U_i$. In each interval $[t_i, t_{i+1}]$, let τ_i^+ (respectively, τ_i^-) denote the total durations of time during which γ_1 is an integral curve of an element in $\mathcal{F}$ (respectively, $-\mathcal{F}$). Let x_1^i denote the first coordinate of $\phi_i(\gamma_1(t_i))$. Then $\gamma_1(t_{i+1})$ is contained in the zero-time orbit whose first coordinate x_1^{i+1} is given by $x_1^{i+1} = x_1^i + \tau_i^+ - \tau_i^-$. Denote $s_0 = 0 < s_1 < s_2 < \cdots < s_k = T_2$, the partition of $[0, T_2]$ such that $\gamma_2(s_i)$ is in the same zero-time orbit as $\gamma_1(t_i)$. Because $\gamma_1(0) = \gamma_2(0)$ and $\gamma_1(T_1) = \gamma_2(T_2)$, it follows that both trajectories originate and terminate on the same zero-time orbits of $\mathcal{F}$. Let μ_i^+ (respectively, μ_i^-) denote the total duration of time in $[s_i, s_{i+1}]$ such that γ_2 is an integral curve of $\mathcal{F}$ (respectively, $-\mathcal{F}$). It follows that $\tau_i^+ - \tau_i^- = \mu_i^+ - \mu_i^-$. Because $T_1^+ = \Sigma \tau_i^+$, $T_1^- = \Sigma \tau_i^-$, $T_2^+ = \Sigma \mu_i^+$, $T_2^- = \Sigma \mu_i^-$, it follows that

$$T_1^+ - T_1^- = \Sigma \tau_i^+ - \tau_i^- = \Sigma \mu_i^+ - \mu_i^- = T_2^+ - T_2^-.$$

Suppose now that $\sigma_1 : [0, T_{\sigma_1}] \to M$ and $\sigma_2 : [0, T_{\sigma_2}] \to M$ are any two trajectories of $\mathcal{F}$ that satisfy $\sigma_1(0) = \sigma_2(0)$ and $\sigma_1(T_{\sigma_1}) = \sigma_2(T_{\sigma_2})$ and that are homotopic to each other. It follows by the preceding argument that if ω is any path near σ_1, and if $\gamma : [0, T_\gamma] \to M$ is any trajectory of $\mathcal{F} \cup -\mathcal{F}$ near ω [subject to $\omega(0) = \sigma_1(0) = \gamma_1(0), \omega(1) = \sigma_1(T_{\sigma_1}) = \gamma(T_\gamma)$], then $T_{\sigma_1}^+ = T_{\sigma_1} = T_\gamma^+ - T_\gamma^-$ (σ_1 is a trajectory of $\mathcal{F}$, and therefore $T_{\sigma_1}^- = 0$). Thus the difference $T_\gamma^+ - T_\gamma^-$ is constant along small deformations and propagates to the other path σ_2. At the other path, $T_{\sigma_1} = T_\gamma^+ - T_\gamma^- = T_{\sigma_2}$. Thus, for any homotopic trajectories of $\mathcal{F}$ that originate and terminate at the same points, the times of travel along each trajectory are equal.

In particular, if $\mathcal{F}$ is controllable, there exists a trajectory $\sigma : [0, T] \to M$ of $\mathcal{F}$ such that $\sigma(0) = \sigma(T) = x$. Let σ^n denote the trajectory of $\mathcal{F}$ on $[0, nT]$ defined by $\sigma^n(t) = \sigma(t - kT)$ for each t in $[kT, (k+1)T]$, $k = 0, 1, \ldots, n-1$.

σ^n cannot be homotopic to σ^m, for $m \neq n$, as follows from the preceding arguments. Hence, the corresponding element $[\sigma]$ in the fundamental group of M based at x is of infinite order (because M is connected, the fundamental group is independent of the base point). So if the fundamental group of M has no elements of infinite order, then $\mathcal{I}_x(\mathcal{F}) = \text{Lie}_x(\mathcal{F})$ for all x in M, and therefore $\mathcal{A}_\mathcal{F}(x, T)$ has a nonempty interior in M, by Theorem 3(b).

To prove part (b), assume that $\mathcal{A}_\mathcal{F}(x, \leq T) = M$ for each $x \in M$ and $T > 0$, and assume that $\mathcal{I}_x(\mathcal{F}) = \text{Lie}_x(\mathcal{F})$ for all $x \in M$. Then $\mathcal{I}_x(\mathcal{F}) = T_x M$ for each $x \in M$, and hence each of $\mathcal{A}_\mathcal{F}(x, T)$ and $\mathcal{A}_{-\mathcal{F}}(x, T)$ will have a nonempty interior in M for each $x \in M$ and each $T > 0$ (Theorem 3).

The first step in the proof is to show that $\mathcal{A}_\mathcal{F}(x, T)$ is dense in M. Let $y \in M$, and let U be an open neighborhood of y. For any $T > 0$, there exists $\hat{T}$ ($\hat{T} \leq T$) and a trajectory defined on $[0, T]$ for which $\gamma(0) = x$ and $\gamma(\hat{T}) = y$. Assume that $\hat{T} < T$. Let $\varepsilon > 0$ be a number sufficiently small that $\varepsilon < T - \hat{T}$ and $\gamma(t) \in U$ for all $t \in [\hat{T} - \varepsilon, \hat{T} + \varepsilon]$.

For any positive number τ and any $\delta > 0$, there exists $\hat{t}$ $(\tau - \delta < \hat{t} < \tau)$ and a trajectory of $\mathcal{F}$ defined on $[0, \hat{t}]$ such that $\hat{\gamma}(0) = \hat{\gamma}(\hat{t}) = x$. Choose $\hat{t}$ such that $T - \hat{T} < \hat{t} < T - \hat{T} + \varepsilon$, and define

$$\sigma(t) = \begin{cases} \hat{\gamma}(t), & t \in [0, \hat{t}], \\ \gamma(t - \hat{t}), & t \in [\hat{t}, \hat{t} + \hat{T}], \end{cases}$$

$\sigma(T) = \gamma(T - \hat{t})$, and hence $\sigma(T) \in U$, since $\hat{T} > T - \hat{t} > \hat{T} - \varepsilon$. Therefore $\mathcal{A}_{\mathcal{F}}(x, T)$ is dense in M.

Now let $y \in M$, and let $T > 0$ be given. For any $\varepsilon > 0$, $\mathcal{A}_{-\mathcal{F}}(y, \varepsilon)$ has a nonempty interior. Because $\mathcal{A}_{\mathcal{F}}(x, T - \varepsilon)$ is dense, $\mathcal{A}_{\mathcal{F}}(x, T - \varepsilon) \cap \mathcal{A}_{-\mathcal{F}}(y, \varepsilon)$ $\neq \phi$. Therefore, $y \in \mathcal{A}_{\mathcal{F}}(x, T)$. The proof is now finished.   ∎

The following examples show that the conditions on the fundamental group are really necessary for the preceding theorem to be true.

Example 12 Let $M = S^1 \times \mathbb{R}^1$, and $\mathcal{F} = \{X + uY : u \in \mathbb{R}^1\}$, with $X(x, y) = y(\partial/\partial x) - x(\partial/\partial y)$ and $Y = \partial/\partial z$, where (x, y, z) denote the coordinates on M, with $x^2 + y^2 = 1$, and z in $\mathbb{R}^1$.

X and Y commute, and therefore the zero-time ideal is one-dimensional and consists of all scalar multiples of Y. $\mathcal{F}$ is controllable, but the reachable set at each instant T is a vertical line (Figure 3.11).

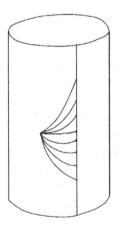

Fig. 3.11.

The next example shows that it may happen that $\mathcal{A}(x, \leq T) = M$ for any x in M and any $T > 0$, and yet $\mathcal{A}(x, T) \neq M$ for all x and all $T > 0$.

Example 13 Consider the differential system

$$\frac{d\theta_1}{dt} = 1 + \alpha u, \qquad \frac{d\theta_2}{dt} = u,$$

on the torus $M = S^1 \times S^1$. α is a fixed irrational number, and u can be any piecewise-constant control function. Any solution $\theta_1(t)$, $\theta_2(t)$ of the foregoing differential equation must satisfy $\theta_1(T) - \alpha\theta_2(T) = T + \theta_1(0) - \alpha\theta_2(0)$. Therefore, there are points that cannot be reached at exactly T units of time. However, $\mathcal{A}((\theta_1^0, \theta_2^0), \leq T) = M$ for any $T > 0$ and any initial angles θ_1^0, θ_2^0. The argument is simple:

$\mathcal{F} = \{(1 + \alpha u)(\partial/\partial\theta_1) + u(\partial/\partial\theta_2) : u \in \mathbb{R}^1\}$. The line $\{u(\alpha(\partial/\partial\theta_1) + (\partial/\partial\theta_2)) : u \in \mathbb{R}^1\}$ is in the strong Lie saturate of $\mathcal{F}$. Hence the entire integral curve of $\alpha(\partial/\partial\theta_1) + (\partial/\partial\theta_2)$ is contained in the closure of $\mathcal{A}_{\mathcal{F}}((\theta_1^0, \theta_2^0), \leq T)$. Because the trajectory of each integral curve $\alpha(\partial/\partial\theta_1) + (\partial/\partial\theta_2)$ is dense, it follows that $\mathrm{cl}(\mathcal{A}_{\mathcal{F}}((\theta_1^0, \theta_2^0), \leq T)) = M$. But then $\mathcal{A}_{\mathcal{F}}((\theta_1^0, \theta_2^0), \leq T) = M$, because the distribution spanned by $\mathcal{F}$ is two-dimensional at each point of T^2.

Problem 2

(a) Let $\mathcal{F}$ be any Lie-determined family of vector fields. Show that every point y of the interior of $\mathcal{A}_{\mathcal{F}}(x, \leq T)$ is normally accessible from x.
(b) Give an example to show that part (a) may be false for an arbitrary family of smooth vector fields.

Problem 3 Consider the linear system $dx/dt = Ax + u(t)b$ introduced in Theorem 10.

(a) Show that the orbit of the corresponding family of vector fields through the origin is a vector space V spanned by $b, Ab, \ldots$.
(b) Show that each reachable $A(x, T) = V$ for any x in V.
(c) Give an example showing that $A(x, T)$ may be a proper subset of the orbit through x when x is not in V.

Problem 4 Consider the Serret-Frenet system defined in Example 11.

(a) Show that the zero-time ideal of the corresponding family of vector fields is equal to $\mathrm{Lie}(\mathcal{F})$.
(b) Show that each reachable set $\mathcal{A}_{\mathcal{F}}(x, T)$ is a proper subset of E_2 and an open neighborhood of the initial point x.

Notes and sources

The material contained in Section 1 is a synthesis from the early papers of Krener (1974), Lobry (1970, 1974), and Sussmann and Jurdjevic (1972). Theorems 1, 2, and 3 are essentially taken from Sussmann and Jurdjevic (1972), with their proofs improved using the ideas of Krener. The concept of a normally accessible point is due to Sussmann (1973).

The idea of using the closure of the reachable sets as the basic invariant object for Lie-determined systems first appeared in a paper by Jurdjevic and Sussmann (1972) concerning control systems on Lie groups. The corresponding equivalence of systems leading up to the notion of the Lie saturate was introduced by Jurdjevic and Kupka (1981a,b) in a study of controllability. Convexification of control systems and the corresponding chattering controls can be traced to the early papers of Warga (1972) and A. F. Fillipov; see Lee and Markus (1967) for further details.

The results presented in Section 3 are fundamentally inspired by the paper of Elliott (1971) and also appear in the Sussmann and Jurdjevic (1972) paper. The proof of Theorem 3 presented here is due to J. P. Gauthier (personal communication). Example 13 was constructed by I. Kupka.

4

Control affine systems

By a "control affine system" we shall mean a differential system on a smooth manifold M of the form

$$\frac{dx}{dt} = X_0(x) + u_1 X_1(x) + \cdots + u_m X_m(x), \tag{1}$$

with $X_0, \ldots, X_m$ smooth vector fields on M, and functions $u_1, \ldots, u_m$ that are the controls. The vector field X_0 is called the drift, and the remaining vector fields $X_1, \ldots, X_m$ are called controlled vector fields.

For the time being, it may be best to consider control functions as parameters, without getting specific about the precise mathematical definitions. Their meaning will be clear from the context. We shall soon discover that the class of control affine systems serves as a kinematic model for a wide range of problems relevant to mechanics, geometry, and control. The purpose of this chapter is to set the general foundation for some specific classes of control affine systems having further mathematical structures, such as systems on Lie groups with symmetries, or polynomial vector fields in $\mathbb{R}^n$.

As described earlier, piecewise-constant controls induce an equivalent family of vector fields that in many applications consists of analytic vector fields and hence is Lie-determined. Therefore, the general formalism of the preceding chapters is applicable and leads to precise accessibility results within this specific context. Although convenient for geometric considerations, piecewise-constant controls are not particularly suitable for problems of optimal control. In this chapter we shall show that points that are in the interiors of sets reachable by piecewise-constant controls are also reachable by smooth trajectories.

All these investigations will be carried out under the additional assumption that the control functions, when regarded as an m-tuple $u = (u_1, \ldots, u_m)$, are constrained to take values in a fixed subset U of $\mathbb{R}^m$, called the control constraint set. It is also customary to refer to control system (1) as the unconstrained control system whenever $U = \mathbb{R}^m$.

In order to be able to reach all points on the boundary of the reachable set defined by piecewise-constant controls, it becomes necessary to enlarge the class of controls. This chapter also contains a closure result when U is a convex and compact set.

Apart from the foregoing matters, this chapter contains an introductory discussion of non-holonomic systems as they relate to geometric control theory. A sphere that rolls without slipping on a stationary or moving platform amply illustrates the non-holonomic aspect of control systems. Partly for that reason, and partly as motivation for further studies of control systems on Lie groups, we shall derive its kinematic equations from the basic principles.

Finally, this chapter contains an introduction to linear systems and some qualitative results arising from the problem of control of a rigid body by means of jets situated on the body. These last examples will provide additional context for the general theory and will serve as links to the material to be presented in the next chapter.

1 Kinematic equations of a rolling sphere

We shall assume that a unit sphere rolls, without slipping, on a horizontal plane in the Euclidean space E^3. The Euclidean metric will facilitate our kinematic description of the movements of the sphere, because we shall be able to express the rotational kinematics of the sphere in terms of a moving orthonormal frame situated on the sphere.

The development is as follows: Let $(\vec{e}_1, \vec{e}_2, \vec{e}_3)$ denote a fixed orthonormal frame in E^3 such that the plane spanned by $\vec{e}_1$ and $\vec{e}_2$ coincides with the plane on which the sphere rolls, and such that $\vec{e}_3$ points to the top.

We shall assume that $(\vec{a}_1, \vec{a}_2, \vec{a}_3)$ denotes an orthonormal frame situated on the sphere that will be called the moving frame. Then each vector $\vec{a}_i$ can be expressed by means of coordinates relative to the basis vectors $\vec{e}_1$, $\vec{e}_2$, and $\vec{e}_3$. R denotes the transition matrix from the frame $(\vec{a}_1, \vec{a}_2, \vec{a}_3)$ to the fixed frame $(\vec{e}_1, \vec{e}_2, \vec{e}_3)$. R is an orthogonal matrix whose columns are made up of coordinates of $\vec{a}_1$, $\vec{a}_2$, and $\vec{a}_3$ relative to the fixed basis $(\vec{e}_1, \vec{e}_2, \vec{e}_3)$.

In order to obtain the kinematic equations for the sphere, further notations are necessary. We shall use Q to denote the coordinates of a point in E^3 relative to the moving frame realized as a column vector in $\mathbb{R}^3$, and q for the coordinate vector relative to the fixed frame. With these notations,

$$q = RQ. \tag{2}$$

Each movement of the sphere is described by the movement of its center and the movement of its frame, which in turn describes its rotation. But the movement of the frame is monitored by the rotation matrix R. So each configuration path is described by the pair of curves $\sigma(t)$ and $R(t)$, the first describing the movement of the center, and the second describing its rotation. Because the e_3 coordinate of the center is constant, σ can be regarded as a curve in E^2. Thus, $M = E^2 \times SO_3(R)$ is the configuration space, because $R(t)$ is a curve in $SO_3(R)$.

Consider now an arbitrary point P on the surface of the sphere. Denote by O the origin of the fixed frame, and by C the origin of the moving frame, as shown in Figure 4.1. We have the following vector diagram:

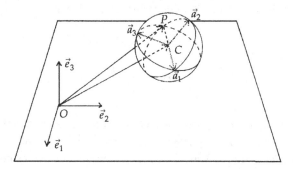

Fig. 4.1.

$$\overrightarrow{OP} = \overrightarrow{OC} + \overrightarrow{PC}.$$

If p, c, and q denote the coordinate vectors of $\overrightarrow{OP}$, $\overrightarrow{OC}$, and $\overrightarrow{PC}$, then $p = c + q$. Corresponding to each path of the sphere, the velocities are given by

$$\frac{dp}{dt} = \frac{dc}{dt} + \frac{dq}{dt}.$$

Because $q(t) = R(t)Q$, with Q equal to the coordinate of $\overrightarrow{CP}$ relative to the moving frame, it follows that $dq/dt = (dR/dt)Q$. The velocity of $\overrightarrow{CP}$ relative to the moving frame is equal to zero, and that fact, combined with an important geometric observation to be explained later, will illustrate the advantage of the moving-frame method.

Because $R(t)$ is a curve in $SO_3(R)$, dR/dt belongs to the tangent space of $SO_3(R)$ at $R(t)$. We shall regard the tangent space at any point R of $SO_3(R)$ as the set of right-translations by R of the elements of its Lie algebra. The Lie algebra of $SO_3(R)$ consists of all 3×3 antisymmetric matrices. We shall use A_1, A_2, and A_3 to denote its standard basis:

$$A_1 = \begin{pmatrix} 0 & 0 & 0 \\ 0 & 0 & -1 \\ 0 & 1 & 0 \end{pmatrix}, \quad A_2 = \begin{pmatrix} 0 & 0 & 1 \\ 0 & 0 & 0 \\ -1 & 0 & 0 \end{pmatrix}, \quad A_3 = \begin{pmatrix} 0 & -1 & 0 \\ 1 & 0 & 0 \\ 0 & 0 & 0 \end{pmatrix}.$$

Then $dR/dt = (\omega_1 A_1 + \omega_2 A_2 + \omega_3 A_3)R$ for some functions $\omega_1(t)$, $\omega_2(t)$, and $\omega_3(t)$. Substituting this expression into $dq/dt = (dR/dt)Q$ leads to $dq/dt = (\omega_1 A_1 + \omega_2 A_2 + \omega_3 A_3)q$. The vector $\vec{\omega} = \omega_1 \vec{e}_1 + \omega_2 \vec{e}_2 + \omega_3 \vec{e}_3$ is called the angular velocity of q. The reader can easily verify that the actions of antisymmetric matrices on column vectors coincide with the cross-product

$$\begin{pmatrix} \omega_2 q_3 - \omega_3 q_2 \\ \omega_3 q_1 - \omega_1 q_3 \\ \omega_1 q_2 - \omega_2 q_1 \end{pmatrix}.$$

Hence, $dq/dt = \omega \times q$.

The assumption that the sphere rolls without slipping means that the velocity of the point of contact of the sphere with the plane is equal to zero, all measured relative to the fixed frame. This constraint provides relationships between the movements of the center of the sphere and the rotational kinematics of the sphere, as follows:

P is the point of contact when

$$Q = \begin{pmatrix} 0 \\ 0 \\ -1 \end{pmatrix} = -e_3.$$

Therefore, $0 = dp/dt = (dc/dt) + \omega \times (-e_3)$, or

$$\frac{dc}{dt} = \omega \times e_3. \tag{3}$$

Denoting the coordinates of c by x_1, x_2, and 1, the preceding relations become

$$\frac{dx_1}{dt} = \omega_2 \quad \text{and} \quad \frac{dx_2}{dt} = -\omega_1. \tag{4}$$

We shall now express the movements of the sphere as the trajectories of a control affine system in $M = E^2 \times SO_3(R)$. The Euclidean plane E^2 is a Lie group with its left- and right-invariant vector fields coinciding with constant vector fields. Being the product of two Lie groups, M itself is a Lie group. Let X_1, X_2, and X_3 denote the right-invariant vector fields on M defined by

$$X_1(x, R) = \vec{e}_1 \oplus A_2 R, \quad X_2(x, R) = \vec{e}_2 \ominus A_1 R, \quad X_3(x, R) = O \oplus A_3 R,$$

for each point (x, R) in M. Denoting the pair $(x(t), R(t))$ by the single construct $g(t)$, it follows that each path in the configuration space is a trajectory of

$$\frac{dg}{dt} = u_1 X_1(g) + u_2 X_2(g) + u_3 X_3(g), \tag{5}$$

with $u_1 = \omega_2$, $u_2 = -\omega_1$, and $u_3 = \omega_3$. Equation (5) can also be written as

$$\frac{dx_1}{dt} = u_1, \quad \frac{dx_2}{dt} = u_2, \quad \frac{dR}{dt} = \begin{pmatrix} 0 & -u_3 & u_1 \\ u_3 & 0 & u_2 \\ -u_1 & -u_2 & 0 \end{pmatrix} R. \tag{5a}$$

System (5) contains no drift term, because the plane on which the sphere rolls is stationary.

The drift term appears when the horizontal plane is not stationary relative to the fixed frame, for then the velocity of the point of contact, instead of being equal to zero, is equal to the velocity of the moving platform. Take, for example, the case where the horizontal plane is revolving about the $\vec{e}_3$ axis with constant angular velocity Ω. Then $dp/dt = \Omega(e_3 \times p)$ whenever p is the point of contact of the sphere with the revolving plane. Then

$$p = \begin{pmatrix} x_1 \\ x_2 \\ 0 \end{pmatrix} \quad \text{and} \quad \frac{dc}{dt} = \omega \times e_3 + \Omega \begin{pmatrix} -x_2 \\ x_1 \\ 0 \end{pmatrix}.$$

Therefore,

$$\frac{dx_1}{dt} = \omega_2 - \Omega x_2 \quad \text{and} \quad \frac{dx_2}{dt} = -\omega_1 + \Omega x_1. \tag{6}$$

The revolving plane introduces a drift vector field $X_0(x, R) = (Ax \oplus 0)$, with

$$A = \Omega \begin{pmatrix} 0 & -1 \\ 1 & 0 \end{pmatrix}.$$

The kinematic equations (5) and (6) allow the vertical component of the angular velocity of the rolling sphere to take any value. In particular, the sphere can spin about the vertical axis and keep its point of contact stationary. The point of contact will undergo circles in the case of a rotating platform (Figure 4.2).

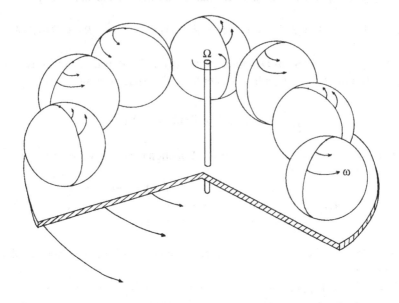

Fig. 4.2.

For some applications, the sphere is rolled in such a way that its angular velocity is always parallel to the horizontal plane. This constraint rules out the trajectories depicted in Figure 4.2, and the corresponding kinematic equations are obtained from equations (5) and (6) by setting $u_3 = 0$.

2 Linear systems

Considering the importance of linear systems in the development of geometric control theory, it would be inappropriate to discuss the general theory of control affine systems without explicit mention of linear systems.

A linear control system is a control affine system on a finite-dimensional vector space M with a linear drift X_0 and each controlled vector field X_i constant. Denoting the constant values of $X_1, \ldots, X_m$ by $b_1, \ldots, b_m$, and the drift term by a linear mapping A (i.e., $X_0(x) = Ax$), the corresponding linear system is given by

$$\frac{dx}{dt} = Ax + \sum_{i=1}^{m} u_i b_i. \tag{7}$$

The preceding system of equations is often written more compactly as

$$\frac{dx}{dt} = Ax + Bu, \tag{7a}$$

with u denoting the column vector of controls

$$\begin{pmatrix} u_1 \\ \vdots \\ u_m \end{pmatrix},$$

and B denoting the linear mapping from $\mathbb{R}^m$ into M, given by

$$Bu = \sum_{i=1}^{m} u_i b_i.$$

The case $m = 1$ is called the single-input case.

Linear systems are intricately connected with the nth-order linear nonhomogeneous differential equation with constant coefficients

$$y^{(n)} + a_1 y^{(n-1)} + \cdots + a_n y = u(t). \tag{8}$$

In this notation, y is a real function, and $y^{(k)}$ denotes its kth derivative. Equation (8) models a mechanical or an electrical linear system controlled by means of an external force u. Equation (8) can be converted into a first-order system in $\mathbb{R}^n$ via the transformation

$$x_1 = y, \qquad x_2 = y^{(1)} \cdots x_n = y^{(n-1)}.$$

Then

$$\frac{dx_1}{dt} = x_2, \qquad \frac{dx_2}{dt} = x_3, \ldots, \qquad \frac{dx_{n-1}}{dt} = x_n,$$
$$\frac{dx_n}{dt} = -a_1 x_n - a_2 x_{n-1} - \cdots - a_n x_1 + u. \tag{9}$$

Denoting by x the column vector

$$\begin{pmatrix} x_1 \\ \vdots \\ x_n \end{pmatrix},$$

the differential system (9) takes the form (7), with $m = 1$, $b = e_n$, and

$$A = \begin{pmatrix} 0 & 1 & \cdots & 0 \\ \vdots & & \ddots & \vdots \\ & & & 1 \\ -a_n & \cdots & & -a_1 \end{pmatrix}.$$

Conversely, for any solution

$$x(t) = \begin{pmatrix} x_1 \\ \vdots \\ x_n \end{pmatrix}$$

of the preceding control affine system, the first coordinate $x_1(t)$ is a solution of the nth-order equation (8).

It is somewhat remarkable that an arbitrary single-input linear system is an nth-order linear equation in disguise, as the following theorem describes.

Theorem 1 *Let $dx/dt = Ax + bu$ be any single-input linear system for which $b, Ab, \ldots, A^{n-1}b$ form a basis for the state space M. Then there exists a linear system of coordinates in M in which A is equal to the matrix*

$$\begin{pmatrix} 0 & 1 & 0 & \cdots & 0 \\ \vdots & & 1 & & \vdots \\ \vdots & & & \ddots & \vdots \\ 0 & \cdots & \cdots & & 1 \\ -a_n & \cdots & \cdots & \cdots & -a_1 \end{pmatrix}, \quad \text{and} \quad b = \begin{pmatrix} 0 \\ \vdots \\ 0 \\ 1 \end{pmatrix}.$$

Proof Let $\lambda^n + a_1 \lambda^{n-1} + \cdots + a_n = 0$ be the characteristic polynomial of A. We shall find a basis $v_1, \ldots, v_n$ for $\mathbb{R}^n$ such that, with respect to this basis, A and b are as given earlier.

Let $v_n = b$ and $v_{n-k} = A^k b + a_1 A^{k-1} b + \cdots + a_k b$ for each $k = 1, 2, \ldots, n - 1$. It follows that $v_1, \ldots, v_n$ are linearly independent and

$$Av_n = Ab = v_{n-1} - a_1 v_n,$$
$$Av_{n-k} = A^{k+1} b + a_1 A^k b + \cdots + a_k Ab$$
$$= v_{n-(k+1)} - a_{k+1} v_n, \qquad 1 \le k < n - 1.$$

With respect to the system of linear coordinates induced by the basis $v_1, \ldots, v_n$, A and b are as claimed in the statement of the theorem, and the proof is finished.

∎

The connection with an nth-order differential equation provides extra information about the structure of the trajectories of a single-input linear controllable case, for it shows that any n-times-differentiable function f is the projection of a trajectory $x(t)$ of

$$\frac{dx}{dt} = Ax + ub$$

for some control $u(t)$. The explicit correspondence is given by $x_1(t) = f(t)$ and $u(t) = f^{(n)}(t) + a_1 f^{(n-1)}(t) + \cdots + a_n f(t)$.

This observation can be used as a basis for an alternative proof of strong controllability, which goes as follows:

Given any initial state x^0, any terminal state x^1, and any interval $[0, T]$, it suffices to find a smooth function $f(t)$, defined on $[0, T]$, that will match the boundary conditions

$$f^{(k)}(0) = x_{k+1}^0 \quad \text{and} \quad f^{(k)}(T) = x_{k+1}^1 \quad \text{for each } k = 0, 1, \ldots, n - 1.$$

It is easy to construct solutions to the preceding boundary-value problem. We leave it to the reader to show the existence of one such solution in the class of polynomial functions.

The multi-input case has a slightly more complicated structure, which will be discussed in the next chapter. Instead, we shall shift attention to another specific class of control affine systems arising from control of the orientation of a rigid body.

3 Control of a rigid body by means of jet torques

We shall begin by recalling the equations that describe the angular velocity $\omega = (\omega_1, \omega_2, \omega_3)$ of a rigid body fixed at its center of gravity and free to move around this pivotal point. For each movement of the body, the equations for $\omega(t)$ are given by the famous equations of Euler:

$$I_1 \frac{d\omega_1}{dt} = (I_2 - I_3)\omega_2\omega_3, \qquad I_2 \frac{d\omega_2}{dt} = (I_3 - I_1)\omega_3\omega_1,$$

$$I_3 \frac{d\omega_3}{dt} = (I_1 - I_2)\omega_1\omega_2.$$

I_1, I_2, and I_3 are called the principal moments of inertia of the body. The angular momentum $x = (x_1, x_2, x_3)$ of the body is given by $x_1 = I_1\omega_1$, $x_2 = I_2\omega_2$, and $x_3 = I_3\omega_3$. The differential equation for ω can also be expressed in terms of the angular momentum, as $dx/dt = x \times \omega$. If there are external forces, then the equation for the momentum must include the momenta due to those forces, whose sum is denoted by Q, and the equation reads $dx/dt = x \times \omega + Q$.

Although Euler's equation is well known, its derivation is based on subtle geometric facts that are often blurred in the classic literature. We shall have occasion to discuss its proof in complete detail in our chapters dealing with variational problems on Lie groups, which will then also clarify the preceding remarks.

For the time being, we shall imagine a pair of jets situated on the body that can exert torques constant in magnitude and opposite in direction. This situation is modeled by $Q = ub$, with b equal to the constant vector corresponding to the direction of the torque, and the control function $u(t)$, which can take only three values $(+1, -1, 0)$. The control values correspond to the possible states of the jets, which can be turned on, reversed, or turned off. We shall now suppose that $b_1, \ldots, b_m$ are m pairs of jets situated on the body, and their actions are controlled by the control functions $u_1, \ldots, u_m$, each of which can take only the values $1, -1$, or 0 (Figure 4.3).

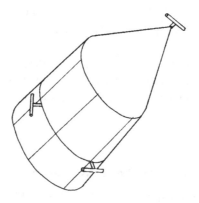

Fig. 4.3.

Denote by

$$\begin{pmatrix} b_i^1 \\ b_i^2 \\ b_i^3 \end{pmatrix}$$

the body coordinates of each moment vector b_i:

$$\frac{dx_1}{dt} = a_1 x_2 x_3 + \sum_{i=1}^{m} u_i b_i^{(1)},$$

$$\frac{dx_2}{dt} = a_2 x_1 x_3 + \sum_{i=1}^{m} u_i b_i^{(2)}, \tag{10}$$

$$\frac{dx_3}{dt} = a_3 x_1 x_2 + \sum_{i=1}^{m} u_i b_i^{(3)},$$

where $a_1 = (I_2 - I_3)/I_2 I_3$, $a_2 = (I_3 - I_1)/I_3 I_1$, and $a_3 = (I_1 - I_2)/I_1 I_2$.

The foregoing equations constitute a particular case of a control affine system in $M = \mathbb{R}^3$, with the drift vector field X_0 a quadratic vector field whose coordinates are given by

$$\begin{pmatrix} a_1 x_2 x_3 \\ a_2 x_1 x_3 \\ a_3 x_1 x_2 \end{pmatrix},$$

and the controlled vector fields $X_1, \ldots, X_m$ equal to the constant vector fields $b_1, \ldots, b_m$. The control constraint set U consists of points u in $\mathbb{R}^m$ whose coordinates can take the values 1, -1, or 0.

The simplest case occurs when all moments of inertia are equal, in which case $X_0 = 0$. This case occurs when the rigid body in question is a sphere of uniform density. Then the minimum number of jets required to control the orientation of the sphere is three, and the jets must not be situated in any common plane passing through the center of the sphere.

We shall return to the problem of controlling the orientation of a rigid body with an arbitrary number of jets, aided with the extra information from the subsequent general theory of control affine systems.

4 Reachability by piecewise-constant controls

Assuming that the control functions take values in a fixed constraint set U, a control affine system (1) defines a family of vector fields $\mathcal{F}(U) = \{X_0 + \sum_{i=1}^{m} u_i X_i : u = (u_1 \ldots, u_m) \in U\}$.

It will be convenient to consider only the constraint subsets U of $\mathbb{R}^m$ that contain m linearly independent points of $\mathbb{R}^m$, for then the Lie algebra generated by $\mathcal{F}(U)$ is independent of U and is generated by the vector fields $X_0, \ldots, X_m$. The proof is simple: The linear span of elements of $\mathcal{F}(U)$ is contained in $\text{Lie}(\mathcal{F}(U))$. But the linear span of elements of $\mathcal{F}(U)$ is equal to $\mathcal{F}(\mathbb{R}^m)$, which in turn is equal to the linear span of $X_0, \ldots, X_m$. We shall use $\text{Lie}\{X_0, \ldots, X_m\}$ to denote the Lie algebra generated by $\{X_0, \ldots, X_m\}$. The zero-time ideal generated by $\mathcal{F}(U)$ is described by the following proposition.

Proposition 1 *The zero-time ideal $\mathcal{I}(\mathcal{F}(U))$ is equal to the smallest ideal in $\text{Lie}\{X_0, \ldots, X_m\}$ that contains $X_1, \ldots, X_m$, provided that U is such that the origin in $\mathbb{R}^m$ is in the convex hull of U.*

Proof $\mathcal{I}(\mathcal{F}(U))$ contains the linear span of the differences of elements in $\mathcal{F}(U)$. Each difference of elements in $\mathcal{F}(U)$ is of the form $\sum_{j=1}^{m}(u_j - v_j)X_j$, with u and v in U.

Let $0 = \sum_{i=1}^{m} \alpha_i u^{(i)}$ for some elements $u^{(1)}, \ldots, u^{(m)}$ in U and numbers $\alpha_1, \ldots, \alpha_n$, with $\sum_{i=1}^{m} \alpha_i = 1$. For any $u \in U$, $\sum_{i=1}^{m} \alpha_i \sum_{j=1}^{m}(u_j - u_j^{(i)})X_j$ is an element of $\mathcal{I}(\mathcal{F}(U))$. Because $\sum_{i=1}^{m} \alpha_i u_j = u_j$, and $\sum_{i=1}^{m} \alpha_i u_j^{(i)} = 0$, this element is equal to $\sum_{j=1}^{m} u_j X_j$. The linear span of all such elements is equal to the linear span of $X_1, \ldots, X_m$, and the proof is finished. ∎

Example Let X_1 and X_2 be any constant linearly independent vector fields in $\mathbb{R}^2$, and let $\mathcal{F}(U) = \{u_1 X_1 + u_2 X_2 : u \in U\}$. Then $\mathcal{I}(\mathcal{F}(U))$ is equal to the linear span of

(a) $X_1 - X_2$ when $U = \{u : u_1 + u_2 = 1\}$,
(b) $2X_1 - X_2$ when $U = \{u : u_1 + 2u_2 = 1\}$,
(c) X_1 and X_2 when $U = \{(1, 0), (2, 0), (0, 1)\}$.

Definition 1 A control affine system is called driftless or symmetric if $X_0 = 0$.

Theorem 2 *Assume that the controlled vector fields $X_1, \ldots, X_m$ generate a Lie algebra $\text{Lie}\{X_1, \ldots, X_m\}$ that satisfies $\text{Lie}_x\{X_1, \ldots, X_m\} = T_x M$ for all x in M. Then*

(a) *the corresponding control affine system is strongly controllable whenever there are no restrictions on the size of controls, and*
(b) *a driftless system remains controllable (but not necessarily strongly controllable) in the presence of constraints, provided that the convex hull of the constraint set U is a neighborhood of the origin in $\mathbb{R}^m$.*

Proof $\mathcal{F} = \{X_0 + \sum_{i=1}^{m} u_i X_i : u \in \mathbb{R}^m\}$ when there are no restrictions on the controls. Each vector field of the form $\frac{1}{\lambda}(X_0 + \lambda u X_i)$ with $\lambda \leq 1$, is in the strong saturate of $\mathcal{F}$ (Chapter 3, Theorem 5 and Theorem 8). Then

$u_i X_i = \lim_{\lambda \to 0}((1/\lambda)(X_0 + \lambda u_i X_i))$, and hence any scalar multiple of X_i is in the strong Lie saturate of $\mathcal{F}$ (Chapter 3, Theorem 1 and Theorem 5). The convex hull of $\{u X_i : u \in \mathbb{R}^1, i = 1, 2, \ldots, m\}$ is equal to the vector space spanned by $X_1, \ldots, X_m$. Therefore the vector space spanned by $X_1, \ldots, X_m$ is in the strong Lie saturate of $\mathcal{F}$, because of Theorem 11 in Chapter 3. But then $\mathrm{Lie}\{X_1, \ldots, X_m\}$ is contained in the strong Lie saturate of $\mathcal{F}$. Because $\mathrm{Lie}_x\{X_1, \ldots, X_m\} = T_x M$, it follows from Theorem 12 in Chapter 3 that $\mathcal{F}$ is strongly controllable.

In order to prove the second statement, let $\mathrm{co}(U)$ denote the convex hull of U. Then $\mathcal{F}(\mathrm{co}(U)) = \mathrm{co}(\mathcal{F}(U))$. Moreover, for each vector field X_i there exists an $\epsilon > 0$ such that the line segment $\{u X_i : |u| \le \varepsilon\}$ is contained in $\mathrm{co}(\mathcal{F}(U))$, because $\mathrm{co}(U)$ is a neighborhood of the origin. Therefore the affine hull of $\mathcal{F}(U)$ is equal to the vector space spanned by $X_1, \ldots, X_m$. But then $\mathrm{Lie}\{X_1, \ldots, X_m\}$ is in the Lie saturate of $\mathcal{F}$. Theorem 12 in Chapter 3 implies that the system is controllable, and the proof is finished. ∎

Corollary *A rolling sphere on a rotating platform is strongly controllable for any angular velocity Ω of the rotating platform.*

Proof We shall prove that the Lie algebra generated by the controlled vector fields X_1 and X_2 spans the tangent space of the configuration space at each point. We recall that X_1 and X_2 are right-invariant vector fields on a five-dimensional Lie group $M = E^2 \times SO_3(R)$. Therefore it suffices to show that X_1 and X_2 generate a five-dimensional Lie algebra.

Denote by e the group element of M, and by L_1 and L_2 the elements of the Lie algebra defined by $X_1(e)$ and $X_2(e)$. It follows that $L_1 = e_1 \oplus A_2$ and $L_2 = e_2 \ominus A_1$. The standard basis A_1, A_2, A_3 of $SO_3(R)$ satisfies the following commutation relations:

$$[A_1, A_2] = A_3, \qquad [A_3, A_1] = A_2, \qquad [A_2, A_3] = A_1.$$

Therefore,

$$[L_1, L_2] = [e_1 \oplus A_2, e_2 \ominus A_1] = [e_1, e_2] \ominus [A_2, A_1] = (0 \oplus A_3),$$
$$[L_1, [L_1, L_2]] = [e_1 \oplus A_2, (0 \oplus A_3)] = (0 \oplus A_1),$$
$$[L_2, [L_1, L_2]] = [e_2 \ominus A_1, (0 \oplus A_3)] = (0 \oplus A_2).$$

But then $e_1 \oplus 0 = L_1 - [L_2, [L_1, L_2]]$, and $e_2 \oplus 0 = L_2 + [L_1, [L_1, L_2]]$, and therefore $\mathrm{Lie}\{L_1, L_2\} = T_e M$. The proof is now finished. ∎

We can combine our preceding results to get the exact controllability at each time T, for Proposition 1 implies that $\mathrm{Lie}\{X_1, X_2\} \subset \mathcal{I}(\mathcal{F})$, and then the exact time controllability follows from Theorem 13 in Chapter 3.

The particular case of a stationary platform on which the sphere is rolled, such that its angular velocity is always parallel to the horizontal plane, was of some interest at Oxford University in the late 1950s.

The Oxford problem Assuming that the sphere is rolled along straight line segments, what is the minimum number of line segments along which the sphere can be rolled from an arbitrary initial configuration into any prescribed terminal configuration? In the language of control theory, the Oxford problem asks for the minimum number of control switches required to bring an arbitrary initial state into an arbitrary terminal state.

Hammersley (1983) recounted the history of this problem at Oxford and showed that three is the minimum number of line segments for the Oxford problem. In that paper, using an easy argument that we shall reproduce later, he pointed out that any two configurations can be connected by four moves. The most difficult situation arises when the initial point of contact coincides with the terminal point of contact, and the terminal rotation differs from the initial rotation by an arbitrary rotation about the e_3 axis.

Hammersley's four-move solution is based on the following observations:

(a) Any two points in the plane can be connected by two rolls without changing the initial orientation of the sphere.
(b) Any initial orientation can be rolled into any terminal rotation by two rolls.

Therefore, to get a four-move solution, first roll for the orientation, and then use the remaining moves to get to the desired terminal point of contact. Figure 4.4 offers pictorial proof of statements (a) and (b).

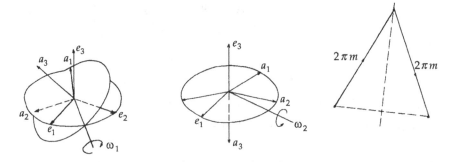

Fig. 4.4.

Problem 1 Take $g_0 = (0, I)$ and $g_1 = (0, e^{\omega A_3})$ for the initial and terminal configurations (Hammersley, 1983).

(i) Show that it is not possible to reach g_1 from g_0 in three moves along line segments for which at least one line segment is an integral multiple of 2π.
(ii) Show that there are three line segments along which g_1 can be reached from g_0.

We return now to the general systems.

Theorem 3 *Assume that a driftless control affine system satisfies* $\mathrm{Lie}_x\{X_1, \ldots, X_m\} = T_x M$ *for all x in M and that the convex hull of the constraint set U is a neighborhood of the origin in $\mathbb{R}^m$. Then for any open connected set Ω in M, any two points of Ω can be transferred to each other along a trajectory of the system that remains in Ω during the entire duration of the transfer.*

Proof Because $\mathrm{co}(\mathcal{F}(U)) = \mathcal{F}(\mathrm{co}(U))$, it follows that there exists an $\epsilon > 0$ such that $G = \{uX_i : |u| \le \varepsilon, \ i = 1, \ldots, m\} \subset \mathrm{co}(\mathcal{F}(U))$. Let G_Ω denote the family of vector fields on Ω consisting of the restriction of elements in G to Ω. The orbit of G_Ω through each point of Ω is equal to Ω, since $\mathrm{Lie}_x(G_\Omega) = T_x\Omega$ for each x in Ω.

But each orbit of G_Ω is also equal to the reachable set of G_Ω from any point of the orbit, because $-G_\Omega = G_\Omega$. It then follows that any two points x and y of Ω can be transferred to each other by a trajectory of G that remains in Ω, since trajectories of G_Ω are also trajectories of G.

Denote by $\mathcal{F}_\Omega(U)$ the family of restrictions to Ω of the vector fields in $\mathcal{F}(U)$. According to Theorem 2 in Chapter 3, the reachable sets of $-\mathcal{F}_\Omega(U)$ from any point y in Ω have a nonempty interior in Ω.

Denote this interior by Δ. If x denotes a point in Ω, then it follows by the preceding argument that there exists a trajectory of G that transfers x to a point in Δ and remains in Ω for the duration of the travel. Because trajectories of $\mathrm{co}(\mathcal{F}(U))$ are uniform limits of the trajectories in $\mathcal{F}(U)$, it follows that there is also a trajectory of $\mathcal{F}(U)$ that transfers x to a point z in Δ and remains in Ω for the entire duration of the transfer. Our proof is now finished, since y can be reached from z by a trajectory of $\mathcal{F}_\Omega(U)$. ∎

Corollary *Under the hypothesis of Theorem 3, for every continuous curve $\sigma: [0, T] \to M$, and every neighborhood Ω of σ, there exists a trajectory $x(t)$ of the control system defined on some interval $[0, S]$ that remains in Ω for each t in $[0, S]$ and further satisfies $x(0) = \sigma(0)$ and $x(S) = \sigma(T)$.*

The proof follows from Figure 4.5.

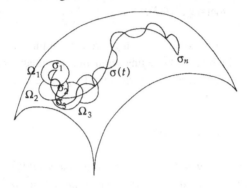

Fig. 4.5.

Problem 2 Show that $dx/dt = \alpha x + u$, with $|u| \leq 1$, is not controllable in $\mathbb{R}^1$ for any $\alpha \neq 0$. Therefore, (a) in Theorem 2 is false if there are bounds on the size of controls.

Problem 3 Let X denote a nonvanishing vector field on a unit circle. Show that $dx/dt = uX$ is not strongly controllable when the control u is restricted to $|u| \leq 1$. Relate to part (b) of Theorem 2.

5 Reachability by smooth controls

It will be convenient to continue using $\mathcal{F}(U)$ as a synonym for system (1) with piecewise-constant controls. Then,

Theorem 4

(a) *Suppose that U is convex and that $\hat{y}$ is normally accessible by $\mathcal{F}$ from $\hat{x}$ in $\hat{T}$ units of time. Then there exists a smooth control $u(t)$ defined on an interval $[0, T]$ such that the corresponding trajectory $x(t)$ of (1) generated by $u(t)$ satisfies $x(0) = \hat{x}$ and $x(T) = \hat{y}$.*

(b) *If $\mathcal{F}(U)$ is a Lie-determined system, then any point in the interior of the reachable set of $\mathcal{F}(U)$ that can be reached from a point $\hat{x}$ is also reachable by a trajectory $x(t)$ of (1) generated by a smooth control $u(t)$. If, furthermore, $\mathcal{F}(U)$ is controllable, then any two points of M can be connected to each other in a positive time by a trajectory of (1) generated by a smooth control.*

To prove (a), assume that $Y_i = X_0 + \sum_{j=1}^{m} u_j^{(i)} X_j$, with $i = 1, 2, \ldots, n$, are elements of $\mathcal{F}$ such that $\hat{y}$ is normally accessible from $\hat{x}$ by $Y_1, \ldots, Y_n$

at $\hat{t} = (\hat{t}_1, \ldots, \hat{t}_n)$. Therefore the mapping $F(t_1, \ldots, t_n) = (\exp t_n Y_n) \circ \cdots \circ (\exp t_1 Y_1)(\hat{x})$ is a local diffeomorphism from a neighborhood of $\hat{t}$ in $\mathbb{R}^n$ onto a neighborhood of $\hat{y}$ in M. Let $t_1(y), \ldots, t_n(y)$ be the local inverse of F.

For each i ($i = 1, 2, \ldots, n$), denote by $\tau_i(y)$ the sum $\sum_{k=1}^{i} t_k(y)$. The piecewise-constant control $u(t)$ that transfers $\hat{x}$ to y is given by

$$u(t) = u^{(1)} \text{ on the interval } [0, \tau_1(y)), \text{ and}$$
$$u(t) = u^{(i)} \text{ on each interval } [\tau_{i-1}(y), \tau_i(y)).$$

For any $\varepsilon > 0$, let $u(t, y, \varepsilon)$ be the smooth control defined by

$$u(t, y, \varepsilon) = \begin{cases} u^{(1)} & \text{for } t \in [0, \tau_1(y) - \varepsilon] \\ (1 - \alpha_i(t, y))u^{(i)} + \alpha_i(t, y)u^{(i+1)}, \\ \quad \text{for } t \in [\tau_i(y) - \varepsilon, \tau_i(y) + \varepsilon], i = 2, \ldots, n-1 \\ u^{(i)} & \text{for } t \in [\tau_{i-1}(y) + \varepsilon, \tau_i(y) - \varepsilon], i = 2, \ldots, n-1 \\ u^{(n)} & \text{for } t \in [\tau_{n-1}(y) + \varepsilon, \tau_n(y)], \end{cases}$$

where each function $\alpha_i(t, y)$ satisfies the following conditions:

(1) $\alpha_i(t, y)$ is a smooth and increasing function of t (for each y) in the interval $[\tau_i(y) - \epsilon, \tau_i(y) + \epsilon]$ with $\alpha_i(\tau_i(y) - \epsilon, y) = 0$ and $\alpha_i(\tau_i(y) + \epsilon, y) = 1$.
(2) $\frac{\partial^k}{\partial t^k}\alpha_i(t, y) = 0$ at both $t = \tau_i(y) - \epsilon$ and $t = \tau_i(y) + \epsilon$ for each $k \geq 1$.

Figure 4.6 illustrates the preceding construction.

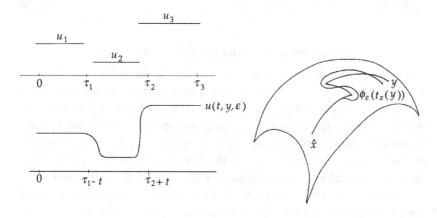

Fig. 4.6.

Let $X_\varepsilon(y, t) = X_0(y) + \sum_{j=1}^{m} u_j(t, y, \varepsilon) X_j(y)$. X_ε is a smooth time-varying vector field defined for each y, and $t \in [0, \tau_n(y)]$, with y belonging to some neighborhood of $\hat{y}$.

Denote the integral curve of X_ε through $\hat{x}$ by $\Phi_\varepsilon(t)$. It follows that $\lim_{\varepsilon \to 0} \Phi_\varepsilon(\tau_n(y)) = y$ for each y in some neighborhood of $\hat{y}$. Our proof will be based on the fact that the mapping $y \to \Phi_\varepsilon(\tau_n(y))$ contains $\hat{y}$ in its image for some $\varepsilon > 0$.

Let (ψ, V) be a local chart at $\hat{y}$ such that V is contained in the domain of the inverse of F. Then for each $y \in V$, $\tau_1(y), \ldots, \tau_n(y)$ is well defined. Denote $\psi(y)$ by $(z_1(y), \ldots, z_n(y))$, and $\psi(\hat{y})$ by $\hat{z}$. Let $\delta > 0$ be small enough that the sphere $\|z - \hat{z}\| \leq \delta$ is contained in $\psi(V)$.

Choose $\varepsilon > 0$ sufficiently small that $\Phi_\varepsilon(\tau_n(\psi^{-1}(z)))$ is contained in V for each z in $\psi(V)$. Then the mapping $g_\varepsilon(z) = \psi \circ \Phi_\varepsilon(\tau_n(\psi^{-1}(z)))$ is smooth as a mapping from $\psi(V)$ into $\psi(V)$.

Because $\lim_{\varepsilon \to 0} g_\varepsilon(z) = z$ for each z, it follows that ε may be further refined so that $\|g_\varepsilon(z) - z\| < \delta$ for each z in the sphere $B = \{z : \|z - \hat{z}\| \leq \delta\}$. But then the mapping $f(z) = z + \hat{z} - g_\varepsilon(z)$ satisfies $\|f(z) - \hat{z}\| = \|z - g_\varepsilon(z)\| \leq \delta$ for each $z \in B$. Therefore f maps B into itself and hence has a fixed point (Dugundji and Granas, 1982, p. 47). If z is the fixed point of f, then $g_\varepsilon(z) = \hat{z}$. But then $\hat{y} = \Phi_\varepsilon(\tau_n(y))$, with $y = \psi^{-1}(z)$. Hence,

$$u(t) = u(t, y, \varepsilon)$$

is the smooth control defined on the interval $[0, \tau_n(y)]$ that transfers $\hat{x}$ to $\hat{y}$.

In order to prove part (b), it suffices to remark that any point $\hat{y}$ contained in the interior of $\mathcal{A}_\mathcal{F}(\hat{x})$ is also normally accessible from $\hat{x}$ by $\mathcal{F}$, provided that $\mathcal{F}$ is a Lie-determined family (Problem 2 in Chapter 3). Therefore part (a) is applicable, and the proof is finished. ∎

6 Recurrent drifts and control of a rigid body

The presence of drifts complicates the dynamics of the system and often accounts for deterioration of its controllability properties. For instance, the problem of controlling the sphere becomes more difficult when the platform begins to rotate, even though the overall system remains strongly controllable. In order to counteract the effects of the drift, it becomes necessary to use large controls even when executing small changes of the state. In some situations, such as that described in Problem 2, controllability is lost as soon as the drift appears.

We shall now return to the problem of controlling the orientation of a rigid body by means of jet torques, with particular emphasis on its controllability properties when the number of jet pairs is less than three. The reader can easily verify that instantaneous controllability is not possible, since the controls are uniformly bounded. We shall show that state-to-state control is possible, even when the number of jet pairs is less than three, provided that the firing of the jets is synchronized with the geometry of the body and with its free motion. The key property is the recurrence of the free motion. The relevant definitions are as follows.

Definition 2 The positive limit set of a complete vector field X on a manifold M through a point x in M consists of all points y in M such that $y = \lim_{n \to \infty}((\exp t_n X)(x))$ for some sequence $\{t_n\}$ of real numbers such that $\lim_{n \to \infty}(t_n) \to \infty$. $\omega(x)$ will be used to denote the positive limit set of X through x.

Each $\omega(x)$ is a closed, possibly empty, subset of M that is also invariant for X; that is, each trajectory of X that originates in $\omega(x)$ remains in $\omega(x)$ for all time.

Definition 3 Let X be a complete vector field on M.

(a) A point x of M is called recurrent for X if $x \in \omega(x)$.
(b) X is called recurrent if there is a dense subset of recurrent points.

Problem 4 Let

$$A_1 = \begin{pmatrix} 0 & \omega_1 \\ -\omega_1 & 0 \end{pmatrix}, \qquad A_2 = \begin{pmatrix} 0 & \omega_2 \\ -\omega_2 & 0 \end{pmatrix}, \qquad A = \begin{pmatrix} A_1 & 0 \\ 0 & A_2 \end{pmatrix}$$

for some real numbers ω_1 and ω_2. Show that the linear vector field X on $\mathbb{R}^4$ given by $X(x) = Ax$ is recurrent.

Problem 5 Suppose that A is any $n \times n$ antisymmetric matrix. Show that the linear vector field $x \to Ax$ is recurrent.

The preceding examples are special cases of the following general situation: Let G be a compact real Lie group that acts on a manifold M. Suppose that X is any vector field on M induced by a left- or right-invariant vector field $\tilde{X}$ on G. That is, if $\theta: G \times M \to M$ denotes the action of G on M, then $(\exp t X)(x) = \theta((\exp t \tilde{X})(x))$. Then X is a recurrent vector field on M. We postpone the proof of this statement to a more opportune place in the chapter dealing with systems on Lie groups.

The quadratic vector field X_0 describing the free momenta of a rigid body is a recurrent vector field in $\mathbb{R}^3$ because of its conservation laws: X_0 has two integrals of motion, the total angular momentum $x_1^2 + x_2^2 + x_3^2$, and the total energy

$$E = \frac{1}{2}\left(\frac{x_1^2}{I_1} + \frac{x_2^2}{I_2} + \frac{x_3^2}{I_3}\right).$$

That is, each integral curve $x(t)$ of X_0 is contained in the intersection of the sphere $\|x(t)\|^2 = $ constant, with the energy ellipsoid $\frac{1}{2}(x_1^2(t)/I_1 + x_2^2(t)/I_2 + x_3^2(t)/I_3) = $ constant. For instance, when the energy ellipsoid E is held fixed, and the radius r of the sphere $\|x\|^2 = r^2$ is varied, one gets the well-known picture for the integral curves of X_0 shown in Figure 4.7.

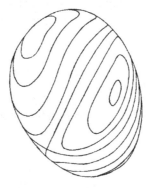

Fig. 4.7.

We now return to systems affine in control, with recurrent drifts. The following theorem is basic.

Theorem 5 *Suppose that $X_0, \ldots, X_m$ are vector fields on M that define a system affine in control, as described by equation (1). Assume that*

(a) *the drift X_0 is recurrent, and*
(b) *the control constraint set U contains the origin of $\mathbb{R}^m$ in the interior of its convex hull. Then the corresponding control system is controllable provided that $\mathrm{Lie}_x\{X_0, \ldots, X_m\} = T_x M$ for each x in M.*

Proof It will be shown that the Lie saturate of the family of vector fields $\mathcal{F}(U)$ that corresponds to the control system (1) is equal to $\mathrm{Lie}\{X_0, \ldots, X_m\}$. Then the statement of this theorem is an immediate consequence of Theorem 12 in

Chapter 3. If co(U) denotes the convex hull of U, then $\mathcal{F}(\mathrm{co}(U))$ is in the Lie saturate of $\mathcal{F}(U)$ because of Theorem 3 in Chapter 3. Because $O \in \mathrm{co}(U)$, it follows that $X_0 \in \mathcal{F}(\mathrm{co}(U))$.

We shall show that $\pm X_0$ is in the Lie saturate of $\mathcal{F}$ because of the recurrence assumption. Let $\hat{x}$ be any point of M, and let $T > 0$. Because $\mathrm{Lie}(\mathcal{F}(U)) = \mathrm{Lie}(\mathcal{F}(\mathrm{co}(U)))$, it follows that $\mathrm{Lie}(\mathcal{F}(U)) = \mathrm{Lie}\{X_0, \ldots, X_m\}$. Assume that $\mathrm{Lie}_x\{X_0, \ldots, X_m\} = T_x M$ for each x in M. Then for any open neighborhood O of $\hat{x}$, $\mathrm{int}\, \mathcal{A}_{\mathcal{F}(\mathrm{co}(U))}(\hat{x}) \cap O$ is not empty. Therefore, there exists a recurrent point x contained in $\mathcal{A}_{\mathcal{F}(\mathrm{co}(U))}(\hat{x}) \cap O$, because the set of recurrent points is dense in M. This fact implies that there is a sequence $\{x_n\}$ of recurrent points of X_0 contained in $\mathrm{int}\, \mathcal{A}_{\mathcal{F}(\mathrm{co}(U))}(\hat{x})$ that converges to $\hat{x}$.

Then $(\exp -T X_0)(\hat{x}) = \lim_{n \to \infty} ((\exp -TX_0)(x_n))$. For each recurrent point x_n, let $\{t_{k,n}\}$ be a sequence of numbers that tends to ∞ as $k \to \infty$, and for which $(\exp t_{k,n} X_0)(x_n) \to x_n$ as $k \to \infty$. Then

$$
\begin{aligned}
(\exp -TX_0)(x_n) &= \lim_{k \to \infty} ((\exp -TX_0)(\exp t_{k,n} X_0)(x_n)) \\
&= \lim_{k \to \infty} ((\exp(t_{k,n} - T)X_0)(x_n)).
\end{aligned}
$$

For large k, $t_{k,n} - T > 0$, and therefore

$$
(\exp(t_{k,n} - T)X_0)(x_n) \in \mathrm{cl}(\mathcal{A}_{\mathcal{F}(\mathrm{co}(U))}(x_n)).
$$

But $\mathcal{A}_{\mathcal{F}(\mathrm{co}(U))}(x_n) \subset \mathcal{A}_{\mathcal{F}(\mathrm{co}(U))}(\hat{x})$, hence $(\exp -TX_0)(x_n) \subset \mathrm{cl}(\mathcal{A}_{\mathcal{F}(\mathrm{co}(U))}(\hat{x}))$ for each n. Therefore, $(\exp -T X_0)(\hat{x}) \in \mathrm{cl}(\mathcal{A}_{\mathcal{F}(\mathrm{co}(U))}(\hat{x}))$.

But this fact implies that $\pm X_0$ is in the Lie saturate of $\mathcal{F}$. On subtracting X_0, we get that $u_1 X_1 + \cdots + u_m X_m$ is in the Lie saturate of $\mathcal{F}$ for any $(u_1, \ldots, u_m)$ in $\mathrm{co}(U)$. Because the origin is contained in the interior of $\mathrm{co}(U)$, it follows that the vector space spanned by $X_1, \ldots, X_m$ is contained in the Lie saturate of $\mathcal{F}(U)$. But then the Lie saturate is equal to $\mathrm{Lie}\{X_0, \ldots, X_m\}$, and the proof is finished. ∎

It follows from the preceding theorem that controllability of a rigid body is determined from the rank of the Lie algebra generated by the drift X_0 and the controlled vector fields $b_1, \ldots, b_m$. This Lie algebra, although infinite-dimensional, admits an easy constructive procedure for determining its rank, as will be explained in more detail in the next chapter. This constructive procedure exploits the fact that the vector fields whose Lie algebra is to be computed are all homogeneous polynomial vector fields, and therefore its minimal rank occurs at the origin.

In the case of a rigid body, the degree of homogeneity of the drift is 2, and it is an easy matter to compute the rank of the appropriate algebra at the origin. For simplicity, this will be done only for the case of one torque.

Theorem 6 *Suppose that $X_0(x)$ is a quadratic vector field in $\mathbb{R}^3$ with co-ordinates $(a_1x_2x_3, a_2x_1x_3, a_3x_1x_2)$ for some constants $a_1, a_2,$ and a_3. Let $X_1 = b = (b_1, b_2, b_3) \neq 0$ be a constant vector field. Then $\text{Lie}_x\{X_0, X_1\}$ has rank 3 at each point $x \in \mathbb{R}^3$ if and only if the plane P spanned by b and $X_0(b)$ is not invariant under X_0.*

Proof Suppose that the plane P spanned by b and $X_0(b)$ is invariant under X_0. Then for each $u \in \mathbb{R}^1$, $X_0 + uX_1$ is tangent to P, and therefore the orbit of $\mathcal{F} = \{X_0 + uX_1 : u \in \mathbb{R}^1\}$ through a point of P must be contained in P. Then $\text{Lie}_x\{X_0, X_1\} \subset P$ for each $x \in P$. Conversely, assume that P is not invariant under X_0. That is, there exists a point c in P such that $X_0(c) \notin P$.

It is easy to verify that for any constant vector field $X(x) = a$, $X_0(a) = \frac{1}{2}(\text{ad}X_0)^2(X)$. If $c = \alpha b + \beta X_0(b)$, then $c = \alpha X_1 + \frac{1}{2}\beta(\text{ad}X_0)^2(X_1)$. Thus both c and $X_0(c)$ are in $\text{Lie}\{X_0, X_1\}$. Because b, $X_0(b)$, and $X_0(c)$ are assumed linearly independent, it follows that $\text{Lie}\{X_0, X_1\}$ contains three linearly independent constant vector fields, which in turn implies that $\text{Lie}_x(X_0, X_1) = \mathbb{R}^3$ for each $x \in \mathbb{R}^3$. The proof is now finished. ∎

Theorem 7 *Let $X_0(x) = (a_1x_2x_3, a_2x_1x_3, a_3x_1x_2)$. If no two coordinates of $a = (a_1, a_2, a_3)$ are equal to zero, then there exists a constant vector field $X_1(x) = b$ such that $\text{Lie}_x(X_0, X_1) = \mathbb{R}^3$ for each $x \in \mathbb{R}^3$.*

Proof For any $b = (b_1, b_2, b_3)$, $X_0(b) = (a_1b_2b_3, a_2b_1b_3, a_3b_1b_2)$. Let $c = \alpha b + \beta X_0(b)$. Then $X_0(c)$ is equal to $\alpha^2 X_0(b) + \beta^2(a_1a_2a_3b_1b_2b_3)b + \alpha\beta d$, with $d = (a_1(a_2b_3^2b_1 + a_3b_1b_2^2), a_2(a_3b_1^2b_2 + a_1b_2b_3^2), a_3(a_1b_2^2b_3 + a_2b_1^2b_3))$. $\text{Lie}_x(X_0, X_1) = \mathbb{R}^3$ for each $x \in \mathbb{R}^3$ if and only if

$$\Delta = \det\begin{pmatrix} b_1 & a_1b_2b_3 & a_1(a_2b_3^2b_1 + a_3b_1b_2^2) \\ b_2 & a_2b_1b_3 & a_2(a_3b_1^2b_2 + a_1b_2b_3^2) \\ b_3 & a_3b_1b_2 & a_3(a_1b_2^2b_3 + a_2b_1^2b_3) \end{pmatrix} \neq 0,$$

$$\Delta = (a_2^2a_3b_1^4b_3^2 - a_2a_3^2b_1^4b_2^2) + (a_1a_3^2b_2^4b_1^2 - a_1^2a_3b_2^4b_3^2)$$
$$+ (a_1^2a_2b_3^4b_2^2 - a_1a_2^2b_3^4b_1^2),$$

or

$$\Delta = b_1^4a_2a_3(a_2b_3^2 - a_3b_2^2) + b_2^4a_1a_3(a_3b_1^2 - a_1b_3^2) + b_3^4a_1a_2(a_1b_2^2 - a_2b_1^2).$$

$\Delta = 0$ whenever any two coordinates of a are equal to zero, independently of b. Otherwise, Δ is a nonzero homogeneous polynomial in b_1, b_2, and b_3, and there always exist b_1, b_2, b_3 such that $\Delta(b_1, b_2, b_3) \neq 0$. The proof is finished. ∎

Theorem 8 *A rigid body with principal moments of inertia not all equal to each other can always be controlled by one jet torque.*

Proof $a_1 = (I_2 - I_3)/I_2 I_3$, $a_2 = (I_3 - I_1)/I_3 I_1$, $a_3 = (I_1 - I_2)/I_1 I_2$. Either $a_1 = a_2 = a_3$ when $I_1 = I_2 = I_3$ or else two coordinates of (a_1, a_2, a_3) are not equal to zero. In the latter case, apply Theorem 7. End of the proof. ∎

7 Compact constraints and the closure of the reachable sets

In order to obtain the closure of the reachable sets, it becomes necessary to extend the class of admissible controls. The appropriate extensions are paraphrases of analogous developments in the theory of the calculus of variations and are as follows:

For any compact interval $[t_0, t_1]$ in $\mathbb{R}^1$, let $L^1(t_0, t_1)$ denote the space of all measurable functions f on the interval $[t_0, t_1]$ for which $\int_{t_0}^{t_1} |f(t)| \, dt < \infty$. Then $L_m^1[t_0, t_1]$ will denote the space of all m-tuples of elements of $L^1[t_0, t_1]$. As is well known, $L_m^1[t_0, t_1]$ is a Banach space, and its dual is equal to $L_m^\infty[t_0, t_1]$, the space of m-tuples of all essentially bounded functions on $[t_0, t_1]$. Because any essentially bounded function on $[t_0, t_1]$ is an element of $L_m^1[t_0, t_1]$, it follows that $L_m^\infty[t_0, t_1] \subset L_m^1[t_0, t_1]$.

For the reader's convenience, we shall recall the relevant facts and theorems from functional analysis that will be required for proof of the main theorem. We begin with the notion of weak convergence.

Definition 4 The *weak topology* in a locally convex linear space E is the smallest topology for which each continuous linear functional $f : E \to \mathbb{R}^1$ remains continuous.

In contrast to the weak topology, the topology of E is called *strong*. It follows by Mazur's theorem that a convex set $A \subset E$ is weakly closed if and only if it is closed. In particular, any Banach space is weakly closed. A sequence $\{x_n\}$ in a Banach space E converges weakly to x if $\lim_{n \to \infty}(f(x_n)) = f(x)$ for any continuous linear functional on E. The basic facts concerning weak convergence are assembled in the following theorem.

Theorem 9 *Let E be a Banach space. Then we have the following:*

(a) *If $\{x_n\}$ is a weakly convergent sequence in E, then $\sup\{\|x_n\| : 1, 2, \ldots\} < \infty$.*

(b) *A subset is weakly compact if and only if it is weakly sequentially compact (Eberlein).*

(c) *If A is weakly compact, then the weak closure of the convex hull of A is also weakly compact (Krein and Smulian).*

We shall use $L_m^1[t_0, t_1, U]$ to denote the set of all elements of $L_m^1[t_0, t_1]$ that take values in the control set U in $\mathbb{R}^m$. When U is a closed and convex subset of $\mathbb{R}^m$, then $L_m^1[t_0, t_1, U]$ is a convex and closed subset of $L_m^1[t_0, t_1]$. It then follows by Mazur's theorem that $L_m^1[t_0, t_1, U]$ is weakly closed. If, in addition, U is compact, then $L_m^1[t_0, t_1, U]$ is weakly compact.

Because every continuous linear functional in $L_m^1[t_0, t_1, U]$ is of the form

$$ u \to \sum_{i=1}^m \int_{t_0}^{t_1} u_i(t) f_i(t)\, dt $$

for some essentially bounded function $f = (f_1, \ldots, f_m)$ in $L_m^1[t_0, t_1]$, it follows that a sequence $\{u^{(n)}\}$ in $L_m^1[t_0, t_1, U]$ converges weakly to u if and only if

$$ \lim_{n \to \infty} \sum_{i=1}^m \int_{t_0}^{t_1} f_i(t) u_i^{(n)}(t)\, dt = \sum_{i=1}^m \int_{t_0}^{t_1} f_i(t) u_i(t)\, dt $$

for each $f = (f_1, \ldots, f_m)$ in $L_m^\infty[t_0, t_1]$.

It can then be shown that any sequence $\{u^{(n)}\}$ in $L_m^1[t_0, t_1, U]$ for which the foregoing limit holds for the characteristic function of any subinterval $[a, b]$ in $[t_0, t_1]$ necessarily weakly converges to u in $L_m^1[t_0, t_1]$.

We now have essentially all the theoretical background required for the proofs. The following result is of central importance for proving compactness of reachable sets.

Theorem 10 *Suppose that $\{u^{(n)}\}$ is a sequence in $L_m^1[t_0, t_1, U]$ that converges weakly to u. Let $Z_n(t, x) = X_0(x) + \sum_{i=1}^m u_i^{(n)}(t) X_i(x)$, and let $Z(t, x) = X_0(x) + \sum_{i=1}^m u_i(t) X_i(x)$. Suppose that $z(t)$ is the integral curve of Z defined on the interval $[t_0, t_1]$ that satisfies $z(t_0) = z^0$. Then, for sufficiently large n, the integral curves $z^{(n)}$ of Z_n, with $z^{(n)}(t_0) = z^0$, are defined for all t in $[t_0, t_1]$, and the sequence $\{z^{(n)}\}$ converges uniformly to z on $[t_0, t_1]$.*

Proof We shall prove that if $z^{(n)}$ is an integral curve of Z_n that satisfies $z^{(n)}(t_0) = \bar{z}^{(n)}$, then $z^{(n)}$ tends uniformly to z whenever $\bar{z}^{(n)}$ tends to z^0. It is enough to prove this statement for t_1 close to t_0, which in turn implies that it suffices to prove the statement in a fixed coordinate chart. Therefore, there is no loss of generality in assuming that all of the vector fields are in $\mathbb{R}^n$.

Let

$$z^{(n)}(t) = \bar{z}^{(n)} + \int_{t_0}^t Z_n(\tau, z^{(n)}(\tau)) \, d\tau, \qquad \text{and}$$

$$z(t) = z^0 + \int_{t_0}^t Z(\tau, z(\tau)) \, d\tau.$$

Let V be a compact neighborhood of z^0 such that $\|X_i(x) - X_i(y)\| \le K \|x - y\|$ for some $K > 0$ for all x and y in V and each $i = 0, 1, \ldots, m$. Then

$$\|z^{(n)}(t) - z(t)\| \le \|\bar{z}^{(n)} - z^0\| + \int_{t_0}^t K\left(1 + \sum_{i=1}^m |u_i^{(n)}(\tau)|\right) \|z^{(n)}(\tau) - z(\tau)\| \, d\tau$$

$$+ \sum_{i=1}^m \left| \int_{t_0}^t (u_i^{(n)}(\tau) - u_i(\tau)) X_i(z(\tau)) d\tau \right|.$$

Because $u^{(n)} \to u$ weakly, it follows by Theorem 9(a) that

$$\sup_n \left\{ \int_{t_0}^t |u_i^{(n)}(t)| \, dt, \quad i = 1, 2, \ldots, m \right\} < \infty,$$

and $\sum_{i=1}^m \int_{t_0}^t (u_i^{(n)}(\tau) - u_i(\tau)) X_i(z(\tau)) \, d\tau \to 0$ as $n \to \infty$. Let $\alpha_n(t) = K(1 + \sum_{i=1}^m |u_i^{(n)}(\tau)|)$, and $\phi_n(\tau) = \|z^{(n)}(\tau) - z(\tau)\|$. Then, for all sufficiently large n,

$$\phi_n(t) \le \varepsilon_n + \int_{t_0}^t \alpha_n(\tau)\phi_n(\tau) \, d\tau,$$

with $\varepsilon_n > 0$ and $\lim \varepsilon_n = 0$. Let $M = \sup_n \{\int_{t_0}^{t_1} \alpha_n(\tau) \, d\tau\}$. It follows from Gronwall's inequality that $\phi_n(t) \le \varepsilon_n e^M$, and the proof is finished. ∎

The results of Theorem 2 can be paraphrased by saying that the trajectories of a control affine system converge uniformly on compact intervals whenever the controls that generate such trajectories converge weakly. This observation leads to the following:

Theorem 11 *Assume that U is a compact and convex subset of $\mathbb{R}^m$, and assume that $T > 0$ is given. Suppose further that for each u in $L_m^1[0, T, U]$ and each x_0 in M the integral curve that corresponds to u and passes through x_0 at $t = 0$ is* *defined for all t in the interval* $[0, T]$. *Then each of the reachable sets* $\mathcal{A}(x_0, t)$ *and* $\mathcal{A}(x_0, \leq t)$ *of* (1) *defined by controls in* $L_m^1(0, t, U)$ *is compact for all* x_0 *in* M *and* $t \leq T$.

Proof Let $x(x_0, u, t)$ denote the integral curve of (1) that corresponds to the control u and that satisfies $x(x_0, u, 0) = x_0$. For any point y in the closure of $\mathcal{A}(x_0, t)$, with $t \leq T$, let $\{u_n\}$ denote a sequence of controls in $L_m^1[0, t, U]$ such that $\lim_{n \to \infty}(x(x_0, u_n, t)) = y$.

Because $L_m^1[0, t, U]$ is weakly compact, it is weakly sequentially compact (Theorem 9(b)). This implies that $\{u_n\}$ has a weakly convergent subsequence $\{u_{n_k}\}$. Let u denote its weak limit. Then $u \in L_m^1[0, T, U]$, and

$$y = \lim_{k \to \infty} \left(x\left(x_0, u_{n_k}, t\right) \right) = x\left(x_0, \lim_{k \to \infty} u_{n_k}, t\right) = x(x_0, u, t).$$

Therefore $\mathcal{A}(x_0, t)$ is closed, and hence compact, because it is an image of a compact set under a continuous map.

The proof of compactness of $\mathcal{A}(x_0, \leq t)$ requires a minor modification of the preceding argument, for if y belongs to the closure of $\mathcal{A}(x_0, \leq t)$, then there exists a sequence of controls $\{u_n\}$ defined on intervals $[0, t_n]$ such that $\lim_{n \to \infty}(x(x_0, u_n, t_n)) = y$. Then first extend each u_n arbitrarily to the entire interval $[0, T]$. Then choose a convergent subsequence $\{t_{n_k}\}$ of $\{t_n\}$, and finally choose a weakly convergent subsequence of the sequence $\{u_{n_k}\}$. Denoting the limit of this subsequence by u, it follows that $y = x(x_0, u, \hat{t})$, where $\hat{t}$ denotes the limit of $\{t_{n_k}\}$. Therefore, y is reachable, and the proof is now finished. ∎

Remark 1 We have already seen (Example 10, Chapter 3) that a convex combination of complete vector fields need not be complete. In particular, the same example shows that even if each of the fields $X_0, \dots, X_m$ in equation (1) is complete, and U is compact and convex, the integral curves of $X_0 + \sum_{i=1}^m u_i X_i$ need not be defined for all t. Therefore, reachable sets need not be compact without the additional assumption concerning forward completeness of the solution curves. ∎

Remark 2 If the control set U is merely closed, and not bounded, then it is easy to see that attainable sets need not be closed. The simplest example is provided by

$$\frac{dx}{dt} = uX,$$

where X is a fixed vector field in M, and u is an arbitrary function. Then for each x_0 and each $T > 0$, the set at T reachable from x_0 is the entire trajectory of X through x_0. The reachable set $\mathcal{A}(x_0, T)$ is not closed whenever the trajectory of X through x_0 has a nonempty limit set not contained on the trajectory (as in the case of a rigid body). ∎

Problem 6 Show that the reachable set at time T for

$$\frac{dx}{dt} = -y^2 + 1, \qquad \frac{dy}{dt} = u, \qquad u(t) \in \{-1, 1\},$$

is not closed.

8 Non-holonomic aspects of control theory

The preceding discussions of the rolling sphere and the control of a rigid body illustrate the interdisciplinary nature of control theory and hint of further connections with mechanics. In order to free the flow of ideas between these subjects, it becomes necessary to recall the fundamental principles of Lagrangian mechanics.

Definition 5 A Lagrangian system is a triple (M, T, V), with M a smooth manifold, T a Riemannian metric on M, and V a smooth function on M.

T generalizes the notion of kinetic energy and is any smooth function on the tangent bundle of M that is positive-definite in each fiber $T_x M$. We shall follow the usual conventions of mechanics and use $q_1, \ldots, q_n, \dot{q}_1, \ldots, \dot{q}_n$ to denote the coordinates of an arbitrary point in TM relative to the basis $\partial/\partial q_1, \ldots, \partial/\partial q_n$. The meaning of the notion of positive definiteness of T is that for each q in M,

$$T(q, \dot{q}) = \sum_{i,j=1}^{n} T_{ij}(q) \dot{q}_i \dot{q}_j$$

is a positive-definite quadratic form in $\dot{q}_1, \ldots, \dot{q}_n$. The coordinates $\dot{q}_1, \ldots, \dot{q}_n$ are called generalized velocities at the generalized position $q_1, \ldots, q_n$. V can be thought of as the potential energy due to a conservative field of force.

Definition 6 Associated with each Lagrangian system (M, T, V), $L = T - V$ is called the Lagrangian.

We shall continue with the notation of mechanics and use $\partial L/\partial \dot{q}$ and $\partial L/\partial q$ to denote the appropriate differentials of L. The reader should note that both of these differentials take values in $T_q^* M$ at each point q in M. A curve $q(t)$ is called a *motion of the system* if it satisfies the d'Alembert-Lagrange principle of virtual work:

$$\sum_{i=1}^{n} \left(\frac{d}{dt} \frac{\partial L}{\partial \dot{q}_i} \left(q(t), \frac{dq}{dt} \right) - \frac{\partial L}{\partial q_i} \left(q(t), \frac{dq}{dt} \right) \right) \xi_i = 0$$

for any generalized velocity $\xi = (\xi_1, \ldots, \xi_n)$ at $q(t)$. Equivalently, $q(t)$ is a solution curve of the Euler-Lagrange equation

$$\frac{d}{dt} \frac{\partial L}{\partial \dot{q}} \left(q(t), \frac{dq}{dt} \right) - \frac{\partial L}{\partial q} \left(q(t), \frac{dq}{dt} \right) = 0$$

because ξ is arbitrary.

Usually M is an embedded submanifold of a Euclidean space E^N. M is called the configuration space, and its tangent bundle TM is called the state space or phase space. For instance, the configuration space for a three-dimensional pendulum is a sphere S^2, and the configuration space for a double pendulum is $M = S^2 \times S^2$ embedded in $E^3 \times E^3$.

The quantity $(d/dt)(\partial L/\partial \dot{q}) - \partial L/\partial q$, which can be defined independently of any coordinate system, is a differential form on E^n and is called the reaction force. The tangent vectors ξ in $T_q M$ are called virtual displacements, and d'Alembert's principle of virtual work states that along any motion $q(t)$ of the system the reaction force does no work along any virtual displacement.

Consider now the motions of the system that are further constrained to a subset S of the phase space; that is, at each point $q \in M$ the velocity is constrained to a subset of the tangent space of M at q.

For simplicity of exposition, assume that S is an embedded submanifold of the tangent bundle of M such that the natural projection $\pi : TM \to M$ restricted to S is surjective. Then for each $q \in M$, $\pi^{-1}(q) \cap S \neq \phi$. Very often S is the zero set of m linearly independent differential forms $\omega_1, \ldots, \omega_m$. These forms then define a distribution $\mathcal{D}$ that at each point q in M consists of the subspace of $T_q M$ equal to the annihilator of $\omega_1, \ldots, \omega_m$.

The constraints are said to be *holonomic* if the distribution $\mathcal{D}$ is integrable. In such a case, each q in M is contained in an $(n - m)$-dimensional submanifold N_q of M whose tangent space at each of its points coincides with the evaluation of $\mathcal{D}$ at these points. The constraints are called *non-holonomic* when $\mathcal{D}$ is a non-integrable distribution.

In terms of local coordinates, S is expressed locally as the zero set of m independent functions $f_1, \ldots, f_m$ defined on some open subset of $\mathbb{R}^{2n}$. The linear independence of differentials $df_1, \ldots, df_m$ implies the existence of coordinates $q_1, \ldots, q_n, \dot{q}_1, \ldots, \dot{q}_n$ in $\mathbb{R}^{2n}$ such that the zero set of $f_1, \ldots, f_m$ can be written as

$$\dot{q}_i = \Phi_i(q_1, \ldots, q_n, \dot{q}_{m+1}, \ldots, \dot{q}_n), \qquad i = 1, \ldots, m,$$

for some functions $\Phi_1, \ldots, \Phi_m$. This means that any conceivable motion $q(t)$ must satisfy the following differential system:

$$\frac{dq_i}{dt} = \Phi_i\left(q_1(t), \ldots, q_n(t), \frac{dq_{m+1}}{dt}, \ldots, \frac{dq_n}{dt}\right), \qquad i = 1, \ldots, m.$$

The preceding differential system can be viewed as a control system:

$$\frac{dq_i}{dt} = \Phi_i(q_1, \ldots, q_n, u_1, \ldots, u_{n-m}), \qquad i = 1, 2, \ldots, m,$$

$$\frac{dq_i}{dt} = u_{i-m}, \qquad m < i \le n.$$

The integral manifolds N_q are $(n-m)$-dimensional if and only if the reachable set of the foregoing control system reachable from q is an $(n-m)$-dimensional manifold. That will be the case if and only if $q_i = $ constant or, equivalently, if and only if $\Phi_i = 0$ for $i = 1, 2, \ldots, m$.

Virtual displacements can take only values in S, in the sense that $\xi_i = \Phi_i(q, \xi_1, \ldots, \xi_{n-m})$, $i = 1, \ldots, m$, and ξ_i is arbitrary for $i > m$.

In terms of these coordinates, the d'Alembert-Lagrange principle of virtual work states that the actual motion must satisfy

$$\sum_{i=1}^{m}\left(\frac{d}{dt}\frac{\partial L}{\partial \dot{q}_i} - \frac{\partial L}{\partial q_i}\right)\Phi_i(q, \xi) + \sum_{i=m+1}^{n}\left(\frac{d}{dt}\frac{\partial L}{\partial \dot{q}_i} - \frac{\partial L}{\partial q_i}\right)\xi_i = 0$$

for all $\xi = (\xi_{m+1}, \ldots, \xi_n)$.

When S corresponds to non-holonomic constraints, $\Phi_i = 0$, and therefore $\sum_{i=m+1}^{n}((d/dt)(\partial L/\partial \dot{q}_i) - (\partial L/\partial q_i))\xi_i = 0$. Thus the motions satisfy

$$\frac{d}{dt}\frac{\partial L}{\partial \dot{q}_i} - \frac{\partial L}{\partial q_i} = 0 \quad \text{for each} \quad i = m+1, \ldots, n.$$

That is, the d'Alembert-Lagrange principle and the Euler-Lagrange equation remain equivalent necessary conditions for the actual motion under holonomic constraints. Stated differently, a Lagrangian system (M, L, S) remains

Lagrangian when restricted to the integral submanifolds of the holonomic constraints.

For non-holonomic systems, the functions Φ_i are not equal to zero, and the d'Alembert-Lagrange principle no longer implies the Euler-Lagrange equation.

A sphere that rolls without slipping on a horizontal plane is a classic example of a non-holonomic system. As we showed earlier, the non-slipping condition defines a three-dimensional distribution in a five-dimensional ambient space, which is non-integrable, because the corresponding control system is controllable. Seen from these perspectives, each controllable system having fewer control functions than the dimension of the ambient manifold can be viewed as a non-holonomic subset S of the tangent bundle.

Notes and sources

Hammersley (1983) uses quaternions to describe the kinematic equations of the rolling sphere. The description of this mechanical system is the same as that of Jurdjevic (1993). The reader may want to compare those descriptions with that of Neimark and Fufaev (1972).

The literature on control of linear systems is vast. I have included only a very selective list of mathematical publications that are of interest for the problems of optimal control. Control of a rigid body by means of external jets is an old problem in optimal control (Brockett, 1972). The proofs given here are the same as those of Lobry (1974).

The results in Section 5 concerning the sets reachable by smooth controls first appeared in the work of Brunovsky and Lobry (1975). The compactness results of Section 7 are classic mathematical results in optimal control (Lee and Markus, 1967). Section 8 is a brief summary of the work of Neimark and Fufaev (1972) adapted to problems of control.

5

Linear and polynomial control systems

The differential systems that are encountered in applications are fundamentally analytic and therefore share a common geometric base. At the same time, such systems have extra-mathematical properties that differentiate one system from another and account for the particular features of their solutions. The theory of such systems is a blend of the general, shared by all Lie-determined systems, and the particular, due to other mathematical structures, and this recognition provides much insight and understanding, even for systems motivated by narrow practical considerations.

In this chapter we shall expand the theory of linear systems for systems with single inputs initiated in Chapter 4. In contrast to the traditional presentation of this theory, which relies entirely on the use of linear algebra and functional analysis, we shall develop the basic theory using the geometric tools introduced in Chapter 3. The geometric point of view will reveal that much of the theory of linear systems follows from considerations independent of the linear properties of the system and therefore extends to larger classes of systems.

Our selection of topics in this theory is motivated partly by the striking nature of the mathematical results and partly by their relevance to the second part of this book, dealing with optimality. Our treatment of linear systems begins with their controllability properties. We shall show that the main theorem has natural Lie-theory interpretations in terms of the Lie saturate of the system. Second, we shall show that a linear controllable system can be decoupled by means of linear feedback into a finite number of independent scalar linear differential equations. This result is known as Brunovsky's normal form. It easily yields another striking theorem of linear theory, that any linear controllable system can be stabilized by means of linear feedback.

Our attention then shifts to linear systems having bounds on the size of controls. The "bang-bang principle" and the global controllability theorem are the main results presented.

The concluding part of the chapter deals with control affine systems having a drift vector field with polynomial coordinates. We show that the controllability theorem of linear systems is a particular case of a polynomial system with the drift having homogeneous coordinates of an odd degree that is controlled by constant vector fields. This chapter also contains a theorem that specifies the maximal length of Lie brackets required for determining the rank of the Lie algebra of the corresponding system.

1 Feedback, controllability, and the structure of linear systems

Continuing with the notation from Chapter 4, a linear system in a real n-dimensional vector space M is denoted by

$$\frac{dx}{dt} = Ax + \sum_{i=1}^{m} u_i(t)b_i. \tag{1}$$

From the geometric point of view, system (1) can be regarded as an affine family of vector fields $\mathcal{F} = A + \mathcal{B}$, with $\mathcal{B}$ equal to the linear subspace of constant vector fields spanned by the controlled vector fields $b_1, \ldots, b_m$. Conversely, any affine family of vector fields $\mathcal{F} = A + \mathcal{B}$, with $\mathcal{B}$ equal to a vector space of constant vector fields, can be parametrized by m control functions $u_1, \ldots, u_m$ corresponding to a choice of basis $b_1, \ldots, b_m$ for $\mathcal{B}$. A different choice of a basis $\hat{b}_1, \ldots, \hat{b}_m$ in $\mathcal{B}$ defines an equivalent linear control system:

$$\frac{dx}{dt} = Ax + \sum_{i=1}^{m} v_i(t)\hat{b}_i. \tag{2}$$

The passage from one system of controls to the other is given by $u_i = \sum_{j=1}^{m} b_{ij} v_j$, with (b_{ij}) equal to the transition matrix from one basis in $\mathcal{B}$ to another.

In order to get to the canonical description of a system $\mathcal{F} = A + \mathcal{B}$, it will be necessary to make use of affine feedback controls.

Definition 1 We shall say that linear systems $\mathcal{F}_1 = A_1 + \mathcal{B}$ and $\mathcal{F}_2 = A_2 + \mathcal{B}$ are feedback-equivalent whenever $A_1 - A_2$ is a linear vector field that takes values in $\mathcal{B}$ for all x in M.

Suppose that two feedback-equivalent systems $\mathcal{F}_1 = A_1 + \mathcal{B}$ and $\mathcal{F}_2 = A_2 + \mathcal{B}$ are parametrized by control functions $u_1, \ldots, u_m$ corresponding to a basis $b_1, \ldots, b_m$ in $\mathcal{B}$. Then

$$A_2 + \sum_{i=1}^{m} u_i b_i = A_1 + (A_2 - A_1) + \sum_{i=1}^{m} u_i b_i.$$

Because the range of $A_2 - A_1$ is contained in $\mathcal{B}$, it follows that there exist linear functions $f_1, \ldots, f_m$ on M such that $A_2 - A_1 = \sum_{i=1}^m f_i b_i$. After substitutions,

$$A_2 + \sum_{i=1}^m u_i b_i = A_1 + \sum_{i=1}^m (f_i + u_i) b_i.$$

Thus, feedback-equivalent systems have the same reachable sets, for if $x_1(t)$ is a trajectory of system $\mathcal{F}_1$ corresponding to the control $u(t) = (u_1(t), \ldots, u_m(t))$, then $x_1(t)$ is a trajectory of system $\mathcal{F}_2$ corresponding to the control $v(t) = (v_1(t), \ldots, v_m(t))$, with $v_i(t) = f_i(x(t)) + u_i(t)$, $i = 1, 2, \ldots, m$. An analogous argument shows that any trajectory of $\mathcal{F}_2$ is also a trajectory of $\mathcal{F}$. Therefore, feedback-equivalent systems have the same trajectories in the state space M.

We have already seen that the Lie algebra generated by a linear system $\mathcal{F} = A + \mathcal{B}$ is equal to the linear span of A and the vector space C of all constant vector fields of the form $A^k b$, with $b \in \mathcal{B}$ and $k \geq 0$. C has a natural filtration $\mathcal{B} = \mathcal{L}_1 \subset \mathcal{L}_2 \subset \cdots \subset \mathcal{L}_{r+1} = C$, with each $\mathcal{L}_k$ defined inductively as $\mathcal{L}_{k+1} = \mathcal{L}_k + [A, \mathcal{L}_k]$. It follows that

$$\mathcal{L}_k = \mathcal{B} + A\mathcal{B} + \cdots + A^{k-1}\mathcal{B}, \qquad k \geq 1. \tag{3}$$

Alternatively, we can think of $\mathcal{L}_k$ as the vector space of constant vector fields in $\text{Lie}(\mathcal{F})$ spanned by all Lie brackets of length less than or equal to $k - 1$. The Lie algebras of feedback-equivalent linear systems $\mathcal{F}_1 = A_1 + \mathcal{B}$ and $\mathcal{F}_2 = A_2 + \mathcal{B}$ differ by the linear term $A_2 - A_1$, and therefore the space of constant vector fields in each Lie algebra remains the same. Moreover, all of the constituent spaces $\mathcal{L}_k$ of C also remain the same for both systems, as can be seen through the following Lie-bracket formula:

$$[A_2, b] = \left[A_1 + \sum_{i=1}^m f_i b_i, \ b \right] = [A_1, b] + \sum_{i=1}^m f_i(b) b_i.$$

Thus, if b is an element of $\mathcal{L}_k$, then $[A_2, b] - [A_1, b]$ belongs to $\mathcal{B}$, and therefore $\mathcal{L}_{k+1}$ is independent of the choice of feedback.

With these observations behind us, we come to the first structure theorem of linear systems. Because M is an abelian Lie group, its tangent space at each point can be identified with M, and therefore the space of constant vector fields can be regarded as a subspace of M.

Theorem 1 *C is an invariant subspace of a linear system* $\mathcal{F} = \mathcal{A} + \mathcal{B}$, *and* $\mathcal{F}$ *is strongly controllable on C. $\mathcal{F}$ is strongly controllable on M if and only if* $M = C$.

Proof Because $[\mathcal{A}, C] \subset C$, it follows that $\mathcal{A}(C) \subset C$. Thus, $\mathcal{A}$ is tangent to C, and therefore $\mathcal{F}$ is tangent to C, because $\mathcal{B} \subset C$. Hence C is an invariant subspace for $\mathcal{F}$.

We shall now show that each subspace $\mathcal{L}_k$ of C is contained in the strong Lie saturate of $\mathcal{F}$. For each b in $\mathcal{B}$,

$$ub = \lim_{\lambda \to \infty} \frac{1}{\lambda}(\mathcal{A} + \lambda ub),$$

and therefore the line $\{ub : u \in \mathbb{R}^1\}$ is contained in the strong Lie saturate of $\mathcal{F}$. Because $\mathcal{B}$ is equal to the convex hull of all lines $\{ub : u \in \mathbb{R}^1\}$, it follows from Theorem 11 in Chapter 3 that $\mathcal{B}$ is contained on $\mathcal{LS}_s(\mathcal{F})$. We shall proceed by induction; so assume that $\mathcal{L}_k \subset \mathcal{LS}_s(\mathcal{F})$.

Let $b \in \mathcal{L}_k$. Then $(\exp vb)_\#(\mathcal{A}) \in \mathcal{LS}_s(\mathcal{F})$ for each real number v. But

$$(\exp vb)_\#(\mathcal{A}) = \mathcal{A} - v(\mathcal{A}b).$$

Therefore,

$$v(\mathcal{A}b) = \lim_{\lambda \to \infty} \frac{1}{\lambda}(\mathcal{A} + \lambda v \mathcal{A}b)$$

is contained in $\mathcal{LS}_s(\mathcal{F})$ for each v. The convex hull of the lines $\{ub : u \in \mathbb{R}^1\}$, $b \in \mathcal{L}_k$, and $\{v\mathcal{A}b : v \in \mathbb{R}^1\}$, $b \in \mathcal{L}_k$, is equal to the vector sum $\mathcal{L}_k + \mathcal{A}(\mathcal{L}_k)$, and therefore $\mathcal{L}_{k+1}$ is contained in $\mathcal{LS}_s(\mathcal{F})$. Hence $C \subset \mathcal{LS}_s(\mathcal{F})$, and therefore $\mathcal{F}$ is strongly controllable on C, by Theorem 12 in Chapter 12.

It now follows that $\mathcal{F}$ is strongly controllable on M when $M = C$. Because C is an invariant subspace of $\mathcal{F}$, it follows that trajectories of $\mathcal{F}$ that originate in C remain in C. Hence, $\mathcal{F}$ cannot be controllable on M whenever C is a proper subspace of M. ■

Remark $\mathcal{F}$ is also exactly time-controllable on C for any $T > 0$, because $\mathcal{A}(0, T) = \mathcal{A}(0, \leq T)$. ■

Corollary *Let E be any subspace of M complementary to C. Then the restriction of $\mathcal{F}$ to E is a single linear vector field equal to the projection of A to E.*

The proof of the corollary is evident and will be omitted.

Problem 1 Suppose that

$$A = \begin{pmatrix} 0 & 1 & 0 \\ -1 & 0 & 0 \\ 0 & 0 & 0 \end{pmatrix} \quad \text{and} \quad b = e_3.$$

Show that the controllability space of $dx/dt = Ax + ub$ is equal to the line $\{\alpha b : \alpha \in \mathbb{R}^1\}$ and that the orbit through each x is a vertical cylinder of radius equal to $\sqrt{x_1^2 + x_2^2}$.

Problem 2 Suppose that

$$A = \begin{pmatrix} 0 & 1 & 0 \\ 1 & 0 & 0 \\ 0 & 0 & 0 \end{pmatrix} \quad \text{and} \quad b = e_3.$$

Show that each orbit is a hyperboloid $\{(x_1, x_2, x_3) : x_1^2 - x_2^2 = \text{constant}\}$, and also show that the sets reachable from any point not on the x_3-axis are proper subsets of the orbit through that point.

1.1 Controllability indices and the feedback-decomposition theorem

Continuing with the preceding notation, we come to the following:

Definition 2 The index of each element b in $\mathcal{B}$ is the smallest integer j with the property that $A^j b$ belongs to $\mathcal{L}_j = \mathcal{B} + A\mathcal{B} + \cdots + A^{j-1}\mathcal{B}$. We shall use $\text{ind}(b)$ to denote this integer.

Thus $j = \text{ind}(b)$ if and only if

$$A^j b = b_1 + Ab_2 + \cdots + A^{j-1}b_j$$

for some elements $b_1, \ldots, b_j$ in $\mathcal{B}$ but $A^k b$ is not contained in $\mathcal{L}_k$ for each k ($k < j$). In particular, $b, Ab, \ldots, A^{j-1}b$ are linearly independent constant vector fields.

Evidently, if there exists an integer r such that $\mathcal{L}_{r+1} = \mathcal{L}_{r+2}$, then $\mathcal{L}_{r+k} = \mathcal{L}_{r+1}$ for all $k \geq 1$. We shall use r to denote the smallest integer such that $\mathcal{L}_{r+1} = \mathcal{L}_{r+2}$. Then every element c of C can be written as $c = b_1 + Ab_2 + \cdots + A^r b_{r+1}$ for some elements $b_1, \ldots, b_{r+1}$ in $\mathcal{B}$, and therefore $\text{ind}(b) < r$ for each b in $\mathcal{B}$.

Definition 3 For each positive integer k, $\mathcal{B}(k)$ will denote the set of all b in $\mathcal{B}$ whose indices are less than or equal to k. $\mathcal{B}(0)$ is defined to be equal to zero.

It follows that each $\mathcal{B}(k)$ is a linear subspace of $\mathcal{B}$ and that the sequence $\{\mathcal{B}(k)\}$ is a nested sequence whose union is equal to $\mathcal{B}$.

Definition 4 Let $i(1) < i(2) < i(3) < \cdots < i(p)$ denote all integers equal to an index of an element b in $\mathcal{B}$. The integers $i(k)$, $k = 1, \ldots, p$, are called the controllability indices of $\mathcal{F}$.

For each controllability index $i(k)$, let E_k denote any subspace of $\mathcal{B}(i(k))$ complementary to $\mathcal{B}(i(k - 1))$, with the understanding that $\mathcal{B}(i(0)) = 0$. Because $\mathcal{B}(i(0)) = 0$, it follows that $E_1 = \mathcal{B}(i(1))$, and therefore

$$\mathcal{B} = E_1 \oplus E_2 \oplus \cdots \oplus E_p.$$

Moreover, the index of each element in E_k is equal to $i(k)$, because E_k is complementary to $\mathcal{B}(i(k - 1))$.

Definition 5 For each choice of the complementary space E_k,

$$M_k = E_k + AE_k + \cdots + A^{i(k)-1} E_k$$

is called the *index tower* over E_k.

It is easy to prove that the vector sum that defines M_k is direct. The argument is as follows: If $b_1 + Ab_2 + \cdots + A^{i(k)-1}b_{i(k)} = 0$ for some elements $b_1, \ldots,$ $b_{i(k)}$ of E_k, then the largest integer j for which $A^j b_j \neq 0$ would imply that $A^j b_j = -\sum_{i=0}^{j-1} A^i b_i$, which in turn would imply that $\mathrm{ind}(b_j) < i(k)$, but that is not possible, because b_j belongs to E_k.

The index tower is determined by the index and the choice of a complementary space E_k. However, the towers that correspond to different complementary spaces are isomorphic, because the complementary spaces are of the same dimension, and the vector sums are direct.

We now come to the main theorem of this section, which will be called the feedback-decomposition theorem. The statement of this theorem requires additional notation. For any choice of a basis $b_1, \ldots, b_m$ in $\mathcal{B}$, B will denote the linear mapping from $\mathbb{R}^m$ into $\mathcal{B}$ given by $\sum_{i=1}^{m} u_i b_i$, for $u \in \mathbb{R}^m$, and $[B]$ will denote the matrix of B relative to the basis $b_1, \ldots, b_m$. Similarly, $[A]$ will denote the matrix of the restriction of A to C relative to a basis in C. The feedback-decomposition theorem is as follows:

Theorem 2 *For each choice of a basis $\{e_{kj}\}$ in $\mathcal{B}$ such that $\{e_{kj},\ j = 1, \ldots,$*

$\dim E_k\}$ *forms a basis for* E_k $(k = 1, \ldots, p)$, *there exists a linear feedback* $F : C \to B$ *and a basis* $\{a_{kji}, \ k = 1, \ldots, p, \ j = 1, \ldots, \dim E_k, i = 1, \ldots, i(k) - 1\}$ *of* C *such that*

(i) $[A + F] = \text{diag}(A_1, \ldots, A_p)$, $[B] = \text{diag}(B_1, \ldots, B_p)$, *and*
(ii) *each of the matrices* A_k *and* B_k *is further decomposed into* $A_k = \text{diag}$ $(A_{k1}, \ldots, A_{kn_k})$, $B_k = \text{diag}(B_{k1}, \ldots, B_{kn_k})$, *with* $n_k = \dim E_k$, *each matrix* A_{kj} *and* B_{kj} *having size* $i(k) \times i(k)$, *given by*

$$
A_{kj} = \begin{pmatrix} 0 & 1 & 0 & \cdots & 0 \\ \vdots & \ddots & \ddots & & \vdots \\ \vdots & & \ddots & \ddots & 0 \\ \vdots & & & \ddots & 1 \\ 0 & \cdots & \cdots & \cdots & 0 \end{pmatrix}, \quad B_{kj} = \begin{pmatrix} 0 & \cdots & \cdots & \cdots & 0 \\ \vdots & 0 & & & \vdots \\ \vdots & & \ddots & & \vdots \\ \vdots & & & \ddots & \vdots \\ 0 & \cdots & \cdots & \cdots & 1 \end{pmatrix},
$$

for each $k = 1, \ldots, p$ *and* $j = 1, \ldots, n_k$.

Proof Let e_{kj} $(j = 1, \ldots, n_k)$ denote any basis for E_k. It follows that each vector field e_{kj} generates $i(k)$ linearly independent elements in the tower M_k, of the form $e_{kj}, Ae_{kj}, \ldots, A^{i(k)-1}e_{kj}$, and that $A^{i(k)}e_{kj} = \sum_{i=1}^{i(k)} A^{i-1}b_{kji}$ for some elements b_{kji} in B.

Let a_{kji} denote the following elements of C:

$$
a_{kji} = A^{i(k)-i}e_{kj} - \left(b_{kj(i+1)} + Ab_{kj(i+2)} + \cdots + A^{i(k)-i-1}b_{kji(k)}\right),
$$

for $i = 1, \ldots, i(k) - 1$. a_{kji} is defined to be equal to e_{kj} for $i = i(k)$.
It follows directly from the definitions of a_{kji} that

$$
a_{kji} = Aa_{kj(i+1)} - b_{kj(i+1)}
$$

for each $k, j, i = 1, \ldots, i(k) - 1$ and that

$$
Aa_{kj1} = A^{i(k)}e_{kj} - Ab_{kj2} - \cdots - A^{i(k)-1}b_{kji(k)} = b_{kj1}.
$$

Assuming that the set $\{a_{kji} : k = 1, \ldots, p, \ j = 1, \ldots, n_k, \ i = 1, \ldots, i(k)\}$ forms a basis for C, the choice of linear feedback is fixed: $F(a_{kji}) = -b_{kji}$. Then

$$
(A + F)a_{kj(i+1)} = a_{kji} \quad \text{for} \quad i = 1, \ldots, i(k) - 1,
$$

and

$$(A + F)a_{kj1} = 0.$$

Relative to the basis a_{kji}, the restriction of $A + F$ to the space $\{a_{kji} : i = 1, \ldots, i(p)\}$ is given by

$$A_{kj} = \begin{pmatrix} 0 & 1 & 0 & \cdots & 0 \\ \vdots & \ddots & \ddots & & \vdots \\ \vdots & & \ddots & \ddots & 0 \\ \vdots & & & \ddots & 1 \\ 0 & \cdots & \cdots & \cdots & 0 \end{pmatrix} \quad \text{and} \quad e_{kj} = \begin{pmatrix} 0 \\ \vdots \\ \vdots \\ 0 \\ 1 \end{pmatrix}.$$

The proof will be complete once we show that the set $\{a_{kji}\}$ forms a basis for C. We shall first show that a_{kji} are linearly independent. If $\Sigma \lambda_{kji} a_{kji} = 0$ for some real numbers λ_{kji}, let ℓ denote the maximal difference $i(k) - i$ for which $\lambda_{kji} \neq 0$ for some $k = 1, \ldots, p$ and $j = 1, \ldots, n_k$. Then let m denote the maximal integer k for which $\lambda_{kji} \neq 0$ for some $j = 1, \ldots, n_k$ and $i(k) - i = \ell$. Then

$$\sum_{j=1}^{n_m} \lambda_{mj(i(m)-\ell)} a_{mj(i(m)-\ell)} = - \sum_{\substack{k \neq m \\ i(k)-i \leq \ell}} \lambda_{kji} a_{kji}. \tag{4}$$

Recall that each a_{kji} is defined by

$$a_{kji} = A^{i(k)-i} e_{kj} - \left(b_{kj(i+1)} + \cdots + A^{i(k)-i-1} b_{kji(k)} \right)$$

for $i = 1, \ldots, i(k) - 1$. Therefore, $a_{kji} = 0 \pmod{\mathcal{L}_\ell}$ if $i(k) - i < \ell$, and $a_{kji} = A^\ell e_{kj} \pmod{\mathcal{L}_\ell}$ if $i(k) - i = \ell$.

It follows that equation (4) reduces to

$$\sum_{j \leq m} \lambda_{mj(i(m)-\ell)} A^\ell e_{mj} = - \sum_{\substack{k < m \\ j \leq n_k}} \lambda_{kj(i(k)-\ell)} A^\ell e_{kj} \pmod{\mathcal{L}_\ell}.$$

After multiplying both sides of the preceding equation by $A^{i(m)-\ell}$, we get that

$$A^{i(m)} \sum_{j=1}^{m} \lambda_{mj(i(m)-\ell)} e_{mj} = \sum_{\substack{k \leq m \\ j \leq n_k}} \lambda_{kj(i(k)-\ell)} A^{i(m)} e_{kj} \pmod{\mathcal{L}_\ell}.$$

Because the index of each e_{kj} with $k < m$ is less than $i(m) - 1$, it follows that $A^{i(m)} e_{kj} = 0 \pmod{\mathcal{L}_{i(m)}}$. Therefore,

$$A^{i(m)} \left(\sum_{j=1}^{m} \lambda_{mj(i(m)-\ell)} e_{mj} \right) = 0 \qquad (\mathrm{mod}\ \mathcal{L}_{i(m)}).$$

The preceding relation implies that $\sum_{j=1}^{m} \lambda_{mj(i(m)-\ell)} e_{mj} = 0$, because A is injective on each space E_k. This conclusion contradicts our initial assumption that $\lambda_{mj(i(m)-\ell)} \neq 0$ for some j, and therefore $\{a_{kji}\}$ is a linearly independent set.

It remains to show that $\{a_{kji}\}$ spans C. Because $(A + F)a_{kj(i+1)} = a_{kji}$ and $(A + F)a_{kj1} = 0$, it follows that $(A + F)$ is tangent to the linear span of $\{a_{kji}\}$. $\mathcal{B}$ is also tangent to this linear span, and therefore the entire family $\mathcal{F}$ is tangent to the linear span $\{a_{kji}\}$. It then follows that the orbit of $\mathcal{F}$ through the origin is contained in the linear span of $\{a_{kji}\}$. Because the orbit of $\mathcal{F}$ through the origin is equal to C, it follows that C is contained in the linear span of $\{a_{kji}\}$, and our proof is finished. ∎

The feedback-decomposition theorem shows that all the trajectories of $\mathcal{F} = (A + F) + \mathcal{B}$ can be parametrized by m functions $f_{11}(t), \ldots, f_{1n_1}(t), \ldots,$ $f_{p1}(t), \ldots, f_{pn_p}(t)$. Recall that $n_1 + \cdots + n_p = m$. That is, each trajectory $x(t)$ of the system $\mathcal{F}$ relative to the coordinates $\{a_{kji}\}$ satisfies $x_{kj1}(t) = f_{kj}(t)$, and the corresponding control u_{kj} satisfies $u_{kj} = d^{i(k)} f_{kj}/(dt)^{i(k)}$. Conversely, any m functions $f_1(t), \ldots, f_m(t)$ can be traced by a trajectory of $\mathcal{F}$. The argument consists of reversing the foregoing steps.

Stated differently, any linear controllable system can be decoupled into p scalar differential equations whose orders are determined by the controllability indices.

1.2 Controllability and the spectrum

As explained in Chapter 4, any nth-order linear differential equation

$$y^{(n)} + a_1 y^{(n-1)} + \cdots + a_n y = u$$

can be converted to its first-order linear control system in $\mathbb{R}^n$ given by $dx/dt = Ax + ub$, with

$$A = \begin{pmatrix} 0 & 1 & 0 & \cdots & 0 \\ \vdots & & \ddots & & \vdots \\ \vdots & & & \ddots & \vdots \\ \vdots & & & & 1 \\ -a_n & \cdots & \cdots & \cdots & -a_1 \end{pmatrix} \quad \text{and} \quad b = \begin{pmatrix} 0 \\ \vdots \\ \vdots \\ 1 \end{pmatrix}.$$

The characteristic polynomial of A is given by

$$\lambda^n + a_1\lambda^{n-1} + \cdots + a_n = 0.$$

When u is a closed-loop linear control, $u(x) = \ell(x)$ for some linear function ℓ. The characteristic polynomial of the feedback-equivalent system $A + \ell \otimes e_n$ is given by

$$\lambda^n + (a_1 - \ell_1)\lambda^{n-1} + \cdots + (a_n - \ell_n) = 0,$$

and therefore each coefficient of the preceding polynomial can be assigned an arbitrary value by a suitable linear feedback. This means that the spectral characteristics of $A + \ell \otimes e_n$ can be controlled by means of linear feedbacks. Theorem 1 in Chapter 4 implies that the same is true for an arbitrary controllable linear system with a single input.

The feedback-decomposition theorem shows that the foregoing results extend to linear systems with multiple inputs, because an arbitrary controllable linear system is equivalent to p independent linear differential equations with orders equal to the controllability indices.

This remarkable theorem shows that any linear system $dx/dt = Ax$, with $A \neq 0$, can always be stabilized by a single control u provided that the controlled constant vector field b is chosen so that $b, Ab, \ldots, A^{n-1}b$ forms a basis for M.

Problem 3 Show that for any linear vector field A with distinct eigenvalues there exists b such that $b, Ab, \ldots, A^{n-1}b$ spans M.

2 Bounded controls and the bang-bang principle

As we saw in Chapter 4, the reachable set at time T for any system affine in control is compact whenever the control constraint set U is compact and convex, provided that any integral curve generated by a control in $L_m^1(0, T, U)$ is defined for all t in $[0, T]$. The latter condition is always satisfied for linear systems, and therefore each reachable set $\mathcal{A}(x, U, T)$ is compact whenever U is compact and convex. It turns out that for linear systems, convexity in U is inessential.

Our proof of this unexpected result will be based on two theorems from convex analysis. The first theorem deals with the existence of extreme points for compact subsets in a locally convex real vector space V. Recall that a point e is called an extreme point of a set A in V if e is not an interior point of any line segment contained in A. Then we have the following:

Proposition 1 *Any nonempty compact subset A of a real, locally convex vector space has an extreme point.*

The second result, due to Lyapunov, concerns the range of vector-valued measures.

Proposition 2 *Let* $f : [0, T] \to \mathbb{R}^m$ *be a bounded and measurable function. For each measurable set* $E \subset [0, T]$, *let* $x_E = \int_E f(t)\, dt$. *Then the set of values* x_E *as* E *ranges over all measurable subsets of* $[0, T]$ *is convex and compact.*

The proofs for both of these propositions can be found in the work of es (1975, p. 74). Proposition 1 is essentially a paraphrase of a famous Krein-Milman theorem stating that the closed convex hull of extreme points of a compact and convex set K is equal to K.

We now return to the main development with the promised theorem.

Theorem 3 *Let* U *be a compact subset of* $\mathbb{R}^m$, *and denote by* $\mathcal{A}(x, U, T)$ *the set reachable from* x *in exactly* T *units of time by the trajectories of* (1) *generated by the measurable controls on* $[0, T]$ *that take values in* U. *Then* $\mathcal{A}(x, U, T)$ *is compact for each* $x \in \mathbb{R}^n$ *and each* $T > 0$.

Proof Because $\mathcal{A}(x, U, T) = (\exp T A)(x) + \mathcal{A}(0, U, T)$, it suffices to prove the theorem for $\mathcal{A}(0, U, T)$. Denote by $\mathrm{co}(U)$ the convex hull of U. $\mathrm{co}(U)$ is compact, and therefore $\mathcal{A}(x, \mathrm{co}(U), T)$ is compact and convex for each $T > 0$ (consequence of Theorem 11, Chapter 4).

Denote by $L_m^1(U, T)$ and $L_m^1(\mathrm{co}(U), T)$ the class of all measurable m-valued controls with values in U and in $\mathrm{co}(U)$, respectively. For $a \in \mathcal{A}(0, \mathrm{co}(U), T)$, denote by $C(a)$ the set of all controls u in $L_m^1(\mathrm{co}(U), T)$ such that $\int_0^T (\exp(T - \tau)A)Bu(\tau)\, d\tau = a$.

$C(a)$ is a compact and convex subset of $L_m^1(\mathrm{co}(U), T)$. According to Proposition 1, $C(a)$ has an extreme point e, which must be an extreme point of $L_m^1(\mathrm{co}(U), T)$, for if not, then $e = (\xi_1 + \xi_2)/2$, with $\xi_i \in L_m^1(\mathrm{co}(U), T)$, $i = 1, 2$, and $\xi_1 \neq \xi_2$. But then, according to Proposition 2, there exists a measurable subset E of $[0, T]$ such that

$$\frac{1}{2} \int_0^T e^{A(T-t)} B\xi_1(t)\, dt = \int_E e^{A(T-t)} B\xi_1(t)\, dt,$$

$$\frac{1}{2} \int_0^T e^{A(T-t)} B\xi_2(t)\, dt = \int_E e^{A(T-t)} B\xi_2(t)\, dt.$$

Let $u(t) = \chi_E(t)\xi_1(t) + (1 - \chi_E(t))\xi_2(t)$, and $v(t) = \chi_E(t)\xi_2(t) + (1 - \chi_E(t))\xi_1(t)$, with χ_E denoting the characteristic function of E. Both u and v belong to $L_m^1(\text{co}(U), T)$, and

$$
\int_0^T e^{A(T-t)} Bu(t)\, dt = \int_E e^{A(T-t)} B\xi_1(t)\, dt + \int_{[0,T]-E} e^{A(T-t)} B\xi_2(t)\, dt
$$

$$
= \frac{1}{2} \int_0^T e^{A(T-t)} B\xi_1(t)\, dt + \frac{1}{2} \int_0^T e^{A(T-t)} B\xi_2(t)\, dt
$$

$$
= \int_0^T e^{A(T-t)} Be(t)\, dt = a.
$$

An analogous argument shows that $\int_0^T e^{A(T-t)} Bv(t)\, dt = a$. It follows that $e = (\xi_1 + \xi_2)/2 = (u + v)/2$, which is a contradiction to the fact that e is an extreme point of $C(a)$. Hence e is an extreme point of $L_m^1(\text{co}(V), T)$, as claimed earlier.

Because each extreme point of $L_m^1(\text{co}(U), T)$ belongs to $L_m^1(U, T)$, it follows that e is an element of $L_m^1(U, T)$. Our proof is now finished. ∎

Definition 6 A control u is called a "bang-bang control" on an interval $[0, T]$ if u is measurable and takes values in the boundary ∂U of a compact control constraint set U.

Theorem 4 (The bang-bang principle) *Let U be any compact set. Then*

$$
\mathcal{A}(x, \text{co}(U), T) = \mathcal{A}(x, U, T) = \mathcal{A}(x, \partial U, T).
$$

That is, points reachable by admissible controls are reachable by bang-bang controls.

Proof According to Theorem 3, $\mathcal{A}(x, \partial U, T) = \mathcal{A}(x, \text{co}(\partial U), T)$. Furthermore, $\mathcal{A}(x, \text{co}(\partial U), T) = \mathcal{A}(x, \text{co}(U), T)$, since $\text{co}(\partial U) = \text{co}(U)$. The proof is now finished. ∎

Corollary *Let $a < b$ be any numbers. The reachable sets of a single-input linear system controlled by measurable controls that take values in $\{a, b\}$ remain the same when the control functions are allowed to take values in the entire interval $[a, b]$.*

The control functions $u(t)$, which can take only the value a or b, must frequently switch between these values, which accounts for their bang-bang name. For systems with multiple inputs, the boundary ∂U of compact sets is often a connected and smooth manifold. For such situations, the controls that steer to the boundary of the reachable sets often are smooth and do not live up to their bang-bang epithet.

3 Controllability of linear systems with bounded controls

As we have demonstrated, strong controllability is not possible whenever the controls are required to take values in a compact set U, for then each of the reachable sets $A(x, U, t)$ is compact. In this section we shall investigate the conditions on A and $b_1, \ldots, b_m$ that guarantee controllability when the controls $u = (u_1, \ldots, u_m)$ take values in a compact set U.

Before going into further details, it will be useful to consider two examples that are typical of the general results.

Example 1 Let

$$A = \begin{pmatrix} 1 & 1 \\ 0 & 0 \end{pmatrix}, \quad \text{and} \quad b = e_2.$$

Then $(b, \ Ab) = (e_1, \ e_2)$, and therefore the controllability condition is satisfied. Suppose that the magnitude of the control u is bounded by some constant K $(K < \infty)$. The control system is given by

$$\frac{dx_1}{dt} = x_1 + x_2, \qquad \frac{dx_2}{dt} = u(t).$$

Suppose that $x_2(0) = 0$. Then $|x_2(t)| \leq Kt$, and therefore $| \int_0^t e^{-s} x_2(s) \, ds | \leq K \int_0^t e^{-s} s \, ds \leq K$. Because

$$x_1(t) = e^t \left(x_1(0) + \int_0^t e^{-s} x_2(s) \, ds \right),$$

it follows that $x_1(t) > 0$ no matter what control u is chosen, provided that $x_1(0) > K$. Therefore, the system is not controllable.

Example 2 Let

$$A = \begin{pmatrix} 0 & 1 \\ 0 & 0 \end{pmatrix} \quad \text{and} \quad b = e_2.$$

Suppose that $\varepsilon > 0$ is any number, and let $U = \{u : |u| \leq \varepsilon\}$. The control system is given by

$$\frac{dx_1}{dt} = x_2, \qquad \frac{dx_2}{dt} = u.$$

If $u(t) = \pm\varepsilon$, then the corresponding solutions are $x_1(t) = x_1(0) + x_2(0)t \pm \varepsilon(t^2/2)$ and $x_2(t) = x_2(0) \pm \varepsilon t$. The corresponding trajectories are parabolas $x_1 - c_1 = \pm(1/2\varepsilon)(x_2^2 - c_2^2)$, where $(c_1, c_2) = (x_1(0), x_2(0))$. It is easy to see that any pair of points in $\mathbb{R}^2$ can be joined to each other by positively oriented arcs of these parabolas, and therefore the system is controllable (Figure 5.1).

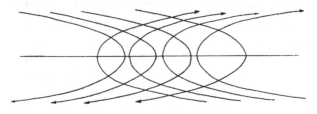

Fig. 5.1.

Theorem 5 *Suppose that the control set U is compact. Then a necessary condition for controllability is that every eigenvalue of A have its real part equal to zero.*

Proof The proof in this theorem is based on the Jordan block decomposition of matrix A. Recall that if $\lambda = \alpha + i\beta$ is an eigenvalue of A, then there exists a subspace V_λ of M such that $A(V_\lambda) \subset V_\lambda$. The structure of V_λ is determined by β. If $\beta = 0$, there exists a basis $v_1, \ldots, v_k$ in V_λ such that, relative to this basis, the restriction of A to V_λ is given by

$$
\begin{pmatrix}
\alpha & \mu & 0 & \cdots & \cdots & 0 \\
0 & \alpha & \mu & 0 & \cdots & 0 \\
\vdots & & & & & \vdots \\
0 & \cdots & \cdots & 0 & \alpha & \mu \\
0 & & \cdots & \cdots & 0 & \alpha
\end{pmatrix},
$$

with μ equal to either 0 or 1. If $\beta \neq 0$, then V_λ is an even-dimensional space, and there exists a basis $v_1, \ldots, v_{2k}$ for V_λ such that, relative to this basis, the restriction of A to V_λ is given by

$$
\begin{pmatrix}
R & \mu & 0 & \cdots & \cdots & 0 \\
0 & R & \mu & 0 & \cdots & 0 \\
\vdots & & & & & \vdots \\
0 & & & R & \mu & 0 \\
0 & & & 0 & R & \mu \\
0 & & \cdots & \cdots & 0 & R
\end{pmatrix}, \quad \text{where } R = \begin{pmatrix} \alpha & \beta \\ -\beta & \alpha \end{pmatrix}
$$

and either μ is

$$\begin{pmatrix} 1 & 0 \\ 0 & 1 \end{pmatrix}$$

or u is the zero matrix.

We shall now show that the system is not controllable whenever $\alpha \neq 0$, regardless of β. Assume first that λ is real. Let x_1 denote the first coordinate in V_λ of a trajectory $x(t)$. Then

$$\frac{dx_1}{dt} = \alpha x_1(t) + \sum_{i=1}^{m} u_i b_{ik}.$$

Because U is compact, it follows that there exists a constant K such that

$$-K < \sum_{i=1}^{m} u_i b_{ik} < K,$$

valid for each $u_1, \ldots, u_m$. If $\alpha > 0$, then choose $x_1(0) > (1/\alpha)K$. Then $dx_1/dt > 0$ for any solution of the system. Hence no point x, with $x_1 < x_1(0)$, can be reached in any positive time. For $\alpha < 0$, the argument is reversed: Points x for which $x_1 > x_1(0)$ cannot be reached from $x(0)$ in a positive time.

When λ is not real, then let x_1 denote the projection of x on the first two coordinates of V_λ. Then

$$\frac{dx_1}{dt} = Rx_1 + \sum_{i=1}^{m} u_i b_{ik},$$

and therefore

$$\frac{d}{dt}\|x_1(t)\|^2 = \frac{d}{dt}(x_1 \cdot x_1) = \left(Rx_1 + \sum u_i b_{ik}, x_1\right)$$

$$= \alpha \|x_1\|^2 + \sum u_i (b_{ik} \cdot x_1).$$

An argument analogous to the real case shows that when $\alpha > 0$, there exists a sphere S such that no points inside of S can be reached from outside of S. If $\alpha < 0$, the argument shows that the reverse holds. The proof is now finished. ∎

The spectral condition that the eigenvalues be confined to the imaginary axis is easily violated when the entries of the matrix are perturbed. Hence, from a practical point of view, controllability is almost never possible when there is a bound on the size of controls. Nevertheless, our next theorem gives mathematical conditions that ensure controllability.

Theorem 6 *Suppose that A is such that all its eigenvalues have real parts equal to zero. Let U be any control set that is a neighborhood of the origin in* $\mathbb{R}^m$. *Then the linear control system with controls in U is controllable whenever* $\bigcup_{k=1}^{n-1}\{A^k b_j : j = 1, \ldots, m\}$ *spans M.*

An arbitrary linear vector field can always be decomposed into a semisimple part A_S and a nilpotent part A_N in such a way that A_S and A_N commute. (A linear vector field is called semisimple if it can be diagonalized over the complex numbers.) The assumption that the eigenvalues of A have zero real parts implies that A_S is a recurrent vector field. When $A_N = 0$, the foregoing theorem becomes a particular case of a control affine system with a recurrent drift, and so there is nothing to prove, because of Theorem 5 in Chapter 4.

The proof of the general case with neither A_S nor A_N equal to zero is a bit messy, so it will be omitted. Instead, we shall prove the theorem for the nilpotent case.

Proof Assume that A is nilpotent, with $A^p \neq 0$ and $A^{p+1} = 0$ for some positive integer p. Because U contains a neighborhood of the origin in $\mathbb{R}^m$, there is no loss of generality in taking U to be a cube $\{u : |u_i| \leq \varepsilon, i = 1, \ldots, m\}$ for some $\varepsilon > 0$. Then the reachable set $\mathcal{A}(0)$ is convex. Any trajectory $x(t)$ that emanates from the origin is given by $x(t) = \int_0^t e^{A(t-s)} Bu(s)\, ds$. Because $A^{p+1} = 0$, $x(t) = \sum_{k=0}^p A^k B \int_0^t ((t-s)^k/k!) u(s)\, ds$.

Taking $u = (0, \ldots, u_j, 0, \ldots)$, with u_j an arbitrary constant, $x(t) = \sum_{k=0}^p (t^{k+1}/(k+1)!) A^k b_j u_j$. If λ is any real number, then there exists $T > 0$ such that $|u_j| = |\lambda|/T^p < \varepsilon$. Then

$$\lim_{T \to \infty} x(T) = \frac{\lambda}{(p+1)!} A^p b_j.$$

Therefore the line through $A^p b_j$ is in the closure of the set reachable from the origin.

Repeat the process with $u_j = \lambda/T^{p-1}$. Because $-(\lambda T/(p+1)!) A^p b_j$ is in the closure of $\mathcal{A}(0)$, and the latter is convex,

$$\frac{1}{2}\left(x(T) - \frac{\lambda T}{(p+1)!} A^p b_j\right) = \frac{\lambda}{2p!} A^{p-1} b_j + \frac{\lambda}{2(p-1)!T} A^{p-2} b_j + \cdots$$

is also contained in the closure of $\mathcal{A}(0)$. Then

$$\lim_{T \to \infty} \frac{1}{2} \left(x(T) - \frac{\lambda T}{(p+1)!} A^p b_j \right) = \frac{\lambda A^{p-1}}{2p!} b_j.$$

Therefore the vector space spanned by $A^p b_j$ and $A^{p-1} b_j$, $j = 1, \ldots, m$, is contained in the closure of $\mathcal{A}(0)$. Obviously this procedure shows that the vector space spanned by $A^k b_j, k = 0, \ldots, p, j = 1, \ldots, m$, is contained in the closure of $\mathcal{A}(0)$. Therefore, $\mathcal{A}(0)$ is a dense subset of M. But then $\mathcal{A}(0) = M$, because linear systems are Lie-determined, and the Lie algebra has full rank.

Because $-A$ is also nilpotent, the preceding argument is applicable to the time-reversed system $dx/dt = -A + \sum_{j=1}^{m} u_j b_j$. Its set reachable from the origin consists of points in M that can be steered to the origin along the trajectories of the original system. This fact, combined with the preceding, shows that any two points of M can be connected by a trajectory of the system. End of the proof. ∎

Problem 4 Show that the linear system corresponding to

$$y^{(n)} + a^2 y = u$$

is not controllable, with $|u| \leq \varepsilon$ for any $\varepsilon > 0$, whenever $n \neq 2$.

Problem 5 Show that the preceding problem remains not controllable even if $a = 0$ provided that the control u is restricted to $u \geq 0$.

4 Polynomial drifts

We now return to the structure theorems of linear systems with a different perspective. A linear system can be represented as a system in $\mathbb{R}^n$ in terms of the coordinates

$$x = \begin{pmatrix} x_1 \\ \vdots \\ x_n \end{pmatrix}$$

relative to a fixed basis $v_1, \ldots, v_n$ in M. Denoting by (a_{ij}) the entries of the matrix of A relative to this basis, then each coordinate $A^{(i)}$ of A is given by $A^{(i)}(x_1, \ldots, x_n) = \sum_{j=1}^{n} a_{ij} x_j$. The induced system in $\mathbb{R}^n$ is given by n differential equations:

$$\frac{dx_i}{dt} = A^{(i)}(x_1, \ldots, x_n) + \sum_{j=1}^{m} u_j(t) b_j^{(i)}, \qquad i = 1, \ldots, n, \qquad (5)$$

with $b_j^{(i)}$ denoting the ith coordinate of the control vector field b_j.

Differential system (5) can be considered a particular case of a system in $\mathbb{R}^n$ in which the coordinates of the drift are polynomials in the variables $x_1, \ldots, x_n$. We shall now reconsider theorems about linear systems in this larger class of polynomial systems.

Definition 7 A vector field $A(x) = (A^{(1)}(x), A^{(2)}(x), \ldots, A^{(n)}(x))$ in $\mathbb{R}^n$ will be called a polynomial vector field if each coordinate $A^{(i)}(x_1, \ldots, x_n)$ is a polynomial in the variables $x_1, \ldots, x_n$.

In this section we shall consider the following differential system:

$$\frac{dx_i}{dt} = A^{(i)}(x_1, \ldots, x_n) + \sum_{j=1}^m u_j b_j^{(i)}, \qquad i = 1, 2, \ldots, n, \qquad (6)$$

with each $A^{(i)}$ a polynomial in the variables $x_1, \ldots, x_n$, and $b_1, \ldots, b_m$ constant vector fields in $\mathbb{R}^n$, with $b_j^{(i)}$ denoting the ith coordinate of b_j. In vector form, the preceding system is written as

$$\frac{dx}{dt} = A(x) + Bu, \qquad (7)$$

with B equal to the matrix whose columns are $b_1, \ldots, b_m$, and u equal to the column vector with components $u_1, \ldots, u_m$.

Let us first recall the basic facts concerning polynomials. A polynomial function $f(x_1, \ldots, x_n)$ is called *homogeneous of order p* if $f(\lambda x_1, \ldots, \lambda x_n) = \lambda^p f(x_1, \ldots, x_n)$. Differentiating f with respect to λ and setting $\lambda = 1$ yields the familiar Euler identity:

$$\sum_{i=1}^n \frac{\partial f}{\partial x_i} x_i = pf(x_1, \ldots, x_n),$$

which can also be expressed as $(\operatorname{grad} f(x), x) = pf(x)$, with $(x, y) = \Sigma x_i y_i$ denoting the usual scalar product on $\mathbb{R}^n$. In particular, homogeneous polynomials of order zero are constant. It follows from the foregoing that if f is a homogeneous polynomial of order p, then each $\partial f / \partial x_i$ is a homogeneous polynomial of order $p - 1$. An arbitrary polynomial $f(x_1, \ldots, x_n)$ can always be written as the sum of homogeneous polynomials of various orders. Each homogeneous polynomial in this sum corresponds to a term in the Taylor-series expansion of f at the origin. Thus,

$$f(x_1, \ldots, x_n) = f_0(x_1, \ldots, x_n) + f_1(x_1, \ldots, x_n) + \cdots + f_p(x_1, \ldots, x_n),$$

with each f_i a homogeneous polynomial of order i. In particular, f_0 is constant, and f_1 is linear.

We shall use $A(x)$ to denote the column vector with components $A^{(1)}, \ldots,$ $A^{(n)}$, and we shall write $A(x) = A_0(x) + A_1(x) + \cdots + A_p(x)$ for its decomposition into homogeneous parts. Each component of the vector A_i is a homogeneous polynomial of degree i.

In terms of the decomposition into the homogeneous polynomial vector fields, equation (7) is given by

$$\frac{dx}{dt} = A_0(x) + A_1(x) + \cdots + A_p(x) + \sum_{j=1}^{m} u_j b_j. \tag{8}$$

Ultimately we shall derive conditions for the controllability of equation (8) in terms of the highest order of homogeneity A_p of A and the controlled vectors $b_1, \ldots, b_m$.

4.1 Homogeneous polynomial vector fields and their Lie algebras

When A is a homogeneous polynomial vector field, the Lie algebra generated by system (7) has a particular structure that accounts for many properties of the system. The following theorems describe this structure in more detail.

Theorem 7 *Let $\mathcal{F}$ be a family of homogeneous polynomial vector fields. Then*

(a) *the evaluation of* $\mathrm{Lie}(\mathcal{F})$ *at the origin coincides with the subalgebra of constant vector fields in* $\mathrm{Lie}(\mathcal{F})$, *and*

(b) $\mathrm{Lie}_x(\mathcal{F}) = \mathbb{R}^n$ *for all x in $\mathbb{R}^n$ if and only if the subalgebra of constant vector fields in* $\mathrm{Lie}(\mathcal{F})$ *is of dimension n.*

Before giving the proof of this theorem, let us note that if X and Y are homogeneous polynomial vector fields, with $\deg(X)$ and $\deg(Y)$ respectively denoting their degrees of homogeneity, then their Lie bracket $[X, Y]$ is a homogeneous polynomial vector field with $\deg([X, Y]) = \deg(X) + \deg(Y) - 1$.

Proof Every element of $\mathrm{Lie}(\mathcal{F})$ is the sum of homogeneous fields each belonging to $\mathrm{Lie}(\mathcal{F})$. Such fields vanish at the origin unless they are constant. Therefore, $\mathrm{Lie}_0(\mathcal{F})$ is equal to the evaluation of all constant fields in $\mathrm{Lie}(\mathcal{F})$.

If $\mathrm{Lie}_0(\mathcal{F}) = \mathbb{R}^n$, then $\mathrm{Lie}_x(\mathcal{F}) = \mathbb{R}^n$ for all x, and the proof is finished. ∎

The next result describes the structure of constant vector fields in a Lie algebra of polynomial vector fields.

Theorem 8 *Let* $\mathcal{F} = \{X, Y_1, \ldots, Y_m\}$ *be a family of vector fields, with* X *a homogeneous polynomial vector field of degree* p ($p \geq 1$), *and each* $Y_1, \ldots, Y_m$ *a constant vector field. Then the subalgebra of all constant vector fields* $\mathcal{L}$ *in* Lie($\mathcal{F}$) *has the following characterization: It is equal to the smallest vector space* S *of constant vector fields that contains vector fields* $Y_1, \ldots, Y_m$ *and that in addition contains all vector fields of the form* ad $V_1 \circ \cdots \circ$ ad $V_p(X)$ *for arbitrary elements* $V_1, \ldots, V_p$ *in* S.

The proof of this theorem is a bit lengthy, and for that reason it will be omitted. The interested reader can find the proof elsewhere (Jurdjevic and Kupka, 1985). Theorem 8 provides a constructive procedure of generating constant vector fields in Lie($\mathcal{F}$) that goes as follows:

Let $\mathcal{L}_1$ denote the linear span of $Y_1, \ldots, Y_m$. For each integer $k \geq 1$, let $\mathcal{L}_{k+1}$ denote the linear span of $\mathcal{L}_k$ and $\{$ad $V_1 \circ \cdots \circ$ ad $V_p(X) : \{V_1, \ldots, V_p\} \subset \mathcal{L}_k\}$. Note that each $\mathcal{L}_k$ is a linear subspace in the space of constant vector fields, as can be easily established by induction on k. Recall that for $[X, Y]$, deg($[X, Y]$) = deg(X) + deg(Y) − 1, and therefore deg(ad $V_p(X)$) = $p - 1$, deg(ad V_{p-1}ad $V_p(X)$) = $p - 2$, and so forth. Therefore, deg(ad $V_1 \circ \cdots \circ$ ad $V_p(X)$) = 0, and the corresponding vector field is constant.

It follows that $\mathcal{L}_1 \subseteq \mathcal{L}_2 \subseteq \cdots = \bigcup_{k \geq 1} \mathcal{L}_k$. Because $\bigcup_{k \geq 1} \mathcal{L}_k$ is contained in the smallest vector space S that satisfies the conditions of Theorem 8, it must be equal to the space of constant vector fields $\mathcal{L}$ in Lie($\mathcal{F}$). The foregoing filtration of $\mathcal{L}$ coincides with the filtration for linear systems when $p = 1$.

Because $\mathcal{L}$ is a linear subspace of $\mathbb{R}^n$, it follows that $\mathcal{L}$ is finite-dimensional. Therefore there exists an integer r such that $\mathcal{L}_{r+1} = \mathcal{L}_r$, and then $\mathcal{L}_k = \mathcal{L}_r$ for any $k > r$. The vector fields in $\mathcal{L}_r$ consist of Lie brackets of orders up to $p(r - 1)$. The following corollary shows further implications of Theorem 8.

Corollary *Let* $\mathcal{F} = \{X, Y_1, \ldots, Y_m\}$ *be a family of vector fields in* $\mathbb{R}^n$, *with* X *a homogeneous polynomial vector field of degree* p ($p \geq 1$) *and each* Y_i *a constant vector field. If* $\mathcal{L}$ *denotes the subalgebra of constant vector fields in* Lie ($\mathcal{F}$), *then a basis for* $\mathcal{L}$ *can be found among the Lie brackets of* $\mathcal{F}$ *of order up to* ($n - 1$)p.

For linear systems, this corollary can be used to show that the controllability space of linear systems is equal to the linear span of $\{A^i b_j : j = 1, \ldots, m, 0 \leq i \leq n - 1\}$, and its proof bypasses the usual arguments based on the use of the Cayley-Hamilton theorem.

The following proposition gives a useful formula that further elucidates the structure of Lie algebras defined by homogeneous vector fields.

Proposition 3 *Let* $X = \sum_{i=1}^{n} A^{(i)} (\partial/\partial x_i)$ *and* $Y = \sum_{i=1}^{n} b^{(i)} (\partial/\partial x_i)$ *be polynomial vector fields, with X homogeneous of degree p, and Y a constant vector field. Then* $(\mathrm{ad}Y)^p (X)$ *is a constant vector field whose coordinates are given by* $(-1)^p p! A^{(i)} (b^{(1)}, \ldots, b^{(n)})$, $i = 1, \ldots, n$.

Proof Use induction on p. $\mathrm{ad}\, Y (X)$ is a homogeneous vector field of degree $p - 1$ whose ith coordinate is given by $-\sum_{j=1}^{n} (\partial A^{(i)}/\partial x_j)(x_1, \ldots, x_n) b^{(j)}$. Because $(\mathrm{ad}Y)^p (X) = ((\mathrm{ad}Y)^{p-1})(\mathrm{ad}\, Y (X))$, it follows by the induction hypothesis that the ith coordinate of $(\mathrm{ad}Y)^p (X)$ is given by $(-1)^{p-1}(p - 1)!$ $(-\sum_{j=1}^{n} (\partial A^{(i)}/\partial x_j)(b^{(1)}, \ldots, b^{(n)})) b^{(j)}$. But the foregoing expression is equal to $(-1)^p p! A^{(i)} (b^{(1)}, \ldots, b^{(n)})$, as can be seen from Euler's formula. End of the proof. ∎

There is a one-to-one correspondence between mappings of $\mathbb{R}^n$ with polynomial coordinates and polynomial vector fields.

Each mapping

$$A = \begin{pmatrix} A^{(1)} \\ \vdots \\ A^{(n)} \end{pmatrix}$$

defines $X_A = \sum_{i=1}^{n} A^{(i)} (\partial/\partial x_i)$. Then a constant mapping $A(x) = b$ defines a constant vector field $Y_b = \sum_{i=1}^{n} b^{(i)} (\partial/\partial x_i)$. In terms of this notation, $(\mathrm{ad}Y_b)^p (X_A) = (-1)^p p! Y_{A(b)}$ is the formula given by the foregoing proposition. Its use is crucial in establishing the following:

Theorem 9 *Let X be a homogeneous polynomial vector field defined by the mapping A. Let $b_1, \ldots, b_m$ be any vectors in $\mathbb{R}^n$, and let $\mathcal{F} = \{X_A, Y_{b_1}, \ldots, Y_{b_m}\}$. Then the algebra of all constant vector fields in $\mathrm{Lie}(\mathcal{F})$ is isomorphic to the smallest vector space in $\mathbb{R}^n$ that contains $b_1, \ldots, b_m$ and is invariant under the mapping A.*

Proof Let $\mathcal{L}_0$ be the algebra of constant vector fields in Lie $(\mathcal{F})$. Let S denote the smallest vector space in $\mathbb{R}^n$ that contains $b_1, \ldots, b_m$ and is invariant under the mapping A. For any Y_b in $\mathcal{L}_0$, $(\mathrm{ad}Y_b)^p (X)$ is in $\mathcal{L}_0$. Because $(\mathrm{ad}Y_b)^p (X) = (-1)^p p! Y_{p(b)}$, the space $\{b : Y_b \in \mathcal{L}_0\}$ is invariant under A, and it contains $b_1, \ldots, b_m$. Therefore, $\{b : Y_b \in \mathcal{L}_0\} \subset S$.

In order to show the opposite inclusion, let a and v be in S. Then $A(a + \lambda v) \in S$ for all numbers λ. By differentiating with respect to λ, it follows that $dA(a)v \in S$. If $\mathrm{ad}\, Y_v(X)(a) = \sum_{j=1}^{n} f_j(a)(\partial/\partial x_j)$, then $f(a) = -dA(a) \cdot v$.

Therefore $f(S) \subset S$. This same reasoning applies to $\operatorname{ad} Y_{v_1} \circ \cdots \circ \operatorname{ad} Y_{v_p}(X_A)$, with $v_1, \ldots, v_p$ in S. But $\operatorname{ad} Y_{v_1} \circ \cdots \circ \operatorname{ad} Y_{v_p}(X_A)$ is constant. Hence if $\operatorname{ad} Y_{v_1} \circ \cdots \circ \operatorname{ad} Y_{v_p}(X) = Y_a$, then $a \in S$.

We have proved that $\operatorname{ad} Y_{v_1} \circ \cdots \circ \operatorname{ad} Y_{v_p}(X_A) \in \{Y_a : a \in S\}$ for any $Y_{v_1}, \ldots, Y_{v_p}$ in $\{Y_v : v \in S\}$. It then follows from Theorem 8 that $\{Y_a : a \in S\} \subset \mathcal{L}$, and our proof is finished. ∎

Theorem 9 shows that the filtration of the space of constant vector fields $\mathcal{L}_1 \subseteq \cdots \subseteq \mathcal{L}_k \subseteq \cdots$ is in exact correspondence with the increasing sequence of linear subspaces $S_1 \subseteq S_2 \subseteq \cdots$ in $\mathbb{R}^n$, defined as follows: S_1 is the linear span of $b_1, \ldots, b_m$, S_{k+1} is the linear span of S_k and $\{A(v) : v \in S_k\}$.

The following examples illustrate the foregoing theory.

Example 3 Consider $X = x_1^3(\partial/\partial x_2)$ and $Y = \partial/\partial x_1$ in $\mathbb{R}^2$. The polynomial mapping A associated with X is given by $A^{(1)} = 0$ and $A^{(2)} = x_1^3$, and

$$b = \begin{pmatrix} 1 \\ 0 \end{pmatrix} = e_1.$$

It follows that $A(b) = e_2$, and therefore $\operatorname{Lie}\{X, Y\}$ has rank 2 at each point in $\mathbb{R}^2$.

Indeed, $(\operatorname{ad} Y)^3(X) = -6(\partial/\partial x_2)$, and therefore $(\operatorname{ad} Y)^3(X)$ and Y are linearly independent. It follows that $\operatorname{Lie}\{X, Y\}$ is an infinite-dimensional Lie algebra, since $(\operatorname{ad} X)^n(Y) = a_n x^{2n}(\partial/\partial x_2)$ for some nonzero constant a_n.

Example 4 The polynomial mapping associated with control of a rigid body is given by

$$A(x_1, x_2, x_3) = \begin{pmatrix} a_1 x_2 x_3 \\ a_2 x_1 x_3 \\ a_3 x_1 x_2 \end{pmatrix}.$$

If

$$b = \begin{pmatrix} b^{(1)} \\ b^{(2)} \\ b^{(3)} \end{pmatrix},$$

then

$$A(b) = \begin{pmatrix} a_1 b^{(2)} b^{(3)} \\ a_2 b^{(1)} b^{(3)} \\ a_3 b^{(2)} b^{(1)} \end{pmatrix} \quad \text{and} \quad A^2(b) = (a_1 a_2 a_3)(b^{(1)} b^{(2)} b^{(3)}) \begin{pmatrix} b^{(1)} \\ b^{(2)} \\ b^{(3)} \end{pmatrix}.$$

Even though b and $A^2(b)$ are linearly dependent, the plane spanned by b and $A(b)$ is generally not invariant under A, because A is not a linear mapping.

4.2 Controllability

The following results further justify some earlier remarks about nonlinear aspects of the theory of linear systems.

Theorem 10 *Suppose that a control system is defined by equation* (7), *with a polynomial drift A whose coordinates are homogeneous polynomials, all of the same degree of homogeneity p, which is odd. Such a system is strongly controllable if and only if the smallest vector space that contains $b_1, \ldots, b_m$ and that is invariant under A is equal to $\mathbb{R}^n$.*

Proof The proof of this theorem is almost identical with the proof of the corresponding result concerning linear systems, and it is based on an inductive argument of the filtration $\mathcal{L}_1 \subset \mathcal{L}_2 \subset \cdots \subset \mathcal{L}_r$ showing that each $\mathcal{L}_k$ is contained in the strong Lie saturate of the system.

Let $\mathcal{LS}_s(\mathcal{F})$ denote the strong Lie saturate of $\mathcal{F} = \{A + \sum_{i=1}^m u_i b_i : (u_1, \ldots, u_m) \in \mathbb{R}^m\}$. Because $ub_j = \lim_{\lambda \to \infty}((1/\lambda)(A + \lambda u b_j))$, and because $\mathcal{LS}_s(\mathcal{F})$ is convex, $\mathcal{L}_1$ is contained in $\mathcal{LS}_s(\mathcal{F})$.

Assume now that $\mathcal{L}_k \subset \mathcal{LS}_s(\mathcal{F})$. Let $V \in S_k$ and $\lambda \in \mathbb{R}^1$. Then $\exp \lambda V$ is a strong normalizer for $\mathcal{F}$, and therefore $(\exp \lambda V)_\#(A) \subset \mathcal{LS}_s(\mathcal{F})$. But

$$(\exp \lambda V)_\#(A) = \sum_{k=0}^{\infty} \frac{\lambda^k}{k!}(\text{ad}V)^k(A) = \sum_{k=0}^{p} \frac{\lambda^k}{k!}(\text{ad}V)^k(A),$$

since $(\text{ad}V)^k(A) = 0$ for $k > p$. Let $\epsilon = \pm 1$ and $\alpha > 0$. Then, because p is odd,

$$\lim_{\epsilon\lambda \to \infty} \frac{\alpha}{|\lambda|^p}(\exp \lambda V)_\#(A) = \varepsilon \frac{\alpha}{p!}(\text{ad}V)^p(A).$$

It now follows from the closure and time-scaling invariance of $\mathcal{LS}_s(\mathcal{F})$ (Theorem 11, Chapter 3) that $(\varepsilon\alpha/p!)(\text{ad}V)^p(A) \in \mathcal{LS}_s(\mathcal{F})$.

Therefore, $\lambda(\text{ad}V)^p(A) \in \mathcal{LS}_s(\mathcal{F})$ for each $\lambda \in \mathbb{R}^1$. Because $\mathcal{LS}_s(\mathcal{F})$ is convex, it follows that the vector space spanned by $S_k \cup \{(\text{ad}V)^p(A) : V \in S_k\}$ is contained in $\mathcal{LS}_s(\mathcal{F})$. The proof is now finished, since S_{k+1} is equal to such a space. ∎

The preceding controllability result easily extends to the general polynomial drifts. If A is an arbitrary polynomial drift, let $A = A_0 + A_1 + \cdots + A_p$ be its decomposition into homogeneous parts, each A_i having i as its degree of homogeneity; p is the highest degree of A. It follows that for any constant vector field Y_b in Lie $(\mathcal{F})$, $(\mathrm{ad}Y_b)^p(A)$ is the constant vector field whose value in $\mathbb{R}^n$ is equal to $(-1)^p p! A_p(b)$. Therefore each constant field Y_b, with b in the linear span of $\{A_p^i(b_j) : 0 \leq i \leq n - 1, \ 1 \leq j \leq m\}$, is contained in Lie($\mathcal{F}$). The methods used in Theorem 10 extend easily to the following situation.

Theorem 11 *Suppose that p is odd and that the smallest vector space that contains $b_1, \ldots, b_m$ and is invariant under A_p is equal to $\mathbb{R}^n$. Then the system*

$$\frac{dx}{dt} = A_0(x) + \cdots + A_p(x) + \sum_{i=1}^{m} u_i b_i$$

is strongly controllable.

Because the proof of this theorem is quite analogous to that for Theorem 10, it will be omitted.

If the constant term A_0 of A is equal to zero, then the reachable set at exactly t units of time is equal to the union of all points reachable from zero in t or fewer units of time. Therefore, in such a case it is evident that strong controllability is the same as controllability at each instant t $(t > 0)$. Theorem 13 in Chapter 3 shows that the same conclusion is valid even when $A_0 \neq 0$.

Problem 6 Show that $dx_1/dt = x_2^3$, $dx_2/dt = u$ is strongly controllable in $\mathbb{R}^2$.

Problem 7 Show that $dx_1/dt = x_2^3 + \varepsilon x_2^4$, $dx_2/dt = u$ is not controllable for any $\varepsilon \neq 0$.

Problem 8 Let $\mathcal{F} = A + \mathcal{B}$ be any linear controllable system. Show the existence of b in $\mathcal{B}$ and a linear feedback F such that $dx/dt = (A + F)x + u(t)b$ is controllable.

Notes and sources

Theorem 1 is due to Kalman, Ho, and Narendra (1962). The matrix $(B\,AB \ldots A^{n-1}B)$ is known as the controllability matrix (Lee and Markus, 1967). Apart from geometric control theory, a Lie-theory interpretation of vectors b, Ab, $\ldots$, $A^{n-1}b$ also appears in the work of Hörmander (1967) in reference to an old paper by Kolmogoroff in 1939 dealing with solutions for the hypoelliptic quadratic operator

$$L(u) = \sum_{i,j=1}^{n} (A_{ij}x_i x_j) \frac{\partial^2}{\partial x_i \partial x_j} u + \sum_{i=1}^{n} b_i \frac{\partial}{\partial x_i}.$$

Theorem 2 is due to Brunovsky (1976). Control of a spectrum by means of linear feedback is known as "pole assignment" in linear control theory (Wonham, 1974). The bang-bang principle seems to have originated with LaSalle (1954). The statement of Theorem 6 appears in the work of Lee and Markus (1967); the proof of this theorem given here is original.

The extension of controllability results to systems having polynomial drifts was published done by Jurdjevic and Kupka (1985). As pointed out by H. J. Sussmann (private communication), some of these results can also be deduced from the work of Brunovsky (1976) on odd systems.

6

Systems on Lie groups and homogeneous spaces

Continuing with the general theme begun in Chapter 5 of amalgamating the basic theory with additional mathematical structures, we shall now consider differential systems possessing group symmetries. Having in mind particular applications in geometry, mechanics, and control of mechanical systems, this chapter focuses on differential systems on Lie groups having either left or right invariance properties. We shall presently show that the basic geometric control theory described in earlier chapters adapts well to systems on Lie groups, and when enriched with additional geometric structure, it provides a substantial theoretical foundation from which various mathematical topics can be effectively pursued. The reader may find it useful to consider, first, several specific situations that have motivated our interest in much of the material in this chapter.

Motions of a rigid body The motions of a rigid body around a fixed point in a Euclidean space E^3 can be viewed as paths in the group of rotations $SO_3(R)$. The correspondence between the motions and the paths is achieved through an orthonormal frame attached to the body, called a moving frame, and an orthonormal frame stationary in the ambient space. The stationary frame is called fixed or absolute. At each instant of time, the position of the body is described by a rotation defined by the displacement of the moving frame relative to the fixed frame.

Associated with each path in the rotation group is its angular velocity. We shall be interested in the motions of a rigid body whose angular velocities are constrained to belong to a fixed subset of $\mathbb{R}^3$. Such situations typically occur in the presence of non-holonomic constraints. For instance, the motions of the rolling sphere that neither slips nor spins around the point of contact with the stationary plane, as in the Oxford problem discussed in Chapter 4, are confined to paths having angular velocities parallel to the horizontal plane. We shall show that the controllability of any such mechanical system is determined by the rank of the Lie algebra induced by the infinitesimal motions of the body. As we shall see, compactness of $SO_3(R)$ is the key property behind this phenomenon.

Framed curves in E^n It is well known that an n-times-differentiable curve in a Euclidean space E^n having its first n derivatives linearly independent can be developed to a curve in the group of motions of E^n. The developed curve consists of the curve itself and an orthonormal frame along the curve, adapted to the curve so that its first vector coincides with the tangent vector of the curve. The frame is called the Serret-Frenet frame. The Serret-Frenet frame can be thought of as the geometric analogue of the body frame in mechanics that measures the amount of curving and twisting of the curve relative to a stationary frame in E^n. In analogy with the motions of a rigid body, each curve defines its angular velocities, which are called the curvatures of the curve. For curves in E^3 these curvatures are called the curvature and the torsion.

We shall consider the curvature functions as the controls and treat the Serret-Frenet differential system as a control system. We shall be interested in the trajectories of such a system, subject to further constraints that some curvature functions are fixed in advance. For instance, we shall be interested in curves in E^3 having the torsion constant.

It turns out that the controllability of this class of systems is also determined by the rank of the associated Lie algebras, even though the underlying group is not compact.

Control systems on affine groups Linear systems can also be lifted to Lie groups through the following reasoning.

Let V denote an n-dimensional vector space, with end(V) and GL(V) denoting the space of all linear endomorphisms on V and the set of all linear automorphisms of V. We shall say that a vector field X on V is an affine vector field if $X(x) = Ax + a$ for some A in end(V) and a in V.

The set of affine vector fields can be viewed as the algebra of vector fields generated by the action of the semidirect product of GL(V) with V. The semidirect product $V * G$ of any group G that acts linearly on V consists of all pairs (v, g) in $V \times G$, with the group structure given by $(v_1, g_1)(v_2, g_2) = (v_1 + g_1 v_2, g_1 g_2)$. The reader can easily verify that $(0, e)$ is the group identity and that $(-g^{-1}v, g^{-1})$ is the group inverse of (v, g).

The semidirect product $V * G$ acts on V as follows: $(v, g)x = gx + v$ for all (v, g) in $V * G$ and all x in V. The Lie algebra of $V * G$ consists of all pairs $L = (a, A)$, with a in V, and A in the Lie algebra of G. The action of $\exp tL$ on V induces a one-parameter group of diffeomorphisms on V whose infinitesimal generator is the affine vector field $X(x) = Ax + a$.

Any finite collection of affine vector fields $X_0, X_1, \ldots, X_m, X_i(x) = A_i x + a_i, i = 0, 1, \ldots, m$, defines a natural control system:

$$\frac{dx}{dt} = X_0(x) + \sum_{i=1}^{m} u_i(t) X_i(x) = (A_0 x + a_0) + \sum_{i=1}^{m} u_i(t)(A_i x + a_i). \quad (1)$$

Linear systems compose a particular subclass of affine systems obtained by setting $a_0 = 0$ and $A_1 = A_2 = \cdots = A_m = 0$. In general, however, affine systems define a much richer class of systems than the linear class, and as we shall see, their controllability properties are essentially governed by their linear parts $A_0, A_1, \ldots, A_m$.

The class of affine systems is a particular class of differential systems on homogeneous spaces that are subordinated to a Lie-group action, as defined in Section 3.5 in Chapter 2. Any differential system that is subordinated to a group action can be lifted to either a left-invariant or a right-invariant system on the group that acts on the space. In particular, differential system (1) lifts to a control affine, right-invariant system on the semidirect product $V * G$. Although the lifted system does not provide new information for linear systems, mainly because the class of linear systems is too small in the class of affine systems, it will provide easy controllability results when G is compact, as in the case of the group of motions.

The principal aim of this chapter is to study the controllability properties of left-invariant and right-invariant systems on Lie groups and to relate their controllability properties to projected systems on homogeneous spaces. Our exposition is guided by the foregoing applications, and so we first consider systems on compact groups. The corresponding theory is illustrated with applications to mechanical systems whose motion is constrained by non-holonomic constraints. The sphere that is free to roll on another sphere without slipping provides a striking example of such a system. We show that the kinematic equations of motion can be lifted to a left-invariant distribution on $SO_3(R) \times SO_3(R)$, and hence to any of its covering groups, from which several easy controllability results can be deduced.

The results for compact groups extend to semidirect products of compact groups and vector spaces, as we shall show in this chapter. These results imply striking controllability properties of framed curves in E^n having prescribed curvature values.

The section on affine systems provides a natural transition between systems on semidirect products and systems on semisimple Lie groups. The chapter ends with a controllability theorem for right-invariant systems on a semisimple Lie group, which is then discussed in specific situations on $SL_n(R)$.

1 Families of right-invariant vector fields on a Lie group

Let G be a connected real Lie group, with its Lie algebra $L(G)$. Recall the basic properties of a right-invariant vector field X on G:

(a) $(\exp tX)(e) = \exp tL$, with $L = X(e)$, and $\exp tL = \sum_{k=0}^{\infty}(t^k/k!)L^k$, where e is the group identity on G.

(b) $(\exp tX)(x) = (\exp tL) \cdot x$ for any $x \in G$. The orbit of X through the identity is an abelian subgroup H of G, and the orbit through an arbitrary point x of G is the right coset Hx. For each $t \in \mathbb{R}^1$, $\exp tX$ is the left group translation by $\exp tL$.

For any family of right-invariant vector fields $\mathcal{F}$, $\Gamma(\mathcal{F})$, or simply Γ when there is no ambiguity, will denote the set $\{X(e): X \in \mathcal{F}\}$. Thus $\Gamma(\mathcal{F}) \subset L(G)$. The orbit of $\mathcal{F}$ through the identity e is the subgroup H consisting of all products of the form

$$\exp t_k A_k \cdots \exp t_1 A_1, \tag{2}$$

with $\{A_1, \ldots, A_k\} \subset \Gamma(\mathcal{F})$ and $(t_1, \ldots, t_k) \in \mathbb{R}^k$. Any other orbit of $\mathcal{F}$ is a right coset of H. In particular, the dimension of each orbit is equal to the dimension of H. H is a Lie subgroup of G. H is equal to G if and only if $\mathrm{Lie}(\Gamma(\mathcal{F})) = L(G)$. Therefore $\mathcal{F}$ has only one orbit if and only if $\mathrm{Lie}(\Gamma(\mathcal{F})) = L(G)$.

Let $A_{\mathcal{F}}(x)$ be the reachable set of $\mathcal{F}$ through x. The reachable set through e consists of all group products of (2) with $t_1 \geq 0$, $t_2 \geq 0, \ldots, t_k \geq 0$. Such products, in general, constitute a semigroup, not a group. We shall use $S(\Gamma)$ to denote such a semigroup. Then

$$A_{\mathcal{F}}(e) = S(\Gamma) \quad \text{and} \quad A_{\mathcal{F}}(x) = S(\Gamma)(x) = \{gx : g \in S(\Gamma)\},$$

and we have the following:

Theorem 1 $\mathcal{F}$ *is controllable if and only if*

(a) $S(\Gamma)$ *is a group and*
(b) $\mathrm{Lie}(\Gamma) = L(G)$.

The proof is evident and will be omitted.

It is important to remark that $S(\Gamma)$ has a nonempty interior in H, but that, in general, e is on the boundary of $S(\Gamma)$. The following result is basic.

Theorem 2 $S(\Gamma) = G$ *if and only if the group identity e is in the interior of* $S(\Gamma)$.

Proof Suppose that $e \in \text{int}(S(\Gamma))$. We shall show that $S(\Gamma)$ is both open and closed, and therefore must be equal to G, since the latter is connected.

To show that $S(\Gamma)$ is open, let $x \in S(\Gamma)$. Denote by V an open neighborhood of the identity contained in $S(\Gamma)$. Then $x \cdot V$ is a neighborhood of x in $S(\Gamma)$, and therefore $S(\Gamma)$ is open.

In order to show that $S(\Gamma)$ is closed, let $\{x_n\}$ be a sequence in $S(\Gamma)$ such that $\lim x_n = x$. Then $\{xx_n^{-1}\}$ is a sequence that converges to e. Therefore there exists an index N such that $xx_N^{-1} \in S(\Gamma)$, because there is a neighborhood of e contained in $S(\Gamma)$. But then $x = xx_N^{-1} \cdot x_N \in S(\Gamma)$, and therefore $S(\Gamma)$ is closed. The proof is now finished. ∎

Corollary 1 *A family of right-invariant vector fields $\mathcal{F}$ is controllable if and only if e is normally accessible from e by some elements $X_1, \ldots, X_k$ in $\mathcal{F}$.*

Proof Assume that $\mathcal{F}$ is controllable. According to Theorem 1 in Chapter 3, there exists a point g in G that is normally accessible from e. Let $X_1, \ldots, X_m$ denote the vector fields in $\mathcal{F}$ along which g is normally accessible from e. Because $\mathcal{F}$ is controllable, there exist vector fields $X_{m+1}, \ldots, X_k$ in $\mathcal{F}$ such that $(\exp t_k X_k) \circ (\exp t_{m+1} X_{m+1})(g) = e$. But then e is normally accessible from e by the fields $X_1, \ldots, X_k$.

The converse follows from Theorem 2, and the proof is finished. ∎

It is an easy consequence of Theorem 2 and Corollary 1 that the controllability of families of right-invariant vector fields is stable under small perturbations.

1.1 Compact Lie groups

Let G be a compact connected Lie group, such as $SO_n(R)$, the group of all $n \times n$ orthogonal matrices with determinants equal to 1. The connection between the orbits of families of invariant vector fields and their reachable sets is particularly simple when G is compact; they are equal to each other. Therefore, reachability questions on compact Lie groups reduce to Lie-algebra computations. The proof is based on a simple lemma.

Lemma 1 *Let $A \in L(G)$, with G compact. Then*

$$\text{cl}\{\exp -tA : t \geq 0\} \subset \text{cl}\{\exp tA : t \geq 0\}.$$

Proof Let $\omega(A) = \bigcap_{n=1}^{\infty} \text{cl}\{\exp tA : t \geq n\}$ denote the positive limit set of the orbit through the identity of the right-invariant vector field X induced by A. Because G is compact, $\omega(A)$ is nonempty. Let $x \in \omega(A)$, and let $\{t_n\}$ be a sequence of positive numbers such that $t_n \geq n$ and $\lim_{n \to \infty} (\exp t_n A) = x$. Without any loss of generality, we can assume that $t_{n+1} - t_n \to \infty$ as $n \to \infty$. But then

$$\lim_{n \to \infty} ((\exp t_{n+1} A)(\exp -t_n A)) = \lim_{n \to \infty} \exp((t_{n+1} - t_n)A) = e,$$

because $\lim_{n \to \infty} (\exp -t_n A) = \lim_{n \to \infty} (\exp t_n A)^{-1} = x^{-1}$. Therefore $e \in \omega(A)$, and then $\exp -tA = e \cdot (\exp -tA) = \lim_{n \to \infty} (\exp[(t_{n+1} - t_n) - t]A)$. Our proof is finished, since for large n, $t_{n+1} - t_n - t > 0$. ∎

It follows immediately from this lemma that any family of right-invariant vector fields is equivalent to $(\mathcal{F} \cup - \mathcal{F})$ and that therefore the orbits of $\mathcal{F}$ agree with the reachable sets. These observations yield the following:

Theorem 3 *A necessary and sufficient condition that a right-invariant family $\mathcal{F}$ be controllable on a compact connected group G is that $\mathrm{Lie}_e(\mathcal{F}) = \mathrm{Lie}(\Gamma(\mathcal{F})) = L(G)$.*

The reader may already have noticed that the proof of Lemma 1 depends on the use of large times, and consequently the time required to transfer one state to another may be large. Whereas for general groups that may be so, for compact groups there is a uniform bound on the length of time required to transfer one state to another. The proof of the preceding statement is as follows:

The interior of $\mathcal{A}(e, \leq T)$ is an open and nonempty set in G. The union of these sets covers G, and because G is compact, there is an instant $T_1 > 0$ such that $\mathcal{A}(e, \leq T_1) = G$. A similar argument applied to $-\mathcal{F}$ shows that there exists $T_2 > 0$ such that $\mathcal{A}_{-\mathcal{F}}(e, \leq T_2) = G$. But then any two states can be connected to each other in a time less than or equal to $T_1 + T_2$.

Let us now illustrate the meaning of Theorem 3 with examples from $SO_3(R)$. Assume that A_1 and A_2 are any 3×3 antisymmetric matrices such that A_1, A_2, and $[A_1, A_2]$ are linearly independent. According to Theorem 3, each rotation in $SO_3(R)$ can be written as the product of exponentials each of the form $\exp tA$, with $t \geq 0$ and A belonging to $\Gamma = \{A_1, A_2\}$. In particular, when the axes of rotation of A_1 and A_2 are orthogonal, then the axis of rotation of $[A_1, A_2]$ is orthogonal to both of those axes, and therefore $A_1, A_2,$ and $[A_1, A_2]$ are linearly independent. In this case, Theorem 3 is very closely related to the classic theorem of Euler, which states not only that every rotation can be written as the product of exponentials but also that the minimal number of exponential terms in each product is three. That is, every rotation g can be written as

$$g = (\exp t_1 A_1)(\exp t_2 A_2)(\exp t_3 A_3),$$

with each t_1, t_2, and t_3 belonging to the interval $[0, 2\pi]$. The preceding numbers t_1, t_2, and t_3 are known as Euler's angles. Figure 6.1 shows a pictorial proof of Euler's theorem. Of course, when the matrices A_1 and A_2 are not orthogonal, the minimum number of exponential terms in each product may be considerably greater than three.

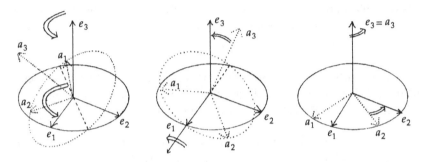

Fig. 6.1.

Problem 1 Let A_1 and A_2 be any antisymmetric 3×3 matrices such that $[A_1, A_2]$ is not contained in the linear span of A_1 and A_2. Show that $\Gamma = \{A_1 + A_2, A_1 - A_2\}$ is controllable on $SO_3(R)$.

We shall now return to the framed curves in E^n mentioned earlier and consider the associated Serret-Frenet equations as control systems, with the curvatures playing the role of controls.

Let $\sigma(t)$ denote any curve in a Euclidean space E^n whose derivatives $d^k\sigma(t)/dt^k$, $k = 1, \ldots, n$, span an n-dimensional vector space at each point along the curve.

The Serret-Frenet frame along the curve σ is described by an orthonormal matrix $R(t)$ in $SO_n(R)$ that relates this frame to a standard orthonormal frame $\vec{e}_1, \vec{e}_2, \ldots, \vec{e}_n$ in E^n and that further satisfies the following differential system on $SO_n(R)$:

$$\frac{dR}{dt} = R(t) \begin{pmatrix} 0 & -k_1(t) & 0 & \cdots & 0 \\ k_1(t) & 0 & -k_2(t) & & \vdots \\ \vdots & k_2(t) & \ddots & & \vdots \\ \vdots & 0 & & & -k_{n-1}(t) \\ 0 & \cdots & & k_{n-1}(t) & 0 \end{pmatrix}, \qquad (3)$$

where $k_1(t), \ldots, k_{n-1}(t)$ are called the curvature functions associated with curve σ. (For curves in E^3, k_2 is called the torsion of σ.) Because the first vector in the Serret-Frenet frame coincides with the tangent vector $d\sigma/dt$, it follows that $d\sigma/dt = R(t)e_1$, with the understanding that $d\sigma/dt$ is the column vector consisting of its coordinates relative to the fixed frame $\vec{e}_1, \ldots, \vec{e}_n$. We shall use $e_1, \ldots, e_n$ to denote the coordinates corresponding to $\vec{e}_1, \ldots, \vec{e}_n$. In particular, e_1 is the column vector

$$\begin{pmatrix} 1 \\ 0 \\ \vdots \\ 0 \end{pmatrix}.$$

The curve

$$\sigma = \begin{pmatrix} \sigma_1 \\ \vdots \\ \sigma_n \end{pmatrix}$$

and the rotation matrix $R(t)$ can be expressed as the curve

$$g(t) = \begin{pmatrix} 1 & 0 \\ \sigma(t) & R(t) \end{pmatrix}$$

in the group of motions E_n, realized as the closed subgroup of $GL_{n+1}(R)$ of all matrices

$$\begin{pmatrix} 1 & 0 \\ x & R \end{pmatrix},$$

with $x \in \mathbb{R}^n$ and $R \in SO_n(R)$.

The system of equations

$$\frac{dR(t)}{dt} = R(t)A(k_1, \ldots, k_{n-1}), \qquad \frac{dx}{dt} = R(t)e_1,$$

with

$$A(k_1, \ldots, k_{n-1}) = \begin{pmatrix} 0 & -k_1 & & & 0 \\ k_1 & 0 & -k_2 & & \\ & k_2 & \ddots & -k_{n-1} \\ 0 & & k_{n-1} & 0 \end{pmatrix},$$

will be regarded as a control affine differential system in E_n given by

$$\frac{dg}{dt} = g(t) \begin{pmatrix} 0 & 0 & \cdots & \cdots & & & 0 \\ 1 & 0 & -k_1 & & & & \\ 0 & k_1 & 0 & -k_2 & & & \vdots \\ \vdots & 0 & k_2 & \ddots & \ddots & & 0 \\ & & & \ddots & \ddots & & -k_{n-1} \\ 0 & \cdots & \cdots & & k_{n-1} & & 0 \end{pmatrix}, \tag{4}$$

with $k_1, \ldots, k_{n-1}$ playing the role of controls.

We should remind the reader that each curvature $k_1, \ldots, k_{n-2}$ of σ is positive, although the last curvature could be of either sign. Of course, system (4), as a control system, has well-defined trajectories for any choice of control functions. However, the projections of the trajectories on E^n define a larger class of curves than those that have well-defined curvatures. For instance, the projection of any trajectory of (4) for which $k_1 = 0$ is a constant curve.

Apart from the planar curves, the geometric significance of the Serret-Frenet frames and the associated curvature parameters seem not to be well documented in the literature on differential geometry. In this context we shall answer only one of the easiest questions that one may ask about the trajectories of system (4), namely, the question of its controllability. However, in order to further emphasize the role of noncommutativity, and the scope of these results, we shall ask our controllability questions under an additional constraint that some of the curvatures in system (4) are fixed in advance. Among many such possibilities, we shall consider only the extreme case where all but one curvature are constant. This assumption reduces the number of controls in equation (4) to a single control and introduces a drift term in the rotational part of the equations corresponding to the constant curvatures.

Under this assumption, equation (3) can be written as $dR/dt = R(t)(A + uB)$, with A defined by the given curvature values $k_1, \ldots, k_{i-1}, k_{i+1}, \ldots, k_{n-1}$, and $B = (e_{i+1} \otimes e_i) - (e_i \otimes e_{i+1})$, where u denotes the remaining curvature, which is free to vary. Here, as well as elsewhere, $(e_j \otimes e_i) - (e_i \otimes e_j)$ denotes the antisymmetric matrix for which $((e_j \otimes e_i) - (e_i \otimes e_j))(x) = (e_i \cdot x)e_j - (e_j \cdot x)e_i$ for all $x \in \mathbb{R}^n$.

Writing $g(t)$ for $R^{-1}(t)$ turns the left-invariant system into a right-invariant system:

$$\frac{dg(t)}{dt} = -(A + uB)g(t), \tag{5}$$

with

$$-A = \begin{pmatrix} A_1 & 0 \\ 0 & A_2 \end{pmatrix}, \qquad A_1 = \begin{pmatrix} 0 & k_1 & 0 & \cdots & & 0 \\ -k_1 & 0 & k_2 & & & \vdots \\ 0 & -k_2 & 0 & \ddots & & 0 \\ \vdots & & \ddots & \ddots & k_{i-1} \\ 0 & \cdots & & 0 & -k_{i-1} & 0 \end{pmatrix},$$

$$A_2 = \begin{pmatrix} 0 & k_{i+1} & 0 & \cdots & & \cdots & 0 \\ -k_{i+1} & 0 & \ddots & & & & \vdots \\ 0 & \ddots & \ddots & & & & \vdots \\ \vdots & & & \ddots & & 0 \\ \vdots & & & & \ddots & k_{n-1} \\ 0 & \cdots & & & 0 & -k_{n-1} & 0 \end{pmatrix},$$

and

$$-B = \begin{pmatrix} 0 & \cdots & & & \cdots & 0 \\ \vdots & \ddots & & & & \vdots \\ & & 0 & 1 & & \\ & & -1 & 0 & & \\ \vdots & & & & \ddots & \vdots \\ 0 & \cdots & & & \cdots & 0 \end{pmatrix}$$

It will be convenient for future reference to name the foregoing system the *stiff Serret-Frenet system*. It then follows from Theorem 3 that the controllability of that system is determined by the Lie algebra generated by $\Gamma = \{-A - uB : u \geq 0\}$, which is the same as the Lie algebra generated by matrices A and B. It should be emphasized that this Lie algebra depends on the given values of $k_1, \ldots, k_{i-1}, k_{i+1}, \ldots, k_{n-1}$ and also depends on the index i. The classification of such Lie algebras makes use of the unitary matrices and the complex structure and necessitates a brief detour into the Euclidean and Hermitian geometry of $\mathbb{C}^n$.

1.2 *Orthogonal and symplectic groups and the unitary group*

Let V be an n-dimensional complex vector space. Because the complex field $\mathbb{C}$ includes the reals, V can also be regarded as a real vector space $\mathcal{R}(V)$. For any basis $a_1, \ldots, a_n$ in V, $a_1, \ldots, a_n, ia_1, \ldots, ia_n$ is a basis for $\mathcal{R}(V)$. Therefore $\mathcal{R}(V)$ is a $2n$-dimensional real vector space.

Any complex linear transformation $L : V \to V$ is also linear as a transformation of $\mathcal{R}(V)$, but the converse may not be true. The precise connection between real linear transformations and complex linear transformations is provided by the operator $J : V \to V$ defined by $J(v) = iv$, $v \in V$. It is easy to check that any linear endomorphism $L : \mathcal{R}(V) \to \mathcal{R}(V)$ is complex linear if and only if L commutes with J.

A complex bilinear form $\langle \, , \, \rangle$ on V is called *Hermitian* if

(a) $\langle \lambda u, v \rangle = \lambda \langle u, v \rangle$ and $\langle u, \lambda v \rangle = \bar{\lambda} \langle u, v \rangle$ for each $\lambda \in \mathbb{C}$ and all u, v in V,
(b) $\langle u, v \rangle = \overline{\langle v, u \rangle}$ for all u, v in V, and
(c) $\langle u, u \rangle \geq 0$ for all $u \in V$, with $\langle u, u \rangle = 0$ if and only if $u = 0$.

Definition 1 The set of all linear transformations of V that leave its Hermitian form invariant is called the unitary group. It is denoted by $U_n(V)$.

If $e_1, \ldots, e_n$ is an orthonormal basis for V relative to the Hermitian product, then

$$\langle u, v \rangle = \sum_{i=1}^{n} u_i \bar{v}_i,$$

with $(u_1, \ldots, u_n)$ and $(v_1, \ldots, v_n)$ denoting the complex coordinates of u and v relative to the foregoing basis. In this basis of V, each element T of $U_n(V)$ can be represented by an $n \times n$ matrix with complex entries. The set of all such matrices will be denoted by $U_n(\mathbb{C})$.

It follows that $T \in U_n(\mathbb{C})$ if and only if T^{-1} is equal to the transpose of the complex conjugate of T. Therefore, the absolute value of the determinant of T is equal to 1, which accounts for the name of the group.

The elements of $U_n(V)$ can also be regarded as linear automorphisms of $\mathcal{R}(V)$. Each element of $U_n(V)$, when represented by its matrix relative to the basis $e_1, \ldots, e_n, ie_1, \ldots, ie_n$, becomes a $2n \times 2n$ matrix with real entries. The set of all such matrices will be denoted by $U_{2n}(R)$.

For each u and v in V, let $x_1, \ldots, x_n, y_1, \ldots, y_n$ and $p_1, \ldots, p_n, q_1, \ldots, q_n$ denote their real coordinates relative to the preceding basis. The complex coordinates of u and v are given by $u_j = x_j + iy_j$ and $v_j = p_j + iq_j$. Hence,

$$\langle u, v \rangle = \sum_{j=1}^{n}(x_j + iy_j)(p_j - iq_j) = \sum_{j=1}^{n}(x_j p_j + y_j q_j) + i \sum_{j=1}^{n} y_j p_j - x_j q_j.$$

Denoting by (,) and [,] the real and the imaginary parts of $\langle\,,\,\rangle$, it follows that

$$(u, v) = \sum_{j=1}^{n} x_j p_j + y_j q_j \quad \text{and} \quad [u, v] = \sum_{j=1}^{n} y_j p_j - x_j q_j.$$

The bilinear form (,) is symmetric and positive-definite, whereas [,] is anti-symmetric and nondegenerate. The subgroups of $\mathrm{GL}_{2n}(R)$ that leave these forms invariant are denoted by $O_{2n}(R)$ and $\mathrm{Sp}_{2n}(R)$, respectively called the orthogonal group and the symplectic group. It then follows from our definitions that

$$U_{2n}(R) = O_{2n}(R) \cap \mathrm{Sp}_{2n}(R).$$

The Lie algebra of $O_{2n}(R)$ consists of all anti-symmetric $2n \times 2n$ matrices, and the Lie algebra of $\mathrm{Sp}_{2n}(R)$ consists of all matrices A whose transposes A^* are equal to JAJ, with

$$J = \begin{pmatrix} 0 & -I \\ I & 0 \end{pmatrix}.$$

(In this notation, J is identified with its matrix relative to the basis $e_1, \ldots, e_n,$ $ie_1, \ldots, ie_n$.)

Conversely, any $2n$-dimensional real vector space V with a positive-definite (,) bilinear form can be considered an n-dimensional complex vector space, with its Hermitian product $\langle\,,\,\rangle$ defined through a complex structure J. A complex structure on V is any linear transformation J such that $J^2 = -I$. The reader may verify that complex structures exist on any even-dimensional real vector space.

Given any complex structure J on V, the complex multiplication on V is defined as $(x + iy)(v) = xv + yJ(v)$ for all $x + iy$ in $\mathbb{C}$ and v in V, and the Hermitian form $\langle\,,\,\rangle$ is defined through the following formula:

$$\langle u, v \rangle = (u, v) + i(u, Jv) \quad \text{for all } u \text{ and } v \text{ in } V.$$

The reader may notice that the preceding groups could have been defined independently of the basis for V, in which case $U(V)$, $O(V)$, and $\mathrm{Sp}(V)$ would be the usual notations for these groups. Because $\mathrm{Sp}(V)$ is a closed subgroup of $\mathrm{GL}(V)$, and because $O(V)$ is a compact group, it follows that $U(V)$ is a compact subgroup of $O(V)$.

The Lie algebra of $U(V)$ can be regarded as the vector space of all $n \times n$ complex matrices A whose transposes are equal to the negatives of their complex conjugates. The dimension of this Lie algebra regarded as a real vector space is equal to n^2, and therefore $U(V)$ is a proper subgroup of $O(V)$ whenever $n > 1$.

Each orthonormal basis $e_1, \ldots, e_n$ in V induces a natural basis for the space of antisymmetric matrices, as follows:

Let A_{ij} denote the antisymmetric matrix $(e_i \otimes e_j) - (e_j \otimes e_i)$. The set $\{A_{ij} : i \neq j\}$ forms a basis for the space of antisymmetric matrices, and

$$[A_{ij}, A_{k\ell}] = [(e_i \otimes e_j) - (e_j \otimes e_i), (e_k \otimes e_\ell) - (e_\ell \otimes e_k)]$$
$$= \delta_{jk} A_{i\ell} - \delta_{j\ell} A_{ik} - \delta_{ik} A_{j\ell} + \delta_{i\ell} A_{jk}$$

describes its Lie-algebra structure.

1.3 Stiff Serret-Frenet frames

With this notation at our disposal, the matrices A and B in equation (5) are given by $A = \sum_{j \neq i}^{n-1} k_j A_{j(j+1)}$ and $B = A_{i(i+1)}$. Our next theorem describes the Lie algebra generated by A and B.

Theorem 4 *Suppose that each fixed curvature k_j in the foregoing expression for A is nonzero. The Lie algebra generated by A and B is equal to the Lie algebra of $SO_n(R)$ in all but one case. The exceptional case occurs when $n = 2m$, $i = m$, and $k_1 = k_2 = \cdots = k_{m-1} = k_{m+1} = \cdots = k$. The Lie algebra in the exceptional case is equal to the Lie algebra of the unitary group $U_m(\mathbb{C}) = U_{2m}(R)$.*

The proof of this theorem is somewhat lengthy and will be omitted. The exceptional case is handled as follows:

There is no loss of generality in taking each k_j equal to 1. Then $A = \sum_{j \neq m}^{2m-1} A_{j(j+1)}$ and $B = A_{m(m+1)}$. Let $J = \sum_{j=1}^{m} (e_i \otimes e_j) - (e_j \otimes e_i)$, with $i + j = 2m + 1$. Then $J(e_j) = e_i$ for $1 \leq j \leq m$, and $J(e_i) = -e_j$, and therefore $J^2 = -I$. Thus J is a complex structure on $\mathbb{R}^{2m}$. It is easy to check that both A and B commute with J. Therefore, both A and B are linear on $\mathbb{C}^m$.

Relative to the complex basis $e_1, \ldots, e_m$, A and B have the following form:

$$A = \begin{pmatrix} 0 & 1 & & & & 0 \\ -1 & 0 & 1 & & & \\ & -1 & 0 & 1 & & \\ & & -1 & & & \\ & & & & & 1 \\ 0 & & & & -1 & 0 \end{pmatrix}, \quad B = \begin{pmatrix} 0 & \cdots & \cdots & \cdots & 0 \\ \vdots & & & & \vdots \\ \vdots & & & & \vdots \\ \vdots & & & & \vdots \\ \vdots & & & 0 & 0 \\ 0 & \cdots & \cdots & 0 & -i \end{pmatrix}.$$

For $m = 2$,

$$\left[\begin{pmatrix} 0 & 1 \\ -1 & 0 \end{pmatrix}, \begin{pmatrix} 0 & 0 \\ 0 & i \end{pmatrix} \right] = \begin{pmatrix} 0 & i \\ i & 0 \end{pmatrix},$$

$$\left[\begin{pmatrix} 0 & 1 \\ -1 & 0 \end{pmatrix}, \begin{pmatrix} 0 & i \\ i & 0 \end{pmatrix} \right] = \begin{pmatrix} 2i & 0 \\ 0 & -2i \end{pmatrix},$$

and

$$\left[\begin{pmatrix} 0 & i \\ i & 0 \end{pmatrix}, \begin{pmatrix} 2i & 0 \\ 0 & -2i \end{pmatrix} \right] = 2 \begin{pmatrix} 0 & 1 \\ -1 & 0 \end{pmatrix}.$$

Therefore, the Lie algebra generated by A and B is four-dimensional and consists of all complex 2×2 matrices

$$\begin{pmatrix} a & b \\ -\bar{b} & c \end{pmatrix}$$

with a and c imaginary. The remaining part of the proof is carried out by induction on m.

Each group $SO_n(R)$ and $U_{2m}(R)$ acts linearly on $\mathbb{R}^n$ by the obvious action $gx = y$, with x written as an n-column vector ($2m = n$ in the case of the unitary group). Each of these groups acts transitively on the spheres in $\mathbb{R}^n$. It is an easy corollary of Theorems 3 and 4 that the bilinear system in $\mathbb{R}^n$,

$$\frac{dx}{dt} = Ax + u(t)Bx, \qquad u(t) \geq 0,$$

is controllable on each sphere $\{x : \|x\| = \|x_0\|\}$. The same result holds for any antisymmetric matrices A and B provided that their Lie algebra is equal to the Lie algebra of either $SO_n(R)$ or $U_{2n}(R)$.

1.4 The Grassmann manifolds

The Grassmann manifolds $G(k, n)$ consist of all k-dimensional subspaces of $\mathbb{R}^n$. We shall presently show that each $SO_n(R)$ acts transitively on $G(k, n)$, from which the controllability properties of systems subordinated to this action can be decided using the Lie-algebraic properties of the lifted system on $SO_n(R)$.

For convenience to the reader, we shall briefly recall the manifold structure of $G(k, n)$. The simplest way to introduce the manifold structure on $G(k, n)$ is to embed it into $O_n(R)$, as follows:

Each k-dimensional subspace S can be identified with an orthogonal reflection P_S, given by $P_S(x) = x$ for $x \in S$ and $P_S(x) = -x$ for x in the orthogonal complement $S^\perp$ of S. $G(k, n)$ is made into a topological space by requiring that the correspondence $S \to P_S$ be a homeomorphism. Because $\{P_S : S \in G(k, n)\}$ is a closed subset of $O_n(R)$, $G(k, n)$ is a compact topological space.

$O_n(R)$ acts on $G(k, n)$ in a natural way: For any $S \in G(k, n)$ and any $T \in O_n(R)$, $TS = \{Tx : x \in S\}$ is an element of $G(k, n)$. This action is transitive, since for any two points S_1 and S_2 in $G(k, n)$ there exists $T \in O_n(R)$ such that $TS_1 = S_2$. In terms of the correspondence $S \to P_S$, the foregoing action is expressed by $TS \to T^*P_ST$, with T^* equal to the transpose of T.

The isotropy group H at S consists of all orthogonal transformations T for which S is an invariant subspace. Evidently $H = O_{n-k}(R) \times O_k(R)$. This argument shows that $G(k, n)$ is equal to the homogeneous space $O_n(R)/H$. Consequently, $G(k, n)$ is an analytic manifold, and its dimension is equal to

$$\dim(O_n(R)) - \dim H = \frac{n(n-1)}{2} - \left(\frac{(n-k)(n-1-k)}{2} + \frac{k(k-1)}{2} \right)$$

$$= \frac{1}{2}(n^2 - n - (n-k)^2 + (n-k) - k^2 + k)$$

$$= \frac{1}{2}(2nk - 2k^2). = k(n-k).$$

For each antisymmetric matrix A, and each S in $G(k, n)$,

$$\frac{d}{dt}(\exp -tA)P_S(\exp tA)|_{t=0} = P_S A - AP_S = [P_S, A].$$

That is, $X_A(S) = [P_S, A]$ is the infinitesimal generator of the one-parameter group of diffeomorphisms induced by A.

Let $\mathcal{P}$ denote the set of all vector fields X_A in $G(k, n)$ with A in $SO_n(R)$. $\mathcal{P}$ is the family of vector fields in $G(k, n)$ subordinated to the group action of $O_n(R)$ on $G(k, n)$. Consider now any sub family $\mathcal{F}$ of $\mathcal{P}$. The basic structure of the reachable sets of $\mathcal{F}$ is described by the following:

Theorem 5

(a) *The reachable set of $\mathcal{F}$ from any point x in $G(k, n)$ is equal to the orbit of $\mathcal{F}$ through x.*

(b) *Let Γ denote the set of all matrices A such that X_A is in $\mathcal{F}$. Then, $\mathcal{A}_{\mathcal{F}}(x) = Hx = \{h^*xh : h \in H\}$, with H equal to the subgroup of $O_n(R)$ generated by $\{\exp tA : A \in \Gamma, t \in R^1\}$*

(c) *$\mathcal{F}$ is controllable on $G(k, n)$ if and only if H acts transitively on $G(k, n)$.*

The proof of this theorem is an easy application of Theorem 7 in Chapter 2 and Theorem 3 in this chapter.

1.5 Motions of a sphere rolling on another sphere

A sphere that rolls on another sphere without slipping defines a non-holonomic system on $S^2 \times SO_3(R)$ whose kinematic properties can be effectively studied using the formalism of Lie groups. The kinematic equations, substantial generalizations of the one-dimensional case (a circle rolling either on the outside or on the inside of another circle), raise challenging questions of control in higher dimensions.

The simplest way to derive the equations of motion is to assume that the rolling sphere is of radius 1 and is equipped with an orthonormal frame fixed at its center and attached to the sphere. The radius of the stationary sphere will be denoted by ρ. Then the orientation of the rolling sphere is expressed by the rotation matrix R that measures the position of the moving frame relative to the fixed orthonormal frame centered at the origin of the stationary sphere, completely analogous to the framework for the sphere rolling on a plane explained in Chapter 4.

Let p denote the coordinates (relative to the stationary frame) of a vector emanating from the origin of E^3 to a point on the sphere. Then, $p = r + q$, with r denoting the coordinates of the center of the rolling sphere, and q denoting the coordinates of the vector from the center of the rolling sphere to the point with coordinates p. The latter vector can also be expressed by the coordinates Q relative to the moving frame, in which case $q = RQ$.

It then follows that $dp/dt = dr/dt + dq/dt = dr/dt + (dR/dt)Q$ along any motions of the rolling sphere. Using $dR/dt = A(\omega)R$ yields $dp/dt = dr/dt + A(\omega)q$.

A non-slipping assumption means that $dp/dt = 0$ whenever p is the point of contact with the rolling sphere. There are two possibilities, depending on whether the sphere rolls inside or outside of the stationary one.

Rolling on the outside of the stationary sphere At the point of contact, $q(t) = -(1/(\rho + 1))r(t)$, and therefore $0 = dr/dt - A(\omega)(1/(\rho + 1))r(t)$. Hence the motion of the center is related to the rotational kinematics through the equation $dr/dt = (1/(\rho + 1))A(\omega)r$.

Rolling on the inside of the stationary sphere In this situation, the point of contact is defined by $q = (1/(\rho - 1))r$, and therefore $0 = dr/dt + A(\omega)((1/(\rho - 1))r)$, or $dr/dt = (1/(1 - \rho))A(\omega)r$.

It will be convenient to project r on the unit sphere by letting $x = (1/(\rho \pm 1))r$, with the sign corresponding to one of the two situations described earlier. In terms of this notation, the kinematic equations are

$$\frac{dx}{dt} = \frac{1}{1 \pm \rho} A(\omega)x, \qquad \frac{dR}{dt} = A(\omega)R. \qquad (6)$$

Equations (6) describe all possible motions on $M = S^2 \times SO_3(R)$ in terms of the angular velocity ω, which is free to take any value in $\mathbb{R}^3$ (Figure 6.2).

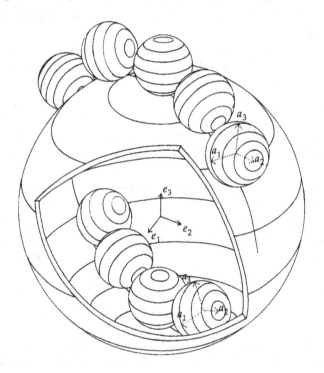

Fig. 6.2.

It is possible to lift equations (6) to $SO_3(R) \times SO_3(R)$ by regarding S^2 as the symmetric space $SO_3(R)/SO_2(R)$: As we have already seen, $SO_3(R)$ acts transitively on S^2 by matrix multiplications from the left. For each point y on S^2, the isotropy group $\{T : T \in SO_3(R), Ty = y\}$ is isomorphic to $SO_2(R)$, and $S^2 = SO_3(R)/SO_2(R)$.

Then each solution curve $x(t)$ is the projection of a curve $T(t)$ in $SO_3(R)$, with $T(t)$ a solution curve of $dT/dt = (1/(1 \pm \rho))A(\omega)T(t)$, and $x(t) = T(t)y$. For the sake of convenience, y will be taken as the north pole e_3.

The overall system in $G = SO_3(R) \times SO_3(R)$ is given by the equations

$$\frac{dT}{dt} = \frac{1}{1 \pm \rho} A(\omega) T(t), \qquad \frac{dR}{dt} = A(\omega) R(t), \qquad (7)$$

which define a three-dimensional right-invariant distribution on a six-dimensional compact Lie group G.

The controllability properties of (7) are determined by the Lie algebra generated by the set of matrices $\Gamma = \{(\alpha A(\omega), A(\omega)) : \omega \in \mathbb{R}^3, \alpha = 1/(1 \pm \rho)\}$. It follows that

$$[(\alpha A(\omega_1), A(\omega_1)), (\alpha A(\omega_2), A(\omega_2))] = (\alpha^2[A(\omega_1), A(\omega_2)], [A(\omega_1), A(\omega_2)]).$$

Let A_i denote the matrix $A(\omega)$, with $\omega = e_i$. It follows that

$$A_1 = \begin{pmatrix} 0 & 0 & 0 \\ 0 & 0 & -1 \\ 0 & 1 & 0 \end{pmatrix}, \qquad A_2 = \begin{pmatrix} 0 & 0 & 1 \\ 0 & 0 & 0 \\ -1 & 0 & 0 \end{pmatrix}, \qquad A_3 = \begin{pmatrix} 0 & -1 & 0 \\ 1 & 0 & 0 \\ 0 & 0 & 0 \end{pmatrix},$$

and $[A_1, A_2] = A_3$, $[A_3, A_1] = A_2$, $[A_2, A_3] = A_1$.

Therefore, $(\alpha A_i, A_i)$ and $(\alpha^2 A_i, A_i)$ belong to $\mathrm{Lie}(\Gamma)$ for each $i = 1, 2, 3$. Their difference is of the form $(\alpha(1-\alpha)A_i, 0)$, and because $\alpha > 0$ and $\alpha \neq 1$, it follows that both $(A_i, 0)$ and $(0, A_i)$ are in the Lie algebra of Γ. This calculation shows that the distribution (7) is controllable on G and that the Lie-bracket structure of $\mathrm{Lie}(\Gamma)$ is of the form

$$\Gamma + [\Gamma, \Gamma] = \mathrm{Lie}(\Gamma).$$

Let us now consider the situation in which the rolling sphere is not allowed to spin around the radial axis through the point of contact. Then the angular velocity ω is perpendicular to x, and therefore it must be of the form $\omega = v \times x$.

There is a convenient choice of variables through which equations (7) can be lifted to a left-invariant distribution on $SO_3(R) \times SO_3(R)$. These variables are defined as follows:

Let $v = Tu$. Then $\omega = Tu \times Te_3 = T(u \times e_3)$. The reader may verify that $A(T(u \times e_3)) = TA(u \times e_3)T^{-1}$ and that therefore

$$\frac{dT}{dt} = \frac{1}{1 \pm \rho} A(\omega) T = \frac{1}{1 \pm \rho} TA(u \times e_3).$$

The transformation $v = Tu$, reminiscent of the passage from a fixed frame to a moving frame in the mechanics of a rigid body, transforms a right-invariant system into a left-invariant one. Therefore, v can be viewed as the angular velocity relative to the moving frame.

The rotational kinematics are now given by $dR/dt = TA(u \times e_3)T^{-1}R$. Let S denote the transformation $S = R^{-1}T$. Then $dS/dt = (dR^{-1}/dt)T(t) + R^{-1}(dT/dt)$. The relation $dR^{-1}/dt = -R^{-1}TA(u \times e_3)T^{-1}$ yields

$$\frac{dS}{dt} = -R^{-1}TA(u \times e_3) + \frac{1}{1 \pm \rho}R^{-1}TA(u \times e_3) = \left(\frac{1}{1 \pm \rho} - 1\right)SA(u \times e_3).$$

The resulting kinematic equations are

$$\frac{dT}{dt} = \alpha TA(u \times e_3), \qquad \frac{dS}{dt} = (\alpha - 1)SA(u \times e_3), \qquad (8)$$

with $\alpha = 1/(1 \pm \rho)$ and

$$A(u \times e_3) = \begin{pmatrix} 0 & 0 & -u_1 \\ 0 & 0 & -u_2 \\ u_1 & u_2 & 0 \end{pmatrix}.$$

Equations (8) define a two-dimensional left-invariant distribution in a six-dimensional compact Lie group G.

The reachable set from the group identity of system (8) is the Lie subgroup H of G, whose Lie algebra is equal to Lie(Γ), with $\Gamma = \{u_2(\alpha A_1, (\alpha - 1)A_1) - u_1(\alpha A_2, (\alpha - 1)A_2) : (u_1, u_2) \in R^2\}$. It follows that

$$[(\alpha A_1, (\alpha - 1)A_1), (\alpha A_2, (\alpha - 1)A_2)] = (\alpha^2 A_3, (\alpha - 1)^2 A_3),$$
$$[(\alpha A_1, (\alpha - 1)A_1), (\alpha^2 A_3, (\alpha - 1)^2 A_3)] = (\alpha^3 A_2, (\alpha - 1)^3 A_2),$$
$$[(\alpha A_2, (\alpha - 1)A_2), (\alpha^2 A_3, (\alpha - 1)^2 A_3)] = -(\alpha^3 A_1, (\alpha - 1)^3 A_1).$$

It is easy to verify that $(\alpha A_2, (\alpha - 1)A_2)$ and $(\alpha^3 A_2, (\alpha - 1)^3 A_2)$ are linearly independent if and only if $\alpha \neq \frac{1}{2}$. The same applies to the linear independence of $(\alpha A_1, (\alpha - 1)A_1)$ and $(\alpha^3 A_1, (\alpha - 1)^3 A_1)$.

We have $\alpha \neq \frac{1}{2}$ whenever $\rho \neq 1$. Thus system (8) is controllable whenever the spheres have unequal curvatures (different radii). The structure of Lie(Γ) is given by

$$\Gamma + [\Gamma, \Gamma] + [\Gamma, [\Gamma, \Gamma]] + [\Gamma, [\Gamma, [\Gamma, \Gamma]]] = \mathcal{L}(G).$$

Lie(Γ) is isomorphic to $SO_3(R)$ when $\rho = 1$. Then, closed curves traced by the point of contact correspond to the closed curves of the rolling sphere. Hence the initial and the terminal rotations differ by a rotation about the radial axis.

The distributions defined by the rolling spheres lift to the covering spaces of $SO_3 \times SO_3$, as outlined in the next section.

1.6 Quaternions and rotations

Let E^4 denote an oriented four-dimensional Euclidean space. We shall use $(\mathbf{v}, \mathbf{w})$ to denote the scalar product of $\mathbf{v}$ and $\mathbf{w}$, and $\|\mathbf{v}\|$ to denote the norm of $\mathbf{v}$. Of course, $\|\mathbf{v}\| = (\mathbf{v}, \mathbf{v})^{1/2}$.

Let $\mathbf{e}$ denote any unit vector in E^4, and let E be the orthogonal complement to the vector space spanned by $\mathbf{e}$. Any element $\mathbf{q}$ in E^4 can be uniquely expressed as $q_0 \mathbf{e} + \mathbf{q}_1$, with $\mathbf{q}_1$ in E, and q_0 in $\mathbb{R}^1$; q_0 is called the scalar part of $\mathbf{q}$, and $\mathbf{q}_1$ is called the vector part of $\mathbf{q}$.

Assuming that E inherits its sense of orientation from E^4, let $\mathbf{i}, \mathbf{j}, \mathbf{k}$ denote any right-handed orthonormal frame in E. Recall that the cross-product $\mathbf{v} \times \mathbf{w}$ of $v = v_1\mathbf{i} + v_2\mathbf{j} + v_3\mathbf{k}$ and $w = w_1\mathbf{i} + w_2\mathbf{j} + w_3\mathbf{k}$ is equal to $(v_2 w_3 - v_3 w_2)\mathbf{i} + (v_3 w_1 - v_1 w_3)\mathbf{j} + (v_1 w_2 - v_2 w_1)\mathbf{k}$.

Definition 2 Quaternions are elements of E^4 endowed with the vector structure of E^4 and the multiplication

$$\mathbf{q} \cdot \mathbf{s} = (q_0 s_0 - (\mathbf{q}_1, \mathbf{s}_1))\,\mathbf{e} + q_0 \mathbf{s}_1 + s_0 \mathbf{q}_1 + \mathbf{q}_1 \times \mathbf{s}_1$$

for any $\mathbf{q} = q_0 \mathbf{e} + \mathbf{q}_1$ and $\mathbf{s} = s_0 \mathbf{e} + \mathbf{s}_1$.

Evidently, $\mathbf{q} \cdot \mathbf{e} = \mathbf{e} \cdot \mathbf{q} = \mathbf{q}$ for all quaternions, and therefore $\mathbf{e}$ is the multiplicative identity in the space of quaternions. The conjugate $\bar{\mathbf{q}}$ of a quaternion $\mathbf{q}$ is defined by $q_0 \mathbf{e} - \mathbf{q}_1$. The following properties are easy to prove:

(a) $\overline{\mathbf{q} + \mathbf{s}} = \bar{\mathbf{q}} + \bar{\mathbf{s}}$,
(b) $\overline{\mathbf{q} \cdot \mathbf{s}} = \bar{\mathbf{s}} \cdot \bar{\mathbf{q}}$, and
(c) $\mathbf{q} \cdot \bar{\mathbf{q}} = (q_0^2 + \|\mathbf{q}_1\|^2)\mathbf{e}$.

Problem 2 By using properties (b) and (c), prove that $\|\mathbf{q} \cdot \mathbf{s}\| = \|\mathbf{q}\|\,\|\mathbf{s}\|$. Also show that $\mathbf{q}^{-1} = (1/\|\mathbf{q}\|^2)\bar{\mathbf{q}}$ is the unique quaternion such that $\mathbf{q}^{-1}\mathbf{q} = \mathbf{e}$.

It is customary to use H to denote the space of quaternions. The equation $x \cdot \mathbf{q} = \mathbf{e}$ is uniquely solvable for any nonzero $\mathbf{q}$, and therefore H is a division algebra. It is not a field, because its elements do not commute.

A quaternion $\mathbf{q}$ is called a pure quaternion if $q_0 = 0$ or if $\mathbf{q}$ belongs to E. Equivalently, $\mathbf{q}$ is a pure quaternion if it is equal to its vector part. The product of pure quaternions is equal to $-(\mathbf{q} \cdot \mathbf{s})\mathbf{e} + \mathbf{q} \times \mathbf{s}$.

Theorem 6 *The sphere $S^3 = \{\mathbf{q} : \|\mathbf{q}\| = 1\}$ is equal to the group of unit quaternions.*

Proof For any unit quaternion, its inverse is equal to its conjugate. Therefore, $1 = \|\mathbf{q} \cdot \bar{\mathbf{q}}\| = \|\mathbf{q}\| \, \|\bar{\mathbf{q}}\| = \|\bar{\mathbf{q}}\|$, and so the unit quaternions form a multiplicative group. End of the proof. ∎

Any complex number $x + iy$ can be regarded as a quaternion $\mathbf{q} = x\mathbf{e} + y\mathbf{i}$. That is, x is the scalar part of $\mathbf{q}$, and $y\mathbf{i}$ is its vector part. Because $\mathbf{i}, \mathbf{j}, \mathbf{k}$ form a right-handed orthonormal frame, their quaternionic multiplications satisfy $\mathbf{i} \cdot \mathbf{j} = \mathbf{k}$, $\mathbf{k} \cdot \mathbf{i} = \mathbf{j}$, and $\mathbf{j} \cdot \mathbf{k} = \mathbf{i}$. Also, $\mathbf{i}^2 = \mathbf{j}^2 = \mathbf{k}^2 = -\mathbf{e}$, as can be easily proved from the definitions. Then for any complex number $x + iy$, $(x\mathbf{e} + y\mathbf{i})\mathbf{j} = x\mathbf{j} + y\mathbf{k}$, and therefore the space of quaternions can be identified with $\mathbb{C}^2$. The identification is given by

$$\mathbf{q} = q_0\mathbf{e} + q_1\mathbf{i} + q_2\mathbf{j} + q_3\mathbf{k} = \mathbf{z}_1 + \mathbf{z}_2 \cdot \mathbf{j}, \tag{9}$$

with $z_1 = q_0 + iq_1$ and $z_2 = q_2 + iq_3$.

Theorem 7 *The space of all unit quaternions is equal to SU_2, the group of all matrices*

$$\begin{pmatrix} z & w \\ -\bar{w} & \bar{z} \end{pmatrix}$$

with complex entries z and w such that $|z|^2 + |w|^2 = 1$.

Proof Let

$$F \begin{pmatrix} z & w \\ -\bar{w} & \bar{z} \end{pmatrix} = \mathbf{z} + \mathbf{w} \cdot \mathbf{j}$$

be the mapping from SU_2 onto S^3. We leave it to the reader to verify that F is an isomorphism. ∎

Each unit quaternion $\mathbf{q}$ induces a rotation $R_{\mathbf{q}}$ on E defined by $R_{\mathbf{q}}(\mathbf{x}) = \mathbf{q}\mathbf{x}\bar{\mathbf{q}}$. Evidently $R_{\mathbf{q}}$ is a rotation, since $\|R_{\mathbf{q}}(\mathbf{x})\| = \|\mathbf{q}\mathbf{x}\bar{\mathbf{q}}\| = \|\mathbf{q}\| \, \|\mathbf{x}\| \, \|\bar{\mathbf{q}}\| = \|\mathbf{x}\|$. $R_{\mathbf{q}}$

is also orientation-preserving, because

$$R_q(x \times y) = R_q(x \cdot y + (x, y)e) = q(x \cdot y + (x, y)e)\bar{q}$$
$$= R_q(x) \cdot R_q(y) + (R_q(x), R_q(y))e = R_q(x) \times R_q(y).$$

It is easy to verify that $q \to R_q$ is a homomorphism from the group of unit quaternions onto $SO(E)$. The kernel of this homomorphism consists of $\pm e$. Therefore, SU_2 is a double cover of $SO(E) = SO_3(R)$. It follows that the Lie algebra of $SO_3(R)$ is isomorphic with the Lie algebra of SU_2.

It will be useful to establish this isomorphism explicitly. Recall that the Lie algebra of SU_2 consists of all complex matrices

$$A = \begin{pmatrix} \alpha & \beta \\ \gamma & \delta \end{pmatrix}$$

such that $e^{tA} = \sum (t^k/k!)A^k$ belongs to SU_2 for all t.

Theorem 8 *The Lie algebra of SU_2 consists of all complex matrices*

$$\begin{pmatrix} \alpha & \beta \\ -\bar{\beta} & \bar{\alpha} \end{pmatrix}$$

with $\alpha + \bar{\alpha} = 0$.

Proof Let

$$A = \begin{pmatrix} \alpha & \beta \\ -\bar{\beta} & \bar{\alpha} \end{pmatrix},$$

with $\alpha + \bar{\alpha} = 0$. Then $A^2 = -\lambda^2 I$, with $\lambda^2 = |\alpha|^2 + |\beta|^2$. It follows that

$$e^{tA} = \sum_{k=0}^{\infty} \frac{t^k}{k!} A^k = \sum_{n=0}^{\infty} \frac{t^{2n} A^{2n}}{(2n)!} + \sum_{n=0}^{\infty} \frac{t^{2n+1} A^{2n+1}}{(2n+1)!}$$

$$= I \sum_{n=0}^{\infty} \frac{(-1)^n (t\lambda)^{2n}}{(2n)!} + \frac{A}{\lambda} \sum_{n=0}^{\infty} (-1)^n \frac{(t\lambda)^{2n+1}}{(2n+1)!}$$

$$= (\cos t\lambda) I + \left(\frac{1}{\lambda} \sin t\lambda \right) A$$

$$= \begin{pmatrix} (\cos t\lambda + \frac{\alpha}{\lambda} \sin t\lambda) & (\frac{\beta}{\lambda} \sin t\lambda) \\ (-\frac{\beta}{\alpha} \sin t\lambda) & (\cos t\lambda + \frac{\bar{\alpha}}{\lambda} \sin t\lambda) \end{pmatrix}.$$

Evidently e^{tA} belongs to SU_2, because $(\cos t\lambda + (\alpha/\lambda)\sin t\lambda)^2 + (|\beta|^2/\lambda^2)$ $(\sin t\lambda)^2 = 1$. Because SU_2 is a three-dimensional real Lie group, its Lie algebra is a three-dimensional real Lie algebra, and therefore must be equal to the algebra of matrices

$$\begin{pmatrix} \alpha & \beta \\ -\bar{\beta} & \bar{\alpha} \end{pmatrix},$$

with $\alpha + \bar{\alpha} = 0$. End of the proof.   ∎

Because $\mathbf{q} \longrightarrow R_\mathbf{q}$ is a homomorphism, it follows that $\mathbf{q}(t) = e^{tA}$ corresponds to a one-parameter group of rotations. Therefore, $R_{\mathbf{q}(t)} = e^{t\hat{A}}$ for some element of the Lie algebra of $SO(E)$. It follows that

$$e^{t\hat{A}}(\mathbf{x}) = \mathbf{q}(t)\mathbf{x}\bar{\mathbf{q}}(t) \quad \text{for all } t.$$

Differentiating the foregoing expression with respect to t and setting $t = 0$ yields

$$\hat{A}(\mathbf{x}) = A\mathbf{x} - \mathbf{x}A \quad \text{for all } \mathbf{x} \text{ in } E.$$

$A\mathbf{x}$ is equal to the quaternion $(\alpha + \beta\mathbf{j})\mathbf{x}$, and $\mathbf{x}A$ is equal to $\mathbf{x}(\alpha + \beta\mathbf{j})$.

Denote by $\omega = \omega_1\mathbf{i} + \omega_2\mathbf{j} + \omega_3\mathbf{k}$ the angular velocity of $\hat{A}$. The matrix of $\hat{A}$ with respect to the frame $\mathbf{i}, \mathbf{j}, \mathbf{k}$ is

$$\begin{pmatrix} 0 & -\omega_3 & \omega_2 \\ \omega_3 & 0 & -\omega_1 \\ -\omega_2 & \omega_1 & 0 \end{pmatrix}.$$

Then

$$\omega_3\mathbf{j} - \omega_2\mathbf{k} = \hat{A}\mathbf{i} = (\alpha + \beta\mathbf{j})\mathbf{i} - \mathbf{i}(\alpha + \beta\mathbf{j}) = -2\beta\mathbf{k}$$
$$= -2((\operatorname{Re}\beta)\mathbf{e} + (\operatorname{Im}\beta)\mathbf{i})\mathbf{k},$$

and therefore $\omega_2 = 2(\operatorname{Re}\beta)$ and $\omega_3 = 2(\operatorname{Im}\beta)$.

Continuing,

$$-\omega_3\mathbf{i} + \omega_1\mathbf{k} = \hat{A}\mathbf{j} = (\alpha + \beta\mathbf{j})\mathbf{j} - \mathbf{j}(\alpha + \beta\mathbf{j}) = (\alpha - \bar{\alpha})\mathbf{j} - (\beta - \bar{\beta}).$$

Recall that $\alpha + \bar{\alpha} = 0$, and so $\alpha - \bar{\alpha} = 2\alpha$, from which we obtain that $\alpha = (i/2)\omega_1$. The overall relations are

$$\alpha = \frac{i}{2}\omega_1 \quad \text{and} \quad \beta = \frac{1}{2}(\omega_2 + i\omega_3).$$

These relations provide an explicit isomorphism between the Lie algebras of SU_2 and $SO_3(R)$:

$$\begin{pmatrix} \alpha & \beta \\ -\bar{\beta} & \bar{\alpha} \end{pmatrix} \quad \text{is identified with} \quad 2\begin{pmatrix} 0 & -\operatorname{Im}\beta & \operatorname{Re}\beta \\ \operatorname{Im}\beta & 0 & \alpha \\ -\operatorname{Re}\beta & -\alpha & 0 \end{pmatrix},$$

and

$$\begin{pmatrix} 0 & -\omega_3 & \omega_2 \\ \omega_3 & 0 & -\omega_1 \\ \omega_2 & \omega_1 & 0 \end{pmatrix} \quad \text{is identified with} \quad \begin{pmatrix} \frac{i}{2}\omega_1 & \frac{1}{2}(\omega_2 + i\omega_3) \\ -\frac{1}{2}(\omega_2 - i\omega_3) & -\frac{i}{2}\omega_1 \end{pmatrix}.$$

Every rotation R can be written as e^A for some antisymmetric matrix A, because the exponential mapping on $SO_3(R)$ is surjective. If $\omega = \omega_1\mathbf{i} + \omega_2\mathbf{j} + \omega_3\mathbf{k}$ is the angular velocity of A, then R is the rotation through the angle $\|\omega\|$ in the counterclockwise direction about ω.

In the foregoing representation, R corresponds to $\pm\exp A$ in SU_2, with

$$A = \begin{pmatrix} \frac{i}{2}\omega_1 & \frac{1}{2}(\omega_2 + i\omega_3) \\ -\frac{1}{2}(\omega_2 - i\omega_3) & -\frac{i}{2}\omega_1 \end{pmatrix}.$$

In terms of the notation of Theorem 8, $\lambda^2 = |\alpha|^2 + |\beta|^2 = \frac{1}{4}\|\omega\|^2$. Hence,

$$e^A = \begin{pmatrix} \cos\frac{1}{2}\|\omega\| + i\frac{\omega_1}{\|\omega\|}\sin\frac{1}{2}\|\omega\| & \frac{(\omega_2 + i\omega_3)}{\|\omega\|}\sin\frac{1}{2}\|\omega\| \\ \frac{-(\omega_2 - i\omega_3)}{\|\omega\|}\sin\frac{1}{2}\|\omega\| & \cos\frac{1}{2}\|\omega\| - i\frac{\omega_1}{\|\omega\|}\sin\frac{1}{2}\|\omega\| \end{pmatrix}$$

which in turn can also be written as the quaternion

$$\mathbf{q} = \cos\frac{1}{2}\|\omega\|\,\mathbf{e} + \frac{\omega_1}{\|\omega\|}\sin\frac{1}{2}\|\omega\|\,i$$
$$+ \left(\frac{\omega_2}{\|\omega\|}\sin\frac{1}{2}\|\omega\|\,\mathbf{e} + \frac{\omega_3}{\|\omega\|}\sin\frac{1}{2}\|\omega\|\,\mathbf{i}\right)\mathbf{j}.$$

Equivalently,

$$\mathbf{q} = \cos\frac{1}{2}\|\omega\|\,\mathbf{e} + \frac{\omega}{\|\omega\|}\sin\frac{1}{2}\|\omega\|.$$

We now turn to the rotations in E^4, with the following theorem.

Theorem 9 $SU_2 \times SU_2$ *is a double cover of* $SO_4(R)$. *Consequently the Lie algebra of* $SU_2 \times SU_2$ *and the Lie algebra of* $SO_4(R)$ *are isomorphic.*

Proof Let R be an element of $SO(E^4)$. Denote $R_1(\mathbf{x}) = R(\mathbf{x})R^{-1}(\mathbf{e})$. Then $R_1(\mathbf{e}) = \mathbf{e}$, and hence R_1 is a rotation of E. It follows from the preceding discussion that $R_1(\mathbf{x}) = \mathbf{q}\mathbf{x}\bar{\mathbf{q}}$ for some unit quaternion $\mathbf{q}$. Therefore,

$$R(\mathbf{x}) = R_1(\mathbf{x})R(\mathbf{e}) = \mathbf{q}\mathbf{x}\mathbf{q}^{-1}R(\mathbf{e}) = \mathbf{q}\mathbf{x}\mathbf{p}^{-1}, \quad \text{with } \mathbf{p} = R^{-1}(\mathbf{e})\mathbf{q}.$$

The steps can be reversed to show that every element R in $SO(E^4)$ can be represented as

$$R(\mathbf{x}) = \mathbf{p}\mathbf{x}\mathbf{q}^{-1}$$

for some unit quaternions $\mathbf{p}$ and $\mathbf{q}$.

If we now denote $R_{(\mathbf{p}_1,\mathbf{q}_1)}(\mathbf{x}) = \mathbf{p}_1\mathbf{x}\mathbf{q}_1^{-1}$, and $R_{(\mathbf{p}_2,\mathbf{q}_2)}(\mathbf{x}) = \mathbf{p}_2\mathbf{x}\mathbf{q}_2^{-1}$, then

$$R_{(\mathbf{p}_2,\mathbf{q}_2)} \circ R_{(\mathbf{p}_1,\mathbf{q}_1)}(\mathbf{x}) = \mathbf{p}_2\mathbf{p}_1\mathbf{x}\mathbf{q}_1^{-1}\mathbf{q}_2^{-1} = (\mathbf{p}_2\mathbf{p}_1)\mathbf{x}(\mathbf{q}_2\mathbf{q}_1)^{-1}.$$

Hence

$$R_{(\mathbf{p}_2\mathbf{p}_1,\mathbf{q}_2\mathbf{q}_1)} = R_{(\mathbf{p}_2,\mathbf{q}_2)} \circ R_{(\mathbf{p}_1,\mathbf{q}_1)},$$

and therefore the correspondence $(\mathbf{p}, \mathbf{q}) \longrightarrow R_{(\mathbf{p},\mathbf{q})}$ is a group homomorphism from $SU_2 \times SU_2$ onto $SO(E^4)$. The kernel of this homomorphism is equal to $\pm(\mathbf{e}, \mathbf{e})$. Therefore $SU_2 \times SU_2$ is a double cover of $SO(E^4)$, and the proof is finished.   ∎

Let us now establish an explicit isomorphism between the Lie algebra of $SU_2 \times SU_2$ and the Lie algebra of $SO(E^4)$. For that purpose, let $p(t) = e^{A_1 t}$ and $q(t) = e^{A_2 t}$. Then $R_{(\mathbf{p}(t),\mathbf{q}(t))}$ is a one-parameter group in $SO(E^4)$; hence $R_{(\mathbf{p}(t),\mathbf{q}(t))} = e^{At}$ for some element A in the Lie algebra of $SO(E^4)$. The desired correspondence is $(A_1, A_2) = A$.

We shall represent A by its matrix relative to the frame $\mathbf{e}$, $\mathbf{i}$, $\mathbf{j}$, $\mathbf{k}$ as follows:

$$A = \begin{pmatrix} 0 & -a_1 & -a_2 & -a_3 \\ a_1 & 0 & -b_3 & b_2 \\ a_2 & b_3 & 0 & -b_1 \\ a_3 & -b_2 & b_1 & 0 \end{pmatrix}.$$

After differentiating $\mathbf{p}(t)\mathbf{x}\mathbf{q}(t)^{-1}$ and setting $t = 0$, we obtain

$$A\mathbf{x} = A_1\mathbf{x} - \mathbf{x}A_2,$$

with the understanding that $A_1\mathbf{x}$ and $\mathbf{x}A_2$ stand for the quaternions $(\alpha_1 + \beta_1\mathbf{j})\mathbf{x}$ and $\mathbf{x}(\alpha_2 + \beta_2\mathbf{j})$, with

$$A_1 = \begin{pmatrix} \alpha_1 & \beta_1 \\ -\bar{\beta}_1 & \bar{\alpha}_1 \end{pmatrix} \quad \text{and} \quad A_2 = \begin{pmatrix} \alpha_2 & \beta_2 \\ -\bar{\beta}_2 & \bar{\alpha}_2 \end{pmatrix}.$$

We then obtain the following relations:

$$a_1\mathbf{i} + a_2\mathbf{j} + a_3\mathbf{k} = A\mathbf{e} = (\alpha_1 + \beta_1\mathbf{j})\mathbf{e} - \mathbf{e}(\alpha_2 + \beta_2\mathbf{j}) = \alpha_1 + \beta_1\mathbf{j} - (\alpha_2 + \beta_2\mathbf{j}).$$

Then

$$\alpha_1 - \alpha_2 + \beta_1\mathbf{j} - \beta_2\mathbf{j} = \alpha_1 - \alpha_2 + (\operatorname{Re}\beta_1 - \operatorname{Re}\beta_2)\mathbf{j} + (\operatorname{Im}\beta_1 - \operatorname{Im}\beta_2)\mathbf{k}$$

leads to

$$a_1 = \operatorname{Im}\alpha_1 - \operatorname{Im}\alpha_2, \qquad a_2 = \operatorname{Re}\beta_1 - \operatorname{Re}\beta_2, \qquad a_3 = \operatorname{Im}\beta_1 - \operatorname{Im}\beta_2.$$

Continuing,

$$-a_1\mathbf{e} + b_3\mathbf{j} - b_2\mathbf{k} = A\mathbf{i} = (\alpha_1 + \beta_1\mathbf{j})\mathbf{i} - \mathbf{i}(\alpha_2 + \beta_2\mathbf{j})$$
$$= (\alpha_1 - \alpha_2)\mathbf{i} + (\operatorname{Im}\beta_1 + \operatorname{Im}\beta_2)\mathbf{j} - (\operatorname{Re}\beta_1 + \operatorname{Re}\beta_2)\mathbf{k},$$

and therefore

$$b_3 = \operatorname{Im}\beta_1 + \operatorname{Im}\beta_2 \quad \text{and} \quad b_2 = \operatorname{Re}\beta_1 + \operatorname{Re}\beta_2.$$

Finally,

$$-a\mathbf{e} - b_3\mathbf{i} + b_1\mathbf{k} = A\mathbf{j} = (\alpha_1 + \beta_1\mathbf{j})\mathbf{j} - \mathbf{j}(\alpha_2 + \beta_2\mathbf{j})$$
$$= \bar{\beta}_2 - \beta_1 + (\operatorname{Im}\alpha_1 + \operatorname{Im}\alpha_2)\mathbf{k},$$

from which it follows that $b_1 = \operatorname{Im}\alpha_1 + \operatorname{Im}\alpha_2$. These relations yield

$$2\operatorname{Im}\alpha_1 = a_1 + b_1, \quad \text{or} \quad \alpha_1 = \frac{i}{2}(a_1 + b_1),$$
$$2\operatorname{Re}\beta_1 = a_2 + b_2 \quad \text{and} \quad 2\operatorname{Im}\beta_1 = b_3 + a_3, \text{ or}$$
$$\beta_1 = \frac{1}{2}((a_2 + b_2) + i(a_3 + b_3)),$$
$$2\operatorname{Re}\beta_2 = b_2 - a_2 \quad \text{and} \quad 2\operatorname{Im}\beta_2 = b_3 - a_3, \text{ or}$$
$$\beta_2 = \frac{1}{2}((b_2 - a_2) + i(b_3 - a_3)),$$
$$2\operatorname{Im}\alpha_2 = b_1 - a_1, \quad \text{or} \quad \alpha_2 = \frac{i}{2}(b_1 - a_1).$$

Summarizing, the isomorphism between the Lie algebras of $SO_4(R)$ and $SU_2 \times SU_2$ is given by the following correspondences:

$$a_1 = \operatorname{Im}\alpha_1 - \operatorname{Im}\alpha_2, \qquad a_2 = \operatorname{Re}\beta_1 - \operatorname{Re}\beta_2, \qquad a_3 = \operatorname{Im}\beta_1 - \operatorname{Im}\beta_2,$$
$$b_1 = \operatorname{Im}\alpha_1 + \operatorname{Im}\alpha_2, \qquad b_2 = \operatorname{Re}\beta_1 + \operatorname{Re}\beta_2, \qquad b_3 = \operatorname{Im}\beta_1 + \operatorname{Im}\beta_2,$$

and

$$\alpha_1 = \frac{i}{2}(a_1 + b_1), \qquad \beta_1 = \frac{1}{2}((a_2 + b_2) + i(a_3 + b_3)),$$
$$\alpha_2 = \frac{i}{2}(b_1 - a_1), \qquad \beta_2 = \frac{1}{2}((b_2 - a_2) + i(b_3 - a_3)).$$

We shall now return to the problem of rolling spheres to show that the distribution defined by equations (8) on $SO_3 \times SO_3$ canonically lifts to a two-dimensional distribution on $SO_4(R)$, described by

$$\frac{dg}{dt} = g(t)A(u_1, u_2),$$

with

$$A(u_1, u_2) = \frac{1}{2}\begin{pmatrix} 0 & -u_2 & u_1 & 0 \\ u_2 & 0 & 0 & -(2\alpha - 1)u_1 \\ -u_1 & 0 & 0 & -(2\alpha - 1)u_2 \\ 0 & (2\alpha - 1)u_1 & (2\alpha - 1)u_2 & 0 \end{pmatrix}.$$

The corresponding distribution in $G = SU_2 \times SU_2$ is given by the pair of differential equations

$$\frac{dg_1}{dt} = g_1 M_1, \qquad \frac{dg_2}{dt} = g_2 M_2,$$

with

$$M_1 = \frac{\alpha}{2}\begin{pmatrix} iu_2 & -u_1 \\ u_1 & -iu_2 \end{pmatrix} \quad \text{and} \quad M_2 = \left(\frac{\alpha - 1}{2}\right)\begin{pmatrix} iu_2 & -u_1 \\ u_1 & -iu_2 \end{pmatrix}.$$

This rich class of non-holonomic systems poses a number of intriguing controllability questions. In particular, the reader may want to examine the controllability properties near the critical case $\rho = 1$ and determine qualitative criteria that will describe the loss of controllability as ρ tends to 1. Such a study would undoubtedly lead to an interesting research paper.

2 Semidirect products of Lie groups

Motivated by the controllability properties of differential systems on the frame bundle of a Euclidean space E^n, we shall now consider differential systems on semidirect products of Lie groups with vector spaces.

The prototype of this situation is the group of motions $E_n = \mathbb{R}^n \ltimes SO_n(R)$ of $\mathbb{R}^n$. Each element of E_n is an isometry for $\mathbb{R}^n$, because

$$\|(kx + v) - (ky + v)\| = \|k(x - y)\| = \|x - y\|$$

for each $(v, k) \in \mathbb{R}^n \ltimes SO_n(R)$. Conversely, any isometry of $\mathbb{R}^n$ is necessarily an element of E_n. But then, E_n can also be regarded as the orthonormal frame bundle of a Euclidean space E^n.

As mentioned in the Introduction, E_n can be realized as the subgroup of $GL_{n+1}(R)$ consisting of all matrices of the form

$$\begin{pmatrix} 1 & 0 \\ v & R \end{pmatrix},$$

with

$$v = \begin{pmatrix} v_1 \\ \vdots \\ v_n \end{pmatrix},$$

and R in $SO_n(R)$. The group of these matrices acts on the affine space

$$\begin{pmatrix} 1 \\ x \end{pmatrix},$$

with x a column vector in $\mathbb{R}^n$, to give

$$\begin{pmatrix} 1 & 0 \\ v & R \end{pmatrix} \begin{pmatrix} 1 \\ x \end{pmatrix} = \begin{pmatrix} 1 \\ Rx + v \end{pmatrix}$$

and realizes $\mathbb{R}^n$ as the homogeneous space $E_n/SO_n(R)$.

The Lie algebra of E_n consists of all matrices M such that $\exp tM \in E_n$ for all t in R. Denote M by

$$\begin{pmatrix} \alpha & b \\ a & A \end{pmatrix}$$

and $\exp tM$ by

$$\begin{pmatrix} 1 & 0 \\ x(t) & R(t) \end{pmatrix}.$$

It then follows, by differentiating, that

$$\begin{pmatrix} 0 & 0 \\ \frac{dx}{dt} & \frac{dR}{dt} \end{pmatrix} = M \exp t M = (\exp t M) M.$$

The foregoing relations imply that $\alpha = 0$ and $b = 0$ and that $dx/dt = Ax(t) + a$ and $dR/dt = AR$, with A an antisymmetric $n \times n$ matrix. Because $\exp tM = I$ when $t = 0$, $x(0) = 0$ and $R(0) = I$.

It follows from the foregoing considerations that the Lie algebra of E_n is isomorphic with the vector space of matrices of the form

$$M = \begin{pmatrix} 0 & 0 \\ a & A \end{pmatrix},$$

with $a \in \mathbb{R}^n$, and A satisfying $A^* = -A$. The commutator $[M_1, M_2]$ of any two such matrices is equal to

$$\begin{pmatrix} 0 & 0 \\ a_1 & A_1 \end{pmatrix} \begin{pmatrix} 0 & 0 \\ a_2 & A_2 \end{pmatrix} - \begin{pmatrix} 0 & 0 \\ a_2 & A_2 \end{pmatrix} \begin{pmatrix} 0 & 0 \\ a_1 & A_1 \end{pmatrix}$$

$$= \begin{pmatrix} 0 & 0 \\ A_1 a_2 - A_2 a_1 & A_1 A_2 - A_2 A_1 \end{pmatrix}.$$

Completely analogous arguments can be used to show that the Lie algebra $\mathcal{L}(G)$ of an arbitrary semidirect product $G = V \ltimes K$ is equal to the set of all pairs $M = (a, A)$, with a in V and A in the Lie algebra of K, and with the Lie bracket $[(a, A), (b, B)] = (Ab - Ba, [A, B])$.

For any element M in $\mathcal{L}(G)$, $\vec{M}$ will denote the right-invariant vector field whose value at the group identity is equal to M. Recall that the curve $g(t)$ of $\vec{M}$ satisfies the following equation:

$$\frac{dg}{dt} = \vec{M}(g(t)) = dR_{g(t)}(M),$$

where dR_g is the differential of the right translation $R_g(x) = xg$. For linear groups, $dR_{g(t)}(M) = M(g(t))$, and the preceding equation takes the familiar form $dg/dt = M(g(t))$.

Each component $(x(t),\, k(t))$ of $g(t)$ satisfies its own differential equations $dx/dt = Ax(t) + a$ and $dk/dt = Ak$.

Remark Integral curves of left-invariant vector fields satisfy $dg/dt = gM$, and therefore $dx/dt = k(t)a$, and $dk/dt = kA$. ∎

We shall be interested in the controllability properties of an arbitrary family of right-invariant vector fields $\mathcal{F}$ on the semidirect product of V with a compact group K. It is somewhat remarkable that the controllability criteria for compact groups extend to the semidirect products under a very minor hypothesis.

Theorem 10 *Suppose that K has no fixed points in V, in the sense that $Kx = x$ holds only for $x = 0$. A family $\mathcal{F}$ of right-invariant vector fields on $G = V \rtimes K$ is controllable if and only if the Lie algebra generated by $\mathcal{F}$ is equal to the Lie algebra of G.*

We shall prove this theorem only for the cases E_n and $G = \mathbb{R}^{2n} \rtimes U_{2n}(R)$.

Proof Let Γ denote the set $\{X(e) : X \in \mathcal{F}\}$. $S(\Gamma)$ is the semigroup generated by $\bigcup_{A \in \Gamma} \{\exp tA : A \in \Gamma,\ t \geq 0\}$. Because $S(\Gamma)$ is equal to the reachable set from the identity, it follows that $S(\Gamma)$ has a nonempty interior in G if and only if $\mathrm{Lie}\,(\Gamma) = \mathcal{L}(G)$.

Assume that $\mathrm{Lie}\,(\Gamma) = \mathcal{L}(G)$. Then the projection of $S(\Gamma)$ on K is equal to K, since the latter is compact and connected.

It follows from Theorem 2 that it is sufficient to show that $e = (0, I)$ is contained in the interior of $S(\Gamma)$. Let (x, k) be a point in the interior of $S(\Gamma)$. Because the projection of $S(\Gamma)$ on K contains K, it follows that there exists $y \in V$ such that (y, k^{-1}) is contained in $S(\Gamma)$. Then $(x, k)(y, k^{-1}) = (x + ky, I)$, and this product is in the interior of $S(\Gamma)$.

Let Ω be a neighborhood of I in K such that $\{x + ky,\ \Omega\} \subset \mathrm{int}\, S(\Gamma)$. Denote $x + ky$ by v.

For any h in Ω and any positive integer n, $(v, h)^n = (v + hv + \cdots + h^{n-1}v,\ h^n)$ is contained in the interior of $S(\Gamma)$. If $h^n = I$, and if $v = hw - w$ for some w in V, then $v + hv + \cdots + h^{n-1}v = 0$, and $(0, I)$ is contained in the interior of $S(\Gamma)$. In either of the two cases [$K = SO_n(R)$, $V = \mathbb{R}^n$, and $K = U_{2m}(R)$, $V = \mathbb{R}^{2m}$] for any $v \in V$ and any neighborhood Ω of I in K there exists an element h in $K \cap \Omega$ such that v is contained in the image of $h - I$, and $h^n = I$ for some positive integer n. Therefore the proof is finished. ∎

Problem 3 Let P denote a plane in $\mathbb{R}^n$ ($n \geq 2$) that contains a given point v of $\mathbb{R}^n$. Show that for any neighborhood Ω of I in $SO_2(P)$ there exists a rotation R in Ω such that $R - I$ is nonsingular and $\mathbb{R}^m = I$ for some positive integer m. Extend R to $SO_n(R)$ by defining it to be equal to the identity on $P^{\perp}$. Then $v \in$ image $(R - I)$, and $\mathbb{R}^m = I$. Thus, show that for any v in $\mathbb{R}^n$ and any neighborhood Ω of I in $SO_n(R)$ there exists R in Ω and a positive integer m such that v is contained in $R - I$ and $\mathbb{R}^m = I$.

Problem 4 Let P be a complex plane in $\mathbb{C}^m$ ($m \geq 2$) that contains a given point v. Repeat Problem 3 with $SO_n(R)$ replaced by $U_m(\mathbb{C})$.

Theorem 10 has far-reaching implications in the theory of curves, as the following examples illustrate: The Serret-Frenet system associated with a curve $x(t)$ in $\mathbb{R}^3$ is given by

$$\frac{dx}{dt} = R(t)e_1, \qquad \frac{dR}{dt} = R \begin{pmatrix} 0 & -k & 0 \\ k & 0 & -\tau \\ 0 & \tau & 0 \end{pmatrix}.$$

If both the curvature and the torsion are constant, then

$$\omega = \begin{pmatrix} \tau \\ 0 \\ k \end{pmatrix}$$

is the axis of rotation for

$$A = \begin{pmatrix} 0 & -k & 0 \\ k & 0 & -\tau \\ 0 & \tau & 0 \end{pmatrix}.$$

Then $\exp t A$ is the rotation about ω through the angle $t\sqrt{\tau^2 + k^2}$, and $x(t)$ is a helix along ω.

Suppose, now, that we consider curves whose curvature $k(t) = $ constant ($\neq 0$) and whose torsion can take only two distinct values: τ_1 and τ_2. Such curves are concatenations of helices along

$$\omega_1 = \begin{pmatrix} \tau_1 \\ 0 \\ k \end{pmatrix}, \quad \text{and} \quad \omega_2 = \begin{pmatrix} \tau_2 \\ 0 \\ k \end{pmatrix}.$$

The corresponding family of left-invariant vector fields on $G = \mathbb{R}^3 \ltimes SO_3(R)$ is defined by $\Gamma = \{(e_1, A), (e_1, B)\}$, with

$$A = \begin{pmatrix} 0 & -k & 0 \\ k & 0 & -\tau_1 \\ 0 & \tau_1 & 0 \end{pmatrix} \quad \text{and} \quad B = \begin{pmatrix} 0 & -k & 0 \\ k & 0 & -\tau_2 \\ 0 & \tau_2 & 0 \end{pmatrix}.$$

It follows that Lie $(\Gamma) = \mathbb{R}^3 \oplus SO_3(R)$ because of the following calculations:

$(e_1, A) - (e_1, B) = (\tau_1 - \tau_2)(0, A_1)$ and $[(e_1, A), (e_1, B)] = k(\tau_1 - \tau_2)(0, A_2)$.

But then $[(0, A_2), (0, A_1)] = (0, A_3)$, and therefore $0 \oplus so_3(R) \subset \text{Lie}(\Gamma)$. Hence $(e_1, 0) \in \text{Lie}(\Gamma)$, and then $[(e_1, 0), (0, so_3(R))] = (\mathbb{R}^3, 0)$.

According to Theorem 10, any initial point x_0 and any initial frame at x_0 can be connected to any terminal point x_1 and any terminal frame at x_1 along the integral curves of the left-invariant family $\mathcal{F} = \{X_A, X_B\}$ in $G = \mathbb{R}^3 \ltimes SO_3(R)$, with X_A and X_B equal to the left-invariant vector fields that coincide with (e_1, A) and (e_1, B) at the group identity.

Analogous results hold for curves in $\mathbb{R}^n$ that have all but one curvature fixed, while the remaining free curvature can take any positive value. According to Theorem 4, the matrices A and B that correspond to such a situation generate either $SO_n(R)$ or the unitary group $U_{2m}(R)$. The corresponding set of matrices Γ in G is given by

$$\Gamma = \{(e_1, A + uB) : u > 0\}.$$

We shall now show that Lie $(\Gamma) = \mathbb{R}^n \ltimes \mathcal{L}(K)$, with $\mathcal{L}(K)$ equal to the Lie algebra of either $K = SO_n(R)$ or $K = U_{2m}(R)$:

Let $\pi : \text{Lie}(\Gamma) \longrightarrow \mathcal{L}(K)$ be the projection $\pi(a, A) = A$. π is a Lie-algebra homomorphism that is onto $\mathcal{L}(K)$, as a consequence of Theorem 4. The kernel of π is a linear subspace of $\mathbb{R}^n$ that is also an ideal in $\mathcal{L}(G)$. It follows that for each $(v, 0)$ in ker π and for each $(0, k)$ with $k \in \mathcal{L}(K)$, $[(v, 0), (0, k)] = -(kv, 0)$. Thus, ker π is a linear subspace of $\mathbb{R}^n$ invariant under $\mathcal{L}(K)$. But ker π cannot be equal to zero, for then Lie (Γ) would be isomorphic to $\mathcal{L}(K)$ (see Problem 5). Then $\mathcal{L}(K)(\ker \pi) = \mathbb{R}^n$ (see Problem 6). Therefore, Theorem 10 is applicable, and the corresponding controllability conclusions follow.

Problem 5 Show that the solutions of

$$\frac{dg}{dt} = g \begin{pmatrix} 0 & 0 \\ e_1 & A + uB \end{pmatrix}, \qquad u > 0,$$

are not contained in any compact subgroup of E_n.

Problem 6 Show that if Lie (Γ) contains an element $(a, A) \in \mathbb{R}^n \oplus so_n(R)$, with $a \neq 0$, and if $\pi(\text{Lie}(\Gamma)) = so_n(R)$, then $\ker \pi = \mathbb{R}^n$.

3 Controllability properties of affine systems

By "an affine system on $\mathbb{R}^n$" we shall mean any differential system of the form

$$\frac{dx}{dt} = (Ax + a_0) + \sum_{i=1}^{m} u_i(t)(A_i x + a_i), \tag{10}$$

where $A_0, \ldots, A_m$ are $n \times n$ matrices, $a_0, \ldots, a_m$ are column vectors in $\mathbb{R}^n$, and $u = (u_1(t), \ldots, u_m(t))$ are arbitrary functions that will be regarded as controls. We shall follow the usual terminology in the control-theory literature and call (10) bilinear whenever each $a_i = 0$ $(i = 0, \ldots, m)$.

We shall presently show that the class of affine systems is substantially richer than the linear class and that its controllability properties are governed by entirely different geometric considerations. To begin with, small perturbations of controllable linear systems in the class of affine systems can result in an uncontrollable system, as the next example shows.

Example 1 Consider the small perturbations $dx/dt = Ax + u(t)(\varepsilon Bx + b)$ of a controllable linear system $dx/dt = Ax + ub$ in $\mathbb{R}^2$. Take

$$A = \begin{pmatrix} \alpha_1 & 0 \\ 0 & \alpha_2 \end{pmatrix},$$

with $\alpha_1 < 0$ and $\alpha_2 < 0$, and assume that b and Ab are linearly independent.
Let

$$B = \begin{pmatrix} \beta_1 & 0 \\ 0 & \beta_2 \end{pmatrix},$$

with $\beta_1 \beta_2 \neq 0$. The trajectories of the affine vector field $Y_\varepsilon(x) = \varepsilon Bx + b$ are the same as the trajectories of the linear field $x \to \varepsilon Bx$ translated to the new origin $c = -(1/\varepsilon)B^{-1}b$. Assuming that $c_2 > 0$, $\beta_1 > 0$, and $\beta_2 < 0$, the phase portrait of Y_ε is as shown in Figure 6.3.

The half-space $x_2 \leq c_2$ is invariant under $\pm Y_\varepsilon$ for any $\varepsilon > 0$. This half-space is also invariant for the drift vector $X(x) = Ax$, and therefore it is also invariant for the system $dx/dt = X(x) + u(t)Y_\varepsilon(x)$. Consequently, the perturbed system is not controllable (Figure 6.4).

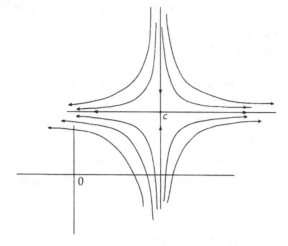

Fig. 6.3.

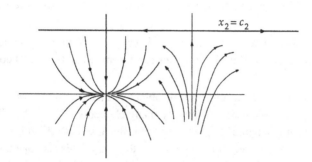

Fig. 6.4.

For large values of x, the perturbed vector field $Y_\varepsilon(x)$ is "far" from the constant vector field, which accounts for the qualitatively different behavior of the perturbed system, and partly explains the loss of controllability in the' foregoing example. The main controllability result concerning affine systems makes use of the controllability properties of their "bilinear parts," as explained in the next theorem.

For any affine vector field $X(x) = Ax + a$, $\vec{X}$ will denote its linear part $\vec{X}(x) = Ax$. If $\mathcal{F}$ is any family of affine vector fields, then $\vec{\mathcal{F}}$ will denote the family of linear fields $\vec{X}$ such that $X \in \mathcal{F}$. An affine family $\mathcal{F}$ is said to have no fixed points if there is no x in $\mathbb{R}^n$ such that $X(x) = 0$ for all $X \in \mathcal{F}$.

Theorem 11 *An affine family of vector fields $\mathcal{F}$ is controllable on $\mathbb{R}^n$ if*

(a) *$\mathcal{F}$ has no fixed points and*
(b) *$\vec{\mathcal{F}}$ is controllable on $\mathbb{R}^n - \{0\}$.*

The first condition, concerning the nonexistence of fixed points, is a necessary condition for controllability, because the set reachable from a fixed point consists of a single point. Theorem 11 applied to system (10) says that it is controllable on $\mathbb{R}^n$ whenever

(a) there is no point x in $\mathbb{R}^n$ such that $A_i x + a_i = 0$, for $i = 0, 1, \ldots, m$, and
(b) the bilinear system $dx/dt = A_0 x + \sum_{i=1}^{m} u_i(t) A_i(x)$ is controllable on $\mathbb{R}^n - \{0\}$.

The proof of Theorem 11 depends on two lemmas, which we state without proofs, in the form of problems for the reader.

Problem 7 Show that any affine family $\mathcal{F}$ that satisfies assumptions (a) and (b) of Theorem 11 contains a finite subfamily that also satisfies (a) and (b).

Problem 8 Show that the reachable set $\mathcal{A}_{\mathcal{F}}(x, <\infty)$ is not bounded for any x in $\mathbb{R}^n$ whenever $\mathcal{F}$ is affine and satisfies conditions (a) and (b) of Theorem 11.

Proof of Theorem 11 According to Problem 7, there is no loss of generality in assuming that $\mathcal{F}$ consists of finitely many elements $X_1, \ldots, X_m$. We shall show that each reachable set $\mathcal{A}_{\mathcal{F}}(x)$ is open. Because both $\mathcal{F}$ and $-\mathcal{F}$ satisfy the hypothesis of the theorem, we then have that $\mathcal{A}_{-\mathcal{F}}(x)$ is also open. But then each reachable set $\mathcal{A}_{\mathcal{F}}(x)$ is also closed, and therefore $\mathcal{A}_{\mathcal{F}}(x) = \mathbb{R}^n$ for each x in $\mathbb{R}^n$.

For each $\lambda > 0$ and each $v \in \mathbb{R}^n$, define $h_{\lambda,v}(x) = v + \lambda(x - v)$ for all $x \in \mathbb{R}^n$. Then $h_{\lambda,v}$ is a transformation of $\mathbb{R}^n$, with $h_{\lambda,v}(v) = v$, and $h_{\lambda,v}^{-1}(x) = v + (1/\lambda)(x - v)$. This transformation transforms any vector field X in $\mathbb{R}^n$ into $X_{\lambda,v}(x) = dh_{\lambda,v} \circ X \circ h_{h,v}^{-1}(x)$. If $X(x) = Ax + a$, then the transformed field $X_{\lambda,v}$ is given by $X_{\lambda,v}(x) = A(x - v) + \lambda(Av + a) = \vec{X}(x - v) + \lambda X(v)$.

We shall use $\mathcal{F}_{\lambda,v}$ to denote the transformed family consisting of $X_{\lambda,v}$, with X in $\mathcal{F}$. $\vec{\mathcal{F}}_{\lambda,v}$ will denote the linear family of $\mathcal{F}_{\lambda,v}$. Because $\vec{\mathcal{F}}$ is controllable on $\mathbb{R}^n - \{0\}$, $\vec{\mathcal{F}}_{\lambda,v}$ is controllable on $\mathbb{R}^n - \{v\}$, and therefore $S(v)$, the sphere of radius 1 centered at v, is contained in the reachable set $\mathcal{A}_{\vec{\mathcal{F}}_{\lambda,v}}(x)$ for each x in $\mathbb{R}^n$. In particular, $S(v) \subset \mathcal{A}_{\vec{\mathcal{F}}_{\lambda,v}}(x)$ for each $x \in S(v)$.

According to a theorem of Sussmann (1976), $S(v)$ is contained in the set reachable from each point x in $S(v)$ for any family of vector fields that is "close" to $\vec{\mathcal{F}}_{\lambda,v}$. In particular, the same conclusion applies to $\mathcal{F}_{\lambda,v}$ for λ sufficiently small; that is, there exists $\lambda > 0$ such that

$$S(v) \subset \mathcal{A}_{\mathcal{F}_{\lambda,v}}(x) \quad \text{for each } x \text{ in } S(v).$$

But then the solid ball $B(v) = \{x : \|x - v\| \le 1\}$ is contained in $\mathcal{A}_{\mathcal{F}_{\lambda,v}}(x)$ for each x in $B(v)$. The argument is as follows: The reachable sets of $\pm\mathcal{F}_{\lambda,v}$ are not bounded (Problem 8). Hence for any x and y in $B(v)$, there exist points x_1 and y_1 on $S(v)$, x_1 reachable from x, and y reachable from y_1. But then it follows from the preceding paragraph that y_1 is reachable from x_1. Hence y is reachable from x.

It remains to interpret these results in terms of the original family $\mathcal{F}$. Because $\mathcal{A}_{\mathcal{F}_{\lambda,v}}(x) = h_{\lambda,v} \circ \mathcal{A}_{\mathcal{F}}(h_{\lambda,v}^{-1}(x))$, it follows that

$$h_{\lambda,v}^{-1} B(h_{\lambda,v}(v)) \subset \mathcal{A}_{\mathcal{F}}(v).$$

But $h_{\lambda,v}^{-1} B(h_{\lambda,v}(v)) = h_{\lambda,v}^{-1} B(v) = (1/\lambda) B(v)$; therefore, the solid ball of radius $1/\lambda$ centered at v is contained in the set reachable from v, and consequently $\mathcal{A}_{\mathcal{F}}(v)$ is open. The proof is now finished. ∎

Theorem 11 marks a sharp departure from the theory of linear control systems and calls for further investigations of the controllability properties of families of linear vector fields on $\mathbb{R}^n - \{0\}$. The task of determining the conditions under which a given family of vector fields is controllable on $\mathbb{R}^n - \{0\}$ is surprisingly difficult. We shall provide partial answers for families of linear vector fields that are subordinated to a linear Lie group G that acts transitively on $\mathbb{R}^n - \{0\}$ by studying the controllability properties of right-invariant families of vector fields on G.

Recall that any linear group G acts, on the left, on points of $\mathbb{R}^n$ regarded as column vectors by the matrix multiplication. Then each right-invariant vector field $X(g) = Ag$ induces a linear vector field $x \to Ax$. Conversely, any linear vector field $x \to Ax$, with A an element of the Lie algebra of G, corresponds to a right-invariant vector field $g \to Ag$ on G. So a linear family of vector fields $\mathcal{F}$ induces a right-invariant family $\mathcal{F}_r$ on G. $\mathcal{F}$ is controllable on $\mathbb{R}^n - \{0\}$ whenever $\mathcal{F}_r$ is controllable on G provided that G acts transitively on $\mathbb{R}^n - \{0\}$.

4 Controllability on semisimple Lie groups

We shall not pursue the problem of classifying all groups that act transitively on $\mathbb{R}^n - \{0\}$. The interested reader can consult Boothby and Wilson (1979) for the list of such groups. Instead, we shall provide some controllability results for right-invariant systems on $SL_n(R)$, the group of $n \times n$ matrices with determinant equal to 1. It is easy to verify that $SL_n(R)$ acts transitively on $\mathbb{R}^n - \{0\}$ and so the results of this section also complement the preceding controllability results for affine systems.

As it happens, our study of controllability on $SL_n(R)$ makes substantial use of the structure of its Lie algebra, in terms of its root system, and is conceptually not much different from a similar study on an arbitrary semisimple Lie group G. Moreover, apart from being concrete, the explanations on $SL_n(R)$ are not any easier. For those reasons we shall present the results on general semisimple Lie groups, rather than on $SL_n(R)$.

It will be convenient to continue with the notation established in Section 1 and denote a right-invariant system by the corresponding set Γ in the Lie algebra defined by the corresponding vectors at the group identity. For the questions of controllability, there is no loss in generality in assuming that Γ is a closed and convex set such that Lie(Γ) is equal to the Lie algebra of G, because the Lie saturate is closed and convex. Although aimed at different goals, the present study shares much common ground with the work of Hilgert et al. (1989). Their study is also concerned with the Lie semigroups $S(\Gamma)$, generated by the closed and convex cones Γ in the Lie algebra of G, which they call wedges.

As tempting as it is to go deeply into this theory, we shall present only the most basic results.

Keeping in mind the reader who is not familiar with the theory of semisimple Lie groups, we shall begin with the basic definitions. For each element L in $\mathcal{L}(G)$, ad L denotes the linear endomorphism $(\text{ad} L)(X) = [L, X]$ on the Lie algebra of G. The correspondence $L \to \text{ad} L$ is linear and also satisfies

$$\text{ad} [L, M] = \text{ad} L \circ \text{ad} M - \text{ad} M \circ \text{ad} L.$$

Therefore, ad is a homomorphism from the Lie algebra of G into the Lie algebra of linear endomorphisms on $\mathcal{L}(G)$, endowed with the Lie bracket equal to the commutator $[A, B] = AB - BA$. The bilinear form trace(ad L ad M) is called the Killing form, usually denoted by $B(L, M)$. G is said to be semisimple if B is nondegenerate, that is, if for every nonzero element L in $\mathcal{L}(G)$ there exists M in $\mathcal{L}(G)$ such that $B(L, M) \neq 0$. The reader can easily deduce the following key properties of the Killing form:

$$B(\theta(L), \theta(M)) = B(L, M) \tag{11}$$

for any Lie-algebra automorphism θ, and

$$B((\operatorname{ad} X)(L), (M)) = -B(L, (\operatorname{ad} X)(M)) \qquad (12)$$

for any elements L, M, and X in $\mathcal{L}(G)$.

Remark The first property follows from the fact that $\theta(\operatorname{ad} L)\theta^{-1} = (\operatorname{ad} \theta)(L)$, and the second property follows from the fact that ad is a homomorphism. ■

G is said to be a simple Lie group if it contains no proper normal Lie subgroup. Equivalently, G is simple if and only if the Lie algebra of G contains no nonzero proper ideals. It is easy to show, using property (12), that for any ideal I of $\mathcal{L}(G)$, the orthogonal complement $I^{\perp}$ induced by the form of Killing is an ideal. An argument based on this observation shows that any semisimple Lie group G is a finite product of simple groups. Consequently, determining the controllability of systems on semisimple Lie groups reduces to a study on each simple group in the foregoing decomposition.

Semisimple Lie groups have other remarkable properties that have a direct bearing on this subject matter. The first property is that each mapping $\operatorname{ad} L: \mathcal{L} \to \mathcal{L}$ is semisimple. That is, the complexification $\mathcal{L}^c$ of $\mathcal{L}$ is a direct sum of eigenspaces of $\operatorname{ad} L$. The maximal dimension of the eigenspace that corresponds to the zero eigenvalue of $\operatorname{ad} L$ as L ranges over $\mathcal{L}$ is called the rank of $\mathcal{L}$. It turns out that any two elements in the zero eigenspace commute. Therefore, the rank of $\mathcal{L}$ can be equivalently defined as the dimension of the maximal commutative subalgebra of $\mathcal{L}^c$. Such an algebra is called a Cartan algebra.

An element L in $\mathcal{L}$ is called *regular* if the zero eigenspace of $\operatorname{ad} L$ has dimension equal to the rank of $\mathcal{L}$. A regular element is called *strongly regular* if every other eigenvalue of $\operatorname{ad} L$ is simple, that is, if each eigenspace is a one-dimensional subspace of $\mathcal{L}^c$. It is known that the set of strongly regular elements in any semisimple Lie algebra is open and dense (Jurdjevic and Kupka, 1981b).

We shall use $\operatorname{sp}(L)$ to denote the spectrum of $\operatorname{ad} L$. For any a in $\operatorname{sp}(L)$, let L_a denote the real eigenspace corresponding to a. That is, L_a is the eigenspace corresponding to a if a is a real number. Otherwise, it is a two-dimensional vector space equal to the linear span of the real and imaginary real vector spaces in the complex eigenspace of a. Each of these spaces is invariant for $\operatorname{ad} L$, and

$$\mathcal{L} = L_0 \oplus \{L_a : a \in \operatorname{sp}(L),\ a \neq 0\}. \qquad (13)$$

In particular, any element M in $\mathcal{L}$ has a corresponding decomposition

$$M = M_0 \oplus \{M_a : a \neq 0, \ a \in \mathrm{sp}(L)\}. \tag{14}$$

The set of all elements M in $\mathcal{L}$ for which each M_a is nonzero, with $a \neq 0$, is an open and dense subset of $\mathcal{L}$, relative to each strongly regular element L.

Theorem 12 *The set C of all pairs (L, M) in $\mathcal{L} \times \mathcal{L}$ for which the Lie algebra generated by $\Gamma = \{L, M\}$ is equal to $\mathcal{L}$ is an open and dense subset of $\mathcal{L} \times \mathcal{L}$.*

Proof If $\mathrm{Lie}\{L, M\} = \mathcal{L}$, then the right-invariant control set $\{\pm\vec{L}, \pm\vec{M}\}$ is controllable on G. $\vec{L}$ and $\vec{M}$ denote the right-invariant vector fields whose values at the group identity are equal to L and M. Because controllable invariant systems remain controllable under small perturbations, it follows that C is open.

To show that C is dense, take L as a strongly regular element, and take M as an element in $\mathcal{L}$ for which $M_a \neq 0$ for $a \neq 0$ in the decomposition (14) of M relative to L. Such pairs form a dense subset of $\mathcal{L} \times \mathcal{L}$, and each pair (L, M) belongs to C. ∎

Corollary *Almost all right-invariant control systems*

$$\frac{dg}{dt} = \vec{X}_0(g) + \sum_{i=1}^{m} u_i(t)\vec{X}_i(g)$$

are controllable on G provided that the number of control functions $m \geq 2$.

Proof For each such system, the vector space spanned by $\{X_1, \ldots, X_m\}$ is contained in the Lie saturate. According to Theorem 12, the set of all m-tuples of right-invariant vector fields that generate $\mathcal{L}$ is open and dense. Each such system is controllable independently of the drift vector field X_0. End of the proof. ∎

Systems with single controls are described by lines Γ of the form $\{A + uB : u \in \mathbb{R}^1\}$, with A and B fixed elements in $\mathcal{L}$. As remarked earlier, determining the controllability reduces to determining the conditions on A and B such that $S(\Gamma) = G$. Γ is equivalent to the closed positive cone spanned by $\{A, \pm B\}$. In order to describe the conditions under which a given system $\Gamma = \{A, \pm B\}$ is controllable on G, we shall assume that the complex numbers are ordered by its lexicographic ordering; that is, $z < w$ if $\mathrm{Re}\,z < \mathrm{Re}\,w$, and if $\mathrm{Re}\,z = \mathrm{Re}\,w$, then $\mathrm{Im}\,z < \mathrm{Im}\,w$.

Theorem 13 *Let G be a real semisimple Lie group with a finite center. Assume that* $\Gamma = \{A, \pm B\}$ *satisfies the following conditions:*

(a) *B is strongly regular.*

(b) Lie $(\Gamma) = \mathcal{L}(G)$.

(c) *The component* A_a *in the decomposition* (14) *of A relative to* ad B *that corresponds to either a maximal or minimal eigenvalue of* ad B *is not equal to zero.*

(d) *If a is a maximal eigenvalue of* ad B, *and if* $r = \mathrm{Re}\, a$ *is also an eigenvalue of* ad B, *then*

$$\mathrm{trace}\,(\mathrm{ad}\, A_r \cdot \mathrm{ad}\, A_{-r}) < 0,$$

with A_r *and* A_{-r} *denoting the components of A in* L_r *and* L_{-r}.

Then $S(\Gamma) = G$.

The proof of Theorem 13 makes intricate use of the root space decomposition of semisimple Lie algebras, whose details would take us too far from the main theme of this book. For that reason we shall omit the proofs. We shall, instead, provide an elementary discussion of the theorem for $G = SL_n(R)$.

The Lie algebra $sl_n(R)$ of $SL_n(R)$ consists of all $n \times n$ matrices with real entries whose traces are equal to zero. It then follows that the eigenvalues of ad B consist of differences of eigenvalues of B. If $\lambda_1, \ldots, \lambda_n$ denote the eigenvalues of B, some of which may be complex, then strongly regular elements satisfy the condition that

$$\lambda_i - \lambda_j \neq \lambda_k - \lambda_\ell$$

for all indices i, j, k, ℓ, with $\{i, j\} \neq \{k, \ell\}$, and $i \neq j$.

The eigenspace L_a corresponding to an eigenvalue $a = \lambda_i - \lambda_j$ consists of all matrices of the form $b_i \otimes c_j$, with b_i and c_j respectively the eigenvectors of B and its transpose B^*. That is, $Bb_i = \lambda_i b_i$ and $B^* c_j = \lambda_j c_j$.

If a is a complex eigenvalue, then $\bar{a}$ is also an eigenvalue, and L_a is the real vector space spanned by $\mathrm{Re}(b_i \otimes c_j)$ and $\mathrm{Im}(b_i \otimes c_j)$. The eigenspace that corresponds to a zero eigenvalue consists of all matrices that commute with B.

Example Let

$$B = \begin{pmatrix} 0 & 1 \\ -1 & 0 \end{pmatrix}.$$

The eigenvalues of B are $\lambda_1 = i$ and $\lambda_2 = -i$. The unit eigenvectors of B are $b_1 = (1/\sqrt{2})(e_1 + ie_2)$ and $b_2 = \bar{b}_1 = (1/\sqrt{2})(e_1 - ie_2)$. Because $B^* = -B$, it follows that $a_1 = b_2$ and $a_2 = b_1$. Therefore,

$$b_1 \otimes a_2 = \frac{1}{2}\begin{pmatrix} 1 & i \\ i & -1 \end{pmatrix} \quad \text{and} \quad b_2 \otimes a_1 = \frac{1}{2}\begin{pmatrix} 1 & -i \\ -i & -1 \end{pmatrix}.$$

Therefore,

$$\mathrm{Re}(b_1 \otimes a_2) = \frac{1}{2}\begin{pmatrix} 1 & 0 \\ 0 & -1 \end{pmatrix} \quad \text{and} \quad \mathrm{Im}(b_1 \otimes a_2) = \frac{1}{2}\begin{pmatrix} 0 & 1 \\ 1 & 0 \end{pmatrix}.$$

It now follows that the linear span of the foregoing matrices is equal to the vector space of 2×2 symmetric matrices having trace equal to zero. Hence, decomposition (13) induced by B coincides with the classic decomposition of matrices into a symmetric part and an antisymmetric part.

If A denotes any matrix in $sl(2)$ that satisfies condition (c) of Theorem 13, then its symmetric part in not zero, and consequently $\Gamma = \{A, B, -B\}$ is controllable. In this case, controllability can be verified directly by elementary means, as follows:

There is no loss of generality in assuming that A is a symmetric matrix

$$a_{11}\begin{pmatrix} 1 & 0 \\ 0 & -1 \end{pmatrix} + a_{12}\begin{pmatrix} 0 & 1 \\ 1 & 0 \end{pmatrix}.$$

Then $(\exp vB)_\#(A) = (\exp vB)A(\exp -vB)$ is contained in the Lie saturate of Γ for any real number v. But

$$(\exp vB)_\#(A) = a_{11}\begin{pmatrix} \cos 2v & -\sin 2v \\ -\sin 2v & -\cos 2v \end{pmatrix} + a_{12}\begin{pmatrix} \sin 2v & \cos 2v \\ \cos 2v & -\sin 2v \end{pmatrix},$$

and, in particular, when $v = \frac{1}{2}\pi$, $(\exp vB)_\#(A) = -A$. Hence, Γ is equivalent to the symmetric system $\{A, -A, B, -B\}$, from which controllability is immediate.

Consider now the case

$$B = \begin{pmatrix} 1 & 0 \\ 0 & -1 \end{pmatrix}.$$

In contrast to the previous case, the eigenvalues of B are real. The corresponding eigenspaces of $\mathrm{ad}\, B$ are one-dimensional and are spanned by

$$e_1 \otimes e_2 = \begin{pmatrix} 0 & 1 \\ 0 & 0 \end{pmatrix} \quad \text{and} \quad e_2 \otimes e_1 = \begin{pmatrix} 0 & 0 \\ 1 & 0 \end{pmatrix}.$$

Using B, $(e_1 \otimes e_2)$, and $(e_2 \otimes e_1)$ as the basis for decomposition (13), then any matrix A can be written as $A = a_{11}B + a_{12}(e_1 \otimes e_2) + a_{21}(e_2 \otimes e_1)$.

According to Theorem 13, $\Gamma = \{A, B, -B\}$ is controllable whenever $a_{12} a_{21} < 0$.

As in the previous case, there is a simple and independent proof of controllability. It goes as follows: There is no loss of generality in taking $a_{11} = 0$. But then the eigenvalues of A are imaginary, and therefore the orbits of A are periodic. Hence, Γ is equivalent to the symmetric system $\{A, -A, B, -B\}$ and is therefore controllable. We leave it to the reader to show that Γ is not controllable when $a_{12}a_{21} \leq 0$.

Returning now to the general case in $s\ell_n(R)$, any strongly regular element B with real eigenvalues can be diagonalized. The eigenspaces of ad B then become scalar multiples of $e_i \otimes e_j$, with $e_1, \ldots, e_n$ denoting the standard basis in $\mathbb{R}^n$. The maximal eigenvalue of ad B is the largest difference between the diagonal elements of B, and the minimal eigenvalue is its negative. Assuming that the diagonal elements of B are written in ascending order $\lambda_1 < \lambda_2 < \cdots < \lambda_n$, the controllability condition (d) in Theorem 13 reduces to $a_{1n}a_{n1} < 0$, with a_{ij} denoting the general entry of the drift matrix A.

Strongly regular elements with real eigenvalues suggest that Theorem 13 can be seen as a distant extension of the controllability result for systems with linear recurrent drift, namely, the case when A is an antisymmetric matrix. From this perspective, the case where both A and B are symmetric matrices is at the other end of the spectrum created by Theorem 13. There is adequate evidence to suggest that such a case is never controllable, neither on $SL_n(R)$ nor on its homogeneous space $\mathbb{R}^n - \{0\}$, but the proof remains an open question. We shall further illustrate the preceding discussion with a controllability result in the space of Jacobi matrices in $s\ell_n(R)$.

A matrix $A = (a_{ij})$ is called a Jacobi matrix if $a_{ij} = 0$ for all i and j, with $|i - j| > 1$. That is, the nonzero entries of any Jacobi matrix A are concentrated along the three main diagonals. We shall use J_n to denote the space of all Jacobi $n \times n$ matrices of trace zero.

J_n contains all diagonal matrices B. B is strongly regular if the differences of its diagonal entries are distinct. The reader may want to show that a strongly regular element B along with an element A in J_n that satisfies $a_{ij} \neq 0$ for $|i - j| = 1$ generates $s\ell_n(R)$. That is, Lie $\{A, B\} = s\ell_n(R)$.

Consider, now, an arbitrary positive closed convex cone Γ in J_n such that

(i) Γ contains all diagonal elements in J_n and
(ii) Γ has a nonempty interior in J_n.

As before, $S(\Gamma)$ denotes the semigroup generated by $\{\exp tA : t \geq 0, A \in \Gamma\}$.

Theorem 14 $S(\Gamma) = G$ if and only if Γ contains an element A having an entry a_{ij}, with $i \neq j$, such that $a_{ij} a_{ji} < 0$.

Proof Assume, first, that there is an element A in Γ having an entry a_{ij} for which $a_{ij} a_{ji} < 0$. Because Γ has a nonempty interior in J_n, we can assume that each entry $a_{k\ell}$, with $|k - \ell| = 1$, is nonzero. Let B be a strongly regular diagonal matrix, with diagonal entries $b_1, \ldots, b_n$, for which $b_i - b_j$ is the maximal difference.

Then $\Gamma_0 = \{A, \pm B\} \subset \Gamma$, and according to Theorem 13, $S(\Gamma_0) = G$. Evidently, $S(\Gamma) = G$, since $S(\Gamma_0) \subset S(\Gamma)$.

Suppose that Γ is any cone in J_n for which $a_{ij} a_{ji} > 0$ for each $|i - j| = 1$ for each matrix A in Γ. Let P be a diagonal matrix, with its diagonal entries $p_1, \ldots, p_n$ satisfying $\det P = 1$.

PAP^{-1} is the matrix with entries given by $a_{ij} p_i p_j^{-1}$. Choose the diagonal elements of P according to the rule $p_i p_j^{-1} = \operatorname{sgn} a_{ij}$ for all i and j such that $|i - j| = 1$. With this choice of P, the conjugate cone $\Gamma_P = P\Gamma P^{-1}$ has all nondiagonal entries positive (relative to J_n). But then $S(\Gamma_P)$ is contained in the set of matrices having nonnegative entries and therefore is a proper semigroup. Our proof is now finished.   ∎

The foregoing rather special result suggests a question directed to a reader interested in the structure of Lie semigroups in a semisimple group G: If Γ is any closed positive convex cone in $\mathcal{L}$ such that $\operatorname{trace}(\operatorname{ad} L \operatorname{ad} M) \geq 0$ for any L and M in Γ, is it always true that $S(\Gamma)$ is a proper semigroup in G?

Problem 9 Let $G = SO(1, n)$ be the closed subgroup of $\mathrm{GL}_{n+1}(R)$ that leaves the quadratic form $-x_0^2 + \sum_{i=1}^n x_i^2$ invariant. Show that the Lie algebra of G is equal to the direct sum $\mathcal{P} \oplus \mathcal{K}$, with $\mathcal{P}$ equal to the n-dimensional space of symmetric matrices A of the form

$$
A = \begin{pmatrix}
0 & a_1 & \cdots & a_n \\
a_1 & 0 & \cdots & 0 \\
\vdots & \vdots & & \vdots \\
a_n & 0 & \cdots & 0
\end{pmatrix},
$$

and $\mathcal{K}$ equal to the algebra of all antisymmetric matrices of the form

$$
B = \begin{pmatrix}
0 & 0 & \cdots & \cdots & 0 \\
0 & 0 & b_{23} & \cdots & b_{2n} \\
\vdots & -b_{23} & 0 & & \vdots \\
\vdots & \vdots & & \ddots & b_{(n-1)n} \\
0 & -b_{2n} & \cdots & -b_{(n-1)n} & 0
\end{pmatrix}.
$$

Problem 10 Let B be an element of $\mathcal{K}$ having n distinct eigenvalues, and let A be any nonzero element of $\mathcal{P}$. Show that $\Gamma = \{A + uB : u \in \mathbb{R}^1\}$ is controllable on G.

Problem 11 Show that $\Gamma = \{A_1 + uA_2 : u \in \mathbb{R}^1\}$ is never controllable on G for any elements A_1 and A_2 in $\mathcal{P}$.

Notes and sources

Controllability results on compact Lie groups were first obtained by Jurdjevic and Sussmann (1972). Extensions of those results to semidirect products of Lie groups were published by Bonnard et al. (1982).

The results of Theorem 5 are apparently true for any semisimple Lie group G, as I have recently learned from San Martin and Tonelli (1994). That is, a subset Γ of $\mathcal{L}(G)$ is controllable on G if and only if the induced family of vector fields $\mathcal{F}_\Gamma$ is controllable on any compact homogeneous space G/H.

Controllability criteria for affine systems were inspired by the work of Bonnard (1981). The version given here is taken from Jurdjevic and Sallet (1984).

The material presented in Section 4 covers a fairly extensive body of work that originated with Jurdjevic and Kupka (1981a, b). Theorem 13, in its present form, was recently proved by El Assoudi, Gauthier, and Kupka (in press).

The papers by Boothby and Wilson (1979) and Kostant (1958) provide a nice complement to the aforementioned papers. Kostant's paper contains criteria useful for calculating the Lie algebras generated by arbitrary subsets of the Lie algebras of classic groups that can be used to shortcut the calculations in Theorem 4.

Apart from the aforementioned literature, the work of Hilgert et al. (1989), dealing with the abstract theory of Lie semigroups, parallels much of the theory presented here.

Part two
Optimal control theory

Much like the unidentified hero figure in classic folk tales who, after having won his position by virtue of his achievements, in the end reveals a hidden but distinguished parentage, so optimal control theory, recognized initially as an engineering subject, reveals, as it reaches maturity, a distinct relationship to classic forebears: the calculus of variations, differential geometry, and mechanics. This distinctive character of optimal control theory can be traced to the mathematical problems of the subject in the mid-1950s dealing with inequality constraints. Faced with the practical, time-optimal control problems of that period, mathematicians and engineers looked to the calculus of variations for answers, but soon discovered, through the papers of Bellman et al. (1956), LaSalle (1954), and Bushaw (1958) on the bang-bang controls, that the answers to their problems were outside the scope of the classic theory and would require different mathematical tools. That realization initiated a search for new necessary conditions for optimality suitable for control problems. That search, further intensified by the space program and the race to the moon, eventually led, in 1959, to the "maximum principle," which answered the practical needs of that period. So strongly was the maximum principle linked to control problems involving bounds on the controls that its significance in a larger context – as an important extension of Weierstrass's excess function in the calculus of variations – went unnoticed for a long time after its discovery.

The treatment here of optimal control problems combines the rich theory of the calculus of variations with control-theory innovations. I believe, as does Young (1969), that a control-theory point of view is more natural for the calculus of variations. I shall further argue that the calculus of variations is fundamentally concerned with problems of optimal control and that much mathematical clarity is gained through this recognition.

With these perspectives in mind, it becomes essential to begin our study of optimality with the definitions that blend the classic calculus of variations with control theory and are at the same time amenable to problems of mechanics and geometry. For those reasons, we shall consider optimal problems defined by a subset C of the tangent bundle of a smooth manifold M and a real-valued function f having C as its domain. C is called the constraint set, and f is called the *cost function*. By "a trajectory of C" we shall mean an absolutely continuous curve $x(t)$ in M such that $(dx/dt)(t) \in C$ for almost all t in the domain of x. If $[0, T]$ denotes the domain of x, then $\int_0^T f((dx/dt)(t))\, dt$ is called the total cost of x. Associated with any two points a and b in M, we shall be interested in the trajectory of C that connects a to b and whose total cost is minimal among all such trajectories of C. The trajectory with minimal total cost will be called *optimal*.

In this exposition we shall consider only constraint sets that admit control-theory descriptions. That is, we shall consider only sets C that admit "sections" of the form $\xi = F(\pi(\xi), u_1, \ldots, u_m)$, where $(u_1, \ldots, u_m)$ takes values in a fixed set U in $\mathbb{R}^m$, π denotes the natural projection from TM onto M, and ξ is an arbitrary point of C. Then the velocity dx/dt of any trajectory $x(t)$ is parametrized by the controls $u_1(t), \ldots, u_m(t)$, and therefore its total cost can be expressed by $f \circ F$ in terms of $x(t)$ and $u(t)$ as

$$\int_0^T c(x(t), u(t))\, dt = \int_0^T f \circ F(x(t), u(t))\, dt, \quad \text{with } c = f \circ F.$$

In a given section of C, the trajectories of C that connect an intial point a to a terminal point b in T units of time coincide with the solution curves $x(t)$ of the differential system

$$\frac{dx}{dt} = F(x(t), u_1(t), \ldots, u_m(t)),$$

with $x(0) = a$ and $x(T) = b$. Under suitable smoothness assumptions on F, each control function $u(t) = (u_1(t), \ldots, u_m(t))$ determines a unique solution curve, and the problem of finding the optimal trajectories of C is converted to one of finding the controls that give rise to the optimal trajectories; that is, the original problem becomes a problem of optimal control.

Any given constraint set can admit different control sections, in which case it will be necessary to assume that there is smooth passage from one set of control parameters to another. That is, different control sections of C differ only by a smooth feedback.

This geometric description of the optimal-control problem allows for natural applications of geometric control theory to variational problems with constraints and to the material in the first part of this book.

We shall begin the subject with the topics that are inseparably connected with optimal control, rather than with the calculus of variations. Apart from its own merits, this choice of material offers clear results that illustrate the innovative aspects of optimal control theory and that also justify our earlier claims about the significance of Lie-determined systems for problems of the calculus of variations.

7

Linear systems with quadratic costs

Minimizing the integral of a quadratic form over the trajectories of a linear control problem, known as the linear quadratic problem, was one of the earliest optimal-control problems (Kalman, 1960). Rather than limit our attention to the positive-definite case, as is usually done in the control-theory literature, we shall consider the most general situation for which the question is well posed. The minimal assumptions under which this problem is treated reveal a rich theory that derives from the classic heritage of the calculus of variations and yet is sufficiently distinctive to describe new phenomena outside the scope of the classic theory. As such, this class of problems is a natural starting point for optimal control theory.

This chapter contains a derivation of the "maximum principle" for this class of problems. The curves that satisfy the maximum principle are called extremal curves. The class of problems for which the Legendre condition holds is called "regular." In the regular case, the maximum principle determines a single Hamiltonian, and the optimal solutions are the projections of the integral curves of the corresponding Hamiltonian vector field. The projections of these extremal curves remain optimal up to the first conjugate point.

Problems in the subclass for which the Legendre condition is not satisfied are called "singular." For singular problems, the maximum principle determines an affine space of quadratic Hamiltonians and a space of linear constraints. The resolution of the corresponding constrained Hamiltonian system reveals a generalized optimal synthesis consisting of turnpike-type solutions. The complete description of these solutions makes use of higher-order Poisson brackets and is sufficiently complex to merit a separate chapter. Their analysis is presented in Chapter 9, following the chapter on the Riccati equation.

The assumptions under which the basic theory is obtained include problems with indefinite cost functions, in which case there may be conjugate points. This chapter contains a complete characterization of conjugate points, along with

precise connections between conjugate points and the optimality properties of
the extremal curves.

It will also be shown that several classic inequalities can be obtained through
the theory of linear quadratic optimal problems. In particular, we shall define
the class of problems that lead to inequalities of the Hardy-Littlewood type. The
corresponding differential systems are called Hardy-Littlewood systems, and
in this chapter we shall establish their basic properties. We shall also consider
a new proof of the famous inequality of Wirtinger by defining an appropriate
linear quadratic optimal problem for which the preceding theory applies.

1 Assumptions and their consequences

As usual, the state space M is an n-dimensional real vector space. Following
our earlier descriptions of linear systems, it is most convenient to think of a
linear system as an affine family of vector fields $A + \mathcal{B}$, with A a linear vector
field, and $\mathcal{B}$ a linear m-dimensional vector space of constant vector fields. As
explained earlier, a basis $b_1, \ldots, b_m$ for $\mathcal{B}$ determines the controls $u_1, \ldots, u_m$,
and trajectories of $A + \mathcal{B}$ are solutions of

$$\frac{dx}{dt} = Ax + \sum_{i=1}^{m} u_i b_i \quad \text{or} \quad \frac{dx}{dt} = Ax + Bu, \tag{1}$$

with

$$u = \begin{pmatrix} u_1 \\ \vdots \\ u_m \end{pmatrix} \quad \text{and} \quad Bu = \sum_{i=1}^{m} u_i b_i.$$

By "a trajectory $x(t)$" we shall mean any absolutely continuous curve $x(t)$
that satisfies (1) almost everywhere for some controls $u_1, \ldots, u_m$ in $L^\infty[0, T]$.
The pair (x, u) is called a trajectory pair. For any pair of states a and b in M, and
any $T > 0$, $\mathrm{Tr}(a, b, T)$ will denote the set of all trajectory pairs defined on the interval $[0, T]$ that satisfy the given boundary conditions $x(0) = a$ and $x(T) = b$.

In addition to these data, it will be assumed that the cost is defined by a real
quadratic form $c(x, u)$ defined on $M \times U$, with $U = \mathbb{R}^m$. As stated in the Introduction, for each trajectory pair (x, u), $\int_0^T c(x(t), u(t)) \, dt$ is called the *total
cost of* (x, u) in the interval $[0, T]$. The problem is to find a trajectory pair $(\bar{x}, \bar{u})$
in $\mathrm{Tr}(a, b, T)$ such that the total cost $\int_0^T c(\bar{x}(t), \bar{u}(t)) \, dt$ is minimal among all
trajectory pairs in $\mathrm{Tr}(a, b, T)$. On occasion we shall refer to this problem as the
basic linear quadratic problem, to differentiate it from the variable-terminal-
point problems and the infinite-horizon problems, both of which also appear
in the literature. Each solution $(\bar{x}, \bar{u})$ will be called optimal, always meaning

optimal relative to the interval $[0, T]$ and its boundary conditions $\bar{x}(0) = a$ and $\bar{x}(T) = b$.

In order to get to the basic assumptions, we shall need additional notation. For any vector space E, E^* denotes its dual. E can be regarded as a subspace of $(E^*)^*$ through the correspondence $e \to g(e)$ for any e in E and g in E^*. When E is finite-dimensional, $E = (E^*)^*$. Recall that a linear mapping $L : E \to E^*$ is said to be symmetric if L is equal to its dual mapping L^*. Using this notation, any quadratic form $c(x, u)$ can be written as $c(x, u) = \frac{1}{2}(Pu, u) + (Qx, u) + \frac{1}{2}(Rx, x)$ for linear mappings $P : U \to U^*$, $Q : M \to U^*$, and $R : M \to M^*$, with P and R symmetric. In this notation, (f, x) means $f(x)$. The required existence assumptions will be extracted from the following theorem.

Theorem 1 *There is an optimal solution $(\bar{x}, \bar{u})$ for some boundary conditions a, b, and T if and only if $\int_0^T c(x(t), u(t))\, dt \geq 0$ for all trajectory pairs (x, u) that satisfy $x(0) = 0 = x(T)$.*

Proof Assume that $(\bar{x}, \bar{u})$ minimizes the total cost among all trajectories in $\mathrm{Tr}(a, b, T)$ for some a, b, and T. Denote by (x_0, u_0) any trajectory pair in $\mathrm{Tr}(0, 0, T)$, and consider $x_\rho = \bar{x} + \rho x_0$ and $u_\rho = \bar{u} + \rho u_0$ for $\rho > 0$. For each ρ, $(x_\rho, u_\rho) \in \mathrm{Tr}(a, b, T)$, and

$$\int_0^T c(x_\rho, u_\rho)\, dt = \rho^2 \int_0^T c(x_0, u_0)\, dt + \rho \int_0^T ((P\bar{u}, u_0)$$

$$+ (Q\bar{x}, u_0) + (Qx_0, \bar{u}) + (R\bar{x}, \bar{x}_0))\, dt + \int_0^T c(\bar{x}, \bar{u})\, dt.$$

Because $\int_0^T c(x_\rho, u_\rho)\, dt \geq \int_0^T c(\bar{x}, \bar{u})\, dt$, it follows that

$$\rho^2 \int_0^T c(x_0, u_0)\, dt + \rho \int_0^T ((P\bar{u}, u_0) + (Q\bar{x}, u_0) + (Qx_0, \bar{u}) + (R\bar{x}, x_0))\, dt \geq 0$$

for all ρ. This can happen only if $\int_0^T c(x_0, u_0)\, dt \geq 0$.

Conversely, if $\int_0^T c(x, u) \geq 0$ for any (x, u) in $\mathrm{Tr}(0, 0, T)$, the zero trajectory $u \equiv 0$, $x \equiv 0$ is a solution, and the proof is finished. ∎

The first assumption we shall make guarantees the existence of solutions. It consists of assuming the existence of an interval $[0, T_{\max}]$, $T_{\max} > 0$, on which there exists at least one trajectory (x, u) that is optimal. According to Theorem 1, this assumption requires that $\int_0^{T_{\max}} c(x, u)\, dt \geq 0$ for all trajectory pairs (x, u) in

Tr$(0, 0, T_{\max})$. Evidently a trajectory that is optimal on $[0, T_{\max}]$ is also optimal on every subinterval $[0, T]$ with $T < T_{\max}$, and therefore $\int_0^T c(x, u)\, dt \geq 0$ for any $(x, u) \in$ Tr$(0, 0, T)$ with $T \leq T_{\max}$.

It will be further assumed that the system is dissipative, in the sense that for every T, $0 < T \leq T_{\max}$, there exists a trajectory pair (x, u) on the interval $[0, T]$ such that $(x, u) \in$ Tr$(0, 0, T)$ and $\int_0^T c(x, u)\, dt > 0$. This assumption implies the following:

Proposition 1 *Suppose that (x, u) is a trajectory in* Tr(a, b, T) *whose total cost $\int_0^T c(x, u)\, dt$ is equal to α. Then for any $\varepsilon > 0$, there exists a trajectory pair $(x_\varepsilon, u_\varepsilon)$ in* Tr(a, b, T) *such that $\int_0^T c(x_\varepsilon, u_\varepsilon)\, dt = \alpha + \varepsilon$.*

Proof Let (x_0, u_0) be a trajectory pair in Tr$(0, 0, T)$ such that $\int_0^T c(x_0, u_0)\, dt > 0$. Consider $x_\rho = x + \rho x_0$, $u_\rho = u + \rho u_0$. Then $\sigma(\rho) = \int_0^T c(x_\rho, u_\rho)\, dt$ is a continuous curve that satisfies $\sigma(0) = \alpha$ and $\lim_{\rho \to \infty} \sigma(\rho) = \infty$. Hence there exists a value ρ such that $\sigma(\rho) = \alpha + \varepsilon$, and the proof is finished. ∎

In addition to the preceding two assumptions, we shall assume that system (1) is controllable. This assumption ensures that the basic problem is well posed, in the sense that Tr(a, b, T) is not empty for any a, b in M and any $T > 0$. We shall presently show that the existence assumption implies that $P \geq 0$ and that $\inf\{\int_0^T c(x, u)\, dt : (x, u) \in$ Tr$(a, b, T)\} > -\infty$ for all a, b in M and for T with $0 < T \leq T_{\max}$.

The second assumption concerning the dissipation property is more convenient than essential; it rules out the possibility that c could be an exact differential over the trajectories of (1), in which case every trajectory would be optimal.

1.1 Optimality and the boundaries of the reachable sets

In the following, $A_{\text{ext}}((0, a), T)$ denotes the reachable set, at time T, of the extended control system in $\mathbb{R}^1 \times M$ defined by

$$\frac{dx_0}{dt} = c(x(t), u(t)) \quad \text{and} \quad \frac{dx}{dt} = Ax + Bu, \quad \text{with } x_0(0) = 0,\ x(0) = a.$$
(2)

The next theorem is fundamental for the following theory.

Theorem 2

(a) *Corresponding to each optimal solution $(\bar{x}, \bar{u})$, the terminal point $(\int_0^T c(\bar{x}, \bar{u})\, dt, \bar{x}(T))$ belongs to the boundary of $A_{\text{ext}}((0, a), T)$.*

(b) *Each point (α, b) of the boundary of $A_{\text{ext}}((0, a), T)$ can be reached only by an optimal trajectory.*

Proof If $(\bar{x}, \bar{u})$ is an optimal-trajectory pair, then its extended terminal point $(\int_0^T c(\bar{x}, \bar{u})) \, dt, \bar{x}(T))$ is the lowest point on the line $\{(\alpha, \bar{x}(T)) \; \alpha \in \mathbb{R}^1\}$ contained in $A_{\text{ext}}((0, a), T)$. Evidently, this point must be on the boundary of $A_{\text{ext}}((0, a), T)$.

To prove part (b), assume that (α, b) is a point on the boundary of $A_{\text{ext}}((0, a), T)$ that can be reached by a trajectory pair $(\bar{x}, \bar{u})$. That is, $\int_0^T c(\bar{x}, \bar{u}) \, dt = \alpha$, $x(0) = a$, and $x(T) = b$. Suppose that $(\bar{x}, \bar{u})$ is not optimal, and let (y, v) be another trajectory such that $y(0) = a$, $y(T) = b$, and $\int_0^T c(y, v) \, dt = \alpha - \varepsilon$ for $\varepsilon > 0$. There is no loss in generality in assuming that v is a smooth control function. Let V_δ denote a neighborhood of all bounded and measurable functions $u : [0, T] \to U$ such that $\|u(t) - v(t)\| < \delta$ for all t in $[0, T]$. For each $u \in V_\delta$, let x be the corresponding trajectory that originates at a (i.e., $x(0) = a$). The mapping $u \to (\int_0^T c(x, u) \, dt, x(T))$ is a continuous mapping on V_δ whose projection on M covers a neighborhood of b. Hence there exists $\delta > 0$ such that $\int_0^T c(x, u) \, dt \leq \alpha - (\varepsilon/2)$ for all u in V_δ. But then it follows from Proposition 1 that each point of Ω can be reached along a trajectory whose total cost is any number β, with $\beta \geq \alpha - (\varepsilon/2)$.

But then $(\alpha - (\varepsilon/2), \alpha + (\varepsilon/2)) \times \Omega$ is an open neighborhood of (α, b) contained in $A_{\text{ext}}((0, a), T)$ that contradicts the initial assumption that (α, b) belongs to the boundary of $A_{\text{ext}}((0, a), T)$. The proof is finished. ∎

Proposition 2 $P \geq 0$.

Proof Assume P indefinite. Because P is symmetric, there exists an invariant splitting $U_1 \oplus U_2$ of U such that the restriction of P to U_1 is negative-definite, and the restriction of P to U_2 is semipositive-definite. Let $K > 0$ be a number such that $(u, Pu) \leq -K \|u\|^2$ for any u in U_1 (Hirsch and Smale, 1974).

It then follows, by elementary estimates, that there exist positive numbers ϵ and C such that for any $\bar{t}$ less than ϵ, $\int_0^{\bar{t}} c(x, u) dt \leq -C(\int_0^t \|u(t)\|^2 dt)^{1/2}$ for any trajectory (x, u) with $x(0) = 0$ and any u with values in U_1. But then there exists a trajectory (x, u) with $x(0) = x(T) = 0$ with $\int_0^T c(x, u) dt < 0$. This contradicts the result of Proposition 1, and our proof is finished. ∎

Remark The preceding result can be regarded as a control-theory version of the classic result of Tonelli concerning the necessity of the Legendre condition. ∎

In general, the remaining terms in the cost function need not be positive, nor is it even necessary for $c(x, u)$ to be non-negative over the trajectories of the system for the preceding assumptions to hold.

Problem 1 Show that $\int_0^\pi (u^2 - x^2)\, dt \geq 0$ over any closed trajectory of $dx/dt = u$ in $M = \mathbb{R}^1$ for which $x(0) = x(\pi) = 0$. Show that there exists a closed trajectory (x, u) on the interval $[0, \pi]$ for which $\int_0^\pi (u^2 - x^2)\, dt < 0$.

Problem 2 A linear quadratic system is said to be convex if $c(\lambda x + (1 - \lambda)y, \lambda u + (1 - \lambda)v) \leq \lambda c(x, u) + (1 - \lambda)c(y, v)$ for any real number λ such that $0 \leq \lambda \leq 1$. Show that convexity implies that $c(x, u) \geq 0$ for all (x, u) in $M \times U$. Show further that $A_{\text{ext}}((0, a), T)$ is a convex subset of $\mathbb{R}^1 \times M$ for each a in M and each $T > 0$.

2 The maximum principle

The maximum principle, a fundamental result of optimal control theory, is a first-order necessary condition for the terminal point $(x_0(T), x(T))$ to be on the boundary of the extended reachable set $A_{\text{ext}}((0, a), T)$. Its formulation takes place on the cotangent bundle of M, and it is expressed in terms of a function on T^*M called the Hamiltonian of the system. In order to arrive at the proper geometric formulation, it will first be necessary to review the basic facts from symplectic geometry.

2.1 Canonical coordinates and Hamiltonian vector fields

Because M is a vector space, the cotangent bundle T^*M can be realized as $M \times M^*$. The standard symplectic form ω on T^*M is given by

$$\omega_{(x,p)}((X, P), (Y, Q)) = Q(X) - P(Y)$$

for all tangent vectors (X, P) and (Y, Q) in $M \times M^*$ at each point (x, p) of T^*M. The reader can easily verify that ω is a nondegenerate, antisymmetric bilinear form at each tangent space of T^*M.

Any function H on T^*M corresponds to a vector field on T^*M called the Hamiltonian vector field of H. We shall use $\vec{H}$ to denote this vector field. H is also called a Hamiltonian of $\vec{H}$. The correspondence between H and $\vec{H}$ is given by the following relation. Denote by $(X(x, p), P(x, p))$ the vector $\vec{H}(x, p)$. Then

$$dH_{(x,p)}(Y, Q) = \omega_{(x,p)}((X(x, p), P(x, p)), (Y, Q))$$

for all tangent vectors (Y, Q) at (x, p).

Corresponding to each basis $a_1, \ldots, a_n$ in M, $a_1^*, \ldots, a_n^*$ denote its dual basis in M^*. Recall that each element a_j^* of this dual basis is defined through $a_j^*(a_i) = \delta_{ij}$ for all $i = 1, 2, \ldots, n$. Each pair a_i and a_i^* is identified with $(a_i, 0)$ and $(0, a_i^*)$. The set of vectors $a_1, \ldots, a_n, a_1^*, \ldots, a_n^*$ is called a *canonical basis* for $M \times M^*$. We shall use $x_1, \ldots, x_n, p_1, \ldots, p_n$ to denote the coordinates relative to a canonical base. Such coordinates are also called canonical. Any vector field (X, P) can be expressed by the canonical coordinates $(X_1, \ldots, X_n, P_1, \ldots, P_n)$. If $(Y_1, \ldots, Y_n, Q_1, \ldots, Q_n)$ are the canonical coordinates of another vector field (Y, Q), then

$$\omega((X, P), (Y, Q)) = \sum_{i=1}^{n} Q_i X_i - P_i Y_i.$$

In particular, if $(X_1, \ldots, X_n, P_1, \ldots, P_n)$ are canonical coordinates of $\vec{H}$, and if $(\partial H/\partial x_1, \ldots, \partial H/\partial x_n, \partial H/\partial p_1, \ldots, \partial H/\partial p_n)$ denote the canonical coordinates of dH, then

$$\sum_{i=1}^{n} \frac{\partial H}{\partial x_i} Y_i + \frac{\partial H}{\partial p_i} Q_i = \sum_{i=1}^{n} Q_i X_i - P_i Y_i$$

for all $(Y_1, \ldots, Y_n, Q_1, \ldots, Q_n)$ in $\mathbb{R}^{2n}$. Therefore,

$$P_i(x_1, \ldots, x_n, p_1, \ldots, p_n) = -\frac{\partial H}{\partial x_i}(x_1, \ldots, x_n, p_1, \ldots, p_n)$$

and

$$X_i(x_1, \ldots, x_n, p_1, \ldots, p_n) = \frac{\partial H}{\partial p_i}(x_1, \ldots, x_n, p_1, \ldots, p_n)$$

$$\text{for all } i = 1, 2, \ldots, n.$$

Each integral curve $(x(t), p(t))$ of $\vec{H}$, when expressed in terms of its symplectic coordinates, satisfies the familiar system of differential equations

$$\frac{dx_i}{dt} = \frac{\partial H}{\partial p_i}(x_1, \ldots, x_n, p_1, \ldots, p_n),$$

$$\frac{dp_i}{dt} = -\frac{\partial H}{\partial x_i}(x_1, \ldots, x_n, p_1, \ldots, p_n).$$

$$(3)$$

Any vector field V on M can be canonically lifted to a Hamiltonian vector field on T^*M via the function $H_V(x, p) = p(V(x))$. In terms of the symplectic coordinates,

$$H_V(x_1, \ldots, x_n, \ p_1, \ldots, p_n) = \sum_{i=1}^{n} p_i a_i^* \left(\sum_{j=1}^{n} V_j(x_1, \ldots, x_n) a_j \right)$$

$$= \sum_{i=1}^{n} p_i V_i(x_1, \ldots, x_n).$$

Therefore each integral curve of the Hamiltonian lift $\vec{H}_V$ of V satisfies

$$\frac{dx_i}{dt} = V_i(x_1, \ldots, x_n) \quad \text{and} \quad \frac{dp_i}{dt} = \sum_{j=1}^{n} p_j \frac{\partial V_j}{\partial x_i}(x_1, \ldots, x_n). \qquad (4)$$

In particular, if V is a linear vector field, then $V_i(x_1, \ldots, x_n) = \sum_{j=1}^{n} V_{ij} x_j$, with (V_{ij}) equal to the matrix of V relative to the basis $a_1, \ldots, a_n$. Hence, (4) becomes

$$\frac{dx_i}{dt} = \sum_{j=1}^{n} V_{ij} x_j \quad \text{and} \quad \frac{dp_i}{dt} = \sum_{j=1}^{n} p_j V_{ji}. \qquad (5)$$

Therefore the Hamiltonian lift of a linear field is linear, and its matrix relative to a canonical base is given by

$$\vec{H} = \begin{pmatrix} V & 0 \\ 0 & -V^* \end{pmatrix}.$$

This formalism extends to control systems: Each control system in M can be naturally lifted to a Hamiltonian control system in T^*M via its Hamiltonian function. The Hamiltonian of linear control system (1) is given by $H(x, p = p(Ax + Bu)$, and the Hamiltonian of the cost extended system is given by $H((p_0, p), (x_0, x)) = p_0 c(x, u) + p(Ax + Bu)$. In each of these cases, the Hamiltonian is to be regarded as a function on the cotangent bundle of the underlying space parametrized by the controls; in terms of the canonical coordinates, the Hamiltonian of the extended system is given by

$$H = p_0 c(x_1, \ldots, x_n, u_1, \ldots, u_m) + \sum_{i=1}^{n} p_i \left(\sum_{j=1}^{n} A_{ij} x_j + \sum_{j=1}^{m} B_{ij} u_j \right).$$

The corresponding differential system is given by

$$
\frac{dx_i}{dt} = \frac{\partial H}{\partial p_i} = \sum_{j=1}^{n} A_{ij} x_j + \sum_{j=1}^{m} B_{ij} u_j, \qquad i = 1, 2, \ldots, n,
$$

$$
\frac{dp_0}{dt} = \frac{\partial H}{\partial x_0} = 0, \tag{6}
$$

$$
\frac{dp_i}{dt} = -\frac{\partial H}{\partial x_i} = -p_0 \frac{\partial c}{\partial x_i} - \sum_{j=1}^{n} A_{ji} p_j, \qquad i = 1, \ldots, n.
$$

Then x_0 is called a cyclic coordinate, because the Hamiltonian H does not explicitly depend on it. For each cyclic coordinate, the corresponding dual variable p_0 is constant along the flow of $\vec{H}$; p_0 can always be normalized to ± 1 or 0.

For linear quadratic problems, under controllability assumptions, p_0 can always be taken as -1, and therefore the Hamiltonian of the cost extended system can always be reduced to

$$
H_u(x, p) = -c(x, u) + pAx + pBu \tag{7}
$$

for all (x, p) in $M \times M^*$.

2.2 Necessary and sufficient conditions of optimality

With the preceding formalism behind us, we come to the main theorem.

Theorem 3 (The maximum principle) *Suppose that $(\bar{x}, \bar{u})$ is an optimal trajectory on an interval $[0, T]$. Then $\bar{x}$ is the projection of an integral curve $(\bar{x}(t), \bar{p}(t))$ of the Hamiltonian vector field $\vec{H}_{\bar{u}}$ defined on $[0, T]$, which further satisfies*

$$
H_{\bar{u}(t)}(\bar{x}(t), \bar{p}(t)) = \sup_{u \in U} H_u(\bar{x}(t), \bar{p}(t))
$$

for almost all t in the interval $[0, T]$.

Proof Assume that $(\bar{x}, \bar{u})$ is an optimal trajectory on $[0, T]$, and denote by (α, b) the terminal point $(\int_0^T c(\bar{x}, \bar{u})\, dt, \bar{x}(T))$. For any bounded and measurable control function u, define a variation $u_\lambda(t) = \bar{u}(t) + \lambda u(t)$ for real λ. Denote by x_λ the trajectory of (1) such that $x_\lambda(0) = a$. Because (1) is a linear system, $x_\lambda = \bar{x} + \lambda x$, with x the solution of (1) corresponding to u, for which $x(0) = 0$.

Let $L^\infty([0, T], U)$ denote the space of essentially bounded and measurable functions on $[0, T]$ with values in U, and consider the mapping F from $L^\infty([0, T], U)$ into $A_{\text{ext}}((0, a), T)$ given by $F(u_\lambda) = (\int_0^T c(x_\lambda(t), u_\lambda(t)) \, dt, x_\lambda(T))$. Let L denote its directional derivative $\lim_{\lambda \to 0}((F(u_\lambda) - F(\bar{u}))/\lambda)$. Because

$$\frac{1}{\lambda} \int_0^T (c(x_\lambda(t), u_\lambda(t)) - c(\bar{x}(t), \bar{u}(t))) \, dt$$

$$= \int_0^T \left(\frac{\partial c}{\partial x}(\bar{x}(t), \bar{u}(t)) x(t) + \frac{\partial c}{\partial u}(\bar{x}(t), \bar{u}(t)) u(t) \right) dt$$

$$+ \lambda \int_0^T c(x(t), u(t)) \, dt,$$

it follows that

$$L(u) = \left(\int_0^T \left(\frac{\partial c}{\partial x}(\bar{x}(t), \bar{u}(t)) x(t) + \frac{\partial c}{\partial u}(\bar{x}(t), \bar{u}(t)) u(t) \right) dt, x(T) \right).$$

The first task in our proof is to show that for each u, $L(u) = (p(x(T)), x(T))$ for some linear function p. To begin with, the range of L cannot be $\mathbb{R}^1 \times M$. The argument is as follows: If the range of L were $\mathbb{R}^1 \times M$, then there would exist u_0 such that $L(u_0) = (-\varepsilon, 0)$ for some $\varepsilon > 0$. Denote by x_0 the trajectory of (1) that corresponds to u_0 and that satisfies $x_0(0) = 0$. Because $L(u_0) = (-\varepsilon, 0)$, it follows that $x_0(T) = 0$, and therefore $x_\lambda(T) = b$ for all λ. Then $F(u_\lambda) = (-\lambda\varepsilon + \alpha + \lambda^2 \int_0^T c(x_0(t), u_0(t)) \, dt, b)$, which would violate the optimality of $(\bar{x}, \bar{u})$, since $\int_0^T c(x_\lambda, u_\lambda) \, dt = -\lambda\varepsilon + \alpha + \lambda^2 \int_0^T c(x_0, u_0) \, dt < \alpha$ for small λ.

As a consequence of the controllability assumption, the projection of the range of L on M is equal to M. Therefore, the range of L is of the same dimension as M, which is equivalent to the existence of a linear form p such that $L(u) = (p(x(T)), x(T))$.

The remaining part of our proof consists in showing that the relation

$$\int_0^T \left(\frac{\partial c}{\partial x}(\bar{x}(t), \bar{u}(t)) x(t) + \frac{\partial c}{\partial u}(\bar{x}(t), \bar{u}(t)) u(t) \right) dt = p(x(T)),$$

for all trajectories (x, u), with $x(0) = 0$, can hold only along the trajectories $(\bar{x}, \bar{u})$ that satisfy the maximum principle.

Let $\bar{p}(t)$ be the curve that satisfies

$$\frac{d\bar{p}}{dt} = -A^* \bar{p}(t) + \frac{\partial c}{\partial x}(\bar{x}(t), \bar{u}(t)) \quad \text{a.e. on } [0, T],$$

and the terminal boundary condition $\bar{p}(T) = p$. Expressing $p(x(T))$ by $\int_0^T ((d\bar{p}/dt)x(t) + \bar{p}(t)(dx/dt))\,dt$ in the preceding equality, we get

$$\int_0^T \left(\frac{\partial c}{\partial x}(\bar{x}(t), \bar{u}(t)) - A^*\bar{p}(t) - \frac{d\bar{p}}{dt} \right) x(t)$$
$$+ \left(\frac{\partial c}{\partial u}(\bar{x}(t), \bar{u}(t)) - B^*\bar{p}(t) \right) u(t)\,dt = 0.$$

Thus, $\int_0^T (\partial c/\partial u)(\bar{x}(t), \bar{u}(t)) - (B^*\bar{p}(t))u(t)\,dt = 0$. Because $u(t)$ is an arbitrary integrable function, $(\partial c/\partial u)(\bar{x}(t), \bar{u}(t)) - B^*\bar{p}(t) = 0$ a.e. on $[0, T]$. This condition can also be expressed as $\partial H/\partial u = 0$ a.e. on$[0, T]$, with $u = \bar{u}$ and $H(\bar{x}(t), \bar{p}(t), u) = -c(\bar{x}(t), u) + \bar{p}(t)(A\bar{x}(t) + Bu)$.

Because $P \geq 0$, H is a concave function of u, and therefore $\partial H/\partial u = 0$, corresponding to its maximum, and our proof is finished. ∎

The maximum principle leads to a special class of curves in T^*M called the extremal curves.

Definition 1 An absolutely continuous curve $\sigma(t) = (x(t), p(t))$ in $M \times M^*$ defined on an interval $[0, T]$ is called an extremal curve if there exists a bounded and measurable control function u on the interval $[0, T]$ such that

$$\frac{d\sigma}{dt} = \vec{H}(\sigma(t), u(t))$$

and

$$H(x(t), p(t), u(t)) = \sup_{v \in U} H(x(t), p(t), v)$$

for almost all t in $[0, T]$. H denotes the Hamiltonian $-c(x, u(t)) + p(Ax + Bu(t))$. The maximality condition can be replaced by an equivalent condition $Pu(t) - B^*p(t) + Qx(t) = 0$ for almost all t in $[0, T]$, corresponding to $\partial H/\partial u = 0$.

According to the maximum principle, each optimal trajectory is the projection of an extremal curve. For the class of linear quadratic problems, the converse is also true.

Theorem 4 *Suppose that* $\sigma(t) = (x(t), p(t))$ *is an extremal curve on an interval* $[0, T]$ *generated by the control* $u(t)$. *Then* (x, u) *is an optimal-trajectory on* $[0, T]$.

Proof The proof consists in reversing the steps in the proof of the maximum principle. These details will be left to the reader.   ∎

Summarizing our findings, we have shown that a trajectory is optimal if and only if it is the projection of an extremal curve and that an absolutely continuous curve $\sigma(t) = (x(t), p(t))$ is an extremal curve on an interval $[0, T]$ if and only if there exists a bounded and measurable control function $u(t)$ on $[0, T]$ such that

$$\frac{d\sigma}{dt} = \vec{H}(\sigma(t), u(t)) \tag{8}$$

and $Pu(t) - B^*p(t) + Qx(t) = 0$ for almost all t in $[0, T]$.

Definition 2 A linear quadratic problem is called *regular* if $P > 0$. Otherwise it is called *singular*.

In the regular case, the equation $Pu - B^*p + Qx = 0$ is uniquely solvable for u in terms of x and p to yield $u(x, p) = P^{-1}(B^*p - Qx)$. This closed-loop solution defines a single Hamiltonian $H_0(x, p)$ equal to $-c(x, u(x, p)) + p(Ax + Bu(x, p))$. It then follows that H_0 is a quadratic function on $M \times M^*$ given by

$$H_0(x, p) = \frac{1}{2}(BP^{-1}B^*p, p) + p(A - BP^{-1}Q)x$$

$$+ \frac{1}{2}(P^{-1}(Qx), Qx) - \frac{1}{2}(Rx, x).$$

In order to avoid cumbersome notation, it will be convenient to use the same symbols for linear objects and their matrices relative to the canonical bases. With this convention, the integral curves $(x(t), p(t))$ are given by the following system of linear differential equations:

$$\frac{dx}{dt} = \frac{\partial H_0}{\partial p} = Ax + Bu(x, p) = P^{-1}B^*p + (A - P^{-1}Q)x,$$

$$\frac{dp}{dt} = -\frac{\partial H_0}{\partial x} = Q^*u(x, p) - A^*p + Rx \tag{9}$$

$$= -(A^* - Q^*P^{-1}B^*)p + Rx - Q^*P^{-1}Qx.$$

The essential properties of H_0 are described by our next theorem.

Theorem 5 *Each extremal curve $(x(t), p(t))$ is an integral curve of $\vec{H}_0$. Conversely, each integral curve $(x(t), p(t))$ of $\vec{H}_0$ corresponds to an extremal curve via the control function $u(t) = P^{-1}(B^*p(t) - Qx(t))$.*

The proof is left to the reader.

In the singular case, the maximum principle does not determine optimal controls, because the equation $Pu - B^*p + Qx = 0$ is not uniquely solvable for u in terms of x and p. In fact, for each v in ker P, the preceding equation reduces to $(-Bv, p) + (Qx, v) = 0$. Thus each extremal curve $(x(t), p(t))$ of the singular problem annihilates the linear function $-(p, Bv) + (Qx, v)$ with v in ker P.

In order to extract the geometric content of the constraint form for the singular case, it is natural to consider the weaker relation

$$Pu - B^*p + Qx = 0 \qquad (\text{mod}(\mathcal{L})), \tag{10}$$

where $\mathcal{L}$ is the vector space of all linear functions $h_v(x, p) = (p, Bv) - (Qx, v)$ with v in ker P.

Definition 3 f is said to be a solution of (10) if f is an affine mapping from $M \times M^*$ into U such that $Pf(x, p) - B^*p + Qx$, as a linear function on $M \times M^*$, belongs to $\mathcal{L}$.

Proposition 3 *Equation (10) has solutions.*

Proof Let E be any subspace of U complementary to ker P. Then $U = (\text{ker } P) \oplus E$. Denoting by A^0 the annihilator of any set A, it follows that $U^* = (\text{ker } P)^0 \oplus E^0$.

For each u in U, Pu belongs to $(\text{ker } P)^0$, and the restriction of P to E is an invertible linear mapping from E onto $(\text{ker } P)^0$. Let π denote the projection from U^* onto $(\text{ker } P)^0$ parallel to E^0.

For each (x, p) in $M \times M^*$, let $u(x, p)$ be the unique element in E such that $Pu(x, p) = \pi(B^*p - Qx)$; $u(x, p)$ is a linear function with values in E. Then $f(x, u) = u(x, p) + v$ is the desired solution for any choice of v in ker P. End of the proof. ∎

The preceding construction can be described more explicitly as follows:

Let $b_1, \ldots, b_m$ denote a particular basis relative to which P is diagonal. Assume that the elements of this basis are so indexed that $b_1, \ldots, b_r$ span ker P. As usual, $b_1^*, \ldots, b_m^*$ denotes the dual basis. Then

$$Pu = \sum_{i=r+1}^{m} u_i b_i^* \quad \text{and} \quad (Pu, u) = \sum_{i=r+1}^{m} u_i^2,$$

where $u_1, \ldots, u_m$ stand for the coordinates of u relative to the foregoing basis.

It follows that $(\ker P)^0$ is spanned by $b_{r+1}^*, \ldots, b_m^*$ and that E^0 is spanned by $b_1^*, \ldots, b_r^*$. Let $h_1, \ldots, h_m$ denote the unique linear functions such that $B^* p - Qx = \sum_{i=1}^{m} h_i(x, p)b_i^*$. Then $h_1, \ldots, h_r$ is a basis for $\mathcal{L}$, and $Pu(x, p) = \pi(B^* p - Qx)$ holds for $u_i(x, p) = h_i(x, p)$, $i > r$. The solution constructed earlier is given by

$$f_i = \begin{cases} v_i, & i = 1, \ldots, r, \\ h_i, & i = r+1, \ldots, m, \end{cases}$$

and $Pf(x, p) - B^* p + Qx = \sum_{i=1}^{r} h_i(x, p)$.

Definition 4 $\mathcal{H}$ denotes the family of all Hamiltonians

$$H_f(x, p) = -c(x, f(x, p)) + p(Ax + Bf(x, p))$$

as f ranges over all solutions of equation (10).

We leave it to the reader to show that any two elements H_g and H_f in $\mathcal{H}$ differ by an element of $\mathcal{L}$.

Definition 5 An absolutely continuous curve $\sigma(t)$ defined on an interval $[0, T]$ is called an integral curve of the family $\mathcal{H}$ if for any $H_f \in \mathcal{H}$ there exist bounded and measurable functions $u_1, \ldots, u_r$ and elements $h_1, \ldots h_r$ in $\mathcal{L}$ such that

$$\frac{d\sigma}{dt} = \vec{H}_f(\sigma(t)) + \sum_{i=1}^{r} u_i(t)\vec{h}_i \quad \text{for almost all } t \in [0, T].$$

For notational simplicity, we shall denote the foregoing equation by the inclusion $d\sigma/dt \in \vec{\mathcal{H}}(\sigma(t))$.

The preceding formalism leads to a natural characterization of the extremal curves for the singular problem, as described by the following:

Theorem 6 *An absolutely continuous curve σ defined on some interval $[0, T]$ is an extremal curve of the singular problem if and only if $d\sigma/dt \in \vec{\mathcal{H}}(\sigma(t))$ for almost all t, and $\mathcal{L}(\sigma(t)) = 0$ for all t.*

Proof Let $b_1^*, \ldots, b_m^*$ denote the dual basis of a basis $b_1, \ldots, b_m$ in U. Let $u = (u_1, \ldots, u_m)$ denote the coordinates relative to this basis. Assume that $b_1, \ldots, b_m$ is so chosen that $b_1, \ldots, b_r$ is a basis for ker P and that $(Pu, u) = \sum_{i=r+1}^m u_i^2$.

Assume now that $(x(t), p(t), u(t))$ is an extremal triple defined on $[0, T]$. Then, $\sigma(t) = (x(t), p(t))$ satisfies $d\sigma/dt = \vec{H}(\sigma(t), u(t))$, and

$$\frac{\partial H}{\partial u}(\sigma(t), u(t)) = Pu(t) - B^*p(t) + Qx(t) = 0$$

for almost all t, with $H(x, p, u) = -c(x, u) + p(Ax + Bu)$. In terms of the coordinates $u_1, \ldots, u_m$,

$$H(x, p, u_1, \ldots, u_m) = H_0(x, p) - \frac{1}{2}\sum_{i=r+1}^m u_i^2 + \sum_{i=1}^m u_i h_i(x, p),$$

with H_0 denoting the part of H independent of the controls, that is, $H_0(x, p) = pAx - \frac{1}{2}(Rx, x)$. Because $Pu(t) - B^*p(t) + Qx(t) = 0$, it follows that $h_1(\sigma(t)) = \cdots = h_r(\sigma(t)) = 0$ and that $u_i(t) = h_i(\sigma(t))$ for $i = r+1, \ldots, m$.

Let $H_f = H_0 + \frac{1}{2}\sum_{i=r+1}^m h_i^2$. Then $\vec{H}_f = \vec{H}_0 + \sum_{i=r+1}^m h_i \vec{h}_i$. Because $\vec{H}(x, p, u(t)) = \vec{H}_0(x, p) + \sum_{i=1}^m u_i \vec{h}_i$, it follows that $d\sigma/dt = \vec{H}_0(\sigma(t)) + \sum_{i=1}^m u_i(t)\vec{h}_i(\sigma(t))$. But $u_i(t) = h_i(\sigma(t))$ for $i = r+1, \ldots, m$, and therefore

$$\frac{d\sigma}{dt} = \vec{H}_0 + \sum_{i=r+1}^m h_i(\sigma(t))\vec{h}_i(\sigma(t)) + \sum_{i=1}^r u_i(t)\vec{h}_i(\sigma(t))$$

$$= \vec{H}_f(\sigma(t)) + \sum_{i=1}^r u_i(t)\vec{h}_i(\sigma(t)).$$

Because $h_1, \ldots, h_r$ span $\mathcal{L}$, it follows that $\mathcal{L}(\sigma(t)) = 0$.

To prove the converse, let $\sigma(t)$ be any absolutely continuous curve such that $d\sigma/dt \in \vec{\mathcal{H}}(\sigma(t))$ for almost all t, and $\mathcal{L}(\sigma(t)) = 0$ for all t. Let $H_f = H_0 + \frac{1}{2}\sum_{i=r+1}^m h_i^2$, with $h_1, \ldots, h_m$ equal to the coordinates of $B^*p - Qx$ relative to the preceding basis. Then there exist bounded and measurable functions $u_1, \ldots, u_r$ such that $d\sigma/dt = \vec{H}_f(\sigma(t)) + \sum_{i=1}^r u_i(t)\vec{h}_i(\sigma(t))$. Define $u_i(t) = h_i(\sigma(t))$ for $i = r + 1, \ldots, m$. Then $(\sigma(t), u(t))$ is an extremal curve, and our proof is finished. ∎

The resolution of the constrained Hamiltonian system $d\sigma/dt \in \vec{\mathcal{H}}(\sigma(t))$, with $\mathcal{L}(\sigma(t)) = 0$, requires a fair bit of mathematical attention and will be deferred to a separate chapter. The remaining part of this chapter addresses other properties of the solutions to the regular problem.

2.3 The Euler-Lagrange equation

From the classic point of view, a linear quadratic problem is known as a problem of Lagrange (Carathéodory, 1935). A Lagrange problem is any variational problem subject to differential equality constraints. In this context, a control system is a submanifold of the tangent bundle of M given by $\dot{x} - Ax - Bu = 0$. A multiplier λ for this problem is a real function whose domain consists of the points in the tangent bundle; that is, it is a differential form (this geometric fact is often overlooked). The basic problem is to select a multiplier (if one exists) so that the critical point of the functional $\int_0^T (c(x, u) + \lambda(\dot{x} - Ax - Bu))\, dt$ will yield the minimum for the original problem. As is well known, the critical value of this functional is a solution curve of the Euler-Lagrange equation. We shall now show that the optimal solutions for the regular problem can also be obtained through the Euler-Lagrange equation.

It may be worthwhile, first, to recall the basic duality between the Hamiltonian and Lagrangian formalisms, obtained through the Legendre transformation. The unconstrained version of the linear quadratic problem in the calculus of variations consists of minimizing $\int_0^T c(x(t), dx/dt)\, dt$ over all absolutely continuous curves $x(t)$ defined on $[0, T]$ that satisfy the prescribed boundary conditions $x(0) = a$ and $x(T) = b$.

Writing $\dot{x}$ for dx/dt, the Euler-Lagrange system of equations for this problem is

$$\frac{d}{dt} \frac{\partial c}{\partial \dot{x}} (x, \dot{x}) - \frac{\partial c}{\partial x} (x, \dot{x}) = 0. \tag{11}$$

Assuming that

$$c(x, \dot{x}) = \frac{1}{2}(P\dot{x}, \dot{x}) + (Qx, \dot{x}) + \frac{1}{2}(Rx, x),$$

$P > 0$ corresponds to the strong Legendre condition; it is a necessary and sufficient condition for the correspondence between the velocity variable $\dot{x}$ and the momentum variable p to be univalent. This correspondence, obtained implicitly through the relation $p = (\partial c/\partial \dot{x})(x, \dot{x})$, is known as the Legendre transformation. In geometric terms, this correspondence transforms the Euler-Lagrange equation (11), as an equation in the tangent bundle of M, into a first-order Hamiltonian system in the cotangent bundle of M. The explicit relations are as follows: The Euler-Lagrange equation is

$$\frac{d}{dt}(P\dot{x} + Qx(t)) - Q^*\dot{x} - Rx = 0.$$

The corresponding Hamiltonian system is obtained through $p(t) = P(dx/dt) + Qx(t)$, to yield

$$\frac{dx}{dt} = P^{-1}(p - Qx),$$

$$\frac{dp}{dt} = \frac{d}{dt}\left(P\frac{dx}{dt} + Qx(t)\right) = Q^*\dot{x} + R = Q^*P^{-1}(p - Qx).$$

This linear system of equations in $M \times M^*$ is generated by the quadratic Hamiltonian

$$H_0(x, p) = \frac{1}{2}(P^{-1}p, p) - (P^{-1}p, Qx) - \frac{1}{2}(Rx, x) + \frac{1}{2}(P^{-1}(Qx), Qx),$$

which is a particular case of the quadratic Hamiltonian described in (8). The duality between Lagrangian and Hamiltonian formalisms extends to any regular linear quadratic problem, as will now be shown. For simplicity, the arguments will be given only for the scalar case.

Let $x_1, \ldots, x_n$ be the coordinates in M in which (1) is in Brunovsky form:

$$\frac{dx_i}{dt} = x_{i+1} \quad \text{and} \quad \frac{dx_n}{dt} = u, \qquad 1 \le i < n.$$

For the regular problem there is no loss in generality in taking $P = 1$. In this system of coordinates the constraint manifold C is equal to $\{(x, \dot{x}) : \dot{x}_i - x_{i+1} = 0, \; i = 1, 2, \ldots, n - 1\}$. The substitution $u = \dot{x}_n$ turns the cost function into a quadratic form on TM. The Euler-Lagrange equation associated with $\int(c(x_1, \ldots, x_n, \dot{x}_n) + \sum_{i=1}^{n-1} \lambda_i(\dot{x}_i - x_{i+1}))\, dt$ is then given by the following system of equations:

$$\frac{d}{dt}\left(\frac{\partial c}{\partial \dot{x}_1} + \lambda_1\right) - \frac{\partial c}{\partial x_1} = 0, \qquad \frac{d}{dt}\left(\frac{\partial c}{\partial \dot{x}_i} + \lambda_i\right) - \frac{\partial c}{\partial x_i} - \lambda_{i-1} = 0,$$

$$i = 1, 2, \ldots, n - 1,$$

and

$$\frac{d}{dt}\frac{\partial c}{\partial \dot{x}_n} - \frac{\partial c}{\partial x_n} - \lambda_{n-1} = 0.$$

Because c does not depend explicitly on $\dot{x}_i$ for $i < n$, $\partial c / \partial \dot{x}_i = 0$. Therefore,

$$\frac{d}{dt}\left(\frac{\partial c}{\partial \dot{x}_i} + \lambda_i\right) - \frac{\partial c_i}{\partial x_i} - \lambda_{i-1} = \frac{d}{dt}(\lambda_i) - \frac{\partial c}{\partial x_i} - \lambda_{i-1} = 0.$$

For $i = n$,

$$\frac{d}{dt}\left(\frac{\partial c}{\partial \dot{x}_n}\right) - \frac{\partial c}{\partial x_n} - \lambda_{n-1} = \frac{d\dot{x}_n}{dt} - \frac{\partial c}{\partial x_n} - \lambda_{n-1} = 0.$$

The multipliers can be eliminated by the following procedure: Let $f(t) = x_1(t)$. Then $x_k(t) = f^{(k-1)}(t)$ for $k > 1$. In particular, $f^{(n+1)} = (d/dt)f^n = (d/dt)\dot{x}_n = \partial c / \partial x_n + \lambda_{n-1}$.

Further differentiations of the preceding relation ultimately lead to

$$f^{(2n)} = \left(\frac{d}{dt}\right)^{n-1}\left(\frac{\partial c}{\partial x_n}\right) + \cdots + \frac{d}{dt}\left(\frac{\partial c}{\partial x_2}\right) + \frac{\partial c}{\partial x_1}. \tag{12}$$

Each term $(d/dt)^{k-1}(\partial c / \partial x_k)$ on the right-hand side of equation (12) is a linear function of $f, \ldots, f^{(2n-1)}$, and therefore (12) is a $2n$th-order linear differential equation in f. Its solutions determine optimal controls, since $u(t) = \dot{x}_n = f^{(n)}(t)$.

The Legendre transformation is given by $p_n = \partial c / \partial \dot{x}_n = \dot{x}_n$. After renaming the remaining multipliers λ_i by p_i, for $i = 1, \ldots, n - 1$, the preceding equations become $dp_1/dt = \partial c / \partial x_1$ and $dp_i/dt = \partial c / \partial x_i + p_{i-1}$, $i = 2, \ldots, n$. Together with $dx_i/dt = x_{i+1}$ and $dx_n/dt = p_n$, these equations are the canonical equations associated with the Hamiltonian $H = -c(x_1, \ldots, x_n, p_n) + \sum_{i=1}^{n-1} p_i x_{i+1} + p_n^2$. This finding coincides with the equations obtained through the maximum principle.

3 Conjugate points for the regular problem

When $P > 0$, then for each sufficiently small T there exists a constant K such that $\int_0^T c(x(t), u(t))\, dt \geq K(\int_0^T \|u(t)\|^2 dt)^{\frac{1}{2}}$ for any trajectory (x, u) in $[0, T]$, with $x(0) = 0$. For such T, both the existence and the dissipative properties are automatically satisfied, and the preceding theory holds. However, as T grows, there is a possibility that the total cost over $[0, T]$ may become sufficiently negative and ultimately destroy the existence of optimal solutions on such intervals. The classic notions of conjugate points and cut-locus theory are related to the bifurcating time interval beyond which there are no optimal solutions, and the main objective of this section is to describe the relevant theory in detail.

Once more it will be necessary to examine the extended reachable sets and extract the relevant information. This time the required information concerns the existence of a largest interval $[0, T]$ for which the extended reachable set contains boundary points.

Theorem 7 *Suppose that* (α, b) *is an interior point of the extended reachable set* $A_{ext}((0, a), T)$. *Then there exists* $\varepsilon > 0$ *such that* (α, b) *remains in the interior of* $A_{ext}((0, a), S)$ *for any* S *in* $(T - \varepsilon, T + \varepsilon)$.

Proof Let $(\bar{x}, \bar{u})$ be a trajectory on $[0, T]$ that satisfies $\bar{x}(0) = a$, $\bar{x}(T) = b$, and $\int_0^T c(\bar{x}, \bar{u}) \, dt = \alpha - 3\delta$ for some $\delta > 0$. It will be convenient to assume that $\bar{u}$ is a continuous (even smooth) control. The existence of δ and of a smooth $\bar{u}$ follows from the assumption that (α, b) is an interior point of $A_{ext}((0, a), T)$ (see Section 5 in Chapter 4). Extend $\bar{u}(t)$ to $[0, T + \varepsilon]$ by any continuous extension. For each $\eta > 0$ and for each $S \in (T - \varepsilon, T + \varepsilon)$, denote by $\mathcal{U}_\eta[0, S]$ the set of all continuous control functions on $[0, S]$ such that $\|\bar{u}(t) - u(t)\| \le \eta$ for all t in $[0, S]$. The mapping $u \mapsto \int_0^S c(x, u) \, dt$ is a continuous mapping from $\mathcal{U}_\eta[0, S]$ into $\mathbb{R}^1$. Therefore there exist $\varepsilon > 0$ and $\eta > 0$ such that $\int_0^S c(x, u) \, dt \le \alpha - \delta$ for all trajectories (x, u), with $u \in \mathcal{U}_\eta[0, S]$ and $x(0) = a$. It can be further arranged, possibly by refining ε and η, that b is in the interior of the set of terminal points $x(S)$ corresponding to such trajectories. Thus, (α, b) is in the interior of $A_{ext}((0, a), S)$ for any S in $(T - \varepsilon, T + \varepsilon)$, and our proof is now finished. ∎

Corollary *Let* T_{max} *be the supremum of all time intervals* $[0, T]$ *for which the total cost is non-negative over the trajectories* (x, u), *with* $x(0) = x(T) = 0$. *Then* $\int_0^{T_{max}} c(x, u) \, dt \ge 0$ *for any trajectory* (x, u) *with* $x(0) = x(T_{max}) = 0$.

Proof If the property did not extend to $[0, T_{max}]$, then there would exist a trajectory (x, u) on this interval such that $x(0) = x(T_{max}) = 0$ and $\int_0^{T_{max}} c(x, u) \, dt < 0$. Therefore, $(\int_0^{T_{max}} c(x, u) \, dt, 0)$ would be an interior point of $A_{ext}((0, 0), T_{max})$. According to Theorem 7, this point would remain in the interior of $A_{ext}((0, 0), T)$ for some T $(T < T_{max})$, which is not possible. ∎

Definition 6 The interval $[0, T_{max}]$ of the preceding corollary will be called the maximal interval of optimality, or the Riccati interval. (The reason for the alternative name will become clear in the chapter on the Riccati equation.)

The following theorem explains the bifurcating significance of the Riccati interval in relation to the existence of optimal trajectories.

Theorem 8

(a) *If $T < T_{max}$, then for each a and b in M there exists a unique optimal trajectory $(\bar{x}, \bar{u})$ on $[0, T]$ such that $\bar{x}(0) = a$ and $\bar{x}(T) = b$. We have $\bar{x}(t)$ as the projection of the extremal curve $(\bar{x}(t), \bar{p}(t))$, and $\bar{u}(t) = P^{-1}(B^* \bar{p}(t) - Q\bar{x}(t))$ for all t in $[0, T]$.*

(b) *For $T = T_{max}$, each extremal trajectory $(x(t), p(t))$ remains optimal for its boundary conditions $x(0) = a$ and $x(T) = b$, but uniqueness is lost. The set of all points b that can be reached by an optimal trajectory in time T_{max} is an affine subvariety of M whose codimension is equal to the dimension of $M_0^* = \{q : (0, q)$ is the initial point for an extremal pair $(x(t), p(t))$ that satisfies $x(T) = 0\}$.*

(c) *There exist no optimal trajectories on intervals $[0, T]$ with $T > T_{max}$.*

Proof Consider a general extremal curve $x(t) = x(a, q, t)$, $p(t) = p(a, q, t)$, with a and q denoting the initial values of x and p, respectively. Point b can be reached by an optimal trajectory in T units of time if and only if b is in the range of the affine mapping $q \rightarrow x(a, q, T)$. Denote by L the linear mapping $q \rightarrow x(0, q, T)$. Then $x(a, q, T) = x(a, 0, T) + L(q)$. We shall now prove that $\ker L = 0$, which in turn will imply that the mapping $q \rightarrow x(a, q, T)$ is onto M.

Point q is in $\ker L$ if and only if $x(0, q, T) = 0$. Let $p(t)$ denote $p(0, q, t)$, and $u(t) = P^{-1}(B^* p(t) - Qx(t))$. Because the total cost $\int_0^T c(x(t), u(t)) \, dt$ is minimal along extremal curves, it follows that $\int_0^T c(x(t), u(t)) \, dt = 0$, because the total cost of the zero trajectory is zero.

Consider now the trajectory $(\bar{x}, \bar{u})$, defined on $[0, T + \varepsilon]$ by

$$\bar{u}(t) = \begin{cases} u(t), & t \in [0, T], \\ 0, & t \in [T, T + \varepsilon], \end{cases} \quad \text{and} \quad \bar{x}(t) = \begin{cases} x(t), & t \in [0, T], \\ 0, & t \in [T, T + \varepsilon]. \end{cases}$$

It follows that $\int_0^{T+\varepsilon} c(\bar{x}, \bar{u}) \, dt = 0$. Because the system admits solutions on $[0, T + \varepsilon]$, for some $\epsilon > 0$ it further follows that $\int_0^{T+\varepsilon} c(x, u) \geq 0$ for (x, u) in $\mathrm{Tr}(0, 0, T + \varepsilon)$, and therefore $(\bar{x}, \bar{u})$ is optimal.

The maximum principle implies that $(\bar{x}, \bar{u})$ is the projection of an extremal curve $(\bar{x}, \bar{p})$. Because the extremal curves are solutions of a linear system, they are analytic, and therefore $\bar{x}(t) = 0$ for all t. But then it follows from (9) that $P^{-1}B^* \bar{p}(t) = 0$ for all t and that $d\bar{p}/dt = A^* \bar{p}(t)$. Successive differentiations of the preceding relations imply that the initial condition q for $\bar{p}(t)$ must be zero because of the controllability assumption. This argument proves that the kernel of the mapping $q \rightarrow x(0, q, T)$ is zero, and the proof of (a) is now finished.

The only fact in (b) that requires a proof is that $M_0^* \neq 0$. If ker L were equal to zero, then the mapping $q \rightarrow x(0, q, T)$ would be nonsingular and therefore would remain nonsingular on an interval $[0, T + \varepsilon]$, which in turn would imply that the problem would admit solutions beyond $T_{\max}$, but that is not possible.

Because part (c) follows from Theorem 1, our proof is finished. ∎

The results obtained in this section can be seen as natural extensions of the classic theory concerning conjugate points and cut-locus points. According to the classic definitions, points t_0 and t_1 are said to be conjugate to each other if there exists a nonzero solution of the Jacobi vector field that initiates at zero at $t = t_0$ and passes through zero at $t = t_1$. Because the extremal field of a linear quadratic problem is linear, the Jacobi field coincides with the extremal system, and therefore $t = 0$ and $t = T_{\max}$ are points conjugate to each other.

The results of Theorem 8 can also be interpreted in terms of a related cut-locus notion from differential geometry. The cut locus of a given point a in M consists of all points on a geodesic curve through a beyond which the geodesic length stops being minimal. Equating the projections of extremal curves with geodesics, the cut locus of a point a is given by all terminal points $x(T_{\max})$ corresponding to the projections of the extremal curves for which $x(0) = a$ (Figure 7.1).

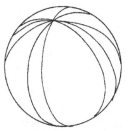

Fig. 7.1.

4 Applications : Wirtinger's inequality and Hardy-Littlewood systems

The first example, well known in the calculus of variations, is introductory.

Example 1 Minimize $\frac{1}{2} \int_0^T ((dx/dt)^2 - x^2)\, dt$ over all absolutely continuous curves x that satisfy $x(0) = a$ and $x(T) = b$. If we let $dx/dt = u$, then the problem, becomes a control problem, with particularly simple dynamics: *Any* *absolutely continuous path from a to b is a trajectory of the system*. The system Hamiltonian $H(x, p)$ is equal to $\frac{1}{2}x^2 + \frac{1}{2}p^2$. The corresponding differential system is

$$\frac{dx}{dt} = \frac{\partial H}{\partial p} = p \quad \text{and} \quad -\frac{dp}{dt} = \frac{\partial H}{\partial x} = x.$$

Hence the extremals are $x(t) = A \sin t + B \cos t$. In particular, the extremals that originate at 0 are given by $x(t) = A \sin t$ for an arbitrary constant A. $T = \pi$ is the first instant such that a nonzero extremal returns to zero. The optimal synthesis is guided by the diagram in Figure 7.2:

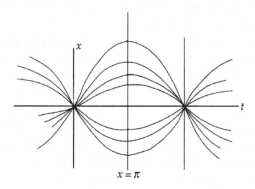

Fig. 7.2.

(a) For $T < \pi$, any point on the vertical line through T is reached by a unique extremal from zero.
(b) At $T = \pi$, only the origin is reached by an extremal from zero. A point b can be reached by an extremal through a if and only if $b = -a$.
(c) At $\pi + \varepsilon$, every point is reached by an extremal from zero, but no extremal is optimal on the interval $[0, \pi + \varepsilon]$.

Example 2 (Wirtinger's problem) Minimize $\frac{1}{2} \int_0^T (u^2 - x_2^2)$ over the trajectories of $dx_1/dt = x_2$ $(dx_2/dt) = u$, relative to the given boundary conditions in $\mathbb{R}^2$. The system Hamiltonian H is given by $H(x_1, x_2, p_1, p_2, u) = -\frac{1}{2}u^2 + \frac{1}{2}x_2^2 + p_1 x_2 + p_2 u$. The maximum of H occurs for $u = p_2$, and therefore the corresponding Hamiltonian H_0 is given by

$$H_0(x_1, x_2, p_1, p_2) = \frac{1}{2}p_2^2 + p_1 x_2 + \frac{1}{2}x_2^2.$$

The associated Hamiltonian system is described by the equations

$$\frac{dx_1}{dt} = \frac{\partial H_0}{\partial p_1} = x_2, \qquad \frac{dx_2}{dt} = \frac{\partial H_0}{\partial p_2} = p_2,$$

$$\frac{dp_1}{dt} = -\frac{\partial H_0}{\partial x_1} = 0, \qquad \frac{dp_2}{dt} = -\frac{\partial H_0}{\partial x_2} = -x_2 - p_1.$$

Any solution curve $p_1(t) = -\alpha$, where α is an arbitrary constant. Then $d^2x_2/dt^2 = dp_2/dt = -x_2 + \alpha$, or $d^2x_2/dt^2 + x_2 = \alpha$. Hence, $x_2(t) = \beta \sin t + \gamma \cos t + \alpha$, where β and γ are arbitrary constants. Because $dx_1/dt = x_2$, it follows that $x_1(t) = -\beta \cos t + \gamma \sin t + \alpha t + \delta$, with δ an arbitrary constant.

The relations $x_1(0) = x_2(0) = 0$ imply that $\gamma = -\alpha$ and $\beta = \delta$. For such extremals, $x_1(t) = 0$ and $x_2(t) = 0$ if and only if

$$\begin{pmatrix} 1 - \cos t & \sin t \\ t - \sin t & 1 - \cos t \end{pmatrix} \begin{pmatrix} \alpha \\ \beta \end{pmatrix} = \begin{pmatrix} 0 \\ 0 \end{pmatrix}$$

for some numbers α and β. This system of equations has nonzero solutions if and only if $(1 - \cos t)^2 - (\sin t)(t - \sin t) = 0$. Because

$$(1 - \cos t)^2 - (\sin t)(t - \sin t) = 2 - 2\cos t - t \sin t$$

$$= 4 \sin^2 \frac{t}{2} - 2t \sin \frac{t}{2} \cos \frac{t}{2}$$

$$= \left(2 \sin \frac{t}{2}\right)\left(2 \sin \frac{t}{2} - t \cos \frac{t}{2}\right),$$

it follows that the preceding determinant is equal to zero at either

$$\sin \frac{t}{2} = 0, \quad \text{or} \quad 2 \sin \frac{t}{2} - t \cos \frac{t}{2} = 0.$$

The second condition means that $\tan(t/2) = t/2$, which has only the zero solution in the interval $[0, 2\pi]$. Hence $T = 2\pi$ is the first positive instant that a nonzero extremal returns to zero.

According to Theorem 8, $\int_0^T (u^2 - x_2^2)\, dt$ admits no optimal trajectories when $T > 2\pi$. For $T < 2\pi$, any two points a and b in $\mathbb{R}^2$ can be connected by a unique extremal.

It remains to examine the case when $T = 2\pi$. Then $x_1(2\pi) = -\beta + 2\pi\alpha + \delta$ and $x_2(2\pi) = \gamma + \alpha$, with $x_1(0) = -\beta + \delta$ and $x_2(0) = \gamma + \alpha$ for any extremal curve $x(t)$. It follows that all extremals satisfy $x_2(0) = x_2(2\pi)$ and

$x_1(2\pi) = x_1(0) + 2\pi\alpha$, with α an arbitrary number. Therefore a can be connected to b by an optimal trajectory in $T = 2\pi$ units of time if and only if $a_2 = b_2$ and $a_1 = b_1 + 2\pi\alpha$. In such a case the optimal cost is given by

$$\frac{1}{2}[x_1(2\pi)p_1(2\pi) + x_2(2\pi)p_2(2\pi) - x_1(0)p_1(0) - x_2(0)p_2(0)] = \pi\alpha^2.$$

In particular, when $\alpha = 0$, both x_1 and x_2 are periodic, and the optimal cost is zero. Therefore, $\int_0^T (u^2 - x_2^2) \, dt \geq 0$ for any periodic trajectory $x(t)$ having a period equal to 2π.

Let $f(t)$ denote any absolutely continuous periodic function with period 2π for which f' belongs to $L^\infty[0, 2\pi]$. Such a function can be traced by a trajectory of the system, with $x_1(t) = \int_0^t f(s) \, ds$, and $x_2(t) = f(t)$ corresponding to $u(t) = f'(t)$. Then $x(t)$ is periodic if and only if $x_1(t) = x_1(t + 2\pi)$ or, equivalently, if $\int_0^{2\pi} f(t) \, dt = 0$. It follows from the preceding inequality that

$$\int_0^T \left(\left(\frac{df}{dt}\right)^2 - f^2(t)\right) dt \geq 0 \quad \text{or that} \quad \int_0^{2\pi} \left(\frac{df}{dt}\right)^2 dt \geq \int_0^{2\pi} f(t)^2 dt$$

for all such functions f, with equality occurring only for $f(t) = r\cos t$, with r an arbitrary number. This inequality, due to Wirtinger (Hardy, Littlewood, and Polya, 1934), was discovered in connection with isoperimetric problems in the calculus of variations.

Hardy-Littlewood systems These systems are motivated by the following problem in analysis: Suppose that f is a function on an interval $[0, T]$ such that its nth derivative $f^{(n)}$ is an element of $L^2[0, T]$. Then f and each intermediate derivative $f^{(k)}$ are also elements of $L^2[0, T]$, and the problem is to determine bounds for $\|f^{(k)}\|$ in terms of $\|f\|$ and $\|f^{(n)}\|$.

A natural way to obtain such bounds is through variational means, which go back to Hardy and Littlewood, and to consider the question of minimizing the functional $\int_0^T f^2(t) - (f^{(k)}(t))^2 + (f^{(n)})^2 dt$ over all functions on the interval $[0, T]$ whose nth derivatives are in $L^2[0, T]$ and that, in addition, may satisfy certain boundary conditions.

This problem has a natural formulation as a linear quadratic problem: Each function f, up to its $n - 1$ derivatives, generates a point x in $\mathbb{R}^n$, with $x_1 = f, x_2 = f^{(1)}, \ldots, x_n = f^{(n-1)}$. Regarding the nth derivative as the control function, f and its intermediate derivatives are trajectories of a linear control system

$$\frac{dx_i}{dt} = x_{i+1}, \quad \text{for } i = 1, 2, \ldots, n-1, \quad \text{and} \quad \frac{dx_n}{dt} = u.$$

The cost functional to be minimized is $\frac{1}{2} \int_0^T (x_1^2 - \alpha x_{k+1}^2 + u^2) \, dt$. (It may be interesting to consider values of α different from 1.)

For each control function u such that $u \in L^2[0, T]$, the corresponding solutions remain in $L^2[0, T]$, and therefore $\int_0^T c(x, u) \, dt$ is a well-defined number. This optimal problem has solutions only if

$$\int_0^T (f^{(2)}(t) - \alpha f^{(k)}(t)^2 + f^{(n)}(t)^2) \, dt \geq 0 \tag{13}$$

for any function f for which $f^{(i)}(0) = f^{(i)}(T) = 0$, $i = 0, 1, \ldots, n-1$.

The largest possible value α for which the inequality (13) holds on an interval $[0, T]$ seems to depend only on the integers n and k, not on T. For instance, when $n = 2$ and $k = 1$, then the maximal value for α is 2, as can be seen through the following argument: $(f')^2 = (d/dt)(ff') - ff''$. For functions f that satisfy the zero boundary conditions, $\int_0^T (f')^2 = -\int_0^T ff''$. Thus,

$$\frac{1}{2} \int_0^T (f^2 - \alpha(f')^2 + (f'')^2) \, dt = \frac{1}{2} \int_0^T \left(\left(f'' + \frac{\alpha}{2} f \right)^2 + f^2 \left(1 - \frac{\alpha^2}{4} \right) \right) dt.$$

Let f be any nonzero solution of $f'' + (\alpha/2)f = 0$ that satisfies the zero boundary conditions. For such an f,

$$\frac{1}{2} \int_0^T (f^2 - \alpha(f')^2 + (f'')^2) \, dt = \left(1 - \frac{\alpha^2}{4} \right) \int_0^T f^2 dt.$$

Therefore the maximal value of α that ensures (13) is 2.

The preceding argument can be adapted to the general case to show that inequality (13) holds for $\alpha = 1$ independently of n and k and T (as proved later in the chapter on the Riccati equation).

The quadratic Hamiltonian H_0 produced by this class of optimal problems is given by

$$H_0(x_1, \ldots, x_n, p_1, \ldots, p_n) = \frac{1}{2} p_n^2 + p_1 x_2 + \cdots + p_{n-1} x_n + \frac{1}{2} (\alpha x_{k+1}^2 - x_1^2)$$

and is obtained by substitution of $u = p_n$. At the risk of confusion in notation, we shall use P to denote the matrix

$$
\begin{pmatrix}
0 & \cdots & \cdots & 0 \\
\vdots & & & \vdots \\
0 & \cdots & \cdots & 1
\end{pmatrix}.
$$

Then H_0 can be simply written as

$$
H_0(x, p) = \frac{1}{2}(Pp, p) + (Ax, p) - \frac{1}{2}(Rx, x),
$$

with

$$
A = \begin{pmatrix}
0 & 1 & 0 & \cdots & 0 \\
0 & 0 & 1 & \cdots & 0 \\
\vdots & & & \ddots & \\
\vdots & & & & 1 \\
0 & \cdots & \cdots & \cdots & 0
\end{pmatrix}
\quad \text{and} \quad
R = \begin{pmatrix}
1 & \cdots & \cdots & \cdots & 0 \\
\vdots & \ddots & & & \vdots \\
\vdots & & -\alpha & & \vdots \\
\vdots & & & \ddots & \vdots \\
0 & \cdots & \cdots & \cdots & 0
\end{pmatrix}.
$$

The symplectic matrix

$$
\vec{H}_0 = \begin{pmatrix} A & P \\ R & -A^* \end{pmatrix}
$$

defines a linear flow $d\xi/dt = \vec{H}_0 \xi$ that coincides with the Hamiltonian flow of H_0. This flow is equivalent to the $2n$th-order linear differential equation

$$
(-1)^n f^{(2n)}(t) + (-1)^{k+1} f^{(2k)}(t) + f(t) = 0, \tag{14}
$$

as can be verified as follows:

Let $f(t) = x_1(t)$. Then $f^{(j)}(t) = x_{j+1}(t)$ for $j < n$, and $f^{(n)}(t) = p_n(t)$. The subsequent derivatives of f are obtained from the differential equations for $p_1, \ldots, p_n$, which are

$$
\frac{dp_1}{dt} = x_1, \qquad \frac{dp_i}{dt} = -p_{i-1}, \qquad 1 < i < k+1,
$$

$$
\frac{dp_{k+1}}{dt} = -\alpha x_{k+1} - p_k,
$$

$$
\frac{dp_i}{dt} = -p_{i-1}, \qquad i > k+1.
$$

Then $f^{(n+i)} = (-1)^i p_{n-i}$ for $n - i > k + 1$. For $i = n - (k+2)$,

$$f^{(2n-(k+2))} = (-1)^{n-(k+2)} p_{k+2},$$

and therefore

$$f^{(2n-k)} = (-1)^{n-(k+1)}(-\alpha f^{(k)} - p_k).$$

Continuing,

$$f^{(2n-(k+i))} = (-1)^{n-(k+1)}(-\alpha f^{(k+i)} - (-1)^i p_{k-i}).$$

In particular, $i = k - 1$ gives $f^{(2n-1)} = (-1)^{n-(k+1)}(-\alpha f^{(2k-1)} - (-1)^{k-1} p_1)$. Another differentiation leads to $f^{(2n)} = (-1)^{n-(k+1)}(-\alpha f^{(2k)} - (-1)^{k-1} f)$, and (14) is verified.

It then follows from (14) that the characteristic polynomial associated with the symplectic matrix

$$\vec{H}_0 = \begin{pmatrix} A & P \\ R & -A^* \end{pmatrix}$$

is given by

$$p(\lambda) = (-1)^n \lambda^{2n} + (-1)^{k+1} \lambda^{2k} + 1.$$

Apart from the fact that for each eigenvalue λ of $\vec{H}_0$, $-\lambda$ is also an eigenvalue of $\vec{H}_0$, which follows from the symplectic structure of $\vec{H}_0$, we have the following:

Theorem 9 $\vec{H}_0$ *has no imaginary eigenvalues for any n and k for which $1 < k < n$ and any α for which $0 \le \alpha \le 1$.*

Proof Suppose that $\lambda = ix$ for some real number x. Then $p(ix) = (-1)^n(ix)^{2n} + (-1)^{k+1}\alpha(ix)^{2k} + 1 = x^{2n} - \alpha x^{2k} + 1$. Let $f(y) = y^n - \alpha y^k + 1$, with $y = x^2$. Then

$$f'(y) = ny^{n-1} - \alpha k y^{k-1} = y^{k-1}(ny^{n-k} - \alpha k).$$

Because $y > 0$, $y_0^{n-k} = \alpha k/n$ corresponds to the minimum of f. It follows that

$$f(y_0) = \left(\frac{\alpha k}{n}\right)^{\frac{n}{n-k}} - \alpha \left(\frac{\alpha k}{n}\right)^{\frac{k}{n-k}} + 1.$$

Because $\alpha k/n < 1$, $f(y_0) > 0$. Therefore $f(y) = 0$ has no real solutions for $y > 0$, and the possibility $p(ix) = 0$ is eliminated.   ■

We shall make further use of the result of Theorem 9 at the end of the next chapter for optimal problems on the interval $[0, \infty)$.

Problem 3 If λ is an eigenvalue of the symplectic matrix

$$\vec{H} = \begin{pmatrix} A & P \\ R & -A^* \end{pmatrix},$$

with $P = P^*$ and $R = R^*$, show that $-\lambda$ is also an eigenvalue of $\vec{H}$.

Problem 4 Let f and g be any smooth functions on $T^*M = M \times M^*$, with M a finite-dimensional real vector space. Let $\vec{f}$ and $\vec{g}$ denote the Hamiltonian vector fields of f and g. The Poisson bracket of f and g denoted by $\{f, g\}$ is a function defined as follows:

$$\{f, g\}(x, p) = \omega_{(x,p)}(\vec{f}(x, p), \vec{g}(x, p)).$$

Show that the Hamiltonian vector field of $\{f, g\}$ is equal to the Lie bracket of $\vec{f}$ and $\vec{g}$.

Notes and sources

The material in this chapter is drawn largely from the work of Kogan (1986) and Jurdjevic and Kogan (1989), although bits and pieces of the theory can be found scattered throughout the literature on control. Most of that literature assumes that $c(x, u) = \frac{1}{2}(Pu, u) + \frac{1}{2}(Rx, x)$, with $P > 0$ and $R \geq 0$. Then the cost extended system is convex, and the maximum principle is obtained through convexity arguments based on the separation theorem for convex sets (Lee and Markus, 1967). For such systems there are no conjugate points, because $c(x, u) \geq 0$.

The results presented in Section 3 are based almost entirely on the work of Jurdjevic and Kogan (1989), and the applications to Hardy-Littlewood systems in Section 4 are original. For $n = 2$, the results are classic, and they have been treated in considerable detail by Hardy et al. (1934). The same source can be consulted for alternative proofs of Wirtinger's inequality.

Even though optimal control theory only follows the long tradition of generating inequalities by variational means, I think that it provides distinctive insights to many classic inequalities and often suggests their generalizations. These mathematical directions need to be investigated further. For the classic literature on norm inequalities, the reader may want to consult Kwong and Zettl (1992).

8

The Riccati equation and quadratic systems

In the early part of the eighteenth century, Count Jacopo Riccati discovered that a second-order differential equation $f(y(x), dy/dx, d^2y/dx^2) = 0$ could be reduced to a first-order equation $f(y, p, p(dp/dy)) = 0$ by regarding dy/dx as a function $p(y)$ of an independent variable y. That discovery led, in particular, to a study of the following differential equation:

$$\frac{dy}{dx} + \psi(x)y^2 + \phi(x)y + \xi(x) = 0, \tag{1}$$

where ϕ, ψ, and ξ are arbitrary functions of the independent variable x. Equation (1), along with its generalizations to other equations of first order containing a quadratic term, came to be known as the Riccati equation. Equation (1) and its generalizations have attracted the attention of many mathematicians, beginning with the Bernoulli family in the first half of the eighteenth century and continuing up to the present. Among its many applications, we shall concentrate on only a narrow aspect connected with its relation to the linear quadratic problem.

The Riccati equation appears naturally in the solutions for linear quadratic problems, as a means of "completing the square" for the quadratic cost $c(x, u) = \frac{1}{2}(Pu, u) + (Qx, u) + \frac{1}{2}(Rx, x)$ over the trajectories of a linear control system $dx/dt = Ax + Bu$. The method of completing the square consists of adding a term of the form $-(d/dt)(S(t)x(t), x(t))$ to $c(x, u)$ in order to show that $\int_0^T c(x, u)\, dt \geq 0$ over the closed trajectories of the system. $S(t)$ is a curve in the space of symmetric matrices, to be determined as follows:

$$\int_0^T c(x, u)\, dt = \int_0^T c(x, u) - \frac{1}{2}\frac{d}{dt}(S(t)x(t), x(t))\, dt$$

$$= \int_0^T \left(\frac{1}{2}(Pu, u) + (Qx, u) + \frac{1}{2}(Rx, x) \right.$$

$$\left. - \frac{1}{2}\left(\frac{dS}{dt}x, x\right) - (S(t)x(t), Ax + Bu) \right) dt$$

228

In the regular case, $P > 0$, and the foregoing integrand can be rewritten as

$$\int_0^T \left(\left(\frac{1}{2} P(u - P^{-1}(B^*S(t)x - Qx)), \; u - P^{-1}(B^*S(t)x - Qx) \right) \right.$$

$$- \left(\frac{1}{2} P^{-1}(B^*S(t)x - Qx), \; B^*S(t)x - Qx \right)$$

$$\left. + \frac{1}{2}(Rx, x) - \frac{1}{2}\left(\frac{dS}{dt}x, x \right) - \left(\frac{A^*S + SA}{2}x, x \right) \right) dt.$$

That integral reduces to

$$\frac{1}{2} \int_0^T (P(u - P^{-1}(B^*S(t) - Qx)), \; u - P^{-1}(B^*S(t)x - Qx)) \, dt,$$

provided that $S(t)$ is chosen as a solution curve of the following matrix Riccati differential equation:

$$\frac{dS}{dt} + (SB - Q^*)P^{-1}(B^*S - Q) + A^*S + SA - R = 0.$$

This equation can be written more simply as

$$\frac{dS}{dt} + S\tilde{P}S + S\tilde{Q} + \tilde{Q}^*S + \tilde{R} = 0, \qquad (2)$$

with $\tilde{P} = BP^{-1}B^*$, $\tilde{Q} = A - BP^{-1}Q$, and $\tilde{R} = Q^*P^{-1}Q - R$.

Evidently, if a solution curve $S(t)$ of equation (2) exists in an interval $[0, T]$, then $u(t) = P^{-1}(B^*S(t)x(t) - Qx(t))$ is the optimal control function that corresponds to a solution $x(t)$ of the following differential system:

$$\frac{dx}{dt} = Ax(t) + B(P^{-1}S(t)x(t) - Qx(t)). \qquad (3)$$

It remains to show that there always exist solutions of (2) and (3) such that the desired boundary conditions $x(0) = a$ and $x(t) = b$ are fulfilled.

Rather than pursuing the details of matching the boundary conditions, it is more natural to establish, first, a geometric base that connects the Riccati equation to the Hamiltonian formalism and thus illuminate the geometric significance behind the method of completing the square.

1 Symplectic vector spaces

Because symplectic geometry is indispensable for understanding the Hamiltonian formalism, it may be worthwhile to take some time to discuss the basic theory.

A bilinear, antisymmetric and nondegenerate form ω defined on a real vector space V is called *symplectic*. Recall that a form is said to be nondegenerate if $\omega(a, v) = 0$ for all v in V implies that $a = 0$. The pair (V, ω) is called a symplectic vector space. A symplectic space must be even-dimensional, for the following reasons: Let $a_1, \ldots, a_m$ be a basis for V, and let $(x_1, \ldots, x_u)$ and $(y_1, \ldots, y_n)$ be the coordinates of v and w in V relative to this basis. Then $\omega(v, w) = \sum_{j=1}^{m} \sum_{i=1}^{m} a_{ij} x_i y_i$ where the matrix (a_{ij}) is an antisymmetric matrix with entries $a_{ij} = \omega(a_i, a_j)$. But then m must be even; otherwise (a_{ij}) would have a zero eigenvalue, which would contradict the nondegeneracy of ω. We shall use $2n$ to denote the dimension of V.

Definition 1 A subspace W is called *isotropic* if $\omega(a, b) = 0$ for all a and b in W.

Every one-dimensional subspace of V is isotropic, as a consequence of the antisymmetry of ω. Corresponding to each subspace W of V, $W^\perp$ denotes its symplectic orthogonal complement; $a \in W^\perp$ if and only if $\omega(a, w) = 0$ for all $w \in W$. Evidently, $W \subset W^\perp$ if and only if W is isotropic.

Definition 2 An isotropic subspace W is called Lagrangian if W is not contained in any larger isotropic subspace of V.

Definition 3 W is called a symplectic subspace of V if the restriction of ω to W is nondegenerate.

Definition 4 A basis $a_1, \ldots, a_n, b_1, \ldots, b_n$ in V is called symplectic if $\omega(b_i, b_j) = \omega(a_i, a_j) = 0$ and $\omega(a_i, b_j) = \delta_{ij}$ for all i and j, with $1 \leq i \leq n$ and $1 \leq j \leq n$. If $a_1, \ldots, a_n, b_1, \ldots, b_n$ is a symplectic basis, then the linear span of $\{a_1, \ldots, a_n\}$ and the linear span of $\{b_1, \ldots, b_n\}$ are both Lagrangian subspaces of V.

Problem 1 Show that if a is any element of V for which $\omega(a, a_j) = 0$, $j = 1, 2, \ldots, n$, then a belongs to the linear span of $\{a_1, \ldots, a_n\}$. Hence, prove that each symplectic basis defines two complementary Lagrangian spaces $\mathcal{L}_1$ and $\mathcal{L}_2$.

Theorem 1 Let $a_1, \ldots, a_k, b_1, \ldots, b_k$ be any set of elements in V with the property that $\omega(a_i, a_j) = \omega(b_i, b_j) = 0$ and $\omega(a_i, b_j) = \delta_{ij}$ for each $i =$

$1, 2, \ldots, k$ *and* $j = 1, \ldots, k$. *Let* W *be the linear span of* $a_1, \ldots, a_k, b_1, \ldots, b_k$. *Then* $V = W \oplus W^\perp$, *and both* W *and* $W^\perp$ *are symplectic subspaces of* V.

Proof Assume that $v \in W \cap W^\perp$. Then $\omega(v, w) = 0$ for all $w \in W$. In particular, $\omega(a_i, v) = \omega(b_i, v) = 0$ for each $i = 1, 2, \ldots, n$. Writing

$$v = \sum_{i=1}^{m} \alpha_i a_i + \beta_i b_i,$$

it follows that $\alpha_i = \omega(v, b_i) = 0$, and $\beta_i = \omega(a_i, v) = 0$. Hence $v = 0$.

Suppose that v is any element in V. Let $v_1 = \sum_{i=1}^{k} -\omega(v, b_i)a_i + \omega(v, a_i)b_i$ and $v_2 = v - v_1$. Then $\omega(v_2, a_i) = \omega(v, a_i) - \omega(v, a_i) = 0$ and $\omega(v_2, b_i) = \omega(v, b_i) - \omega(v, b_i) = 0$ for each i; therefore v_2 belongs to $W^\perp$. Hence $V = W \oplus W^\perp$. Evidently $W^\perp$ is symplectic, and the proof is finished. ∎

Corollary 1 *If* $k < n$, *then there exists a vector* a_{k+1} *such that* $\omega(a_{k+1}, a_i) = 0$ *for each* $i = 1, 2, \ldots, k$.

Proof If $k < n$, then $W^\perp \neq \{0\}$. Let a_{k+1} be any nonzero element of $W^\perp$. ∎

Theorem 2 *Let* $a_1, \ldots, a_k$ *be any basis for an isotropic subspace* W. *There exist vectors* $b_1, \ldots, b_k$ *in* V *such that* $\omega(a_i, b_j) = \delta_{ij}$ *and* $\omega(b_i, b_j) = 0$ *for* $i = 1, \ldots, k$ *and* $j = 1, \ldots, k$.

Proof Use induction on k. If $\dim W = 1$, then there exists $b \in V$ such that $\omega(a_1, b) \neq 0$. Let $b_1 = (1/\omega(a_1, b))b$.

Assume that the theorem is true for any k-dimensional isotropic subspace of V. Let W be any $(k+1)$-dimensional isotropic subspace of V, with a basis consisting of vectors $a_1, \ldots, a_{k+1}$.

Let W_k be the linear span of $a_1, \ldots, a_k$. According to the induction hypothesis, there exist vectors $b_1, \ldots, b_k$ that satisfy the conditions of the theorem.

Denote by X the linear span of $a_1, \ldots, a_k, b_1, \ldots, b_k$. According to Theorem 1, $V = X \oplus X^\perp$. Let $a_{k+1} = v_1 + v_2$, with $v_1 \in X$ and $v_2 \in X^\perp$. Then $v_1 = \sum_{i=1}^{k} \alpha_i a_i + \beta_i b_i$ for some coefficients $\alpha_1, \ldots, \alpha_n$ and $\beta_1, \ldots, \beta_n$. Because $\omega(a_{k+1}, a_j) = 0$, and $\omega(a_{k+1}, a_j) = \omega(v_1, a_j)$, it follows that each $\alpha_j = 0$ for $j = 1, 2, \ldots, k$. But then

$$\omega(a_{k+1}, b_i) = \omega(v_1, b_i) + \omega(v_2, b_i) = 0 + 0 = 0.$$

The foregoing argument shows that $a_{k+1} \in X^\perp$. Because $X^\perp$ is a symplectic subspace of V, there exists b in $X^\perp$ such that $\omega(a_{k+1}, b) \neq 0$. Let $b_{k+1} =$

$(1/\omega(a_{k+1}, b))b$. Evidently, $a_1, \ldots, a_{k+1}, b_1, \ldots, b_{k+1}$ satisfies the conditions of the theorem, and the proof is finished.   ∎

Corollary 2 *Each Lagrangian space is n-dimensional.*

Proof Let $a_1, \ldots, a_k$ denote a basis of a Lagrangian space L. According to Theorem 2, there exist vectors $b_1, \ldots, b_k$ such that $\omega(b_i, b_j) = 0$ and $\omega(a_i, b_j) = \delta_{ij}$. Then k must be n, for otherwise L would be contained in a larger isotropic space (Corollary 1). End of the proof.   ∎

The coordinates $(x_1, \ldots, x_n, p_1, \ldots, p_n)$ of a point v relative to a symplectic basis $a_1, \ldots, a_n, b_1, \ldots, b_n$ are called symplectic. If $(y_1, \ldots, y_n, q_1, \ldots, q_n)$ are the symplectic coordinates of another point w, then

$$\omega(v, w) = \sum_{i=1}^{n} x_i q_i - y_i p_i.$$

The preceding expression for w brings us back to Section 1.2 in Chapter 6: If we write $((x, p), (y, q)) = x \cdot y + p \cdot q$, and if $[(x, p), (y, q)] = x \cdot q - y \cdot p$, then

$$[(x, p), (y, q)] = ((x, p), J(y, q)) \quad \text{for } J = \begin{pmatrix} 0 & I \\ -I & 0 \end{pmatrix}.$$

J is a complex structure for $\mathbb{R}^{2n}$. As we saw in Chapter 6, $\mathbb{R}^{2n}$ can be viewed as the complex space $\mathbb{C}^n$ endowed with its Hermitian product $\langle \, , \, \rangle$:

$$\langle x + ip, \ y + iq \rangle = ((x, p), \ (y, q)) + i[(x, p), \ (y, q)].$$

1.1 The geometry of linear Lagrangians

We shall use $\text{Grass}_n(V)$ to denote the Grassmann manifold of all n-dimensional subspaces of V. Following the earlier discussion of Grassmann manifolds in Section 1.4 in Chapter 6, $\text{Grass}_n(V)$ can be realized as the homogeneous space $O_{2n}(R)/O_n(R) \times O_n(R)$.

The space of all Lagrangians, denoted by $\text{Lg}(V)$, is a closed subset of $\text{Grass}_n(V)$. $\text{Lg}(V)$ can also be realized as a homogeneous space through the following arguments: Let $\mathcal{L}_h$ be any Lagrangian subspace of V, and let $a_1, \ldots, a_n$ be any basis of $\mathcal{L}_h$. Extend $a_1, \ldots, a_n$ into a symplectic basis $a_1, \ldots, a_n, b_1, \ldots, b_n$. Denote by $\mathcal{L}_v$ the Lagrangian spanned by $b_1, \ldots, b_n$. Then $V = \mathcal{L}_h \oplus \mathcal{L}_v$.

Theorem 3 *The space of all Lagrangians that are transversal to $\mathcal{L}_v$ is in one-to-one correspondence with the space of all symmetric linear mappings on $\mathcal{L}_h$.*

Proof We shall work with a fixed symplectic basis $a_1, \ldots, a_n, b_1, \ldots, b_n$. Each symmetric linear mapping S on $\mathcal{L}_h$ is identified with a symmetric $n \times n$ matrix relative to the basis $a_1, \ldots, a_n$, and each point x in $\mathcal{L}_h$ is identified with its coordinate vector x in $\mathbb{R}^n$. The graph $\{(x, Sx), \ x \in \mathbb{R}^n\}$ is a Lagrangian subspace of $\mathbb{R}^{2n}$ that is transversal to $\{(0, x) : \ x \in \mathbb{R}^n\}$. The proof is simple: $[(x, Sx), (y, Sy)] = (x \cdot Sy) - (y \cdot Sx) = 0$, because S is symmetric. Evidently $\dim\{(x, Sx) : \ x \in \mathbb{R}^n\} = n$. Therefore the graph of a symmetric matrix is a Lagrangian subspace of $\mathbb{R}^{2n}$.

Conversely, suppose that L is a Lagrangian space in $\mathbb{R}^{2n}$ transversal to $L_v = \{(0, x) : \ x \in \mathbb{R}^n\}$. For each (x, y) in L, let $Sx = y$. The mapping S is well defined, because $Sx = y_1$ and $Sx = y_2$ imply that $(0, y_1, -y_2) \in L \cap L_v$, which implies that $y_1 = y_2$, because L is transversal to L_v. It is easy to check that the isotropic property is equivalent to S being symmetric. The proof is now finished. ∎

The set of all $n \times n$ symmetric matrices is a real vector space of dimension $n(n+1)/2$. We shall regard $\mathrm{Lg}(V)$ as an $(n(n+1)/2)$-dimensional manifold with a coordinate neighborhood at each L_h given by the space of all $n \times n$ symmetric matrices.

Recall that $T : V \to V$ is a symplectic transformation if $\omega(Tv, Tw) = \omega(v, w)$ for all v and w in V. The group of all symplectic transformations on V will be denoted by $\mathrm{Sp}(V)$.

Relative to any symplectic basis in V, $\mathrm{Sp}(V)$ can be realized by matrices, in which case it can be regarded as a closed subgroup of $\mathrm{GL}_{2n}(R)$ consisting of all nonsingular matrices T such that $T^*JT = J$. The Lie algebra of $\mathrm{Sp}_{2n}(R)$ consists of all matrices A such that $e^{A^*t}Je^{At} = J$ for all $t \in R$.

Problem 2 Show that A is in the Lie algebra of $\mathrm{Sp}_{2n}(R)$ if and only if $A^*J + JA = 0$.

Problem 3 Show that $A^*J + JA = 0$ if and only if

$$A = \begin{pmatrix} Q & P \\ R & -Q^* \end{pmatrix},$$

with P and R symmetric $n \times n$ matrices, and Q an arbitrary $n \times n$ matrix.

Problem 4 Show that T is a symplectic transformation if and only if T transforms any symplectic basis into a symplectic basis.

Problem 5 Show that $TL = \{Tv : v \in L\}$ is a Lagrangian for each $T \in \mathrm{Sp}(V)$ and each Lagrangian L.

$(T, L) \to TL$ defines an action of the symplectic group on the space of all Lagrangian spaces. It is easy to prove that this action is transitive: Let L_1 and L_2 be any two Lagrangians. Let $a_1^i, \ldots, a_n^i$, $b_1^i, \ldots, b_n^i$, $i = 1, 2$, be the symplectic bases in V, which correspond to the bases $a_1^i, \ldots, a_n^i$ in L_i. Let T be the symplectic transformation such that $Ta_i^1 = a_i^2$ and $Tb_i^1 = b_i^2$ for $i = 1, 2, \ldots, n$. Then $TL_1 = L_2$.

The isotropy group H consists of all symplectic transformations T such that $TL_h = L_h$ for some fixed Lagrangian L_h. L_h can be identified with $\mathbb{R}^n$. A matrix A is an element of the Lie algebra of H if and only if

$$A \begin{pmatrix} x \\ 0 \end{pmatrix} = \begin{pmatrix} y \\ 0 \end{pmatrix}$$

for all $x \in \mathbb{R}^n$. Using the notation in Problem 3, we get that $Rx = 0$ for each $x \in \mathbb{R}^n$. Therefore, H can be identified with all matrices in $\mathrm{Sp}_{2n}(R)$ of the form

$$\begin{pmatrix} Q & P \\ 0 & -Q^* \end{pmatrix},$$

with $Q \in \mathrm{GL}_n(R)$. The preceding arguments prove that the space of Lagrangians is a homogeneous space equal to $\mathrm{Sp}(V)/H$.

2 Lagrangians and the Riccati equation

As we saw in Chapter 7, $V = M \times M^*$ is a symplectic space, with its symplectic form ω equal to $q(x) - p(y)$ for all (x, p) and (y, q) in $M \times M^*$. After identifying M with $M \times \{0\}$ and M^* with $\{0\} \times M^*$, it follows that both M and M^* are Lagrangian subspaces of V. M is called the horizontal Lagrangian, and M^* is called the vertical Lagrangian. Any basis $a_1, \ldots, a_n$ in M, along with its dual basis $a_1^*, \ldots, a_n^*$, forms a symplectic basis for V.

Each symplectic matrix A is of the form

$$\begin{pmatrix} Q & P \\ -R & -Q^* \end{pmatrix},$$

with P and R symmetric (Problem 3). The corresponding linear differential system

$$\frac{dx}{dt} = Qx + Pp, \qquad \frac{dp}{dt} = -Rx - Q^*p,$$

is Hamiltonian, induced by $H = \frac{1}{2}\sum_{i,j=1}^n P_{ij}p_ip_j + \sum_{i,j=1}^n Q_{ij}x_jp_i + \frac{1}{2}\sum_{i,j=1}^n R_{ij}x_ix_j$, with

$$x = \begin{pmatrix} x_1 \\ \vdots \\ x_n \end{pmatrix} \quad \text{and} \quad p = \begin{pmatrix} p_1 \\ \vdots \\ p_n \end{pmatrix}$$

equal to the symplectic coordinates of point v in V. It will be more convenient to write the foregoing quadratic Hamiltonian as $H = \frac{1}{2}(Pp \cdot p) + (Qx \cdot p) + \frac{1}{2}(Rx \cdot x)$.

Definition 5 A differential equation in the space of $n \times n$ symmetric matrices of the form

$$\frac{dS}{dt} + SPS + Q^*S + SQ + R = 0,$$

with P, Q, and R given $n \times n$ matrices, P semipositive-definite, and R symmetric, is called a matrix Riccati equation.

Evidently there is an exact correspondence between Riccati equations and quadratic Hamiltonians with $P \geq 0$. It follows from the basic theory of differential equations that for each symmetric matrix K, there is a unique interval $[0, T)$ and a unique solution curve $S(t)$ of the matrix Riccati equation defined on $[0, T)$ such that $S(0) = K$. T could be equal to ∞, in which case $S(t)$ is defined on the entire interval $[0, \infty)$.

Each solution curve $S(t)$ lifts to a unique curve $L(t)$ in the space of Lagrangians, with $L(t)$ equal to the graph of $S(t)$. If $S(t)$ is defined on $[0, \infty)$, then $L(t)$ is always transversal to L_v. Suppose, now, that the maximal interval $[0, T)$ on which $S(t)$ exists is finite. Then $\lim_{t \to T} \|S(t)\| = \infty$, for otherwise $S(t)$ could be extended to an interval $[0, \hat{T})$, with $\hat{T} > T$.

It will be shown in the next theorem that the Lagrangians $L(t)$ converge to a Lagrangian $L(T)$ as t tends to T. $L(T)$ cannot be transversal to L_v, for the following reasons:

Because $\|S(t)\| = \sup_{\|x\|=1} \|S(t)x\|$, it follows that there exist a sequence $\{t_n\}$ with $\lim_{n\to\infty} t_n = T$ and a sequence $\{x_n\}$ on the unit sphere in M such that

$\lim_{n \to \infty} \| S(t_n) x_n \| = \infty$. There is no loss of generality in assuming that $\{x_n\}$ converges to a point x, because of the compactness of the unit sphere.

Let $y_n = S(t_n) x_n / \| S(t_n) x_n \|$. Because $\| y_n \| = 1$, we may as well assume that y_n converges to a point y. But then $(0, y)$ is a point in $L(T)$, because $(1/\| S(t_n) x_n \|)(x_n, S(t_n) x_n) \in L(t_n)$ and $(0, y) = \lim_{n \to \infty} (1/\| S(t_n) x_n \|)(x_n, S(t_n) x_n)$. Figure 8.1 depicts the preceding arguments.

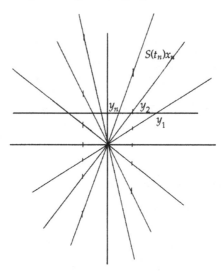

Fig. 8.1.

A quadratic Hamiltonian H induces a linear symplectic flow $\exp t \vec{H}$, with $\vec{H}$ equal to the Hamiltonian vector field defined by H. In terms of symplectic coordinates, $\exp t \vec{H} = \exp t A$, with

$$ A = \begin{pmatrix} Q & P \\ -R & -Q^* \end{pmatrix}. $$

The action of the symplectic group on the space of Lagrangians induces a flow $\{ \Phi_t : t \in \mathbb{R}^1 \}$ on the space of Lagrangians defined by $\Phi_t(L) = \{ (\exp t \vec{H})(x) : x \in L \}$.

The next theorem shows the precise connection between the induced flow on the space of Lagrangians and the Riccati equation.

Theorem 4

(a) *Let $L(t)$ be the graph of a solution curve $S(t)$ of the Riccati equation. Then*

$$ \Phi_t(L(0)) = L(t) $$

for each t for which $S(t)$ is defined.

(b) *Assume that L_0 is transversal to the vertical Lagrangian L_v. Denote by*
   K the symmetric matrix whose graph is L_0. Let $L(t) = \Phi_t(L_0)$. For each
   t for which $L(t)$ is transversal to L_v, $L(t)$ is the graph of a symmetric
   matrix $S(t)$. $S(t)$ is the solution of the matrix Riccati equation with $S(0) =$
   K.
   The maximal interval $[0, T)$ for which $S(t)$ exists coincides with the first
   instant T such that $L(T)$ is not transversal to L_v.

Proof There is no loss of generality in assuming that all the data are given in
terms of the matrices relative to a symplectic basis. Let

$$\begin{pmatrix} \phi_{11}(t) & \phi_{12}(t) \\ \phi_{21}(t) & \phi_{22}(t) \end{pmatrix}$$

denote the partition of $\exp tA$ into $n \times n$ block matrices. Because $(d/dt)\exp tA = A \exp tA$, it follows that

$$\frac{d\phi_{11}}{dt} = Q\phi_{11} + P\phi_{21}, \qquad \frac{d\phi_{12}}{dt} = Q\phi_{12} + P\phi_{22},$$

$$\frac{d\phi_{21}}{dt} = -R\phi_{11} - Q^*\phi_{21}, \qquad \frac{d\phi_{22}}{dt} = -R\phi_{12} - Q^*\phi_{22}.$$

Let L be any Lagrangian transversal to L_v. Because L is transversal to L_v,
L is the graph of a symmetric matrix K:

$$e^{At}(\text{graph}(K)) = L(t) = \left\{ \begin{pmatrix} \phi_{11}v + \phi_{12}Kv \\ \phi_{21}v + \phi_{22}Kv \end{pmatrix} : v \in \mathbb{R}^n \right\}.$$

As long as $L(t)$ is transversal to L_v, the mapping $v \to \phi_{11}(t)v + \phi_{12}(t)Kv$ is
invertible, for if $\phi_{11}v + \phi_{12}Kv = 0$ for some $v \neq 0$, then $\phi_{21}v + \phi_{22}Kv \neq 0$,
because e^{At} is a nonsingular matrix, which contradicts the transversality of $L(t)$
and L_v.

Let $S(t) = (\phi_{21}(t) + \phi_{22}(t)K)(\phi_{11}(t) + \phi_{12}(t)K)^{-1}$. Evidently, $\text{graph}(S(t))$
$= L(t)$. On differentiating $S(t)(\phi_{11}(t) + \phi_{12}(t)K) = (\phi_{21}(t) + \phi_{22}(t))K$, we
get

$$\frac{dS}{dt}(\phi_{11} + \phi_{12}K) + S(Q\phi_{11} + P\phi_{21} + (Q\phi_{12} + P\phi_{22})K)$$

$$= -(R\phi_{11} + Q^*\phi_{21}) - (R\phi_{12} + Q^*\phi_{22})K.$$

This equality can be rewritten as

$$\frac{dS}{dt}(\phi_{11} + \phi_{12}K) + S(Q(\phi_{11} + \phi_{12}K) + P(\phi_{21} + \phi_{22}K))$$
$$= -R(\phi_{11} + \phi_{12}K) - Q^*(\phi_{21} + \phi_{22}K),$$

and on post-multiplying from the right by $(\phi_{11} + \phi_{12}K)^{-1}$, we get that $dS/dt + SQ + SPS + Q^*S + R = 0$.

The foregoing argument proves part (b). Now suppose that $L(t)$ is the graph of any solution curve $S(t)$ of the matrix Riccati equation. Denote by K the initial symmetric matrix for $S(t)$; that is, $K = S(0)$. Denote by $\tilde{S}(t)$ the symmetric matrix whose graph is equal to $(\exp t A)(\text{graph}(K))$. Then it follows from the foregoing construction that $\tilde{S}(t)$ is a solution of the matrix Riccati equation with $\tilde{S}(0) = K$. The uniqueness of solutions for the initial-value problem for differential equations guarantees that $S(t) = \tilde{S}(t)$. Therefore, (a) is also valid, and the proof is now finished. ∎

Theorem 4 shows that the space of symmetric matrices can be identified with an open and dense subset of Lagrangians that are transversal to the vertical Lagrangian L_v. $\text{Lg}(V)$, being a closed subset of $\text{Grass}_n(V)$, is compact and therefore can be viewed as a compactification of the space of symmetric matrices.

For each initial Lagrangian L_0 that is transversal to L_v, $(\exp t \vec{H})(L_0) = L(t)$ corresponds to a solution of the Riccati equation. Although $(\exp t \vec{H})(L_0)$ is a well-defined Lagrangian for each instant of time t, the correspondence with a solution of the Riccati equation is valid only up to the first instant T that $L(T)$ is not transversal to L_v.

Problem 6 (Wirtinger's problem) Consider the matrix Riccati equation in $\mathbb{R}^2$ corresponding to $H = \frac{1}{2}p_2^2 + p_1 x_2 + \frac{1}{2}x_2^2$. Show that $[0, 2\pi]$ is the maximal interval on which the matrix Riccati equation admits a solution.

The following theorem concerning the existence of solutions of the Riccati equation answers the question raised in the introductory part of this chapter concerning the matching of the boundary conditions.

Theorem 5 *Suppose that there is a solution curve $S(t)$ of the Riccati equation defined on an interval $[0, T]$. If $(\exp t \vec{H})(L_v)$ is transversal to L_v for all t in some interval $(0, \varepsilon)$, then $(\exp t \vec{H})(L_v)$ is transversal to L_v for all t in $(0, T]$.*

Conversely, assume that $(\exp t\vec{H})(L_v)$ *is transversal to* L_v *for all* t *in some interval* $[0, T]$. *Then there exists a solution* $S(t)$ *of the Riccati equation defined on the entire interval* $[0, T]$.

Proof Assume that $S(t)$ is a solution of the Riccati equation defined on an interval $[0, T]$. Let $L(t)$ denote its graph. According to Theorem 4,

$$L(t) = (\exp tA)(L(0)), \quad \text{with } A = \begin{pmatrix} Q & P \\ -R & -Q^* \end{pmatrix}.$$

Let

$$\exp tA = \begin{pmatrix} \phi_{11} & \phi_{12} \\ \phi_{21} & \phi_{22} \end{pmatrix} \quad \text{and} \quad S(0) = K.$$

Then $L(t)$ is the graph of $U(t) = \phi_{11}(t) + \phi_{12}(t)K$. Because $L(t)$ is transversal to L_v, $U(t)$ is nonsingular for each t in $[0, T]$. $(\exp tA)(L_v)$ is transversal to L_v if and only if $\phi_{12}(t)$ is nonsingular. So it suffices to show that $\phi_{12}(t)$ is nonsingular for all t in $(0, T]$.

First note that for each t for which $(\exp t\vec{H})(L_v)$ is transversal to L_v, $\phi_{12}(t) = U(t)\phi(t)$ for some time-varying matrix $\phi(t)$. The argument is as follows:

$$(\exp tA)(L_v) = \left\{ \begin{pmatrix} \phi_{12}p \\ \phi_{22}p \end{pmatrix} : p \in \mathbb{R}^n \right\} \quad \text{and} \quad (\exp tA)(\text{graph }(K))$$

$$= \left\{ \begin{pmatrix} Ux \\ \phi_{21}x + \phi_{22}Kx \end{pmatrix} : x \in \mathbb{R}^n \right\}.$$

Both of these Lagrangians are transversal to L_v, and therefore for each $p \in \mathbb{R}^n$ there is a unique $x_t(p)$ in $\mathbb{R}^n$ such that $U(t)x_t(p) = \phi_{12}(t)p$. Moreover, the correspondence $p \to x_t(p)$ is linear and can be written as $x_t(p) = \phi(t)p$. Because p is arbitrary, $U(t)\phi(t) = \phi_{12}(t)$ (Figure 8.2).

The explicit form for $\phi(t)$ is obtained by differentiation:

$$\frac{dU}{dt}\phi + U\frac{d\phi}{dt} = \frac{d\phi_{12}}{dt} = Q\phi_{12} + P\phi_{22} = QU\phi + P\phi_{22},$$

$$\frac{dU}{dt} = \frac{d\phi_{11}}{dt} + \frac{d\phi_{12}}{dt}K = Q\phi_{11} + P\phi_{21} + (Q\phi_{12} + P\phi_{22})K$$

$$= Q(\phi_{11} + \phi_{12}K) + P(\phi_{21} + \phi_{22}K) = QU + PV, \qquad \text{with}$$

$$V = \phi_{21} + \phi_{22}K.$$

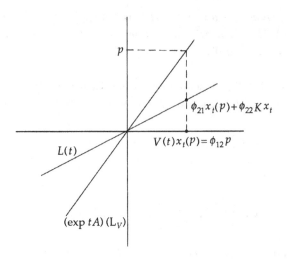

Fig. 8.2.

Therefore,

$$QU\phi + PV\phi + U\frac{d\phi}{dt} = QU\phi + P\phi_{22},$$

and hence

$$\frac{d\phi}{dt} = U^{-1}P(\phi_{22} - V\phi).$$

The fact that $\exp tA \in \mathrm{Sp}_{2n}(R)$ implies that

$$\phi_{11}^*\phi_{21} - \phi_{21}^*\phi_{11} = 0, \qquad \phi_{12}^*\phi_{22} - \phi_{22}^*\phi_{12} = 0, \qquad \phi_{11}^*\phi_{22} - \phi_{21}^*\phi_{12} = I.$$

Because

$$
\begin{aligned}
\phi_{11}^*\phi_{22} - \phi_{21}^*\phi_{12} &= (U^* - K\phi_{12}^*)\phi_{22} - (V^* - K\phi_{22}^*)\phi_{12} \\
&= U^*\phi_{22} - V^*\phi_{12} - K(\phi_{12}^*\phi_{22} - \phi_{22}^*\phi_{12}) \\
&= U^*\phi_{22} - V^*\phi_{12},
\end{aligned}
$$

it follows that $U^*\phi_{22} - V^*\phi_{12} = I$. Therefore $\phi_{22} = (U^*)^{-1}(I + V^*\phi_{12}) = (U^*)^{-1}(I + V^*U\phi)$. That $\phi_{11}^*\phi_{21} - \phi_{21}^*\phi_{11} = 0$ implies that $V^*U - U^*V = 0$. So $\phi_{22} = (U^*)^{-1}(I + U^*V\phi) = (U^*)^{-1} + V\phi$. Therefore,

$$\frac{d\phi}{dt} = U^{-1}P(\Phi_{22} - V\phi) = U^{-1}P((U^*)^{-1} + V\phi - V\phi) = U^{-1}P(U^*)^{-1},$$

and hence $\phi(t) = \phi(0) + \int_0^t U^{-1}(\tau)PU^*(\tau)^{-1}d\tau$. Because $\phi_{12}(0) = 0$ and $U(0)=I$, it follows that $\phi(0) = 0$. So $\phi_{12}(t)=U(t)\int_0^t U^{-1}(\tau)P(U^{-1}(\tau))^*d\tau$.

Suppose, now, that $\phi_{12}(t)x = 0$ for some $t > 0$ and some x in $\mathbb{R}^n$. Because $U(t)$ is nonsingular, $x = U^*(t)y$ for some y. Then

$$\phi_{12}(t)x = \int_0^t (U(t)U^{-1}(\tau))P(U^{-1}(\tau))^*U^*(t)y \, d\tau.$$

Being semipositive-definite, P has a well-defined square root. Therefore, $P^{\frac{1}{2}}U^{-1}(\tau)^*U(t)y = 0$ for each τ ($0 \leq \tau \leq t$). Equivalently, $P^{\frac{1}{2}}U^{-1}$ $(\tau)^*x = 0$. But then $\phi_{12}(t)$ is singular for all t, which contradicts the transversality assumption for small time. Therefore $\phi_{12}(t)$ is nonsingular for all t in $(0, T]$, and the first part is proved.

The second statement follows from the stability of the transversality property under small perturbations. The set of all Lagrangians that are transversal to L_v is open and dense in $\mathrm{Lg}(V)$. Let $\hat{t} > 0$ be an instant for which $(\exp \hat{t}\vec{H})(L_v)$ is transversal to L_v. Therefore there exists a Lagrangian $\hat{L}$ sufficiently close to $(\exp \hat{t}\vec{H})(L_v)$ that is transversal to L_v and for which the entire curve $t \to (\exp((t-\hat{t})\vec{H}))(\hat{L})$ is transversal to L_v for each time t such that $0 \leq (t-\hat{t}) \leq T$. According to Theorem 4, such a curve is the graph of a solution curve $S(t)$ of the Riccati equation. End of the proof. ∎

Corollary 3 *The maximal interval of optimality* $[0, T_{\max}]$ *is equal to* $\sup\{T :$ *there exists a solution of the Riccati equation on the interval* $[0, T)\}$.

Proof On any interval $[0, T]$ on which there exists a solution of the Riccati equation, $(\exp t\vec{H})(L_v)$ is transversal to L_v, and therefore $[0, T]$ is free of conjugate points. ∎

Problem 7 Show that any regular linear quadratic problem has the property that $(\exp t\vec{H})(L_v) \cap L_v = 0$ for each sufficiently small but positive time t. Show that the conjugate point occurs at the first instant of time t for which $(\exp t\vec{H})(L_v)$ is not transversal to L_v.

3 The algebraic Riccati equation

The algebraic Riccati equation is an equation of the form

$$SPS + Q^*S + SQ + R = 0,$$

defined by the linear mappings $P : V \to V^*$, $Q : V \to V$, and $R : V \to V^*$, with P and R symmetric, and $P \geq 0$.

The set of solutions of the algebraic Riccati equation coincides with the set of equilibrium points of the Riccati equation. These solutions admit a simple characterization in the space of Lagrangians and the flow Φ_t induced by $\vec{H}$.

Theorem 6 $\hat{S}$ *is a solution of the algebraic Riccati equation if and only if* graph($\hat{S}$) *is invariant under the flow* Φ_t *induced by* $\vec{H}$.

Proof Assume that $\hat{S}$ is a solution of the algebraic Riccati equation. Let $S(t) = \hat{S}$. Then $S(t)$ is an equilibrium solution of the Riccati equation. According to Theorem 4, the graph of $S(t)$ is given by $L(t) = (\exp tA)(\text{graph}(\hat{S}))$. Hence, graph($\hat{S}$) = $(\exp tA)(\text{graph}(\hat{S}))$ for all t. The converse is proved by reversing the steps. ∎

As we shall see in the next section, the algebraic Riccati equation plays an important role for optimal problems on the semi-infinite time interval $[0, \infty)$. For that reason, it will be convenient to single out the relevant properties of $\vec{H}$-invariant Lagrangians. A Lagrangian L is called $\vec{H}$-invariant if $\vec{H}(L) \subset L$. In symplectic coordinates, $\vec{H}$ is given by the matrix A. Thus, L is $\vec{H}$-invariant if and only if

$$A \begin{pmatrix} x \\ p \end{pmatrix} \in L, \quad \text{with} \quad x = \begin{pmatrix} x_1 \\ \vdots \\ x_n \end{pmatrix} \quad \text{and} \quad p = \begin{pmatrix} p_1 \\ \vdots \\ p_n \end{pmatrix}$$

denoting the symplectic coordinates of an arbitrary point in L.

Theorem 7

(a) *Suppose that L is an $\vec{H}$-invariant Lagrangian. Then $L \cap L_v$ is $\vec{H}$-invariant, with L_v equal to the vertical Lagrangian $\{0\} \times M^*$.*
(b) *If H corresponds to a regular linear quadratic problem, then every $\vec{H}$-invariant Lagrangian is transversal to L_v.*

Proof Assume that L is an $\vec{H}$-invariant Lagrangian. In a symplectic basis,

$$\vec{H} = \begin{pmatrix} Q & P \\ -R & -Q^* \end{pmatrix} \quad \text{and} \quad w((x, p), (y, q)) = x \cdot q - y \cdot p.$$

Because L is isotropic, $\omega((x, p), (y, q)) = 0$ for all (x, p) and (y, q) in L. Moreover, $\omega((x, p), \vec{H}(y, q)) = 0$ for all (x, p) and (y, q) in L because of the invariance property of $\vec{H}$. In particular, if $(y, q) \in L \cap L_v$, then $y = 0$. Hence

$$\vec{H}\begin{pmatrix} 0 \\ q \end{pmatrix} = \begin{pmatrix} Pq \\ -Q^*q \end{pmatrix},$$

and therefore

$$0 = \omega((x, p), \vec{H}(0, q)) = -x \cdot Q^*q - Pq \cdot p.$$

Setting $(x, p) = (0, q)$, we get that $Pq \cdot q = 0$. But then $Pq = 0$, because $P \geq 0$. Therefore,

$$A\begin{pmatrix} 0 \\ q \end{pmatrix} = \begin{pmatrix} 0 \\ -Q^*q \end{pmatrix},$$

and hence $\vec{H}(L \cap L_v) \subset L \cap L_v$. Part (a) is now proved.

In order to prove (b), assume that L is any Lagrangian that is $\vec{H}$-invariant. By part (a), $L \cap L_v$ is $\vec{H}$-invariant, and therefore for any point $(0, q)$ in $L \cap L_v$, $(\exp t\vec{H})(0, q) \in L \cap L_v$. Consequently,

$$(\exp tA)\begin{pmatrix} 0 \\ q \end{pmatrix} = \begin{pmatrix} \phi_{12}q \\ \phi_{22}q \end{pmatrix} = \begin{pmatrix} 0 \\ \phi_{22}q \end{pmatrix} \quad \text{for all } t.$$

Denoting by $(x(t), p(t))$ the extremal curve that satisfies $x(0) = 0$, $p(0) = q$, the preceding argument shows that $x(t) = 0$. For the regular optimal problem, $x(t) = 0$ can obtain only if $p(t) = 0$. But then $q = 0$, and therefore $L \cap L_v = 0$. End of the proof. ∎

Theorem 8 *Suppose that $\vec{H}$ has no eigenvalues on the imaginary axis. Let L^+ and L^- denote the unstable and the stable subspaces $\vec{H}$ in $M \times M^*$. Then both L^+ and L^- are $\vec{H}$-invariant Lagrangian spaces.*

The notion of stable and unstable spaces is taken from the theory of dynamical systems. For their definitions, see Hirsch and Smale (1974). It may be appropriate, first, to remind the reader of some basic facts in symplectic linear algebra:

(i) If λ is an eigenvalue of a symplectic matrix A, then $-\lambda$ is also an eigenvalue of A. The argument is as follows:
Because A is symplectic, $A^*J + JA = 0$. If $Av = \lambda v$, then $A^*(Jv) = -\lambda Jv$. Therefore $-\lambda$ is an eigenvalue for the transpose of A. Hence $-\lambda$ is an eigenvalue for A, because the eigenvalues of A and A^* coincide.

(ii) Suppose that λ and μ are any eigenvalues of A, with corresponding eigen-vectors v_λ and v_μ. Then $(\exp tA)(v_\lambda) = e^{\lambda t}v_\lambda$, and $(\exp tA)(v_\mu) = e^{\mu t}v_\mu$, and therefore

$$\omega(v_\lambda, v_\mu) = \omega((\exp tA)(v_\lambda), \text{ then } (\exp tA)(v_\mu))$$
$$= \omega(e^{\lambda t}v_\lambda, e^{\mu t}v_\mu) = e^{(\lambda+\mu)t}\omega(v_\lambda, v_\mu).$$

So $\omega(v_\lambda, v_\mu) = 0$ unless $\lambda = -\mu$.

This argument would suffice if the eigenvalues of A were real and distinct. In general, A may not be a semisimple matrix with real eigenvalues, and the proof will be more involved.

Proof Because A has no imaginary eigenvalues, e^{tA} is a linear hyperbolic flow. That means that there are A-invariant linear subspaces V_s and V_u of V, with the following properties:

(i) $V = V_s \oplus V_u$, and
(ii) the restriction of A to V_s is a contraction, and the restriction of A to V_u is an expansion.

The reader can find the proof of this theorem in the work of Hirsch and Smale (1974). It follows from (i) that all V_s and V_u have the same dimension. Therefore $\dim V_s = \dim V_u = \dim M$. It remains to show that each of these spaces is isotropic.

Let v and w be any points in V_s. Then $\omega(v, w) = \omega((\exp tA)(v), (\exp tA)(w))$, because A is symplectic. But then

$$\omega(v, w) = \lim_{t \to \infty} \omega((\exp tA)(v), (\exp tA)(w)) = \omega(0, 0) = 0.$$

The same argument applies to the elements in V_u, but with t replaced by $-t$. End of the proof. ∎

4 Infinite-horizon optimal problems

For the remaining part of this section we shall consider the problem of mini-mizing an integral quadratic cost on the interval $[0, \infty)$ over the trajectories of a linear system. As in Chapter 7, $c(x, u) = \frac{1}{2}(Pu, u) + (Qx, u) + \frac{1}{2}(Rx, x)$, and $dx/dt = Ax + Bu$. It will be assumed that $P > 0$. Then H will denote the quadratic Hamiltonian

$$H(x, p) = \frac{1}{2}(\tilde{P}p, p) + (\tilde{Q}x, p) + \frac{1}{2}(\tilde{R}x, x)$$

obtained by setting $u = P^{-1}(B^* p - Qx)$. The corresponding Riccati equation is given by

$$\frac{dS}{dt} + S\tilde{P}S + \tilde{Q}^*S + SQ + \tilde{R} = 0.$$

Definition 6 A trajectory (x, u) defined on $[0, \infty)$ will be called admissible if both

$$u \in L^2([0, \infty), U) \quad \text{and} \quad x \in L^2([0, \infty), M)$$

and the improper integral $\int_0^\infty c(x(t), u(t)) \, dt$ exists and is finite.

Definition 7 An admissible trajectory $(\bar{x}, \bar{u})$ is optimal if

$$\int_0^\infty c(\bar{x}(t), \bar{u}(t)) \, dt \le \int_0^\infty c(x(t), u(t)) \, dt$$

for any other admissible trajectory (x, u) that satisfies $x(0) = \bar{x}(0)$.

The problem of determining optimal trajectories on $[0, \infty)$ will be called the *infinite-horizon problem*. The next theorem provides conditions for the existence of solutions.

Theorem 9

(a) *The infinite-horizon problem admits solutions if and only if $\int_0^\infty c(x(t), u(t))$ $dt \ge 0$ for any admissible trajectory (x, u) such that $x(0) = 0$.*

(b) *$\int_0^\infty c(x(t), u(t)) \, dt \ge 0$ for any admissible trajectory (x, u), with $x(0) = 0$, if and only if the two-point boundary-value problem has solutions on every finite interval $[0, T]$.*

Proof If $\int_0^\infty c(x(t), u(t)) \, dt \ge 0$ for all admissible trajectories with $x(0) = 0$, then $x = 0$, $u = 0$ is an optimal solution. Conversely, if $(\bar{x}, \bar{u})$ is an optimal solution and if (x, u) is any other admissible solution such that $x(0) = 0$, then $(\bar{x} + \rho x, \bar{u} + \rho u)$ is also admissible for each number ρ. As in the proof of Theorem 1 in Chapter 7, large values for ρ force $\int_0^\infty c(x(t), u(t)) \, dt \ge 0$.

We shall now prove (b). If $\int_0^T c(x(t), u(t)) \, dt < 0$ for some trajectory (x, u) on a finite interval $[0, T]$, with $x(0) = x(T) = 0$, then this trajectory concatenated with the zero trajectory on the interval $[T, \infty)$ produces an admissible trajectory on $[0, \infty)$ that originates at zero and whose total cost is negative.

To prove the converse, assume that $\int_0^\infty c(x(t), u(t)) \, dt = -\varepsilon$ for some admissible trajectory (x, u) with $x(0) = 0$ and $\epsilon > 0$. Let Ω be a neighborhood of the origin such that for any point y in Ω there exists a trajectory (x_y, u_y) defined on an interval $[0, \delta]$ such that $x_y(0) = y$, $x_y(\delta) = 0$, and $\int_0^\delta c(x_y, u_y) \, dt < \varepsilon/2$. We can assume that $\delta > 0$ is a fixed number. Let $T > 0$ be any number such that $x(T) \in \Omega$ and $\int_0^T c(x(t), u(t)) \, dt < -\varepsilon/2$.

Denote by (x_T, u_T) the trajectory in the interval $[0, \delta]$ such that $x_T(0) = x(T), x_T(\delta) = 0$, and $\int_0^\delta c(x_T(t), \, u_T(t)) \, dt < \varepsilon/2$. Define the trajectory (y, v) on $[0, T + \delta]$ such that $v(t) = u(t)$ for $0 \le t \le T$, $v(t) = u_T(t - T)$ for t in $[T, T+\delta], y(t) = x(t)$ for $0 \le t \le T$, and $y(t) = x_T(t - T)$ for $T \le t \le T+\delta$. Then

$$\int_0^{T+\delta} c(y(t), v(t)) \, dt < 0 \quad \text{and} \quad y(0) = 0 = y(T + \delta).$$

Hence the two-point boundary-value problem does not admit solutions in the interval $[0, T + \delta]$. End of the proof. ∎

Corollary 4 *The infinite-horizon problem admits solutions if and only if the two-point boundary-value problem has no conjugate points on any finite interval* $[0, T]$.

Theorem 10 *Assume that the infinite-horizon problem admits solutions, and also assume that $\vec{H}$ has no eigenvalues on the imaginary axis.*

(a) *Then for each $a \in M$ there is a unique admissible optimal trajectory $(\bar{x}, \bar{u})$ such that $\bar{x}(0) = a$.*

(b) *$\bar{x}(t)$ is the projection of a unique extremal curve $(\bar{x}(t), \bar{p}(t))$ of $\vec{H}$ such that $(a, \bar{p}(0))$ belongs to the stable Lagrangian L^+, and $\bar{u}(t) = P^{-1}(B^* \bar{p}(t) - Q\bar{x}(t))$ for all t.*

(c) *The optimal cost is given by $\int_0^\infty c(\bar{x}(t), \bar{u}(t)) \, dt = -\frac{1}{2}(\bar{p}(0), \bar{x}(0)) = -\frac{1}{2}(\hat{S}a, a)$. $\hat{S}$ is the solution of the algebraic Riccati equation whose graph is equal to the stable Lagrangian L^+.*

Proof L^+ is an $\vec{H}$-invariant Lagrangian that is transversal to L_v. If π denotes the projection of L^+ on M, then $\ker \pi = 0$ (because of transversality with L_v). Thus π is a bijection.

If a in M is given, let $\sigma(t) = (\bar{x}(t), \bar{p}(t))$ be the integral curve of $\vec{H}$, with $\sigma(0) = \pi^{-1}(a)$, and let $\bar{u}(t) = P^{-1}(B^* \bar{p}(t) - Q\bar{x}(t))$. Then for each finite time T,

$$\int_0^T c(\bar{x}(t), \bar{p}(t))\, dt = \frac{1}{2}\bar{p}(T)\bar{x}(T) - \frac{1}{2}\bar{p}(0)\bar{x}(0),$$

and therefore

$$\lim_{T\to\infty} \int_0^T c(\bar{x}(t), \bar{p}(t))\, dt = -\frac{1}{2}\bar{p}(0)\bar{x}(0).$$

Because L^+ is transversal to L_v, there exists a symmetric matrix $\hat{S}$ whose graph is equal to L^+. Then $\pi^{-1}(a) = (a, \hat{S}a)$, and therefore $\bar{p}(0) = \hat{S}a$. Thus, $\int_0^\infty c(\bar{x}, \bar{u})\, dt = -\frac{1}{2}(\hat{S}a, a)$.

It remains to show that $(\bar{x}, \bar{u})$ is optimal on $[0, \infty)$: Let (x, u) be any admissible trajectory, with $x(0) = a$. Then

$$\int_0^T c(x, u)\, dt - \int_0^T c(\bar{x}, \bar{u})\, dt = \frac{1}{2} \int_0^T (P(u(t) - P^{-1}(B^*\hat{S} - Q)x(t)),$$

$$u(t) - P^{-1}(B^*\hat{S} - Q)x(t))\, dt$$

$$+ \frac{1}{2}(\hat{S}x(T), x(T)) - \frac{1}{2}(\hat{S}\bar{x}(T), \bar{x}(T)).$$

Evidently,

$$\lim_{T\to\infty} \left(\int_0^T c(x(t), u(t))\, dt - \int_0^T c(\bar{x}(t), \bar{u}(t))\, dt \right) > 0$$

unless $u(t) = P^{-1}(B^*\hat{S} - Q)x(t)$. But then $x(t) = \bar{x}(t)$, because $\bar{x} - x$ is the solution of a linear symplectic differential system with zero initial condition. The proof is now finished. ∎

It may be worthwhile to summarize the preceding results in terms of the following notions.

Definition 8 We denote $\inf\{\int_0^T c(x(t), u(t))\, dt : (x, u) \in \mathrm{Tr}(a, b, T)\}$ by $v(a, b, T)$, where v is called the optimal-value function for the time interval T. The optimal-value function for the infinite-horizon problem will be denoted by $v_\infty(a)$. It is equal to $\inf\{\int_0^\infty c(x(t), u(t)) : (x, u)$ admissible and $x(0) = a\}$.

The next theorem summarizes the precise relationships between the solutions of the Riccati equation, the value functions, and the extremal curves under the conditions of Theorem 10 (for which it is also implicitly assumed that $P > 0$).

Theorem 11

(a) $v(a, b, T) = \frac{1}{2}p(T)x(T) - \frac{1}{2}p(0)x(0)$, with $(x(t), p(t))$ denoting the extremal curve that satisfies $x(0) = a$, $x(T) = b$.

(b) Let $S(t)$ be any solution of the Riccati equation defined on an interval $[0, T]$. Then $v(a, b, T) = \frac{1}{2}(S(T)b, b) - \frac{1}{2}(S(0)a, a)$ for each boundary-value problem for which the corresponding extremal curve $p(t)$ satisfies $p(t) = S(t)x(t)$.

(c) $\lim_{T \to \infty} v(a, 0, T) = v_\infty(a)$ for all a in M.

Proof The first statement depends on the homogeneity properties of the system Hamiltonian $H(x, p, u) = -c(x, u) + p(Ax + Bu)$. The degree of homogeneity of H is 2, and therefore $2H(x, p, u) = (\partial H/\partial x)x + (\partial H/\partial p)p + (\partial H/\partial u)u$, by the Euler identity. Along an extremal triple, $\partial H/\partial u = 0, dp/dt = -\partial H/\partial x$, and $dx/dt = \partial H/\partial p$. Therefore,

$$\int_0^T c(x(t), u(t))\, dt$$

$$= \int_0^T (p(t)(Ax(t) + Bu(t)) - H(x(t), p(t), u(t)))\, dt$$

$$= \int_0^T \left(p(t)\frac{dx}{dt} - \frac{1}{2}\left(\frac{\partial H}{\partial x}x(t) + \frac{\partial H}{\partial p}p(t) + \frac{\partial H}{\partial u}u(t) \right) \right) dt$$

$$= \int_0^T \left(p(t)\frac{dx}{dt} + \frac{1}{2}\frac{dp}{dt}x(t) - \frac{1}{2}p(t)\frac{dx}{dt} \right) dt$$

$$= \frac{1}{2}\int_0^T \left(p(t)\frac{dx}{dt} + \frac{dp}{dt}x(t) \right) dt = \frac{1}{2}\int_0^T \frac{d}{dt}(p(t)x(t))\, dt$$

$$= \frac{1}{2}p(t)x(t) - \frac{1}{2}p(0)x(0).$$

Statement (b) follows from (a) by the substitution $p(t) = S(t)x(t)$, and statement (c) follows by an argument similar to the one used in the proof of Theorem 9. End of the proof. ∎

The Riccati equation also makes a link with another cornerstone of the classic calculus of variations called the Hamilton-Jacobi theory. The Hamilton-Jacobi theory and its relevance to optimality will be discussed in greater detail later in the text, but this seems to be a natural place to introduce the main ideas.

Theorem 12

(a) $(\partial/\partial t)v(a, b, t) = -H(b, p_t(t))$ *is valid for any $t > 0$, with the understanding that $p_t(s)$ is the projection of the extremal curve $(x_t(s), (p_t(s))$ such that $x_t(0) = a$ and $x_t(t) = b$.*

(b) *The function $V(x, t) = v(a, x, t)$ is a solution of the Hamilton-Jacobi partial differential equation $\partial V/\partial t + H(x, \partial V/\partial x) = 0$ in a neighborhood of any point a in M with $t > 0$.*

Proof Let $x_t(s) = x(a, q(t), s)$ and $p_t(s) = p(a, q(t), s)$ denote the extremal curve such that

$$x(a, q(t), t) = b.$$

Then

$$2v(a, b, t) = p_t(t)x_t(t) - p_t(0)x_t(0) = p_t(t)b - q(t)a,$$

and therefore

$$2\frac{\partial}{\partial t}v(a, b, t) = -\frac{\partial H}{\partial x}(b, p_t(t))b + p\left(0, \frac{dq}{dt}, t\right)b - \frac{dq}{dt}a.$$

Differentiating $x(a, q(t), t) = b$ leads to

$$\frac{\partial H}{\partial p}(b, p_t(t)) + x\left(0, \frac{dq}{dt}, t\right) = 0$$

or

$$p(t)\frac{\partial H}{\partial p}(b, p_t(t)) = -p(t)x\left(0, \frac{dq}{dt}, t\right).$$

Because the symplectic form is invariant under symplectic transformations, $\omega((\exp t\vec{H})(\xi), (\exp t\vec{H})(\eta)) = \omega(\xi, \eta)$ for all ξ and η in $M \times M^*$. Taking $\xi = (a, q(t))$ and $\eta = (0, dq/dt)$, we get

$$p\left(0, \frac{dq}{dt}, t\right)b - p(t)x\left(0, \frac{dq}{dt}, t\right) = \frac{dq}{dt}a.$$

Thus,

$$p\left(0, \frac{dq}{dt}, t\right)b - \frac{dq}{dt}a = p(t)x\left(0, \frac{dq}{dt}, t\right) = -p(t)\frac{\partial H}{\partial p}(b, p_t(t)),$$

and therefore

$$2\frac{\partial v}{\partial t}(a, b, t) = -\frac{\partial H}{\partial x}b - \frac{\partial H}{\partial p}p = -2H(b, p_t(t)),$$

and part (a) is proved.

In order to prove (b), let $\bar{x}_{(x,t)}(s)$ and $\bar{p}_{(x,t)}(s)$ denote the extremal curves for which $\bar{x}_{(x,t)}(0) = a$ and $\bar{x}_{(x,t)}(t) = x$. We shall show that $(\partial V/\partial x)(x, t) = \bar{p}_{(x,t)}(t)$, from which statement (b) becomes an obvious consequence of (a). Denote $\exp t\vec{H}$ by block matrices

$$\begin{pmatrix} \Phi_{11}(t) & \Phi_{12}(t) \\ \Phi_{21}(t) & \Phi_{22}(t) \end{pmatrix}.$$

Then

$$x = \Phi_{11}(t)a + \Phi_{12}(t)q \quad \text{and} \quad \bar{p}_{(x,t)}(t) = \Phi_{21}(t)a + \Phi_{22}(t)q,$$

and therefore

$$I = \Phi_{12}(t)\frac{\partial q}{\partial x}, \qquad \frac{\partial \bar{p}_{(x,t)}}{\partial x}(t) = \Phi_{22}(t)\frac{\partial q}{\partial x},$$

$$\frac{\partial V}{\partial x} = \frac{\partial}{\partial x}v(a, x, t) = \frac{\partial}{\partial x}\left(\frac{1}{2}\bar{p}_{(x,t)}(t)x - \frac{1}{2}qa\right)$$

$$= \frac{1}{2}\left(\frac{\partial \bar{p}_{(x,t)}}{\partial x}\right)^* x + \frac{1}{2}\bar{p}_{(x,t)}(t) - \frac{1}{2}\left(\frac{\partial q}{\partial x}\right)^* a.$$

It follows from the foregoing that $\partial \bar{p}_{(x,t)}/\partial x = \Phi_{22}\Phi_{12}^{-1}$, and therefore $\frac{1}{2}(\partial \bar{p}_{(x,t)}/\partial x)^*x = \frac{1}{2}(\Phi_{12}^*)^{-1}\Phi_{22}^*(\Phi_{11}a + \Phi_{12}q)$. Being symplectic, matrices Φ_{ij} satisfy

$$\Phi_{11}^*\Phi_{21} - \Phi_{21}^*\Phi_{11} = 0, \quad \Phi_{11}^*\Phi_{22} - \Phi_{21}^*\Phi_{12} = I, \quad \Phi_{12}^*\Phi_{22} - \Phi_{22}^*\Phi_{12} = 0.$$

Therefore,

$$(\Phi_{12}^*)^{-1}\Phi_{22}^*\Phi_{11} = (\Phi_{12}^*)^{-1}(I + \Phi_{12}^*\Phi_{21}) = (\Phi_{12}^*)^{-1} + \Phi_{21},$$

and $(\Phi_{12}^*)^{-1}\Phi_{22}^*\Phi_{12} = \Phi_{22}$. But then $\frac{1}{2}(\partial \bar{p}_{(x,t)}/\partial x)^*x + \frac{1}{2}\bar{p}_{(x,t)} - \frac{1}{2}(\partial q/\partial x)a = \bar{p}_{(x,t)}(t)$, and our proof is finished. ∎

5 Hardy-Littlewood inequalities

The classic problem of interpolating the $L^2([0, \infty))$ norm of an intermediate derivative $f^{(k)}$ in terms of the $L^2([0, \infty))$ norms of f and of the highest derivative $f^{(n)}$ has a natural formulation as a linear quadratic infinite-horizon problem, and its solutions illustrate both the power and the beauty of the preceding theory. The classic results, which we shall investigate by control-theory means, are given by the following theorem.

Theorem 13 *Suppose that f is any function in $L^2([0, \infty))$ such that its nth derivative $f^{(n)}$ is also in $L^2([0, \infty))$. Then each intermediate derivative $f^{(k)}$ is in $L^2([0, \infty))$, and there exists a constant $M_{n,k}$ such that*

$$\| f^{(k)} \| \leq M_{n,k} \| f \|^{((n-k)/n)} \| f^{(n)} \|^{(k/n)}.$$

In this notation, $\| \cdot \|$ denotes the norm in $L^2([0, \infty))$.

These results derive from the bounds on the quadratic form $\int_0^\infty (f^2 - (f^{(k)})^2 + (f^{(n)})^2) \, dt$.

As we saw in Chapter 7, each function f and its successive derivatives $f^{(1)}, \ldots, f^{(n-1)}$ define a curve $x(t)$ in $\mathbb{R}^n$ whose ith coordinate x_i is equal to $f^{(i-1)}$. Regarding the highest derivative $f^{(n)}$ as the control function, then $x(t)$ is a solution curve of a single-input linear system in Brunovsky form:

$$\frac{dx}{dt} = Ax + bu, \quad \text{with} \quad A = \begin{pmatrix} 0 & 1 & \cdots & \cdots & 0 & 0 \\ 0 & 0 & 1 & \cdots & \cdots & \vdots \\ \vdots & & & \ddots & & \vdots \\ \vdots & & & & & 1 \\ 0 & \cdots & \cdots & \cdots & \cdots & 0 \end{pmatrix}, \quad b = \begin{pmatrix} 0 \\ \vdots \\ 0 \\ 1 \end{pmatrix}.$$

The integral $\frac{1}{2} \int_0^T (f^2 - (f^{(k)})^2 + (f^{(n)})^2) \, dt$ is the same as the integral $\frac{1}{2} \int_0^T (u^2 - x_{k+1}^2 + x_1^2) \, dt$. We shall now consider the associated infinite-horizon problem defined by the foregoing data, which will be referred to as the Hardy-Littlewood problem.

As was shown in Chapter 7, the Hardy-Littlewood problem leads to the Hamiltonian

$$H = \frac{1}{2} p_n^2 + p_1 x_2 + \cdots + p_{n-1} x_n + \frac{1}{2} (x_{k+1}^2 - x_1^2).$$

The associated Hamiltonian vector field $\vec{H}$ is described by the matrix

$$\begin{pmatrix} A & \tilde{P} \\ -\tilde{R} & -A^* \end{pmatrix},$$

with

$$\tilde{P} = \begin{pmatrix} 0 & \cdots & 0 \\ \vdots & & \vdots \\ 0 & \cdots & 1 \end{pmatrix} \quad \text{and} \quad \tilde{R} = \begin{pmatrix} 1 & 0 & \cdots & \cdots & \cdots & 0 \\ 0 & 0 & & & & \vdots \\ \vdots & & \ddots & & & \vdots \\ 0 & \cdots & \cdots & -1 & & \vdots \\ \vdots & & & & 0 & \vdots \\ 0 & \cdots & \cdots & \cdots & \cdots & 0 \end{pmatrix}.$$

Recall that the characteristic polynomial of $\vec{H}$ is given by $(-1)^n\lambda^{2n}+(-1)^{k+1}\lambda^{2k}$ $+1$ and that it has no imaginary roots (Theorem 9, Chapter 7). The Hardy-Littlewood problem is well defined because of the following result.

Theorem 14 $\int_0^T (f^2 - (f^{(k)})^2 + (f^{(n)})^2)\, dt \geq 0$ for each $T > 0$ and each $x(t) = (f(t), \ldots, f^{(n-1)}(t))$ such that $x(0) = x(T) = 0$.

Proof Because

$$f^{(k-i)} f^{(k+i)} = \frac{d}{dt}(f^{(k-1-i)} f^{(k+i)}) - f^{(k-i-1)} f^{(k+i+1)},$$

$$\int_0^T (f^{(k-i)} f^{(k+i)})\, dt = -\int_0^T (f^{(k-i-1)} f^{(k+i+1)})\, dt.$$

Putting $i = 0$ in the foregoing expression gives

$$\int_0^T (f^{(k)})^2\, dt = -\int_0^T (f^{(k-1)} f^{(k+1)})\, dt.$$

Proceeding inductively, we arrive at the following formula:

$$\int_0^T (f^{(k)})^2\, dt = (-1)^i \int_0^T (f^{(k-i)} f^{(k+i)})\, dt.$$

If $k \leq n/2$, then

$$\int_0^T (f^{(k)})^2\, dt = (-1)^k \int_0^T (ff^{(2k)})\, dt,$$

and if $k > n/2$, then

$$\int_0^T (f^{(k)})^2 \, dt = (-1)^{n-k} \int_0^T (f^{(2k-n)} f^{(n)}) \, dt.$$

In the first case,

$$\int_0^T (f^2 - (f^{(k)})^2 + (f^{(n)})^2) \, dt$$

$$= \int_0^T (f^2 + (f^{(k)})^2 - 2(-1)^k f f^{(2k)} + (f^{(n)})^2) \, dt$$

$$= \int_0^T (f - (-1)^k f^{(2k)})^2 \, dt + \int_0^T ((f^{(k)})^2 - (f^{(2k)})^2 + (f^{(n)})^2) \, dt,$$

whereas in the second case,

$$\int_0^T (f^2 - (f^{(k)})^2 + (f^{(n)})^2) \, dt$$

$$= \int_0^T (f^2 + (f^{(k)})^2 - 2(-1)^{n-k} f^{(2k-n)} + (f^{(n)})^2) \, dt$$

$$= \int_0^T (f^{(n)} - (-1)^{n-k} f^{(2k-n)})^2 \, dt + \int_0^T (f^2 - (f^{(2k-n)})^2 + (f^{(k)})^2) \, dt.$$

The preceding formulas set up an easy inductive argument on n. We shall leave the details to the reader.   ∎

Theorem 15 *The infinite-horizon problem associated with the Hardy-Littlewood system has a unique solution $(x(t), u(t))$ for each initial point a in $\mathbb{R}^n$. The optimal value $v_\infty(a)$ is given by $-\frac{1}{2}(\hat{S}a, a)$, with $\hat{S}$ equal to the symmetric matrix whose graph is equal to the stable Lagrangian L^+.*

Proof Use Theorems 9 and 10.   ∎

We shall now use Theorem 15 to recover the famous inequality of Hardy and Littlewood: $\|f^{(1)}\|^2 \leq 2\|f\| \, \|f^{(2)}\|$ for any function f in $L^2([0, \infty))$ whose second derivative $f^{(2)}$ is also in $L^2([0, \infty))$.

The characteristic polynomial associated with the Hardy-Littlewood system with $n = 2$ and $k = 1$ is given by $p(\lambda) = \lambda^4 + \lambda^2 + 1$. Its roots are equal to $\lambda, \bar{\lambda}, -\lambda, -\bar{\lambda}$, with $\lambda = -\frac{1}{2} + i\sqrt{3}/2$. It will be convenient to write $\lambda = e^{i\theta}$, $\theta = 2\pi/3$.

We leave it to the reader to show that the normalized eigenvector v_λ is given by

$$v_\lambda = \begin{pmatrix} 1 \\ \lambda \\ \frac{1}{\lambda} \\ \lambda^2 \end{pmatrix}.$$

Because $|\lambda| = 1$, it follows that $1/\lambda = \bar{\lambda}$. Moreover, $\lambda^2 = \bar{\lambda}$, because $\lambda^3 = 1$. Therefore,

$$v_\lambda = \begin{pmatrix} 1 \\ \lambda \\ \bar{\lambda} \\ \bar{\lambda} \end{pmatrix}.$$

The stable Lagrangian L^+ is equal to all points (x_1, x_2, p_1, p_2) in $\mathbb{R}^4$ such that $p_1 = p_2 = -(x_1 + x_2)$. The argument is as follows: L^+ is the real part of the complex linear span of v_λ and $v_{\bar{\lambda}}$. That is,

$$x_1 = \text{Re}(\alpha + \beta), \qquad x_2 = \text{Re}(\alpha\lambda + \beta\bar{\lambda}), \qquad p_1 = p_2 = \text{Re}(\alpha\bar{\lambda} + \beta\lambda)$$

for complex numbers $\alpha = \alpha_1 + i\alpha_2$ and $\beta = \beta_1 + i\beta_2$. Therefore,

$$x_1 = \alpha_1 + \beta_1, \qquad x_2 = -\frac{\alpha_1}{2} - \frac{\alpha_2}{2}\sqrt{3} - \frac{\beta_1}{2} + \frac{\beta_2}{2}\sqrt{3}$$

$$= -\frac{1}{2}(\alpha_1 + \beta_1) - \frac{\sqrt{3}}{2}(\alpha_2 - \beta_2),$$

$$p_1 = -\frac{\alpha_1}{2} + \frac{\alpha_2}{2}\sqrt{3} - \frac{\beta_1}{2} - \frac{\beta_2}{2}\sqrt{3} = -\frac{1}{2}(\alpha_1 + \beta_1) + \frac{\sqrt{3}}{2}(\alpha_2 - \beta_2).$$

It follows that $-(\sqrt{3}/2)(\alpha_2 - \beta_2) = x_2 + \frac{1}{2}x_1$, and therefore

$$p_1 = -\frac{x_1}{2} - \left(x_2 - \frac{1}{2}x_1\right) = -(x_1 + x_2).$$

Let $\hat{S}$ denote the symmetric matrix whose graph is equal to L^+. Then,

$$\hat{S}\begin{pmatrix} x_1 \\ x_2 \end{pmatrix} = -(x_1 + x_2)\begin{pmatrix} 1 \\ 1 \end{pmatrix} \qquad \text{for all } (x_1, x_2) \text{ in } \mathbb{R}^2,$$

and therefore

$$\hat{S} = \begin{pmatrix} -1 & -1 \\ -1 & -1 \end{pmatrix}.$$

Evidently $\hat{S}$ is negative-semidefinite, with its kernel along the line $x_1 = -x_2$.
Theorem 11 implies that for any admissible trajectory (x, u) on $[0, \infty)$,

$$\frac{1}{2} \int_0^\infty \left(u^2 - x_2^2 + x_1^2 \right) dt \geq \frac{1}{2} (x_1(0) + x_2(0))^2.$$

The minimum value of the preceding integral occurs for the optimal solution
that originates along the line $x_1(0) = -x_2(0)$. Because then $p_1(0) = p_2(0) = -(x_1(0) + x_2(0))$, it follows that $p_1(0) = p_2(0) = 0$.

All solutions that originate in L^+ are of the form $\sigma(t) = \mathrm{Re}(\alpha e^{\lambda t} v_\lambda + \beta e^{\bar{\lambda} t} v_{\bar{\lambda}})$
for some complex numbers α and β. In particular,

$$x_1(t) = \mathrm{Re}(\sigma_1(t)) = \mathrm{Re}\left(e^{-\frac{1}{2}t} (\alpha(\cos\theta t + i\sin\theta t) + \beta(\cos\theta t - i\sin\theta(t))) \right)$$

$$= e^{-\frac{1}{2}t} ((\alpha_1 + \beta_1)\cos\theta t + (\beta_2 - \alpha_2)\sin\theta t)$$

$$= e^{-\frac{1}{2}t} \left(x_1(0)\cos\theta t + \frac{2}{\sqrt{3}} \left(x_2(0) + \frac{1}{2} x_1(0) \right) \sin\theta t \right).$$

For $x_1(0) = -x_2(0)$, $x_2(0) + \frac{1}{2}x_1(0) = -\frac{1}{2}x_1(0)$, and $x_1(t) = e^{-\frac{1}{2}t}(\cos\theta t - (1/\sqrt{3})\sin\theta t)x_1(0)$. Along this solution, the total cost is zero.

The interpolation formula that appears in Theorem 13 is a consequence of
another symmetry consideration:

$$g_\rho(t) = f(\rho t) \quad \text{is in } L_2([0, \infty))$$

for any function f in $L_2([0, \infty))$ for each $\rho > 0$. Hence

$$\int_0^\infty \left(\left(g_\rho^{(2)}(t) \right)^2 - \left(g_\rho^{(1)}(t) \right)^2 + g_\rho(t) \right)^2 dt \geq 0 \quad \text{for any } \rho > 0.$$

It follows that $g_\rho^{(1)}(t) = \rho f^{(1)}(\rho t)$ and $g_\rho^{(2)}(t) = \rho^2 f^{(2)}(\rho t)$. Hence

$$\int_0^\infty (\rho^4 f^{(2)}(\rho t) - \rho^2 f^{(1)}(\rho t) + f^2(\rho t)) \, dt \geq 0,$$

or

$$\rho^3 \int_0^\infty (f^{(2)}(t))^2 \, dt - \rho \int_0^\infty (f^{(1)}(t))^2 \, dt + \frac{1}{\rho} \int_0^\infty f^2(t) \, dt \geq 0.$$

Therefore,

$$\rho^4 \|f^{(2)}\|^2 - \rho^2 \|f^{(1)}\|^2 + \|f\|^2 \geq 0 \quad \text{for any } \rho > 0.$$

The minimum value of $\rho^4 \|f^{(2)}\|^2 - \rho^2 \|f^{(1)}\|^2 + \|f\|^2$ occurs for $\rho^2 = \frac{1}{2}(\|f^{(1)}\|^2 / \|f^{(2)}\|^2)$. For this value of ρ,

$$\rho^4 \|f^{(2)}\|^2 - \rho^2 \|f^{(1)}\|^2 + \|f\|^2 = \frac{1}{4} \frac{\|f^{(1)}\|^4}{\|f^{(2)}\|^2} - \frac{1}{2} \frac{\|f^{(1)}\|^4}{\|f^{(2)}\|^2} + \|f\|^2$$

$$= -\frac{1}{4} \frac{\|f^{(1)}\|^4}{\|f^{(2)}\|^2} + \|f\|^2.$$

Therefore,

$$\|f^{(1)}\|^4 \leq 4 \|f\|^2 \|f^{(2)}\|^2$$

or

$$\|f^{(1)}\|^2 \leq 2 \|f\| \, \|f^{(2)}\|.$$

The equality occurs for $f(t) = Ke^{-\frac{1}{2}t}(\cos\theta t - (1/\sqrt{3})\sin\theta t)$, with K an arbitrary constant and $\theta = 2\pi/3$. This conclusion agrees with the famous result of Hardy-Littlewood, except for one small detail: It still remains to show that $f^{(1)}$ is in $L^2([0,\infty))$ whenever both f and $f^{(2)}$ belong to $L^2([0,\infty))$. Equivalently, it remains to show that $x(t) = (x_1(t), x_2(t))$ belongs to $L^2([0,\infty))$ if and only if x_1 belongs to $L^2([0,\infty))$ (assuming that $dx_2/dt = u$ and that $u \in L^2([0,\infty))$). The remaining argument is as follows:

$$x_1^2(T) - x_1^2(0) = \int_0^T \left(\frac{d}{dt} x_1^2(t)\right) dt = 2\int_0^T \left(x_1(t)\frac{dx_1}{dt}\right) dt$$

$$= 2\int_0^T (x_1(t)x_2(t)) \, dt.$$

Because $x_1 \in L^2([0,\infty))$, $\lim_{T\to\infty} x_1^2(T)$ exists (and is equal to zero). Therefore, $\lim_{T\to\infty} \int_0^T (x_1(t)x_2(t)) \, dt$ exists, and hence $\lim_{t\to\infty} x_1(t)x_2(t) = 0$. But then $\lim_{T\to\infty} \int_0^T (x_1^2(t)) \, dt$ exists, as can be easily seen through the expression $\int_0^T (x_2^2(t)) \, dt = x_1(t)x_2(t)|_{t=0}^{t=T} - \int_0^T (x_1(t)u(t)) \, dt$.

Returning now to the general case $\int_0^\infty ((f^{(n)}(t))^2 - (f^{(k)}(t))^2 + f^2(t)) \, dt$, it follows by an analogous argument that

$$\rho^{2n} \|f^{(n)}\|^2 - \rho^{2k} \|f^{(k)}\|^2 + \|f\|^2 \geq 0$$

for each $\rho > 0$ and each $f \in L^2([0, \infty))$ for which $f^{(n)} \in L^2([0, \infty))$, provided that the Riccati solution $\hat{S}$, which corresponds to the stable Lagrangian L^+, is negative-semidefinite.

Denoting $\rho^{2n} \| f^{(n)} \|^2 - \rho^{2k} \| f^{(k)} \|^2 + \| f \|^2$ by $F(\rho)$, we get that $F'(\rho) = 2n\rho^{2n-1} \| f^{(n)} \|^2 - 2k\rho^{2k-1} \| f^{(k)} \|^2$. $F'(\rho) = 0$ occurs at

$$\rho_{\min}^{2(n-k)} = \frac{k}{n} \frac{\| f^{(k)} \|^2}{\| f^{(n)} \|^2},$$

or

$$\rho_{\min} = \left(\frac{k}{n} \right)^{\frac{1}{2(n-k)}} \left(\frac{\| f^{(k)} \|}{\| f^{(n)} \|} \right)^{\frac{1}{n-k}}$$

Letting $A = \| f \|^2$, $B = \| f^{(k)} \|^2$, and $C = \| f^{(n)} \|^2$, the minimum value for F is given by

$$F(\rho_{\min}) = \left(\frac{k}{n} \right)^{\frac{n}{n-k}} \frac{B^{\frac{n}{n-k}}}{C^{\frac{n}{n-k}}} C - \left(\frac{k}{n} \right)^{\frac{k}{n-k}} \frac{B^{\frac{k}{n-k}}}{C^{\frac{k}{n-k}}} B + A$$

or

$$F(\rho_{\min}) = \left(\frac{k}{n} \right)^{\frac{n}{n-k}} \frac{B^{\frac{n}{n-k}}}{C^{\frac{k}{n-k}}} - \left(\frac{k}{n} \right)^{\frac{k}{n-k}} \frac{B^{\frac{n}{n-k}}}{C^{\frac{k}{n-k}}} + A.$$

Therefore,

$$\frac{B^{\frac{n}{n-k}}}{C^{\frac{k}{n-k}}} \left(\left(\frac{k}{n} \right)^{\frac{k}{n-k}} - \left(\frac{k}{n} \right)^{\frac{n}{n-k}} \right) \leq A.$$

Denoting

$$\left(\left(\frac{k}{n} \right)^{\frac{k}{n-k}} - \left(\frac{k}{n} \right)^{\frac{n}{n-k}} \right)^{-\left(\frac{n-k}{n} \right)}$$

by $M_{n,k}^2$ leads to

$$B \leq M_{n,k}^2 A^{\frac{n-k}{n}} C^{\left(\frac{k}{n-k} \right)\left(\frac{n-k}{n} \right)} = M_{n,k}^2 A^{\frac{n-k}{n}} C^{\frac{k}{n}}.$$

Taking the square root for both sides in the foregoing inequality leads to

$$\| f^{(k)} \| \leq M_{n,k} \| f \|^{\frac{n-k}{n}} \| f^{(n)} \|^{\frac{n-k}{n}}.$$

The equality occurs for functions whose derivatives originate in the kernel of $\hat{S}$.

It seems very likely that the stable Lagrangian of any Hardy-Littlewood system is the graph of a negative-semidefinite matrix, although I do not have a definite proof. The reader may want to investigate this situation in more detail.

Problem 8 Show that each eigenvector v_λ of the Hardy-Littlewood problem, normalized so that $x_1 = 1$, is given by $v_\lambda = (1, x_2, \ldots, x_n, p_1, \ldots, p_n)$, with

$$x_i = \lambda^{i-1}, \qquad\qquad i = 2, \ldots, n,$$
$$p_i = (-1)^{i+1}\lambda^{-i}, \qquad i = 1, 2, \ldots, k,$$

and

$$p_{k+1+i} = (-1)^{n+k+1+i}\lambda^{2n-(k+1+i)} \quad \text{for} \quad 0 \leq i \leq n - (k+1).$$

Problem 9 Find an example of a linear quadratic problem that satisfies the conditions of Theorem 11 and has an indefinite optimal cost $v_\infty(a)$.

Notes and sources

The geometry of Lagrangian Grassmannians is an integral part of symplectic geometry. The reader may consult Agrachev and Gamkrelidze (1978) for additional sources on this topic, as well as for further uses of Lagrangians for control-theory generalizations of Maslov's index. The connection between Lagrangian Grassmannians and the Riccati equation has been known for a long time, although I have not been able to trace all the references. Their use in control dates back to Hermann and Martin (1982). See also Shayman (1986).

The discussion of the infinite-horizon optimal problem is essentially a synthesis of known control-theory results, enriched with symplectic geometry and the material from Chapter 7. The applications to Hardy-Littlewood systems are original. As stated earlier, these applications should be studied in greater detail, with the aim of obtaining sharp estimates (best inequality constants) for higher dimensions.

9

Singular linear quadratic problems

Singular linear quadratic problems have additional invariants, not encountered in the regular case, that need to be incorporated into the theory in order to describe the structure of its optimal trajectories. These invariants are obtained from the qualitative properties of the reachable sets of the cost extended system in $\mathbb{R}^1 \times M$:

$$\frac{dx_0}{dt} = c(x, u), \qquad \frac{dx}{dt} = Ax + Bu. \tag{1}$$

Continuing with the notation from the preceding chapters, the cost $c(x, u)$ will be written as $\frac{1}{2}(Pu, u) + (Qx, u) + \frac{1}{2}(Rx, x)$, with $P \geq 0$. We shall maintain the same assumptions as in Chapter 7, and assume that (1) admits optimal trajectories on an interval $[0, T_{max}]$ and that the linear system $dx/dt = Ax + Bu$ is controllable. Recall that $\ker P \neq 0$ defines the singular case. For such systems, differential system (1) contains a distinguished subsystem

$$\frac{dx_0}{dt} = (Qx, u) + \frac{1}{2}(Rx, x), \qquad \frac{dx}{dt} = Ax + Bu, \qquad u \in \ker P. \tag{2}$$

System (2) is a control affine system; large controls in $\ker P$ override the effect of the quadratic drift vector and in the limit produce trajectories that are integral curves of the control symmetric system

$$\frac{dx_0}{dt} = (Qx, u), \qquad \frac{dx}{dt} = Bu, \qquad u \in \ker P. \tag{3}$$

Even though (3) may not be a subsystem of (2), its trajectories are contained in the closure of the reachable sets of (1). The fact that (3) is a time-reversible system provides extra information about system (1), and its effects need to be systematically exploited and incorporated into the analysis of the extremal trajectories.

In order to take full account of these symmetries, it becomes necessary to
find the largest time-reversible system whose trajectories are in the closure of
the reachable set of the original system, and this task essentially amounts to
computing the strong Lie saturate of (1).

The reader should recall that the strong Lie saturate of a family of vector
fields $\mathcal{F}$, denoted by $\mathcal{LS}_s(\mathcal{F})$, is the largest family of vector fields $\hat{\mathcal{F}}$ contained
in Lie($\mathcal{F}$) such that $\mathrm{cl}(\mathcal{A}_{\hat{\mathcal{F}}}(x, \leq T)) \subseteq \mathrm{cl}(\mathcal{A}_{\mathcal{F}}(x, \leq T))$ for all x in M and all
$T > 0$. The key property of $\mathcal{LS}_s(\mathcal{F})$ is that it is a closed convex body, stable
under its normalizer, and that it contains the Lie algebra generated by any of its
linear subspaces of vector fields (Theorem 11 in Chapter 3).

In the present context, $\mathcal{F}$ is the family of vector fields in $\mathbb{R}^1 \times M$ induced by
(1); that is, for each point $\tilde{x} = (x_0, x)$ in $\mathbb{R}^1 \times M$,

$$\mathcal{F}(x_0, x) = \{c(x, u), Ax + Bu : u \in U\}.$$

Because $\mathcal{F}$ does not depend explicitly on the component x_0, it will be natural to write $\mathcal{F}(x)$ in place of $\mathcal{F}(x_0, x)$. Evidently, each control function u in
$L^\infty([0, T])$ is a limit of piecewise-constant functions, and therefore each tra-
jectory $(x_0(t), x(t))$ of (1) satisfies $(x_0(t), x(t)) \in \mathrm{cl}(\mathcal{A}_{\mathcal{F}}((x_0(0), x(0)), t))$ for
all t. In order not to make any distinction between $\mathcal{F}$ and (1), we need to extend
the notion of a trajectory for a family of vector fields to include all absolutely
continuous curves whose derivative $d\tilde{x}/dt$ belongs to $\mathcal{F}(\tilde{x}(t))$ for almost all t.

Beautiful, but involved, this conceptual framework leads to all key ingredients
required for full understanding of the optimal synthesis for the singular problem.
Apart from their effectiveness for this class of problems, these methods extend
to arbitrary Lie-determined systems and suggest a global geometric procedure
for determining the optimal trajectories for variational problems for which the
strong Legendre condition is not satisfied. For that reason, the material in this
chapter is too important to be ignored.

In spite of the diverse ideas involved, the final results are simple, and as a
way of orienting the reader in the right direction, we shall outline the results
that wait at the end of the chapter.

It will be shown that $\mathcal{LS}_s(\mathcal{F})$ contains a commutative Lie algebra $\mathcal{J}$ of vector
fields V of the form $V(x) = (f(x), d)$, with f a linear function on M, and d
a constant vector field on M. $\mathcal{J}$ is called the space of *jump fields* and includes
all vector fields in (3). The projections of elements in $\mathcal{J}$ on M are called *jump
directions* and form a linear subspace of M (provided that vectors in M are
identified with points of M). This subspace is denoted by J.

The space of jump fields extends the original system (1) to a larger linear
quadratic system called the saturated system. The essential property of the

saturated system is that its optimal-value function coincides with the optimal-value function of the original system. Recall that the optimal-value function $v(a, b, T)$ is equal to inf $\{\int_0^T c(x, u)\, dt : (x, u) \in \text{Tr}(a, b, T)\}$. In contrast to the original system, the saturated system admits optimal solutions for any boundary conditions a, b, and T. The optimal solutions of the saturated systems are the projections of the extremal curves σ_s: Each σ_s is the sum of a regular extremal curve σ and an absolutely continuous curve σ_j that takes values in the vector space of Hamiltonian vector fields defined by the Hamiltonian functions of the jump fields.

Denoting the projections of the initial points of $\sigma_s(0)$ and $\sigma(0)$ on M by a and $\bar{a}$, and the projections of the terminal points $\sigma_s(T)$ and $\sigma(T)$ by b and $\bar{b}$, the optimal-value function is given by

$$v(a, b, T) = \frac{1}{2} S(a - \bar{a})(a + \bar{a}) + \frac{1}{2} S(\bar{b} - b)(b + \bar{b}) + v(\bar{a}, \bar{b}, T)$$

for some symmetric linear mapping $S : J \to M^*$.

The preceding expression for the optimal-value function holds all the information required for the optimal synthesis, at least in the case in which the optimal trajectories are unique, for then we shall show that $\bar{a}$ and $\bar{b}$ are unique points in the cosets $a + J$ and $b + J$, which can be connected by an optimal trajectory of the original system in T units of time, and that $v(\bar{a}, \bar{b}, T)$ is equal to the total cost of this trajectory. $\frac{1}{2} S(a - \bar{a})(a + \bar{a})$ is the cost of the initial jump from a to $\bar{a}$, and $\frac{1}{2} S(\bar{b} - b)(b + \bar{b})$ is the cost of the terminal jump from $\bar{b}$ to b.

This kind of synthesis resembles one discovered in mathematical economics some time ago, known as the synthesis of "turnpike type": The optimal trajectory from $\bar{a}$ to $\bar{b}$ is the appropriate turnpike for the prescribed travel data defined by a, b, and T; $\bar{a} - a$ and $b - \bar{b}$ are its access routes.

1 The structure of the strong Lie saturate

Vector fields V in $\mathbb{R}^1 \times M$ will be denoted by pairs (f, X), with the understanding that at each (x_0, x) in $\mathbb{R}^1 \times M$, $f(x_0, x)$ is the projection of $V(x_0, x)$ on the tangent space of $\mathbb{R}^1$ at x_0, while $X(x_0, x)$ is the projection on $T_x M$. Because the vector fields that define (1) do not depend explicitly on the coordinate x_0, the subsequent discussion will be restricted to those vector fields that are constant over $\mathbb{R}^1$ for each x in M. That means that X can be regarded as a vector field in M and that f is a tangent vector in $\mathbb{R}^1$ for each x in M. On identifying f with $f(x)(\partial / \partial x_0)$, f will be regarded as a function on M. With this identification, the Lie bracket of any vector fields $V(x) = (f(x), X(x))$ and $W(x) = (g(x), Y(x))$ in $\mathbb{R}^1 \times M$ is given by

$$[V, W] = (Yf - Xg, \ [X, Y]).$$

In this notation, $[X, Y]$ denotes the Lie bracket of vector fields in M, and Yf and Xg are the derivations of f along Y and g along X. In particular, if $V(x) = (c(x, u), \ Ax + Bu)$ and $W(x) = (c(x, v), \ Ax + Bv)$, then

$$[V, W](x) = ((Ax + Bv, Q^*u) - (Ax + Bu, Q^*v), AB(v - u)).$$

The function $h(x) = (Ax + Bv, \ Q^*u) - (Ax + Bu, \ Q^*v) = (QAx, u - v) + (QBv, u) - (QBu, v)$ is an affine function on M, and $AB(v - u)$ is a constant vector field on M, provided that both u and v are constant controls.

Therefore $[\mathcal{F}, \mathcal{F}]$ is contained in G_1, the class of all vector fields (f, X) in $\mathbb{R}^1 \times M$ with f an affine function on M, and X a constant vector field on M. The derived algebra of G_1, denoted by G_2, consists of vector fields (f, X) with f a constant function and $X = 0$. Then

$$[\mathcal{F}, G_1] \subset G_1,$$

and therefore

$$[\mathcal{F}, [\mathcal{F}, \mathcal{F}]] \subset G_1.$$

Consequently,

$$[\mathcal{F}, [\mathcal{F}, [\mathcal{F}, \mathcal{F}]]] \subset G_2.$$

Because $[G_2, G_2] = 0$, it follows that each element of $\mathrm{Lie}(\mathcal{F})$ is a linear combination of elements in $\mathcal{F}$ and elements in G_1. The same holds for $\mathcal{LS}_s(\mathcal{F})$, since it is contained in $\mathrm{Lie}(\mathcal{F})$.

Definition 1 The largest vector space contained in the strong Lie saturate of $\mathcal{F}$ is called the space of *jump fields*. The projections of jump fields on the state space M are called *jump directions*. $\mathcal{J}$ will denote the space of jump fields, and J the space of jump directions.

There are several noteworthy properties of jump fields that need to be established first to illuminate the geometric significance of jump fields and their relations to the optimal synthesis.

Theorem 1 *Let V be any jump field of $\mathcal{F}$. Suppose that $\tilde{a}$ and $\tilde{b}$ are arbitrary points on an integral curve of V. Then*

(a) *for any ε > 0 and any neighborhood U of $\tilde{b}$ there exists a trajectory $\tilde{x}$ of*
   $\mathcal{F}$ defined on some interval [0, T], with T ≤ ε, such that $\tilde{x}(0) = \tilde{a}$ and
   $\tilde{x}(T) \in U$, and

(b) *for any ε > 0 and any neighborhood U of $\tilde{a}$ there exists a trajectory $\tilde{x}$*
   of $\mathcal{F}$ defined on an interval [0, T], with T ≤ ε, such that $\tilde{x}(0) \in U$ and
   $\tilde{x}(T) = \tilde{b}$.

Proof Let $\tilde{b} = (\exp \lambda V)(\tilde{a})$ for some real number λ. If $\varepsilon > 0$ is given, then
$(\lambda/\varepsilon)V$ is a jump field, and therefore $\tilde{b} = (\exp \varepsilon((\lambda/\varepsilon)V))(\tilde{a}) \in \mathrm{cl}(\mathcal{A}_{\mathcal{F}}(\tilde{a}, \leq\varepsilon))$,
by the definition of the strong Lie saturate. This means that for any neighborhood
U of $\tilde{b}$ there is a trajectory $\tilde{x}$ of $\mathcal{F}$ defined on some interval $[0, T]$, with $T \leq \varepsilon$,
such that $\tilde{x}(0) = \tilde{a}$ and $\tilde{x}(T) \in U$. Part (a) is proved.

Property (b) is essentially equivalent to saying that $\mathcal{J}$ is also the space of
jump fields for the "time-reversed" family $-\mathcal{F}$. Its proof is as follows:

Let $\varepsilon > 0$ be given, and let U be any neighborhood of $\tilde{a}$. Then $(\exp \lambda V)(U)$
is a neighborhood of $\tilde{b}$. Let $\tilde{c}$ be a point in the interior of $\mathcal{A}_{-\mathcal{F}}(\tilde{b}, \leq\varepsilon/2) \cap$
$(\exp \lambda V)(U)$. It follows from (a) that there exists a trajectory $\tilde{x}$ defined on an
interval $[0, T]$, with $T \leq \varepsilon/2$, such that $\tilde{x}(0) = (\exp -\lambda V)(\tilde{c})$ and $\tilde{x}(T) \in$
$\mathcal{A}_{-\mathcal{F}}(\tilde{b}, \leq\varepsilon/2) \cap (\exp \lambda V)(U)$. Every point of $\mathcal{A}_{-\mathcal{F}}(\tilde{b}, \leq\varepsilon/2)$ can be steered to
$\tilde{b}$ by a trajectory of $\mathcal{F}$ in $\varepsilon/2$ units of time or less, and therefore a concatenation
of such a trajectory with $\tilde{x}$ steers $\tilde{x}(0)$ to $\tilde{b}$ in T units of time, with $T \leq \varepsilon$. The
proof is now finished. ∎

Theorem 1 shows that any two points on an orbit of a jump field can be "almost"
reached by a trajectory of $\mathcal{F}$ in an amount of time that is arbitrarily short
(Figure 9.1).

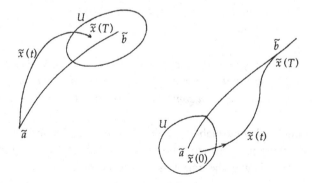

Fig. 9.1.

Our next theorem shows that the space of jump fields is not equal to zero and that it may contain vector fields that are not members of the original family $\mathcal{F}$.

Theorem 2 *Each vector field $((Qx, u), Bu)$, $u \in \ker P$, is a jump field for $\mathcal{F}$.*

Proof $\mathcal{L}S_s(\mathcal{F})$ is a closed convex body, and therefore $(1/|\lambda|)(c(x, u_\lambda), Ax + Bu_\lambda)$ is an element of $\mathcal{L}S_s(\mathcal{F})$ for any $|\lambda| \geq 1$ and any sequence u_λ in U. In particular, if $u \in \ker P$ and $u_\lambda = \lambda u$, then

$$\frac{1}{|\lambda|}(c(x, u_\lambda), \; Ax + Bu_\lambda) = \frac{1}{|\lambda|}(\lambda(Qx, u) + \frac{1}{2}(Rx, x), \; Ax + \lambda Bu).$$

The limit of this one-parameter subfamily of vector fields as $\lambda \to \pm\infty$ is in $\mathcal{L}S_s(\mathcal{F})$, since $\mathcal{L}S_s(\mathcal{F})$ is closed. Therefore, $\pm((Qx, u), Bu)$ is an element of $\mathcal{L}S_s(\mathcal{F})$ for each u in $\ker P$. Because the convex hull of all elements $\pm((Qx, u), Bu)$, with $x \in M$ and $u \in \ker P$, coincides with the vector space spanned by these elements, it follows that each vector field $V(x) = ((Qx, u), Bu)$, $u \in \ker P$, is an element of $\mathcal{I}$. End of the proof. $\blacksquare$

Our next theorem, based on the local controllability properties of linear systems, provides a basic result for "fusing" jump trajectories with the trajectories of the original system. The relevant controllability properties are as follows: Denote by $U_\eta[0, T]$ the space of all control functions u such that $\|u(t)\| \leq \eta$ for almost all t in $[0, T]$. Suppose that $(\bar{x}, \bar{u})$ is any trajectory of a linear controllable system defined on an interval $[0, T]$. Then we have the following:

Property 1 *For any $\eta > 0$ there exist $\alpha > 0$ and a neighborhood Ω of the terminal point $\bar{x}(T)$ with the following property: Any point of Ω can be reached from $\bar{x}(0)$ at any time $T - t$, with $|t| \leq \alpha$, by a trajectory (x, u) for which $(u - \bar{u}) \in U_\eta[0, T - t]$.*

Property 2 *For any $\eta > 0$ there exist $\alpha > 0$ and a neighborhood Ω of the initial point $\bar{x}(0)$ such that any point of Ω can be steered to $\bar{x}(T)$ in $T - t$ units of time, as long as $|t| \leq \alpha$, by a control u that satisfies $(u - \bar{u}) \in U_\eta[0, T - t]$.*

The preceding local controllability properties are illustrated in Figure 9.2. We shall now use these properties to approximate the concatenations of the trajectories of jump fields with the trajectories of the original system. If $V = (f, d)$ and if $(v_0, x) = (\exp V)(0, a)$, then we say that v_0 is the cost of the jump from a to x. Written more explicitly, $v_0 = \int_0^1 f(a + tx)\, dt$, and $x = a + d$.

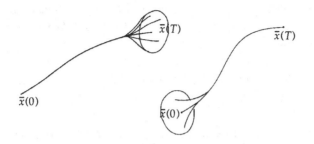

Fig. 9.2.

The following theorem is basic.

Theorem 3 *Let $(\bar{x}_0(t), \bar{x}(t), \bar{u}(t))$ be any trajectory of system (1) defined on an interval $[0, T]$; that is, $\bar{x}_0(T)$ is the total cost of $(\bar{x}(t), \bar{u}(t))$. Suppose that V and W are any jump fields and that $(\exp V)(0, a) = (v_0, \bar{x}(0))$ and $(\exp W)(0, \bar{x}(T)) = (w_0, b)$ for some a and b in M. Then for any $\varepsilon > 0$ and any $\delta > 0$ there exists a trajectory $(x_0(t), x(t), u(t))$ of (1) defined in the interval $[0, T]$ such that*

(a) $x(0) = a, x(T) = b$,
(b) $x(t) = \bar{x}(t)$ *in the interval* $[\delta, T - \delta]$, *and*
(c) $|x_0(T) - (v_0 + w_0 + \bar{x}_0(T))| < \varepsilon$.

Proof Let $\alpha > 0$ and $\eta > 0$, and let Ω_0 be an open neighborhood of the origin in M such that

$$\left| \int_0^t c(\bar{x}(s), \bar{u}(s)) \, ds \right| < \frac{\varepsilon}{8} \quad \text{for all } t \leq \alpha$$

and

$$\left| \int_{t_1}^{t_2} c(x(s), u(s)) \, ds - c(\bar{x}(s), \bar{u}(s)) \, ds \right| < \frac{\varepsilon}{8} \quad \text{for any } t_1 \leq t_2 \leq \alpha$$

and any trajectory (x, u) on the interval $[t_1, t_2]$ such that $(x(t_1) - \bar{x}(t_1)) \in \Omega_0$ and $(u - \bar{u}) \in U_\eta[t, \alpha]$.

Let $\Omega = \bar{x}(0) + \Omega_0$. If necessary, modify α and Ω so that they satisfy the foregoing Property 2 for $(\bar{x}, \bar{u})$ in the interval $[0, \delta]$. According to Theorem 1, there exists a trajectory $(z_0(t), z(t), u_z(t))$ of $\mathcal{F}$ defined on an interval $[0, T_z]$, with $T_z \leq \alpha$, such that $z_0(0) = 0$, $|z_0(T_z) - v_0| < \varepsilon/4$, $z(0) = a$, and $z(T_z) \in \Omega$.

Let $(\hat{x}, \hat{u})$ be a trajectory on the interval $[0, \delta - T_z]$ such that $\hat{x}(0) = z(T_z)$ and $\hat{x}(\delta - T_z) = \bar{x}(\delta)$, with $(u - \bar{u}) \in U_\eta[0, \delta - T_z]$. Define (x, u) on the interval $[0, \delta]$ by

$$x(t) = z(t), \qquad 0 \le t \le T_z, \qquad x(t) = \hat{x}(t - T_z), \qquad T_z \le t \le \delta,$$
$$u(t) = u_z(t), \qquad 0 \le t \le T_z, \qquad u(t) = \hat{u}(t - T_z), \qquad T_z \le t \le \delta.$$

It follows that $x(0) = z(0) = a$ and $x(\delta) = \hat{x}(t - T_z) = \bar{x}(\delta)$. Moreover,

$$\left| \int_0^\delta c(x, u)\, dt - \left(\int_0^\delta c(\bar{x}, \bar{u})\, dt + v_0 \right) \right| \le \left| \int_0^{T_z} c(x, u)\, dt - v_0 \right|$$

$$+ \left| \int_{T_z}^\delta c(x, u)\, dt - \int_{T_z}^\delta c(\bar{x}, \bar{u})\, dt \right| + \left| \int_0^{T_z} c(\bar{x}, \bar{u})\, dt \right| \le \frac{\varepsilon}{4} + \frac{\varepsilon}{8} + \frac{\varepsilon}{8} = \frac{\varepsilon}{2}.$$

Now extend (x, u) to the interval $[0, T - \delta]$ by requiring that $x = \bar{x}$ and $u = \bar{u}$ for all t in $[\delta, T - \delta]$, and then make an analogous argument using Property 1 in the interval $[T - \delta, T]$. These details will be left to the reader. ∎

Theorem 3 justifies this terminology: Any concatenation of a jump trajectory with a regular trajectory can be approximated by a regular trajectory whose total cost is arbitrarily close to the cost of the jump plus the total cost of the regular trajectory (Figure 9.3).

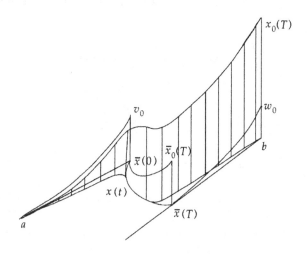

Fig. 9.3.

1.1 The structure of jump fields

We shall ultimately show that each optimal trajectory remains optimal for the saturated system, and this property will be fundamental for the results that follow. We shall first show that the zero trajectory remains optimal for the strong Lie saturate.

Theorem 4 *Let* $\tilde{x}(t) = (x_0(t), x(t))$ *denote a trajectory of the strong Lie saturate of* $\mathcal{F}$ *defined on an interval* $[0, T]$, *with* $T \leq T_{\max}$. *Assume that* $x_0(0) = 0$, $x(0) = 0$, *and* $x(T) = 0$. *Then* $x_0(T) \geq 0$.

Proof Suppose that $x_0(T) = -\alpha$ for a positive number α. There is no loss of generality in assuming that $T < T_{\max}$. Because $\tilde{x}(T) \in \text{cl}(\mathcal{A}_{\mathcal{F}}((x_0(0), x(0)), \leq T))$, it follows that for any neighborhood U of $\tilde{x}(T)$ in $\mathbb{R}^1 \times M$ there exists a trajectory $\tilde{y}$ of $\mathcal{F}$ defined on some interval $[0, \hat{T}]$, with $\hat{T} \leq T$, such that $\tilde{y}(0) = \tilde{x}(0)$ and $\tilde{y}(\hat{T}) \in U$.

Let $\epsilon > 0$ be any positive number, and let Ω_1 be a neighborhood of zero in M such that for any $a \in \Omega_1$ there exists a trajectory (z, u) of $\mathcal{F}$ defined on $[0, \epsilon]$ such that $z(0) = a$, $z(\epsilon) = 0$, and $\int_0^\epsilon c(z, u)\, dt < \alpha/2$. This is a consequence of the local controllability of linear systems.

Let $U = I \times \Omega_1$, with I equal to $\{r : |r + \alpha| < \alpha/2\}$. If $\tilde{y}(t) = (y_0(t), y(t))$ denotes the trajectory of $\mathcal{F}$ such that $\tilde{y}(0) = \tilde{x}(0)$ and $\tilde{y}(T) \in U$, then let (z, u) denote the trajectory of $\mathcal{F}$ defined on $[0, \varepsilon]$ such that $z(0) = y(\hat{T})$, $z(\varepsilon) = 0$, and $\int_0^\varepsilon c(z, u)\, dt < \alpha/2$. Finally, define $w(t) = y(t)$ for t in $[0, \hat{T}]$, and $w(t) = z(t - \hat{T})$ for t in $[\hat{T}, \hat{T} + \varepsilon]$. It follows that $w(0) = w(\hat{T} + \varepsilon) = 0$, and its total cost $w_0(\hat{T} + \varepsilon) = y_0(\hat{T}) + \int_0^\varepsilon c(z, u)\, dt$. Therefore $w_0(\hat{T} + \varepsilon) < 0$ because $|y_0(\hat{T}) + \alpha| < \alpha/2$. This finding contradicts the existence of optimal solutions for (1) in the interval $[0, \hat{T} + \varepsilon]$, and our proof is finished. ∎

Theorem 5 *Any jump field* V *is of the form* (f, d), *with* f *a linear function on* M, *and* d *a vector in* M.

The proof of this theorem is lengthy and will be omitted. Certainly the vector fields that appear in Theorem 2 are of the foregoing form. Because the subsequent constructions will involve only such fields, the lack of a proof for Theorem 5 will not interfere with any of the subsequent results.

Theorem 6 *The space of jump fields of* $\mathcal{F}$ *is a commutative algebra.*

Proof Let $X = (f, a)$ and $Y = (g, b)$ be any jump fields. Assuming that f and g are linear and a and b are constant, then $[X, Y] = (f(b) - g(a), 0)$. Denote $\alpha = f(b) - g(a)$. Because $\rho[X, Y] \in \mathcal{LS}_s(\mathcal{F})$ for any $\rho \in \mathbb{R}^1$, it follows that

$x_0(t) = -\alpha^2 t$, $x(t) = 0$ is a trajectory of the strong Lie saturate whose cost is negative unless $\alpha = 0$. But then the results of Theorem 4 imply that $\alpha = 0$, and our proof is finished. ∎

Theorem 7 *There exists a unique symmetric mapping $S : J \to M^*$ such that each jump field (f, d) is of the form $f = Sd$.*

Proof Let π denote the natural projection from the space of jump fields $\mathcal{J}$ onto the space of jump vectors J. π is a linear mapping whose kernel is zero. The argument is as follows: Suppose that $X = (f, 0)$ is a jump field, with $f(x_0) \neq 0$ for some x_0 in M. But then λX is a jump field for each $\lambda \in \mathbb{R}^1$. Therefore there is no loss of generality in assuming that $f(x_0) < 0$. Let $y_0(t) = \rho^2 f(x_0) t$, $y(t) = x_0$. Then $(y_0(t), y(t))$ is the trajectory of $\rho^2 X$ through $y_0(0) = 0$, $y(0) = x_0$.

Take any regular trajectory (x_1, u_1) that steers the origin to x_0 in time T_1. Also take any regular trajectory (x_2, u_2) that steers x_0 back to the origin in T_2 units of time. The concatenated trajectory

$$x(t) = x_1(t), \qquad 0 \le t \le T_1, \qquad x(t) = x_0, \qquad T_1 \le t \le \varepsilon + T_1,$$

$$x(t) = x_2(t - (T_1 + \varepsilon)), \qquad T_1 + \varepsilon \le t \le T_1 + T_2 + \varepsilon,$$

is a trajectory of the saturated system that satisfies $x(0) = x(T_1 + T_2 + \varepsilon) = 0$. The total cost of this trajectory is given by $\int_0^{T_1} c(x_1, u_1)\, dt + \int_0^{T_2} c_2(x_2, u_2)\, dt + \rho^2 f(x_0)\varepsilon$. This cost is negative for ρ large enough, which is not possible, because of Theorem 4. Therefore $f(x_0) = 0$. Thus $\ker \pi = 0$.

Because π is an injection onto J, it follows that there exists a unique mapping $S : J \to M^*$ such that $Sd = f$ if and only if $(f, d) \in \mathcal{J}$. S must be symmetric, since $\mathcal{J}$ is a commutative Lie algebra. That is, $[(f, c), (g, d)] = (f(d) - g(c), 0) = 0$ if and only if $(Sd, e) = (Se, d)$ for any c and d in J. The proof is now finished. ∎

Corollary 1 *Suppose that $(x_0(t), x(t))$ is an integral curve of a jump field (f, d). Then*

$$x_0(t) - x_0(0) = \frac{1}{2}(S(x(t) - x(0)), \ x(t) + x(0)).$$

Proof

$$x_0(t) - x_0(0) = \int_0^t f(x(0) + sd)\, ds$$

$$= f(x(0))t + \frac{t^2}{2} f(d) = \left(Sd, x(0)t + \frac{t^2}{2} d \right)$$

$$= \left(S dt, x(0) + \frac{t}{2}d\right) = S(x(t) - x(0)), \; x(0) + \frac{x(t) - x(0)}{2}$$

$$= \frac{1}{2}(S(x(t) - x(0)), \; x(t) + x(0))). \qquad \blacksquare$$

Definition 2 For any states a and b in M such that $a - b \in J$, we shall refer to $\frac{1}{2}(S(b - a), a + b)$ as the jump cost from a to b.

Evidently, the jump cost from b to a is equal to the negative of the jump cost from a to b.

Problem 1 Let $c(x, u) = \frac{1}{2}(x_1^2 - x_2 x_4 - x_3^2)$ be the cost associated with $dx_1/dt = x_2$, $dx_2/dt = x_3$, $dx_3/dt = x_4$, and $dx_4/dt = u$. Show that $J = \mathbb{R}^4$ and that the jump cost from any point a to any other point b in $\mathbb{R}^4$ is given by $-\frac{1}{2}(b_2 b_3 - a_2 a_3)$.

Theorem 8

(a) $(Rv, v) - 2(Sv, Av) \geq 0$ *for any jump direction* v.
(b) *If* $(Rv, v) - 2(Sv, Av) = 0$ *for some jump direction* v, *then* Av *is a jump direction. In such a case,* $(Sv, Bu) = (Qv, u)$ *for all* u *in* U.

Proof Let $X_u(x) = (c(x, u), Ax + Bu)$ denote the extended vector field of (1) that corresponds to a constant control u. Denote by V the jump field whose jump direction is v. Then for each real number λ, λV is a jump field, and $(\exp \lambda V)_\#(X_u)$ belongs to the strong Lie saturate of $\mathcal{F}$ for each λ. Written more explicitly,

$$(\exp \lambda V)_\#(X_u)(x) = (\exp \lambda V)_*(X_u)(\exp -\lambda V)(x)$$
$$= (c(x - \lambda v, u) + (S\lambda v, A(x - \lambda v) + Bu),$$
$$A(x - \lambda v) + Bu).$$

The corresponding differential system

$$\frac{dx_0}{dt} = c(x - \lambda v, u) + (S\lambda v, A(x - \lambda v) + Bu),$$
$$\frac{dx}{dt} = Ax - \lambda Av + Bu, \tag{4}$$

is a linear quadratic subsystem of $\mathcal{LS}_s(\mathcal{F})$ with respect to controls λ and u. According to Theorem 4, the total cost of any trajectory of (4) that originates and terminates at zero is not negative, and consequently Tonelli's condition holds for system (4), by Proposition 3 in Chapter 7; that is, $c(-\lambda v, u) + (S\lambda v, -A\lambda v) \geq 0$ for all λ and u. In particular, for $u = 0$ the foregoing inequality reduces to $\lambda^2(\frac{1}{2}(Rv, v) - (Sv, Av)) \geq 0$. Thus (a) is proved.

Assume now that $\frac{1}{2}(Rv, v) - (Sv, Av) = 0$. Then

$$(\exp \lambda V)_\#(X_u)(x)$$

$$= \left(\frac{1}{2}(Pu, u), \ Ax + Bu) + \lambda((Q(x - v), u) + (Sv, Ax + Bu), -Av\right).$$

Because

$$\frac{1}{\lambda}(\exp \alpha \lambda V)_\#(X_u)$$

belongs to $\mathcal{L}S_s(\mathcal{F})$ for $\lambda \geq 1$ and any real number α, $\lim_{\lambda \to \infty} \frac{1}{\lambda}(\exp \alpha \lambda V)_\#(X_u)$ also belongs to $\mathcal{L}S_s(\mathcal{F})$, because the latter is closed.

Thus any scalar multiple of

$$(-(Qv, u) - (Rx, v) + (Sv, Ax + Bu), -Av)$$

belongs to $\mathcal{L}S_s(\mathcal{F})$, and consequently $(-(Qv, u) - (Rx, v) + (Sv, Ax + Bu), -Av)$ is a jump field for any u in U. According to the proof of Theorem 6, the space of jump fields does not contain any elements of the form $(c, 0)$ with c a nonzero constant. This fact easily implies that $(Sv, Bu) - (Qv, u) = 0$ for any u in U, and the proof is finished.   ■

Definition 3 For the remaining part of this chapter, β denotes the quadratic symmetric form on M given by

$$\beta(v, w) = \frac{1}{2}(Rv, w) - \frac{1}{2}((Sv, Aw) + (Sw, Av)).$$

β is called *the extended Legendre form.*

β restricted to J is semipositive-definite, and it vanishes on jump vectors v for which Av is also a jump vector. The results of Theorem 8 suggest a natural filtration for the space of jump directions defined as follows:

Definition 4 For each u in $\ker P$, its index is the smallest integer n such that $\beta(A^n Bu, A^n Bu) = 0$ and $\beta(A^{n+1} Bu, A^{n+1} Bu) > 0$. Then $\mathrm{ind}(u)$ will denote the index of u, with the understanding that $\text{ind}(u) = \infty$ if $\beta(A^k B u, A^k B u) = 0$ for all $k \geq 0$, and $\text{ind}(u) = 0$ if $\beta(Bu, Bu) > 0$.

It follows from Theorem 8 that for each $u \in \ker P$, $\{A^k B u : 0 \leq k \leq \text{ind}(u)\} \subset J$. The index system induces a filtration on $B(\ker P)$ defined as follows: Let i_1 be the smallest finite index. Define $E_1 = \{Bu : u \in \ker P$ such that $\text{ind}(u) > i_1\}$. If i_2 is the next finite index, then $E_2 = \{Bu : u \in \ker P$ such that $\text{ind}(u) > i_2\}$.

For each finite index i_k, define E_k as the set of all $B u$ vectors with u in $\ker P$ such that $\text{ind}(u) > i_k$. E_∞ is the set of all Bu vectors whose indices are infinite. Because β is semipositive-definite on the space of jump vectors, it follows that each E_k is a vector subspace. The argument is as follows:

Let $\sum_{i=1}^r \alpha_i B u_i$ be a linear combination of elements in E_k. Then for each $i = 1, \ldots, r$ and each $0 \leq j \leq i_k + 1$, $A^j B u_i$ is a jump vector, and $\beta(A^j B u_i, A^j B u_i) = 0$. Being semipositive-definite on the space of jump vectors, β satisfies the Cauchy-Schwarz inequality $|\beta(v, w)| \leq \beta(v, v)^{1/2} \beta(w, w)^{1/2}$. Therefore,

$$\left| \beta \left(A^j \sum_{i=1}^r \alpha_i B u_i, \ A^j \sum_{i=1}^r \alpha_i B u_i \right) \right|$$

$$= \left| \beta \left(\sum_{i=1}^r \alpha_i A^j B u_i, \ \sum_{i=1}^r \alpha_i A^j B u_i \right) \right| = \left| \sum_{i,f \leq r} \alpha_i \alpha_f \beta(A^j B u_i, A^j B u_f) \right|$$

$$\leq \sum_{i,f \leq r} |\alpha_i \alpha_f| \beta(A^j B u_i, A^j B u_i)^{\frac{1}{2}} \beta(A^j B u_f, A^j B u_f)^{\frac{1}{2}}$$

$$= 0.$$

Hence $\sum_{r=1}^r \alpha_i B u_i$ belongs to E_k.

The reader should note that the set of elements all having the same index need not be a vector space. The spaces E_k are nested, in the sense that

$$B(\ker P) \supset E_1 \supset E_2 \supset \cdots \supset E_m \supset E_\infty.$$

The results of Theorem 8 can be used to define yet another filtration based on the following definition.

Definition 5 J_∞ is the smallest subspace V of vectors in M that contains $B(\ker P)$ and satisfies $A(V \cap \ker \beta) \subset V$.

It is a corollary of Theorem 8 that $J_\infty \subseteq J$, although we shall ultimately show that $J_\infty = J$. J_∞ is the union of an increasing sequence of subspaces J_k defined inductively as follows:

$J_0 = B(\ker P)$, and J_{k+1} is the linear span of J_k and $\{A(v) : v \in J_k$, and $\beta(v, v) = 0\}$.

We shall come back to these spaces later in this chapter, after having established some more background material.

1.2 The saturated linear quadratic system

The space of jump fields $\mathcal{J}$ defines a natural linear quadratic family $\mathcal{F}_{\text{sat}}$ given by $(\exp \mathcal{J})_\#(\mathcal{F}) + \mathcal{J}$. Such a family will be called the *saturated system of* $\mathcal{F}$. It follows from Theorem 11 in Chapter 3 that $\mathcal{F}_{\text{sat}} \subset \mathcal{LS}_s(\mathcal{F})$. Recall that $(\exp V)_\#(X_u)(x) = (c(x-v), u) + (Sv, A(x-v) + Bu), Ax - Av + Bu)$ for any jump field $V = (Sv, v)$ and any extended vector field $X_u(x) = (c(x, u), Ax + Bu)$ in $\mathcal{F}$. Thus any integral curve $(x_0(t), x(t))$ of $\mathcal{F}_{\text{sat}}$ is a solution of the following differential system:

$$\frac{dx_0}{dt} = c(x(t) - v(t), u(t))$$
$$+ (Sv(t), A(x(t) - v(t)) + Bu(t)) + (Sw(t), x(t)), \quad (5a)$$
$$\frac{dx}{dt} = Ax(t) - Av(t) + w(t) + Bu(t), \quad (5b)$$

for some bounded and integrable controls $u(t)$, $v(t)$, and $w(t)$, with u taking values in U, and v and w taking values in the space of jump vectors J. We shall refer to equations (5) as the saturated system.

The differential system (5) is another linear quadratic system that reduces to the original system when $v = w = 0$. We shall use c_s to denote the quadratic cost of the saturated system:

$$c_s(x, u, v, w) = c((x - v), u) + (Sv, A(x - v) + Bu) + (Sw, x).$$

It follows from Theorem 4 that $\int_0^T c_s(x(t), u(t), v(t), w(t))\, dt \geq 0$ for any trajectory $(x(t), u(t), v(t), w(t))$ of (5b) that satisfies $x(0) = x(T) = 0$, provided that $T \leq T_{\max}$. That is, $x = 0$, $u = 0$, $v = 0$, $w = 0$ is an optimal trajectory for the saturated system. We shall show that the saturated system admits optimal solutions for all the boundary conditions for which the original system did not.

Because the existence of solutions implies Tonelli's condition, it follows that

$$c_s(0, u, v, w) \geq 0 \quad \text{for all} \quad u \in U \quad \text{and } v, \ w \text{ in } J.$$

This means that

$$\frac{1}{2}(Pu, u) - (Qv, u) + \frac{1}{2}(Rv, v) - (Sv, Av) + (Sv, Bu) \geq 0$$

for all u in U and for v and w in J.

In particular, when $u \in \ker P$, $((Qx, u), Bu)$ is a jump field, and therefore $Q^*u = SBu$. Thus $(Qv, u) = (Sv, Bu)$, and the quadratic form c_s reduces to the extended Lagrange form β defined in Section 1.1.

Note that any absolutely continuous curve $z(t)$, such that $dz/dt \in \mathcal{J}(z(t))$, is a trajectory of (5b) that corresponds to $u(t) = 0$ and $v(t) = z(t)$.

Theorem 9 *The saturated system defines the same optimal-value function as the original system. Consequently, each regular optimal curve remains optimal for the saturated system.*

Proof Let $v_s(a, b, T)$ denote the optimal-value function for the saturated system. Evidently, $v_s(a, b, T) \leq v(a, b, T)$. We shall now show that $v_s(a, b, T) \geq v(a, b, T)$. Let a, b be given points in M, and let $T > 0$. Assume first that $(x_s(t), u(t), v(t), w(t))$ is a trajectory of the saturated system that satisfies $x_s(0) = a$ and $x_s(T) = b$ generated by $w(t) = 0$, $v(t)$ a piecewise-constant jump control with discontinuities at $0 < t_0 < t_1 < \cdots < t_n < T$ that takes constant values v_i in each interval $[t_{i-1}, t_i)$, and $u(t)$ an arbitrary integrable control function on $[0, T]$.

In each subinterval (t_{i-1}, t_i), $x_i(t) = x_s(t) - v_i$ is a regular trajectory, since $(dx_i/dt)(t) = Ax_i(t) + Bu(t)$. Extend this trajectory to the entire interval $[t_{i-1}, t_i]$ by requiring that $x_i(t) = x_s(t) - v_i$ hold for all t in that interval. Then, both differences $x_s(t_i) - x_i(t_i)$ and $x_s(t_{i-1}) - x_i(t_{i-1})$ are in the v_i direction. The jump cost associated with the jump from $x_s(t_{i-1})$ to $x_i(t_{i-1})$ is given by $-\frac{1}{2}(Sv_i, x_s(t_{i-1}) + x_i(t_{i-1}))$, and the jump cost associated with the jump from $x_i(t_i)$ back to $x_s(t_i)$ is equal to $\frac{1}{2}(Sv_i, x_s(t_i) + x_i(t_i))$. It follows that the sum of these jump costs is equal to $(Sv, x_s(t_i) - x_s(t_{i-1}))$, because $x_i(t_i) - x_i(t_{i-1}) = x_s(t_i) - x_s(t_{i-1})$.

The cost of x_s in the interval $[t_{i-1}, t_i]$ is given by

$$\int_{t_{i-1}}^{t_i} c_s(x_s(t), u(t), v(t), 0) \, dt = \int_{t_{i-1}}^{t_i} c(x_i(t), u(t)) \, dt + (Sv_i, x_s(t_i) - x_s(t_{i-1})).$$

Therefore, the total cost of x_s in each interval $[t_{i-1}, t_i]$ is equal to the total cost of the regular trajectory (x_i, u) plus the sum of jump costs from $x_s(t_{i-1})$ to $x_i(t_{i-1})$ and from $x_i(t_i)$ back to $x_s(t_i)$. But then it follows from Theorem 3 that there exists a regular trajectory $x(t)$ defined on the interval $[t_{i-1}, t_i]$ that matches the boundary conditions $x(t_{i-1}) = x_s(t_{i-1})$ and $x(t_i) = x_s(t_i)$ and whose total cost is arbitrarily close to $\int_{t_{i-1}}^{t_i} c_s(x_s(t), u(t), v_i, 0) \, dt$.

The concatenation of these regular trajectories produces a regular trajectory on the entire interval $[0, T]$ whose total cost is arbitrarily close to the total cost of x_s. Therefore,

$$\int_0^T c_s(x_s(t), u(t), v(t), 0) \, dt \geq v(a, b, T).$$

We now take an arbitrary jump control v and a regular control u and consider a trajectory x_s of the saturated system that satisfies $x_s(0) = a$ and $x_s(T) = b$. We take any sequence of step functions $\{v_n\}$ such that $\lim_{n \to \infty} v_n(t) = v(t)$ for almost all t in $[0, T]$. The corresponding trajectories $x_n(t)$ generated by u and v_n converge uniformly to the trajectory x_s, and the corresponding costs $\int_0^T c_s(x_n, u, v_n, 0) \, dt$ converge to $\int_0^T c_s(x_s, u, v, 0) \, dt$.

We cannot pass directly to the limit because the approximating trajectories $x_n(t)$ do not necessarily satisfy $x_n(T) = b$ for all n. We shall first make a small perturbation $u + u_n$ of u such that the solution $\bar{x}_n(t)$ of

$$\frac{dx(t)}{dt} = Ax(t) + B(u(t) + u_n(t)) - Av_n(t),$$

with $\bar{x}(0) = a$, satisfies $\bar{x}_n(T) = b$. The existence of such controls follows from the controllability properties of the foregoing system. Then for each n,

$$\int_0^T c_s(\bar{x}_n, u + u_n, v_n, 0) \, dt \geq v(a, b, T),$$

by the preceding arguments. We can now pass to the limit and obtain

$$\int_0^T c_s(x_s(t), u(t), v(t), 0) \, dt \geq v(a, b, T).$$

We leave it to the reader to show that the same arguments can be extended to trajectories generated by an arbitrary jump control w.   ■

Theorem 9 forms a foundation for the optimal synthesis for singular problems. We shall now go to the maximum principle for the remaining information.

2 The maximum principle and its consequences

It was shown in Chapter 7 that the maximum principle for the singular problem determines a family of Hamiltonians $\mathcal{H}$ and the linear space of constraints $\mathcal{L}$ such that $\sigma(t)$ is an extremal curve for the singular problem if and only if $d\sigma/dt \in \vec{\mathcal{H}}(\sigma(t))$ and $\mathcal{L}(\sigma(t)) = 0$ (Theorem 6).

$\mathcal{L}$ is equal to the space of linear forms h_v parametrized by the values v in ker P such that $h_v(x, p) = (p, Bv) - (Qx, v)$ for all (x, p) in $M \times M^*$. Also recall that each element H of $\mathcal{H}$ is of the form $-c(x, f(x, p)) + p(Ax + Bf(x, p))$ for some affine solution $f(x, p)$ of $Pu - B^*p + Qx = 0$, mod($\mathcal{L}$), and finally recall that any two elements in $\mathcal{H}$ differ by a term $\sum \alpha_i h_i$, with each α_i an affine function on $M \times M^*$, and each h_i in $\mathcal{L}$.

Definition 6 Let Ω denote the subset of $M \times M^*$ consisting of all points (x_0, p_0) for which there exists an extremal curve $\sigma(t) = (x(t), p(t))$ defined on some interval $[0, T]$ for which $x(0) = x_0$ and $p(0) = p_0$. Ω is called the *optimal space*.

The optimal space Ω is a linear subspace of $M \times M^*$, because each extremal curve is an integral curve of a linear differential system (see Definition 5 and Theorem 6 in Chapter 7). Our ultimate objective is to obtain the structure of Ω. It will be convenient to assume that all the variables are expressed in terms of the symplectic coordinates relative to a fixed symplectic basis in $M \times M^*$.

Theorem 10

(a) *Suppose that $\sigma(t) = (x(t), p(t))$ is an extremal curve defined on an interval $[0, T]$, with $T \le T_{\max}$. Then*

$$v(x(0), x(T), T) = \frac{1}{2}p(T)x(t) - \frac{1}{2}p(0)x(0).$$

(b) *If $\sigma_1(t) = (x_1(t), p_1(t))$ and $\sigma_2(t) = (x_2(t), p_2(t))$ are any extremal curves on $[0, T]$ for which $x_1(t) = x_2(t)$ for all t in $[0, T]$, then $\sigma_1(t) = \sigma_2(t)$ for all t.*

Proof The proof of (a) is the same as the proof for Theorem 11 in Chapter 8 and will be omitted.

To prove (b), note that the differences of extremal curves are also extremal curves. Therefore, it suffices to prove that any extremal curve $\sigma(t) = (x(t), p(t))$ for which $x(t) = 0$ on the entire interval $[0, T]$ coincides with the zero extremal curve.

Recall that the extremal triples $(x(t), p(t), u(t))$ satisfy the following conditions:

$$\frac{dx}{dt} = Ax(t) + Bu(t), \qquad \frac{dp}{dt} = Rx(t) - A^*p(t) + Q^*u,$$

and

$$-Pu(t) - Qx(t) + B^*p(t) = 0.$$

If $x(t) = 0$, then $dx/dt = 0$, and therefore $Bu(t) = 0$ for almost all t.

Because $-Pu(t) - Qx(t) + B^*p(t) = 0$, it follows that $(u(t), Pu(t)) = (u(t), B^*p(t)) = 0$. This relation implies that $(u(t), Pu(t)) = 0$. Therefore $Pu(t) = 0$ for almost all t, because $P \geq 0$. Because $u(t) \in \ker P$, it follows that $Q^*u(t) = SBu(t)$, as a consequence of Theorem 7. Thus $Q^*u(t) = 0$.

Therefore, $dp/dt = Rx(t) - A^*p(t) + Q^*u(t) = -A^*p(t)$, and $B^*p(t) = 0$. Successive differentiations of $B^*p(t)$ show that $p(0)$ is annihilated by all powers $(A^*)^k B^*$ and therefore must be zero, by the controllability assumption. But then $p(t) = 0$ for all t, and the proof is finished. ∎

Because the saturated system satisfies the existence assumption $\int_0^T c_s(x, u, v, w)\, dt \geq 0$ for any of its trajectories for which $x(0) = x(T) = 0$ for any T $(T \leq T_{\max})$, it follows that its optimal trajectories coincide with the projections of the saturated extremal curves. Our immediate objective is to investigate the structure of these extremals.

The fact that the regular optimal trajectories remain optimal for the saturated system implies that each regular optimal trajectory is the projection of an extremal curve of the saturated system. The corresponding extremal curve of the saturated system coincides with the regular extremal curve, as a consequence of Theorem 10. We shall state this fact explicitly by the following corollary.

Corollary 2 *Suppose that (x, u) is a regular optimal trajectory on an interval $[0, T]$. Let $\sigma(t) = (x(t), p(t))$ denote the corresponding regular extremal curve, and let $\sigma_s(t) = (x(t), p_s(t))$ denote the corresponding saturated extremal curve. Then $\sigma = \sigma_s$.*

Proof

$$\frac{dp}{dt} = \frac{\partial c}{\partial x}(x(t), u(t)) - A^*p(t),$$

and

$$\frac{dp_s}{dt} = \frac{\partial c_s}{\partial x}(x(t), u(t), 0, 0) - A^*p_s(t),$$

subject to the maximality conditions $-Pu(t) - Qx(t) + B^*p(t) = 0$ and
$-Pu(t) - Qx(t) + B^*p_s(t) = 0$. Therefore,

$$\frac{d}{dt}(p_s(t) - p(t)) = A^*(p_s(t) - p(t)),$$

subject to $B^*(p_s(t) - p(t)) = 0$. Thus, $p_s(t) - p(t) = 0$, and consequently
$\sigma = \sigma_s$. ∎

The information provided by the maximum principle for the saturated system
is most naturally extracted through the language of symplectic geometry, and
for that reason it will be necessary to expand the geometric base obtained in the
preceding chapters.

Recall the symplectic form ω_ξ, which at each point ξ in $M \times M^*$ is given by
$\omega_\xi((X, P), (Y, Q)) = \sum_{i=1}^n Q_i X_i - P_i Y_i$.

We shall continue with our convention in which $\vec{H}$ denotes the Hamiltonian
vector field that corresponds to a function H on $M \times M^*$. We then have the
following definition.

Definition 7 The Poisson bracket of two functions f and g on T^*M is a function
$\{f, g\}$ on T^*M defined by $\{f, g\}(\xi) = \omega_\xi(\vec{f}, \vec{g})$ for each ξ in T^*M.

In terms of symplectic coordinates $x_1, \ldots, x_n, \ p_1, \ldots, p_n$,

$$\{f, g\} = \sum_{i=1}^n \frac{\partial f}{\partial x_i} \frac{\partial g}{\partial p_i} - \frac{\partial f}{\partial p_i} \frac{\partial g}{\partial x_i}.$$

That formula shows that the Poisson bracket of homogeneous polynomials f
and g is a homogeneous polynomial whose degree of homogeneity is equal to
$\deg(f) + \deg(g) - 2$. In particular, the Poisson bracket of quadratic functions is
quadratic, and the Poisson bracket of a quadratic function with a linear function
is linear.

The following theorem assembles the classic properties of the Poisson bracket
that will be required for subsequent developments.

Theorem 11 *Suppose that H and G are reduced Hamiltonians of extended
vector fields X and Y on $\mathbb{R}^1 \times M$. Denote by F the reduced Hamiltonian of the
Lie bracket $[X, Y]$. Then*

(a) $F = \{H, G\}$, *and* $\vec{F} = [\vec{H}, \vec{G}]$, *and*

(b) *if* $\sigma(t)$ *is an integral curve of* $\vec{H}$, *then* $(d/dt)G(\sigma(t)) = \{G, H\}(\sigma(t))$ *for all* t.

We shall defer the proof of this theorem to the chapter dealing with the maximum principle on arbitrary manifolds.

Recall that cost extended vector fields are vector fields in $\mathbb{R}^1 \times M$ of the form $V(x) = (f(x), X(x))$. Their reduced Hamiltonians are functions $-V(x) + pX(x)$ for (x, p) in $M \times M^*$. The space of reduced Hamiltonians that correspond to jump fields will be denoted by $\text{Ham}(J)$. Because the form of each jump field V is such that X is constant and f is linear, it follows that its Hamiltonian is a linear function on $M \times M^*$. Elements of $\text{Ham}(J)$ will be called jump Hamiltonians. In particular, the space of all jump Hamiltonians that correspond to $V(x) = ((Qx, u), Bu)$ for some u in $\ker P$ coincides with the linear forms $\mathcal{L}$ that appear as constraints for the regular extremal curves.

Theorem 12 $\text{Ham}(J)$ *is a commutative Poisson algebra contained in the space of linear functions on* $M \times M^*$.

Proof According to Theorem 6, jump fields commute. Therefore their Hamiltonian functions Poisson-commute, as a consequence of (a) in Theorem 11. End of the proof. ∎

We shall use $\text{Ham}(J_\infty)$ to denote the Hamiltonians of jump fields corresponding to J_∞ (Definition 5). This notational distinction reflects our lack of complete a priori knowledge of the structure of the Lie saturate. Ultimately these distinctions are unnecessary, as it will be shown that $J_\infty = J$.

The Hamiltonian lift of the saturated system is obtained through the Hamiltonian

$$H_s(x, p, u, v, w) = -c_s(x, u, v, w) + p(Ax - Av + Bu + w).$$

Recall that the saturated cost c_s is given by

$$c_s(x, u, v, w) = c(x - v, u) + (Sv, A(x - v) + Bu) + (Sw, x),$$

and therefore H_s can be written as $H_s = H(x - v, p - Sv, u) + h_w$, with $H(x, p, u) = -c(x, u) + p(Ax + Bu)$ and $h_w(x, p) = -(Sw, x) + p \cdot w$.

It may also be convenient to write

$$c_s(x, u_s) = \frac{1}{2}(P_s u_s, u_s) + (Q_s x, u_s) + \frac{1}{2}(R_s u_s, u_s)$$

with $u_s = u \oplus v \oplus w$. Then

$$\frac{1}{2}(P_s u_s, u_s) = H_s(0, 0, u_s) = H(-v, -Sv, u).$$

The dynamics of the saturated system are described by

$$\frac{dx(t)}{dt} = Ax(t) + B_s u_s(t), \quad \text{with } B_s u_s(t) = -Av(t) + Bu(t) + w(t).$$

Because the saturated system satisfies the same assumptions as the original system, it follows from Theorem 6 that there exists a family $\mathcal{H}_s$ of quadratic Hamiltonians and a linear space $\mathcal{L}_s$ of linear forms on $M \times M^*$ such that an absolutely continuous curve $\sigma(t)$ in $M \times M^*$ is an extremal curve of the saturated system if and only if $d\sigma(t)/dt \in \vec{\mathcal{H}}_s(\sigma(t))$ for almost all t, and $\mathcal{L}_s(\sigma(t)) = 0$ for all t. Each element of $\mathcal{L}_s$ is of the form $h_{u_s}(x, p) = (p, B_s u_s) - (Q_s x, u_s)$ for some u_s in ker P_s, and $\mathcal{H}_s$ consists of quadratic polynomials $\hat{H}_{f_s} = -c_s(x, f_s) + p(Ax + B_s f_s)$, with f_s a solution of

$$P_s u_s - B_s^* p + Q_s x = 0, \quad \mod(\mathcal{L}_s).$$

Our immediate objective is to relate $\mathcal{L}_s$ and $\mathcal{H}_s$ to the original extremal system $(\mathcal{H}, \mathcal{L})$.

First note that $\mathcal{L}_s = \text{Ham}(J)$, because of the following argument: Each element of $\mathcal{L}_s$ is the Hamiltonian of a jump field of the saturated system. Because the jump fields of the saturated system coincide with the jumps of the original system, if follows that $\mathcal{L}_s \subset \text{Ham}(J)$. On the other hand, every jump direction w is in ker P_s, and therefore the Hamiltonian of the corresponding jump field is in $\mathcal{L}_s$. Hence $\text{Ham}(J) \subset \mathcal{L}_s$, and that, together with the opposite inclusion, gives $\text{Ham}(J) = \mathcal{L}_s$.

The road to $\mathcal{H}_s$ is a bit longer and requires additional notation: For any function f on $M \times M^*$, $\text{ad} f$ will denote the linear operation on the space of smooth functions given by $(\text{ad} f)(g) = \{f, g\}$. In view of Theorem 11, the Hamiltonian vector field of $(\text{ad} f)(g)$ can be written as $\overrightarrow{(\text{ad} f)(g)} = \text{ad} \vec{f}(\vec{g})$.

The foregoing notation, with further reference to Theorem 11, allows for an alternative description of H_s:

$$H_s = H_u + \text{ad} h_v(H_u) + \frac{1}{2}(\text{ad} h_v)^2(H_u),$$

with $H_u = -c(x, u) + p(Ax + Bu)$ and $h_v = -(Sv, x) + pv$. This description follows from the fact that

$$(\exp V)_\#(X_u) = (\exp V)_*(X_u)(\exp -V) = X_u + [V, X_u] + \frac{1}{2}[V, [V, X_u]]$$

and that H_s is the Hamiltonian lift of $(\exp V)_\#(X_u)$.

We have already remarked that $P_s \geq 0$, and therefore

$$-\frac{1}{2}(P_s u_s, u_s) = H_s(0, 0, u, v) = H_u(-v, -Sv) \leq 0$$

for any u in U and v in $\mathrm{Ham}(J)$. In particular, $H_f(v, Sv) \leq 0$ for any Hamiltonian H_f in the original family $\mathcal{H}$, because H_f is a homogeneous polynomial of degree 2, and therefore $H_f(v, Sv) = H_f(-v, -Sv)$.

We leave it to the reader to verify that $\mathrm{ad}\, h_v(H_f) = H_f(x - v, p - Sv)$ and that $(\mathrm{ad}\, h_v)^2(H_f) = 2H_f(v, Sv)$. These observations show that the quadratic form

$$(h_v, h_w) \to \mathrm{ad}\, h_w \mathrm{ad}\, h_v(H_f)$$

is a negative-semidefinite symmetric bilinear form on the space of the Hamiltonians corresponding to jump fields.

Problem 2 Show that $\mathrm{ad}\, h_w \mathrm{ad}\, h_v(H_f) = \mathrm{ad}\, h_v \mathrm{ad}\, h_w(H_f)$, and also show that $\mathrm{ad}\, h_v \mathrm{ad}\, h_w(H_f) = \mathrm{ad}\, h_v \mathrm{ad}\, h_w(H_g)$ for any H_f and H_g in $\mathcal{H}$.

The following theorem shows a close connection between the foregoing quadratic form and the extended Legendre form encountered earlier in this chapter.

Theorem 13 $\mathrm{ad}\, h_v(H_f)$ belongs to $\mathrm{Ham}(J)$ for any h_v in $\mathrm{Ham}(J)$ whenever $(\mathrm{ad}\, h_v)^2(H_f) = 0$. For such a jump direction v, the extended Legendre form β is equal to zero.

Proof Let $V = ((Sv, x), v)$ be the jump field that lifts to h_v. For any H_f in $\mathcal{H}$, let $u = f(v, Sv)$. Then

$$(\exp \lambda V)_\#(X_u) = X_u + \lambda[V, X_u] + \frac{1}{2}\lambda^2[V, [V, X_u]],$$

with $X_u = (c, A+Bu)$. Therefore, the reduced Hamiltonian lift of the foregoing vector field is given by

$$H_s = H_f + \lambda\, \mathrm{ad}\, h_v(H_f) + \frac{1}{2}(\mathrm{ad}\, h_v)^2(H_f) = H_f + \lambda\, \mathrm{ad}\, h_v(H_f).$$

The function $(1/|\lambda|)H_s$ tends to $\pm\mathrm{ad}\, h_v(H_f)$ as λ tends to $\pm\infty$. Remember that λV is a jump field for any real number λ. But then $(1/|\lambda|)(\exp \lambda V)_\#(X_u)$ tends to $\pm[V, X_u]$. Because $\pm[V, X_u]$ is in the strong Lie saturate and the latter is convex, it follows that $[V, X_u]$ is a jump field. The jump direction of $[V, X_u]$ is equal to $[v, A + Bu] = -Av$. The cost of $[V, X_u]$ is given by

$$(Sv, Ax + Bu) - \left(\frac{\partial c}{\partial x}, v\right) = (Sv, Ax + Bu) - (Q^*u + Rx, v).$$

But then $(Sv, Ax+Bu)-(Qv, u)-(Rx, v) = -(SAv, x)$, because of Theorem 7. We can now use Theorem 8 to conclude that $(Sv, Bu) - (Qv, u) = 0$. But then

$$(Sv, Av) - (Rv, v) = -(SAv, v), \quad \text{or} \quad 2(Sv, Av) - (Rv, v) = 0,$$

and our proof is finished. ∎

Remark Theorem 13 is a Hamiltonian analogue of Theorem 8 and will be essential for calculation of the extremal curves. ∎

We now have sufficient information to get to the quadratic Hamiltonians that characterize the extremal curves of the saturated system. Every element $\hat{H}$ of $\mathcal{H}_s$ is given by $H_s(x, p, v(x, p), u(x, p))$, with $v(x, p)$ and $u(x, p)$ solutions of

$$P_s u_s + (Q_s x, u_s) - B_s^* p = 0, \qquad \text{mod}(\text{Ham}(J)). \tag{6}$$

There is a basis in $U \times J$ that makes the calculations simple: To begin with, we shall use the basis $e_1, \ldots, e_m$ in U relative to which $(Pu, u) = \frac{1}{2} \sum_{i=r+1}^m u_i^2$. The elements $e_1, \ldots, e_r$ span ker P. We know that their Hamiltonians $-(Qx, e_i) + pBe_i$ form a basis for $\mathcal{L}$, and we know that $\mathcal{L} \subset \text{Ham}(J)$.

As we showed earlier, the quadratic form $\text{ad}\,h_v\text{ad}\,h_w(H_f)$ is negative-semi-definite on $\text{Ham}(J)$. Therefore there exists a basis $h_1, \ldots, h_\ell$ in $\text{Ham}(J)$ such that with respect to this basis, $\text{ad}\,h_v\text{ad}\,h_v(H_f) = -\frac{1}{2} \sum_{i=q+1}^\ell v_i^2$. That is, $h_1, \ldots, h_q$ form a basis for the kernel of this form. It then follows from Theorem 13 that $\text{ad}\,h_i(H_f) \in \text{Ham}(J)$ for $i = 1, \ldots, q$, because $(\text{ad}\,h_i)^2(H_f) = 0$ for $i = 1, \ldots, q$.

Let $d_1, \ldots, d_\ell$ denote the elements in J defined by $h_1, \ldots, h_\ell$. The solutions $u(x, p)$ and $v(x, p)$ of (6) are characterized by

$$\frac{\partial H_s}{\partial u_i} \in \text{Ham}(J) \quad \text{and} \quad \frac{\partial H_s}{\partial v_j} \in \text{Ham}(J)$$

for all indices $i = 1, \ldots, r$ and $j = 1, \ldots, \ell$.

For $i \le q$,

$$\frac{\partial H_s}{\partial u_i} = \frac{\partial}{\partial u_i} H(x - v, p - Sv, u)$$

$$= \frac{\partial}{\partial u_i}\left(-\frac{1}{2}\sum_{i=r+1}^{m} u_i^2 + \sum_{i=1}^{m} u_i(-Q(x-v), e_i)\right.$$

$$\left. + (p - Sv, Be_i)\right) + H_0(x, p)$$

$$= -(Q(x-v), e_i) + (p - Sv, Be_i) = -(Qx, e_i) + pBe_i,$$

because $Q^*e_i = SBe_i$. Thus $\partial H_s/\partial u_i \in \text{Ham}(J)$ for any choice of $u_{r+1}, \ldots, u_m$ and v in J.

For $i > r$,

$$\frac{\partial H_s}{\partial u_i} = -u_i + (-Q(x-v), e_i) + (p - Sv, Be_i),$$

so we take $u_i = -(Q(x-v), e_i) + (p - Sv, Be_i)$. Such a choice for u corresponds to $u = f(x - v, p - Sv)$, with $H_f = H_0 + \frac{1}{2}\sum_{i=r+1}^{m}(-Qx, e_i) + (p, Be_i)^2$. As was shown earlier, $H_f \in \mathcal{H}$. Then

$$H_s(x, p, f, u) = H_f(x - v, p - Sv)$$

$$= H_f(x, p) + \sum_{i=1}^{\ell} v_i \text{ad} h_i(H_f) + \frac{1}{2}\sum_{i,j=1}^{\ell} v_i v_j \text{ad} h_i \text{ad} h_j(H_f)$$

$$= H_f(x, p) + \sum_{i=1}^{\ell} v_i \text{ad} h_i(H_f) - \frac{1}{2}\sum_{i=q+1}^{\ell} v_i^2.$$

So now,

$$\frac{\partial H_s}{\partial v_i} = \text{ad} h_i(H_f) \quad \text{for } i \le q,$$

$$\frac{\partial H_s}{\partial v_i} = \text{ad} h_i(H_f) - v_i \quad \text{for } i > q.$$

For $i \le q$, $\text{ad} h_i(H_f)$ is in $\text{Ham}(J)$, because $(\text{ad} h_i)^2(H_f) = 0$. For $i > q$, choose $v_i = \text{ad} h_i(H_f)$. The preceding argument shows that

$$\hat{H}_f = H_f + \frac{1}{2}\sum_{i=q+1}^{\ell}(\text{ad} h_i(H_f))^2, \quad \text{with } H_f = H_0 + \frac{1}{2}\sum_{i=r}^{m} f_i^2,$$

is an element of $\mathcal{H}_s$. We leave it to the reader to show that for any $\mathcal{H}_f \in \mathcal{H}$, $\hat{H}_f$, given by the foregoing expression, is also an element of $\mathcal{H}_s$.

The extremals of the saturated system are then easily described by the following theorem.

Theorem 14 *An absolutely continuous curve $\sigma_s(t)$ in $M \times M^*$ is an extremal curve of the saturated system if and only if $h(\sigma_s(0)) = 0$ for any h in $\mathrm{Ham}(J)$ and $d\sigma_s/dt = \vec{\hat{H}}_f + \sum_{i=1}^{\ell} u_i(t)\vec{h}_i$ for some integrable functions $u_1(t), \ldots, u_\ell(t)$.*

Proof We need to show that any solution curve of $d\sigma_s/dt = \vec{\hat{H}}_f(\sigma_s(t)) + \sum_{i=1}^{\ell} u_i(t)\vec{h}_i$ for which $h_i(\sigma_s(0)) = 0$ for all $i = 1, \ldots, \ell$ satisfies $h_i(\sigma_s(t)) = 0$ for all t.

Let $\mathcal{B} = \overline{\mathrm{Ham}(J)}$ denote the vector space of all Hamiltonian vector fields corresponding to the elements in $\mathrm{Ham}(J)$. Because ad $\hat{H}_f(h_i) = $ ad $h_i(H_f)$ for $i \leq q$, and ad $\hat{H}_f(h_i) = 0$ for $i > q$, it follows that ad $\hat{H}_f(\mathrm{Ham}(J)) \subset \mathrm{Ham}(J)$. Therefore, the linear field $A = \vec{\hat{H}}_f$ satisfies $A(\mathcal{B}) \subset \mathcal{B}$, as can be seen from Theorem 11.

Each solution curve $d\sigma_s/dt = A\sigma_s(t) + \sum u_i(t)\vec{h}_i$ is of the form $\sigma_s(t) = (\exp tA)\sigma_s(0) + \sum_{i=1}^{\ell} \int_0^t ((\exp(t-s)A)u_i(s)\vec{h}_i)\, ds$. Because $A(\mathcal{B}) \subset \mathcal{B}$, it follows that both of the terms $\sum_{i=1}^{\ell} \int_0^t ((\exp(t-s)A)u_i(s)\vec{h}_i)\, ds$ and $(\exp tA)\sigma_s(0)$ belong to $\overline{\mathrm{Ham}(J)}$. But then $h_i(\sigma(t)) = 0$ for each i and all t, because $\mathrm{Ham}(J)$ is a commutative algebra. The proof is now finished. ∎

We have already observed that each regular extremal curve is also an extremal curve of the saturated system, and therefore it is annihilated not only by the elements of $\mathcal{L}$ but also by every element in $\mathrm{Ham}(J)$. The resolution of constraints that affect the optimal trajectories is equivalent to calculating the annihilator of the optimal space Ω, which will be our next objective.

3 The reduction procedure

We shall use Theorem 13 to define the indices of jump fields through the properties of their Hamiltonians. Recall that $\mathcal{L}$ denotes the Hamiltonians of the form $-(Qx, u) + pBw$, with u in ker P. Then for each h in $\mathcal{L}$, we define its index as the least integer n such that $(\mathrm{ad}\, H_f)^{n+1}(h)$ is not a jump Hamiltonian. H_f is an arbitrary element of $\mathcal{H}$. We shall leave it to the reader to show that the index is independent of the choice of a representative H_f. We shall use $\mathrm{ind}(h)$ to denote the index of h, with the understanding that $\mathrm{ind}(h) = \infty$ whenever

$(\text{ad } H_f)^n(h) \in \text{Ham}(J)$ for all positive integers n. The reader should note that each jump Hamiltonian has an infinite index relative to the saturated system, because $\text{ad } \hat{H}_f(\text{Ham}(J)) \subset \text{Ham}(J)$ for any $\hat{H}_f$ in $\mathcal{H}_s$.

Definition 8 $\mathcal{L}_k$ will denote the vector space of all elements in $\mathcal{L}$ whose indices are greater than i_κ. $\mathcal{L}_\infty = \{h : (\text{ad } H_f)^k(h) \in \text{Ham}(J) \text{ for each positive integer } k\}$.

The projection of each $\mathcal{L}_k$ on M coincides with the space E_k defined earlier, and we have the following filtration:

$$\mathcal{L} = \mathcal{L}_0 \supset \mathcal{L}_1 \supset \cdots \supset \mathcal{L}_k \supset \cdots \supset \mathcal{L}_\infty$$

$$\downarrow \qquad\qquad \downarrow \qquad\qquad \downarrow \qquad\qquad \downarrow$$

$$B(\ker P) = E_0 \supset E_1 \supset \cdots \supset E_k \supset \cdots \supset E_\infty.$$

The following theorem will enable us to better understand the foregoing correspondence.

Theorem 15 *Let H be a quadratic function on $M \times M^*$, and let $\mathcal{L}$ be a set of linear functions such that $\{(\text{ad } H)^j(f), \, 0 \leq j \leq n, \, f \in \mathcal{L}\}$ Poisson-commute. Then*

(a) $\{(\text{ad } H)^j(f), (\text{ad } H)^{n+1}(g)\} = 0$ *for any f and g in $\mathcal{L}$ and any integers i and j for which $0 \leq i + j \leq n$, and*

(b) $\{(\text{ad } H)^j(f), (\text{ad } H)^{2n+1-j}(g)\} = (-1)^j \, \text{ad } f((\text{ad } H)^{2n+1}(g))$ *for any functions f and g, and $j \leq n$.*

Problem 3 Prove Theorem 15. Make repetitive use of the Jacobi identity.

Although the set of elements having the same index does not form a vector space, the following observation based on Theorem 15 is quite useful: Let $h_1, \ldots, h_m$ denote any basis in $\mathcal{L}_{k-1}$ for which the elements $h_1, \ldots, h_r$ form a basis for $\mathcal{L}_k$. Then each element h in the linear span of the remaining elements $h_{r+1}, \ldots, h_m$ has index equal to i_k. That means that $\{(\text{ad } H_f)^{i_k}(h), \{(\text{ad } H_f)^{i_k}(h), H_f\}\} < 0$ for any element H_f in $\mathcal{H}$. But then

$$\{(\text{ad } H_f)^{i_k}(h), \{(\text{ad } H_f)^{i_k}(h), H_f\}\} = -\{(\text{ad } H_f)^{i_k}(h), (\text{ad } H_f)^{i_k+1}(h)\}.$$

It further follows from Theorem 15, part (b), that

$$\{(\operatorname{ad} H_f)^{i_k}(h), (\operatorname{ad} H_f)^{i_k+1}(h)\} = (-1)^{i_k} \operatorname{ad} h((\operatorname{ad} H_f)^{2i_k+1}(h)).$$

Thus the matrix with entries $\operatorname{ad} h_i((\operatorname{ad} H_f)^{2i_k+1}(h_j))$, with $r+1 \leq i \leq m$ and $r+1 \leq j \leq m$, is either positive-definite or negative-definite, depending on whether i_k is odd or even.

We shall use this fact to show that the optimal space Ω is determined through a sequence of steps numbered by the indices $i_1 < i_2 < \cdots < i_m$; the kth step eliminates the constraints, with index i_k, and modifies the remaining Hamiltonians, through linear feedback, in such a way that the structure for the next step remains the same. Because the subsequent arguments make substantial use of symplectic geometry, it will be convenient to summarize the relevant notions discussed in Chapter 8:

A subspace W of a symplectic space V is called *isotropic* if the symplectic form ω vanishes on W. Isotropic subspaces of maximal dimension are called *Lagrangian*. Lagrangian subspaces of V all have the same dimension, which is equal to $\frac{1}{2}\dim V$. The dual, V^*, of a symplectic space is symplectic, with its symplectic form equal to the Poisson bracket. Isotropic subspaces of V^* agree with Poisson-commutative subspaces of V^*, and Lagrangian subspaces are Poisson-commutative spaces whose dimension is equal to $\frac{1}{2}\dim V$. A subspace W of V^* is symplectic if and only if for each $w \in W$ there is an element $\hat{w}$ in W such that $\{w, \hat{w}\} \neq 0$.

The next theorem prepares the stage for our inductive scheme.

Theorem 16 *Let i_k be any finite index, and H an arbitrary element of $\mathcal{H}$. Denote by Ω_k the zero set of all linear forms $(\operatorname{ad} H)^j(h)$, with $0 \leq j \leq 2\operatorname{ind}(h)+1$, and $\operatorname{ind}(h) \leq i_k$. Then Ω_k is a symplectic subspace of $M \times M^*$, and the linear span of $\{(\operatorname{ad} H)^j(h) : 0 \leq j \leq \operatorname{ind}(h), \operatorname{ind}(h) \leq i_k\}$ is a Lagrangian subspace of the annihilator Ω_k^0 of Ω_k.*

Proof We shall use the fact that Ω is a symplectic subspace of $M \times M^*$ if and only if its annihilator is symplectic. It follows from our earlier remarks that Ω_k^0 is symplectic if and only if for any f in Ω_k^0 there exists $\hat{f}$ in Ω_k^0 such that $\{f, \hat{f}\} \neq 0$.

Let $f = \sum_{m=1}^k \sum_{j=0}^{2i_m+1} \alpha_{mj}(\operatorname{ad} H)^j(h_{mj})$ be an arbitrary element of Ω_k^0, with elements h_{mj} in $\mathcal{L}$ whose index is equal to i_j, with $i_j \leq i_k$. Suppose that ℓ denotes the maximal integer m in the foregoing sum for which not all coefficients α_{mj} are equal to zero. Then let n denote the maximal integer j such that $\alpha_{mj} \neq 0$. Denote $\hat{f} = (\operatorname{ad} H)^{2i_\ell+1-n}(h_{\ell n})$. It follows from

Theorem 16 that $\{\hat{f}, (\text{ad}\, H)^j(h_{mj})\} = 0$ for all $m \leq \ell$ and $j < n$, and $\{\hat{f}, (\text{ad}\, H)^n(h_{\ell n})\} = (-1)^{i_\ell}\text{ad}\, h_{\ell n}((\text{ad}\, H)^{2i_\ell+1}(h_{\ell n}))$. The last term is equal to $\{(\text{ad}\, H)^{i_\ell}(h_{\ell n}), (\text{ad}\, H)^{i_\ell+1}(h_{\ell n})\}$, which cannot be zero, because i_ℓ is the index of $h_{\ell n}$. Thus $\{\hat{f}, f\} \neq 0$, and Ω_k^0 is symplectic.

Now recall that the linear span of $(\text{ad}\, H)^j(h)$, with $0 \leq j \leq \text{ind}(h) \leq i_k$, is a commutative algebra, because each $(\text{ad}\, H)^j(h)$ is a jump Hamiltonian. Therefore their span is an isotropic subspace of Ω_k^0. Suppose that $f = \sum_{m=1}^k \sum_{j=0}^{2i_m+1} \alpha_{mj}$ $(\text{ad}\, H)^j(h_{mj})$ is an element of Ω_k^0 that commutes with each $(\text{ad})^j H(h)$, with $0 \leq j \leq \text{ind}(h)$ and $\text{ind}(h) \leq i_k$. As in the preceding argument, let ℓ denote the maximal index, and n the maximal exponent, in the foregoing sum for which $\alpha_{mj} \neq 0$.

If $n > i_\ell$, then $2i_\ell - n < i_\ell$, and therefore $(\text{ad}\, H)^{2i_\ell-n}(h_{\ell n}) \in \text{Ham}(J)$. But then

$$\{(\text{ad}\, H)^{2i_\ell-n}(h_{\ell n}), f\} = (-1)^{i_\ell}\text{ad}\, h_{\ell n}(\text{ad}\, H)^{2i_\ell+1}(h_{\ell n}) \neq 0.$$

This contradicts the assumption that f commutes with every element $(\text{ad}\, H)^j(h)$, with $0 \leq j \leq \text{ind}(h)$. Hence the linear span of $(\text{ad}\, H)^j(h)$, with $j \leq \text{ind}(h)$ and $\text{ind}(h) \leq i_k$, is a maximal isotropic subspace of Ω_k^0, and hence Lagrangian. Our proof is now finished. ∎

The next theorem describes the resolution of extremal curves.

Theorem 17 *For each finite index i_k, there exists an element H_k in $\mathcal{H}$ with the following properties:*

(a) Ω_k^0, *the linear span of* $\{(\text{ad}\, H_k)^j(h) : 0 \leq j \leq 2\,\text{ind}(h) + 1, \text{ind}(h) \leq i_k\}$, *is a subspace of the annihilator of the optimal space Ω. That is, $\Omega \subset \Omega_k$.*

(b) *An absolutely continuous curve $\sigma(t)$ is an extremal curve if and only if $\sigma(t) \in \Omega_k$ and there exist integrable functions $u_1(t), \ldots, u_m(t)$ and elements $h_1, \ldots, h_m$ in $\mathcal{L}_k$ such that*

$$\frac{d\sigma}{dt}(t) = \vec{H}_k(\sigma(t)) + \sum_{i=1}^m u_i(t)\vec{h}_i \quad \text{and} \quad \mathcal{L}_k(\sigma(t)) = 0$$

for all t. (Remember that $\mathcal{L}_k$ is the space of all h in $\mathcal{L}$ with index greater than i_k.)

Proof By induction on k. It will be convenient to begin the induction with $\Omega_0 = M \times M^*$, with H_0 any element in $\mathcal{H}$, and $\mathcal{L}_0 = \mathcal{L}$. Then $\Omega_0^0 = \{0\}$, and (a) is automatically satisfied, and (b) follows from Theorem 6 in Chapter 7.

Now assume that Ω_{k-1} and H_{k-1} satisfy the conditions of the theorem. Let $h_1, \ldots, h_m$ be any basis for $\mathcal{L}_{k-1}$ such that $h_1, \ldots, h_r$ form a basis for $\mathcal{L}_k$. As we have already remarked, each linear combination of $h_{r+1}, \ldots, h_m$ has index equal to i_k. The system of linear equations

$$0 = (\mathrm{ad}\, H_{k-1})^{2i_k+2}(h_i) + \sum_{j=r+1}^{m} \alpha_j \mathrm{ad}\, h_j (\mathrm{ad}\, H_{k-1})^{2i_k+1}(h_i), \qquad (7)$$

for $i > r$, has a unique solution $\alpha_{r+1}, \ldots, \alpha_m$, because the matrix $M_{ij} = (\mathrm{ad}\, h_i (\mathrm{ad}\, H_{k-1})^{2i_k+1}(h_j))$ is either positive-definite or negative-definite, depending on whether i_k is odd or even. Let $H_k = H_{k-1} + \sum_{i=r+1}^{m} \alpha_i h_i$, with $\alpha_{i+1}, \ldots, \alpha_m$ the solutions of (7).

We shall first show that $\Omega_k^0 \subset \Omega^0$. Let h be any element of $\mathcal{L}$ with index less than or equal to i_k. If $\mathrm{ind}(h) < i_k$, then $(\mathrm{ad}\, H_k)^j(h) \in \Omega_{k-1}^0$ for each $0 \le j \le 2\,\mathrm{ind}(h)+1$, because $(\mathrm{ad}\, H_k)^j(h)$ is a sum of terms $(\mathrm{ad}\, H_{k-1})^i(h)$, with $0 \le i \le j$. Because $\Omega_{k-1}^0 \subset \Omega^0$, it follows that $(\mathrm{ad}\, H_k)^j(h) \in \Omega^0$. It remains to show that if $\mathrm{ind}(h) = i_k$, and if $0 \le j \le 2i_k + 1$, then $(\mathrm{ad}\, H_k)^j(h) \in \Omega^0$.

Let $\sigma(t)$ be an extremal curve of the original system. According to the induction hypothesis,

$$\frac{d\sigma}{dt} = \vec{H}_{k-1}(\sigma(t)) + \sum_{i=1}^{m} u_i(t)\vec{h}_i$$

for some integrable functions $u_1, \ldots, u_m$, and $h_i(\sigma(t)) = 0$ for all t and each $i = 1, \ldots, m$. It follows from Theorem 15 that

$$\{(\mathrm{ad}\, H_{k-1})^\ell(h_i), (\mathrm{ad}\, H_{k-1})^m(h_j)\} = 0 \quad \text{for } 0 \le \ell + m \le 2i_k.$$

Moreover, if both i and j are less than or equal to r, then $\{(\mathrm{ad}\, H_{k-1})^\ell(h_i), (\mathrm{ad}\, H_{k-1})^m(h_j)\} = 0$ for $0 < \ell + m \le 2i_k + 2$, because both $\mathrm{ind}(h_i)$ and $\mathrm{ind}(h_j)$ are larger than i_k.

Each derivative of $h_i(\sigma(t))$ is equal to zero, because $h_i(\sigma(t)) = 0$ for all t. It follows from the preceding formula that

$$0 = \frac{d^j}{dt^j} h_i(\sigma(t)) = (\mathrm{ad}\, H_{k-1})^j(h_i)(\sigma(t))$$

for all j, with $0 \le j \le 2i_k$, and all $i = 1, \ldots, m$. Therefore $(\mathrm{ad}\, H_{k-1})^j(h_i) \in \Omega^0$ for all j with $0 \le j \le 2i_k + 1$. Because $(\mathrm{ad}\, H_k)^j(h_i)$ is a linear combination of terms of the form $(\mathrm{ad}\, H_{k-1})^\ell(h_s)$, with $0 \le \ell \le 2i_k + 1$ and $1 \le s \le m$, it follows that $(\mathrm{ad}\, H_k)^j(h_i) \in \Omega^0$ for $0 \le j \le 2i_k + 1$. Hence $\Omega_k^0 \subset \Omega^0$.

Differentiation of $(d^{2i_k+1}/dt^{2i_k+1})h_i(\sigma(t)) = (\operatorname{ad} H_{k-1})^{2i_k+1}(h_i)$ gives

$$0 = \frac{d^{2i_k+2}}{dt^{2i_k+2}} h_i(\sigma(t))$$

$$= (\operatorname{ad} H_{k-1})^{2i_k+2}(h_i)(\sigma(t)) + \sum_{j=r+1}^{m} u_j(t)\{(\operatorname{ad} H_{k-1})^{2i_k+1}(h_i), h_j\}$$

$$= (\operatorname{ad} H_{k-1})^{2i_k+2}(h_i)(\sigma(t)) + \sum_{j=r+1}^{m} u_j(t)\operatorname{ad} h_j (\operatorname{ad} H_{k-1})^{2i_k+1}(h_i).$$

Therefore $u_j(t) = \alpha_j(\sigma(t))$ for $j = r+1, \ldots, m$, and consequently σ is an integral curve of $H_k + \sum_{i=1}^{r} u_i(t)h_i$ that remains in Ω_k. That is,

$$\frac{d\sigma}{dt} = \vec{H}_k(\sigma(t)) + \sum_{i=1}^{r} u_i(t)\vec{h}_i. \tag{8}$$

Conversely, assume now that $u_1, \ldots, u_r$ are any integrable functions and that σ is any solution curve of (8) that remains in Ω_k for all t and satisfies $h_i(\sigma(t)) = 0$ for each $i = 1, \ldots, r$. Evidently,

$$\frac{d\sigma}{dt} = \vec{H}_{k-1}(\sigma(t)) + \sum_{i=1}^{m} u_i(t)\vec{h}_i, \quad \text{with } u_i(t) = \alpha_i(\sigma(t))$$

$$\text{for } i = r+1, \ldots, m.$$

Because $\sigma(t)$ is contained in Ω_k, and $\Omega_k \subset \Omega_{k-1}$, it follows that $\sigma(t) \subset \Omega_{k-1}$, and $h_i(\sigma(t)) = 0$ for $i = r+1, \ldots, m$. But then, by the induction hypothesis, $\sigma(t)$ is an extremal curve, and our proof is finished. ∎

Theorem 17 has several important corollaries that we shall address next. For the remainder of this chapter, i_∞ will denote the maximal finite index, and $\mathcal{L}_\infty$ will denote the subspace of $\mathcal{L}$ whose index is infinite. $\mathcal{L}_\infty$ is also equal to the set of all elements in $\mathcal{L}$ whose indices are greater than i_∞. We shall also use Ω_∞ and H_∞ to denote the elements described by Theorem 17. That is,

$$\Omega_\infty^0 = \{(\operatorname{ad} H_\infty)^j(h) : 0 \le j \le 2\operatorname{ind}(h) + 1, \quad \operatorname{ind}(h) \le i_\infty\},$$

and $(\Omega_\infty, H_\infty, \mathcal{L}_\infty)$ characterize the extremal curves as all absolutely continuous curves in Ω_∞ that satisfy

$$\frac{d\sigma}{dt} = \vec{H}_\infty(\sigma(t)) + \sum_{i=1}^{m} u_i(t)\vec{h}_i \tag{9}$$

for some integrable functions $u_1(t), \ldots, u_m(t)$ and some elements $h_1, \ldots, h_m$ in $\mathcal{L}_\infty$.

Corollary 3 $\Omega = \Omega_\infty \cap \mathcal{L}_\infty^0$.

Proof We know that $\operatorname{ad} H_\infty(\mathcal{L}_\infty) \subset \mathcal{L}_\infty$, because $(\operatorname{ad} H_\infty)^j(h)$ belongs to $\operatorname{Ham}(J)$ for each positive integer j. But then $(\operatorname{ad} \vec{H}_\infty)(\vec{\mathcal{L}}_\infty) \subset \vec{\mathcal{L}}_\infty$ (Theorem 13). Consequently, any solution curve of (9) that originates in $\mathcal{L}_\infty^0 \cap \Omega_\infty$ remains in $\mathcal{L}_\infty^0 \cap \Omega_\infty$ for all t and is therefore an extremal curve for the original system. Hence $\Omega = \mathcal{L}_\infty^0 \cap \Omega_\infty$. ∎

Corollary 4 Ω *is a symplectic subspace of* $M \times M^*$ *if and only if* $\mathcal{L}_\infty = 0$.

Proof $\Omega = \Omega_\infty$ whenever $\mathcal{L}_\infty = 0$. According to Theorem 16, Ω_∞ is symplectic. Conversely, assume that Ω is symplectic. Then for any f in Ω^0 there exists $\hat{f}$ in Ω^0 such that $\{f, \hat{f}\} \neq 0$. $\mathcal{L}_\infty \subset \Omega^0$ as a consequence of Corollary 3. But then $\{f, \hat{f}\} = 0$ for any f in $\mathcal{L}_\infty$ and any $\hat{f}$ in Ω^0. So $\mathcal{L}_\infty$ must be the zero space. ∎

Definition 9 System (1) is said to have the uniqueness property if $x(t) = 0$, $u(t) = 0$ is the only optimal solution for $x(0) = 0$, $x(T) = 0$.

Corollary 5 *System* (1) *has unique optimal solutions if and only if* $\mathcal{L}_\infty = 0$. *In such a case,* Ω *is a symplectic invariant subspace of* $\vec{H}_\infty$, *and the extremal curves coincide with the integral curves of* $\vec{H}_\infty$ *that originate in* Ω.

Proof It follows from Corollary 3 that any solution curve of (9) that originates at zero is an extremal curve and therefore is optimal. The reachable set defined by the terminal points of these curves is zero if and only if $\mathcal{L}_\infty = 0$. We leave it to the reader to show that the uniqueness property is satisfied if and only if $\mathcal{L}_\infty = 0$.

Then $\Omega = \Omega_\infty$, by Corollary 3, and Ω is symplectic, by Corollary 4. But then equations (8) and (9) reduce to $d\sigma/dt = \vec{H}_\infty(\sigma(t))$, $\sigma(0) \in \Omega_\infty$, and therefore Ω is invariant for $\vec{H}_\infty$. ∎

Corollary 6 $\operatorname{Ham}(J) = \operatorname{Ham}(J_\infty)$ *for systems having the uniqueness property.*

Proof Ω_∞^0 is a symplectic space in which $\operatorname{Ham}(J_\infty)$ is a Lagrangian subspace (Theorem 17). It follows from Corollary 5 that $\Omega^0 = \Omega_\infty^0$. Each element h of

Ham(J) belongs to Ω^0 and commutes with Ham(J_∞). Therefore, Ham(J) $\subset$ Ham(J_∞), by the maximality property of Lagrangian subspaces.   ∎

4 The optimal synthesis

The optimal synthesis for the singular problem is inseparably connected with the optimal synthesis of the saturated system, as will be soon demonstrated. For simplicity of exposition it will be assumed that the original system has the uniqueness property.

Recall (Theorem 14) that an absolutely continuous curve is an extremal curve of the saturated system if and only if there exist integrable functions $u_1, \ldots, u_\ell$ such that

$$\frac{d\sigma_s}{dt} = \vec{H}_f + \sum_{i=q+1}^{\ell} \text{ad}\, h_i \overrightarrow{(H_f)} \text{ad}\, h_i(\vec{H}_f) + \sum_{i=1}^{\ell} u_i(t)\vec{h}_i, \qquad (10)$$

and $h(\sigma_s(0)) = 0$ for any element in Ham(J). $h_1, \ldots, h_\ell$ is a basis in Ham(J) relative to which the quadratic form $\text{ad}\, h$ ad $h(H_f)$ is given by $-\frac{1}{2}\sum_{i=q+1}^{\ell} v_i^2$.

Theorem 18 *The optimal space Ω_s for the saturated system is equal to $\Omega \oplus$ $\overrightarrow{\text{Ham}(J)}$. Every extremal curve σ_s of the saturated system can be written as $\sigma_s = \sigma + \sigma_j$, with σ a regular extremal curve, and σ_j an absolutely continuous curve in $\overrightarrow{\text{Ham}(J)}$. Moreover, this decomposition is unique.*

Proof Let H_∞ be any element of $\mathcal{H}$ such that $\vec{H}_\infty(\Omega) \subset \Omega$. Such an element exists by Corollary 5 following Theorem 17. Then every extremal curve σ_s of the saturated system is described by equations (10), with $H_f = H_\infty$. Denote $H_\infty + \frac{1}{2}\sum_{i=q+1}^{\ell} (\text{ad}\, h_i(H_\infty))^2$ by $\hat{H}_\infty$.

As we noted earlier, ad $\vec{\hat{H}}_\infty(\overrightarrow{\text{Ham}(J)}) \subset \overrightarrow{\text{Ham}(J)}$, and therefore every absolutely continuous curve in $\overrightarrow{\text{Ham}(J)}$ is an extremal curve of the saturated system. We leave it to the reader to verify that $h(\vec{g}) = 0$ for any elements h and g in Ham(J).

Because every regular extremal curve is also an extremal curve of the saturated system, $\Omega \subset \Omega_s$, and therefore $\Omega + \overrightarrow{\text{Ham}(J)} \subset \Omega_s$. The sum is direct, because Ham(J) $\subset \Omega^0$. Note that dim $(\overrightarrow{\text{Ham}(J)}) = \dim(\text{Ham}(J))$, because Ham($J$) contains no constant functions.

The remaining part of the argument consists of counting the appropriate dimensions: Let k denote the dimension of Ham(J). Then $\dim \Omega^0 = 2k$, because

Ham(J) is a Lagrangian subspace of Ω^0. Because $\dim\Omega_s = \dim(\text{Ham}(J)^0)$, it follows that $\dim\Omega_s = 2\dim M - k = 2\dim M - 2k + k = \dim\Omega + \dim(\overrightarrow{\text{Ham}(J)})$. This argument proves that $\Omega_s = \Omega \oplus \overrightarrow{\text{Ham}(J)}$.

To prove the second statement, let σ_s denote an arbitrary extremal curve of the saturated system. Then $\sigma_s(0) = \sigma(0) + \sigma_j(0)$, with $\sigma(0) \in \Omega$ and $\sigma_j(0) \in \overrightarrow{\text{Ham}(J)}$. Let $\sigma(t)$ be the regular extremal curve, with its initial value equal to $\sigma(0)$. Then $\sigma_s(t) - \sigma(t)$ is a solution curve of (10) that belongs to $\overrightarrow{\text{Ham}(J)}$ for all t. The proof is now finished. ∎

Corollary 7 *Suppose that $x(t)$ is an optimal trajectory of the original system for which $x(0) = a$ and $x(T) = b$. Then for any jump directions w_1 and w_2 there exists an optimal trajectory x_s of the saturated system that satisfies $x_s(0) = a + w_1 = \bar{a}$, $x_s(T) = b + w_2 = \bar{b}$ and whose total cost is equal to*

$$\frac{1}{2}S(\bar{b} - b)(\bar{b} + b) - \frac{1}{2}S(\bar{a} - a)(a + \bar{a}) + v(a, b, T).$$

Consequently,

$$v(a + w_1, b + w_2, T) = \frac{1}{2}Sw_2(2b + w_2) - \frac{1}{2}Sw_1(2a + w_1) + v(a, b, T).$$

Proof Let $\sigma(t) = (x(t), p(t))$ denote the extremal curve that projects onto the optimal trajectory $x(t)$. Consider the associated extremal curve $\sigma_s = (x_s, p_s)$ of the saturated system defined by $\sigma_s = \sigma + \sigma_j$, with $\sigma_j(t) = (w_1 + (t/T)(w_2 - w_1), S(w_1 + (t/T)(w_2 - w_1)))$ for all t in $[0, T]$. The projection of σ_s on M is optimal, and according to Theorem 10 its total cost is given by $\frac{1}{2}(p_s(T)x_s(T) - p_s(0)x_s(0))$, where $p_s(T) = p(T) + Sw_2$, $x_s(T) = x(T) + w_2 = b + w_2$, $p_s(0) = p(0) + Sw_1$, and $x_s(0) = a + w_1$. Therefore,

$$\frac{1}{2}p_s(T)x_s(T) - \frac{1}{2}p_s(0)x_s(0) = \frac{1}{2}p(T)x(T) - \frac{1}{2}p(0)x(0) + \frac{1}{2}Sw_2(x(T))$$

$$+ \frac{1}{2}p(T)w_2 - \frac{1}{2}Sw_1(x(0)) - \frac{1}{2}p(0)w_1$$

$$+ \frac{1}{2}Sw_2(w_2) - \frac{1}{2}Sw_1(w_1).$$

It follows from Theorem 10 that $v(a, b, T) = \frac{1}{2}p(T)x(T) - \frac{1}{2}p(0)x(0)$; because $\text{Ham}(J) \subset \Omega^0$, $h_{w_1}(\sigma(t)) = h_{w_2}(\sigma(t)) = 0$ along the jump Hamiltonians $h_{w_1} = -(Sw_1, x) + pw_1$ and $h_{w_2}(x, p) = -(Sw_2, x) + pw_2$. Therefore,

$p(T)w_2 = Sw_2(x(T))$, and $p(0)w_1 = Sw_1(x(0))$. Thus,

$$\frac{1}{2}p_s x_s \Big|_{t=0}^{t=T} = v(a, b, T) + Sw_2(x(T)) - Sw_1(x(0))$$

$$+ \frac{1}{2}Sw_2(w_2) - \frac{1}{2}Sw_1(w_1)$$

$$= v(a, b, T) + Sw_2\left(x(T) + \frac{w_2}{2}\right) - Sw_1\left(x(0) + \frac{w_1}{2}\right)$$

$$= v(a, b, T) + \frac{1}{2}Sw_2(2b + w_2) - \frac{1}{2}Sw_1(2a + w_1)$$

$$= v(a, b, T) + \frac{1}{2}S(\bar{b} - b)(b + \bar{b}) - \frac{1}{2}S(a - \bar{a})(a + \bar{a}),$$

with $\bar{b} = b + w_2$ and $\bar{a} = a + w_1$. So the total cost of x_s is equal to the sum of the total cost of σ, the jump cost from b to $\bar{b}$, and the jump cost from $\bar{a}$ to a. End of the proof.   ∎

Corollary 8 *The only regular optimal trajectory* (x, u) *that satisfies* $x(0) \in J$ *and* $x(T) \in J$ *is the zero trajectory.*

Proof Let $a = x(0)$ and $b = x(T)$. By Corollary 7, $v(a, b, T) = \frac{1}{2}Sb(b) - \frac{1}{2}Sa(a)$. Let x_s be the following trajectory of the saturated system: $x_s(t) = x(t)$ on $[0, T]$, and $x_s(t) = b$ for t in $[T, T + \epsilon]$; ϵ is small enough so that $T + \varepsilon < T_{max}$. The total cost of this trajectory is equal to the total cost in the interval $[0, T]$ plus the total cost in the interval $[T, T + \varepsilon]$. Because the total cost in the interval $[T, T + \varepsilon]$ is zero, it follows that the total cost of x_s is equal to the total cost of (x, u). Therefore, x_s is an optimal trajectory of the saturated system.

Let $\sigma_s = (x_s, p_s)$ denote the corresponding extremal curve of the saturated system. It follows from Theorem 10 that σ_s coincides with the regular extremal curve that projects onto x in the interval $[0, T]$. Thus $\sigma_s(0) \in \Omega$, and $\sigma_s(T) \in \Omega$. In the interval $[T, T + \varepsilon]$, $\sigma_s(t) = (b, Sb)$ (i.e., it is constant). Because σ_s is a continuous curve on $[0, T + \varepsilon]$, $\sigma_s(T) = (b, Sb)$. Therefore, $\sigma_s(T) \in \overrightarrow{\text{Ham}(J)} \cap \Omega$. But $\overrightarrow{\text{Ham}(J)} \cap \Omega = \{0\}$. Consequently $b = 0$.

An analogous argument applied at the initial point $a = x(0)$ shows that $a = 0$. Thus, $x(t)$ is an optimal curve that connects zero to zero and is therefore identically equal to zero.   ∎

Definition 10 Let π denote the natural projection from $M \times M^*$ onto M. The vector subspace of Ω defined by $\pi^{-1}(J) \cap \Omega$ will be denoted by $\text{Lg}(J)$.

Theorem 19 $\mathrm{Lg}(J)$ *is a Lagrangian subspace of* Ω.

Proof Let $E^\perp$ denote the symplectic orthogonal space of a symplectic subspace E of $M \times M^*$; that is, $(v, p) \in E^\perp$ if and only if $p(w) - q(v) = 0$ for all (w, q) in E. Recall that a space E is Lagrangian if and only if $E^\perp = E$.

We shall show that $(\mathrm{Lg}(J))^\perp \cap \Omega = \mathrm{Lg}(J)$, from which it will follow that $\mathrm{Lg}(J)$ is a Lagrangian subspace of Ω. We can write

$$\mathrm{Lg}(J) = (\overrightarrow{\mathrm{Ham}(J)} + M^*) \cap \Omega,$$

with the understanding that $M^* = \{0\} \times M^*$. Then

$$(\mathrm{Lg}(J))^\perp = (\overrightarrow{\mathrm{Ham}(J)} + M^*)^\perp + \Omega^\perp = (\overrightarrow{\mathrm{Ham}(J)})^\perp \cap (M^*)^\perp + \Omega^\perp.$$

M^* is a Lagrangian space, and therefore $(M^*)^\perp = M^*$. Because $\mathrm{Ham}(J)$ is a Lagrangian subspace of Ω^0, $(\overrightarrow{\mathrm{Ham}(J)})^\perp = \overrightarrow{\mathrm{Ham}(J)} + \Omega$. Finally, $\vec\Omega^0 = \Omega^\perp$.

Therefore, $(\mathrm{Lg}(J))^\perp = (\overrightarrow{\mathrm{Ham}(J)} + \Omega) \cap M^* + \vec\Omega^0$. Using the fact from linear geometry that

$$A + (B + C) \cap D = A + (B + D) \cap C = A + (B + C) \cap (B + D)$$

for any linear spaces A, B, C, and D, we get

$$(\mathrm{Lg}(J))^\perp = \vec\Omega^0 + (\overrightarrow{\mathrm{Ham}(J)} + M^*) \cap \Omega = \vec\Omega^0 + \mathrm{Lg}(J).$$

Because Ω is symplectic, $\Omega^\perp \cap \Omega = 0$. Therefore, $\vec\Omega^0 \cap \Omega = \Omega^\perp \cap \Omega = 0$, which implies that

$$(\mathrm{Lg}(J))^\perp \cap \Omega = \vec\Omega^0 \cap \Omega + \mathrm{Lg}(J) \cap \Omega = \mathrm{Lg}(J).$$

Our proof is now finished. ∎

Corollary 9 *Each point x in M has a unique lifting to the saturated optimal space Ω_s.*

Proof Because $\Omega_s = \Omega \oplus \overrightarrow{\mathrm{Ham}(J)}$, it follows that $\dim\Omega_s = \dim\Omega + \dim (\overrightarrow{\mathrm{Ham}(J)}) = \dim\Omega + \dim(\mathrm{Ham}(J))$. But $\dim(\mathrm{Ham}(J)) = \dim\Omega_s^0 = \dim(M \times$

$M^*) - \dim\Omega_s = 2\dim M - \dim\Omega_s$. Therefore, $\dim\Omega_s = \dim\Omega + 2\dim M - \dim\Omega_s$, or $2\dim\Omega_s = 2\dim M + \dim\Omega$.

Now let π_s denote the restriction of π to Ω_s, where π is the projection from $M \times M^*$ onto M. Any point (o, p) in $\ker \pi_s$ admits a unique decomposition $(o, p) = (v, \bar{p}) - (v, Sv)$, with v in J. Thus $\ker \pi_s$ is in a one-to-one correspondence with $\text{Lg}(J)$. Because $\text{Lg}(J)$ is a Lagrangian subspace of Ω, $\dim(\text{Lg}(J)) = \frac{1}{2}\dim\Omega$; therefore, $\dim(\ker \pi_s) = \frac{1}{2}\dim\Omega$.

Using the well-known fact from linear algebra that $\dim(\text{range } \pi_s) = \dim\Omega_s - \dim(\ker \pi_s)$, we get that

$$\dim(\text{range } \pi_s) = \dim M + \frac{1}{2}\dim\Omega - \frac{1}{2}\dim\Omega = \dim M.$$

Hence $\pi_s : \Omega_s \to M$ is an isomorphism, and our proof is finished.   ∎

Our elaborate journey through the symplectic geometry of the singular problem comes to its end with the following theorem.

Theorem 20 (The optimal synthesis theorem) *Assume that a and b are arbitrary points in M and that $T > 0$ is a fixed number such that $T < T_{\max}$. Then*

(a) *there exists a unique regular optimal trajectory $x(t)$ on the interval $[0, T]$ such that $x(0) \in a + J$ and $x(T) \in b + J$. Denoting $x(0) = \bar{a}$ and $x(T) = \bar{b}$, then*

(b) *the optimal-value function $v(a, b, T)$ is given by*

$$v(a, b, T) = v(\bar{a}, \bar{b}, T) + \frac{1}{2}S(b - \bar{b})(b + \bar{b}) + \frac{1}{2}S(a - \bar{a})(a + \bar{a}).$$

Proof Let H_∞ denote any element in $\mathcal{H}$ whose Hamiltonian vector field $\vec{H}_\infty$ leaves the optimal space Ω invariant. As we have already seen, such an element exists as a consequence of Corollary 5.

It follows from Corollary 8 following Theorem 18 that $(\exp T \vec{H}_\infty)(\text{Lg}(J))$ and $\text{Lg}(J)$ are transversal subspaces of Ω, because if $(\exp T \vec{H}_\infty)(\sigma_1) = \sigma_2$, with σ_1 and σ_2 in $\text{Lg}(J)$, then the projection $x(t)$ of the extremal curve $\sigma(t) = (\exp t \vec{H}_\infty)(\sigma_1)$ connects the initial point $x(0)$ in J to a point $x(T)$ in J, which can happen only if $x(0) = x(T) = 0$. But then $\sigma_1 = 0$, and therefore $\sigma(t) = 0$.

Because $\text{Lg}(J)$ and $(\exp T \vec{H}_\infty)(\text{Lg}(J))$ are transversal Lagrangian subspaces of Ω, Ω is a direct sum of $\text{Lg}(J)$ and $(\exp T \vec{H}_\infty)(\text{Lg}(J))$. We shall presently show that for any σ_1 and σ_2 in Ω, there exist unique points $\bar{\sigma}_1$ in $\sigma_1 + \text{Lg}(J)$ and $\bar{\sigma}_2$ in $\sigma_2 + \text{Lg}(J)$ such that $(\exp T \vec{H}_\infty)(\bar{\sigma}_1) = \bar{\sigma}_2$. Let

$$(\exp T \vec{H}_\infty)(\sigma_1) = \xi_1 + (\exp T \vec{H}_\infty)(\eta_1) \quad \text{and} \quad \sigma_2 = \xi_2 + (\exp T \vec{H}_\infty)(\eta_2)$$

for some points ξ_i and η_i in $\mathrm{Lg}(J)$, $i = 1, 2$. Then

$$
\begin{aligned}
(\exp T\,\vec{H}_\infty)(\sigma_1 + \eta_2 - \eta_1) \\
= (\exp T\,\vec{H}_\infty)(\sigma_1) + (\exp T\,\vec{H}_\infty)(\eta_2) - (\exp T\,\vec{H}_\infty)(\eta_1) \\
= \xi_1 + (\exp T\,\vec{H}_\infty)(\eta_1) + (\exp T\,\vec{H}_\infty)(\eta_2) - (\exp T\,\vec{H}_\infty)(\eta_1) \\
= \xi_1 + (\exp T\,\vec{H}_\infty)(\eta_2) = \sigma_2 - \xi_2 + \xi_1.
\end{aligned}
$$

Let $\bar{\sigma}_1 = \sigma_1 + \eta_2 - \eta_1$ and $\bar{\sigma}_2 = \sigma_2 + \xi_1 - \xi_2$. Both $\eta_2 - \eta_1$ and $\xi_1 - \xi_2$ belong to $\mathrm{Lg}(J)$, and therefore $\sigma_i - \bar{\sigma}_i$ belongs to $\mathrm{Lg}(J)$ for $i = 1, 2$, and $(\exp T\,\vec{H}_\infty)(\bar{\sigma}_1) = \bar{\sigma}_2$. These points are unique, because any difference of such solutions is a solution that originates and terminates in $\mathrm{Lg}(J)$.

It follows from Corollary 9 following Theorem 19 that each point x has a unique lifting to a point in Ω_s. Denote the liftings by $\pi_s^{-1}(a) = \sigma_1 + \vec{h}_1$ and $\pi_s^{-1}(b) = \sigma_2 + \vec{h}_2$, with $\vec{h}_1$ and $\vec{h}_2$ in $\overrightarrow{\mathrm{Ham}(J)}$, and σ_1 and σ_2 in Ω.

Let $\bar{\sigma}_1$ and $\bar{\sigma}_2$ denote the unique points in $\sigma_1 + \mathrm{Lg}(J)$ and $\sigma_2 + \mathrm{Lg}(J)$ for which $(\exp T\,\vec{H}_\infty)(\bar{\sigma}_1) = \bar{\sigma}_2$. The projection $\bar{x}(t)$ of the extremal curve $\sigma(t) = (\exp t\vec{H}_\infty)(\bar{\sigma}_1)$ satisfies condition (a) of the theorem, with $\pi(\bar{\sigma}_1) = \bar{a}_1$ and $\pi(\bar{\sigma}_2) = \bar{a}_2$. Evidently, $a - \bar{a}$ and $b - \bar{b}$ belong to J.

Denote by $\vec{h}_{\bar{a}}$ and $\vec{h}_{\bar{b}}$ the Hamiltonian vector fields associated with the jumps $a - \bar{a}$ and $b - \bar{b}$. It follows that $\bar{\sigma}_1 + \vec{h}_{\bar{a}}$ and $\bar{\sigma}_2 + \vec{h}_{\bar{b}}$ are points of the saturated optimal space, and according to Corollary 7 following Theorem 18 the extremal curve $\sigma_s(t) = \sigma(t) + \vec{h}_{\bar{a}} + (t/T)(\vec{h}_{\bar{b}} - \vec{h}_{\bar{a}})$ of the saturated system connects $\bar{\sigma}_1 + \vec{h}_{\bar{a}}$ to $\bar{\sigma}_2 + \vec{h}_{\bar{b}}$, with the total cost equal to

$$
\frac{1}{2} S(b - \bar{b})(b + \bar{b}) - \frac{1}{2} S(\bar{a} - a)(a + \bar{a}) + v(\bar{a}, \bar{b}, T)
$$

$$
= \frac{1}{2} S(b - \bar{b})(b + \bar{b}) + \frac{1}{2} S(a - \bar{a})(a + \bar{a}) + v(\bar{a}, \bar{b}, T).
$$

The proof is now finished. ∎

Figure 9.4 illustrates the geometric construction in the preceding proof.

Corollary 10 *Any points a and b of M can be connected by an optimal trajectory $x_s(t)$ of the saturated system in any interval $[0, T]$, with $T < T_{\max}$. Any two optimal trajectories of the saturated system that conform to the same boundary conditions differ by an absolutely continuous curve $z_j(t)$ with values in J for which $z_j(0) = z_j(T) = 0$.*

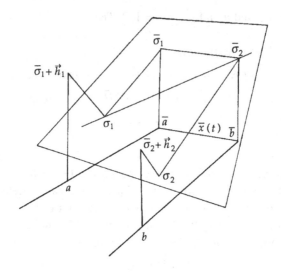

Fig. 9.4.

Proof The existence of optimal trajectories follows from the preceding theorem: $x_s(t)$ is the projection of the extremal curve $\sigma_s(t)$ constructed in the proof of Theorem 20.

Suppose now that $x_s(t)$ and $y_s(t)$ are any optimal trajectories of the saturated system such that $y_s(0) = x_s(0)$ and $y_s(T) = x_s(T)$. Let σ_s and ξ_s denote the corresponding extremal curves of the saturated system. According to Theorem 18, $\sigma_s = \sigma + \sigma_j$, and $\xi_s = \xi + \xi_j$, with σ_j and ξ_j absolutely continuous curves in $\overrightarrow{\text{Ham}(J)}$. Let x and x_j denote the projections of σ and σ_j on M, and let y and y_j denote the projections of ξ and ξ_j on M. Then

$$x_s - y_s = (x - y) + (x_j - y_j).$$

Because $x_s(0) = y_s(0)$ and $x_s(T) = y_s(T)$, it follows that $x(0) - y(0)$ and $x(T) - y(T)$ belong to J. Therefore, $x(t) = y(t)$ for all t, as a consequence of Corollary 8. Consequently, $x_s(t) = y_s(t) + z_j(t)$, with $z_j(t) = x_j(t) - y_j(t)$. End of the proof.   ∎

Alternatively, the solutions to the singular optimal problem can be described by discontinuous or generalized trajectories, as follows: A generalized trajectory $x_g(t)$ of (1) is a piecewise absolutely continuous curve on an interval $[0, T]$ having finitely many jump discontinuities at $0 \le t_0 < t_1 < \cdots < t_n \le T$ such that the restriction of $x_g(t)$ to each subinterval $[t_i, t_{i+1})$ is a regular trajectory, and such that each difference $x_g(t_{i+1}) - \lim_{t \to t_{i+1}} (x_g(t))$ is a jump direction. The total cost of a generalized trajectory is equal to the sum of the costs of regular pieces and the costs of the jumps (Figure 9.5). Each generalized trajectory may be shadowed by a regular trajectory whose cost is arbitrarily close to the total cost of the generalized trajectory (Figure 9.6). The results of Theorem 20 can be paraphrased as follows: For any points a, b in M and any $T > 0$, with $T < T_{\max}$, there exists a unique generalized optimal trajectory x_g such that $x_g(0) = a$ and $x_g(T) = b$; x_g consists of at most two jumps, one at $t = 0$ and the other at $t = T$.

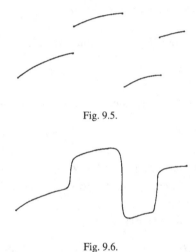

Fig. 9.5.

Fig. 9.6.

Example Consider the problem of minimizing $\frac{1}{2} \int_0^T (x_1^2 - x_2^2 + x_3^2) \, dt$ over the trajectories of $dx_1/dt = x_2, dx_2/dt = x_3, dx_3/dt = u(t)$. This problem corresponds to a slight variation of the Hardy-Littlewood systems discussed in preceding chapters. It corresponds to minimizing the integral $\frac{1}{2} \int_0^T (f^2 - (f^{(1)})^2 + (f^{(2)})^2) \, dt$ over all functions f in $[0, T]$ for which both f and $f^{(3)}$ belong to $L^2([0, T])$. It follows by the arguments used earlier that $\frac{1}{2} \int_0^T (x_1^2 - x_2^2 + x_3^2) \, dt \geq 0$ over the trajectories that satisfy the zero boundary conditions. Therefore the theory of this chapter applies.

According to the maximum principle, the extremal curves $(x(t), p(t))$ attain the maximum of

$$-\frac{1}{2}\left(x_1^2(t) - x_2^2(t) + x_3^2(t)\right) + p_1(t)x_2(t) + p_2(t)x_3(t) + up_3(t)$$

relative to the control variable u. This is possible only if $p_3(t) = 0$. Therefore, $h = p_3$ is a jump Hamiltonian.

It follows that $\mathcal{H} = \{H_0 + uh : u \in \mathbb{R}^1\}$, with $H_0 = -\frac{1}{2}(x_1^2 - x_2^2 + x_3^2) + p_1x_2 + p_2x_3$. Then ad $H_0(h) = -x_3 + p_2$, and $(\text{ad } h)^2(H_0) = 1$. According to Theorem 13, ad $H_0(h)$ is not a jump Hamiltonian. Therefore, ind$(h) = 0$, and Ham(J) is equal to the line containing h. The space of jump directions is equal to the x_3 axis.

It follows from Theorem 17 that Ω is equal to the zero set of h and ad $H_0(h)$. We shall now choose H_∞ in $\mathcal{H}$ such that ad $H_\infty(\Omega^0) \subset \Omega^0$.

Let $H_\infty = H_0 + \alpha h$ for some affine function α that needs to be determined. Then ad $H_\infty(h) = $ ad $H_0(h) + h\{\alpha, h\}$. Because $\{\alpha, h\}$ is constant, it follows that ad $H_\infty(h) \in \Omega^0$ no matter what α. Continuing,

$$(\text{ad } H_\infty)(\text{ad } H_0(h)) = (\text{ad } H_0)^2(h) + \alpha(\text{ad } h)(\text{ad } H_0(h)) + h\{\alpha, \text{ad } H_0(h)\}.$$

Then

$$(\text{ad } H_\infty)(\text{ad } H_0(h)) \in \Omega^0 \quad \text{provided that } \alpha = (\text{ad } H_0)^2(h), \qquad \text{mod}(\Omega^0).$$

That is, $\alpha = (\text{ad } H_0)^2(h) + \lambda_1 h + \lambda_2 \text{ad } H_0(h)$ for arbitrary constants λ_1 and λ_2.

Any such choice of α determines a Hamiltonian H_∞ in $\mathcal{H}$. The restriction of any such Hamiltonian to Ω is equal to $H = \frac{1}{2}p_2^2 + p_1x_2 + \frac{1}{2}x_2^2 - \frac{1}{2}x_1^2$, in the symplectic coordinates x_1, x_2, p_1, p_2 of Ω. H is obtained from H_0 by substituting $x_3 = p_2$.

The extremal curves then satisfy

$$\frac{dx_1}{dt} = x_2, \qquad \frac{dx_2}{dt} = p_2, \qquad \frac{dp_1}{dt} = x_1, \qquad \frac{dp_2}{dt} = -p_1 - x_2.$$

Problem 4 Consider the problem of minimizing $\frac{1}{2}\int_0^T (2x_1^2 + 2x_1x_3 + x_3^2 + x_4^2)\, dt$ over the trajectories of

$$\frac{dx_1}{dt} = u_1, \qquad \frac{dx_2}{dt} = u_2, \qquad \frac{dx_3}{dt} = x_2, \qquad \frac{dx_4}{dt} = x_1.$$

(a) Show that $h_1 = p_1$ and $h_2 = p_2$ are jump Hamiltonians with ind$(h_1) = 0$ and ind$(h_2) = 1$.

(b) Show that the optimal space Ω is determined by the following equations:

$$p_1 = 0, \qquad -2x_1 - x_3 + p_4 = 0, \qquad p_2 = 0, \qquad p_3 = 0,$$
$$x_3 + p_4 = 0, \qquad x_2 + x_4 = 0.$$

(c) Show that each optimal trajectory in the canonical coordinates (x_4, p_4) of Ω satisfies $dx_4/dt = p_4$ and $dp_4/dt = x_4$.

Problem 5 Show that $\lim_{T\to 0} v(a, b, T) < \infty$ if and only if $(b - a) \in J$.

Problem 6 Show that $T_{\max}$ can be defined by the following property: $T_{\max}$ is the smallest time T with the property that there exists a regular trajectory $(x(t), u(t))$ defined on the interval $[0, T]$ such that both $x(0)$ and $x(T)$ belong to J.

Notes and sources

This treatment of the singular problem is essentially taken from an unpublished study by Jurdjevic and Kupka ("Linear systems with singular quadratic cost," a preprint for a 1992 meeting at the University of Toronto). For a problem with a convex cost function, the cost of any jump is equal to zero. For such problems, but particularly for problems with $c = \frac{1}{2}(Pu, u)$, there is a considerable literature under the rubric of singular controls or "cheap" controls; see, for instance, Kitapcu, Silverman, and Willems (1986) for additional sources. The asymptotic nature of jumps and their approximations by regular trajectories have also been recognized in the aircraft industry (Kelly and Edelbaum, 1970).

The existence of Dirac's work (1950) on generalized Hamiltonian mechanics was brought to my attention by R. Hermann (personal communication). Dirac classifies all constraints as primary and secondary. His primary constraints correspond to the jump Hamiltonians in this presentation. The reduction procedure used in this chapter to resolve the constraints bears some similarity to the methods used by van der Schaft (1987).

10
Time-optimal problems and Fuller's phenomenon

The process of transferring one state into another along a trajectory of a given differential system such that the time of transfer is minimal is known as the *minimal-time problem*, and it is one of the basic concerns of optimal control theory.

Minimal-time problems go back to the beginning of the calculus of variations. John Bernoulli's solution of the brachistochrone problem in 1697 was based on Fermat's principle of least time, which postulates that light traverses any medium in the least possible time. According to Goldstine's account (1980) of the history of the calculus of variations, Fermat announced that principle in 1662 in his collected works by saying that "nature operates by means and ways that are the easiest and fastest," and he further differentiated that statement from the statement that "nature always acts along shortest paths" by citing an example from Galileo concerning the paths of particles moving under the action of gravity. Since then, time-optimal problems have remained important sources of inspiration during the growth of the calculus of variations.

In spite of the extensive literature on the subject, control theorists in the early 1950s believed that the classic theory did not adequately confront optimal problems that involved inequalities and was not applicable to problems of optimal control. Their early papers on time-optimal control problems paved the way to the maximum principle as a necessary condition for optimality.

This chapter begins with linear time-optimal problems and provides a self-contained characterization of their time-optimal trajectories. Looking ahead to a more general version of the maximum principle, we consider the brachistochrone problem and Zermelo's navigation problem as time-optimal control problems and obtain their classic solutions by control-theory methods.

In contrast to the aforementioned problems in which optimal controls always take values on the boundary of the control constraint set, there are variational problems for which optimal controls take values in the interior of the constraint set. Such controls are called *singular*. The existence of singular solutions can have dramatic effects on the nature of optimal synthesis, causing drastic oscillations in the solutions, with switching times having accumulation points. Such behavior goes by the name of Fuller's phenomenon.

This chapter ends with a complete discussion of the solutions for Fuller's original problem. Rather than treating Fuller's problem in isolation, we first consider the class of variational problems that are likely to exhibit Fuller's phenomenon; this class is the class of singular linear quadratic problems with convex cost and convex constraints. We show that the maximum principle for this class of problems provides the necessary and sufficient conditions for optimality. It then follows from the material in Chapter 9 that these systems always have singular solutions, because any solution of the singular linear quadratic problem remains optimal for the problem with constraints, provided that the corresponding control does not violate the constraints. Of course, bounds on controls introduce new solutions. Fuller's phenomenon generally occurs when it becomes necessary to fuse a boundary-valued optimal control with a singular one.

1 Linear time-optimal problems: the maximum principle

Time-optimal control theory begins with linear time-invariant systems,

$$\frac{dx}{dt} = Ax + Bu, \tag{1}$$

defined in a real, finite-dimensional vector space M in which the control functions are restricted to a compact and convex neighborhood U_c of the origin in a finite-dimensional control space U. Throughout this section, it will be assumed that (1) is controllable and that control functions are measurable. As in the preceding chapters, the pair (x, u), in which x is an absolutely continuous curve on some interval $[0, T]$ that satisfies (1) almost everywhere in $[0, T]$, will be called a trajectory pair, or simply a trajectory if there is no danger of ambiguity.

Definition 1 A trajectory (x, u) is called time-optimal on an interval $[0, T]$ if for any other trajectory (y, v) of (1) defined on its interval $[0, S]$, for which $y(0) = x(0)$ and $y(S) = x(T)$, S is larger than or equal to T.

Theorem 1

(a) *For any time-optimal trajectory (x, u) on an interval $[0, T]$, the terminal point $x(T)$ belongs to the boundary $\partial \mathcal{A}(x(0), T)$ of the set of reachable points from $x(0)$ at $t = T$ of system (1).*

(b) *Any point b that belongs to the boundary of the set reachable from the origin at time T is the terminal point of a time-optimal trajectory on the interval* $[0, T]$.

Proof If $x(T)$ belonged to the interior of $\mathcal{A}(x(0), T)$, then $x(T)$ would also belong to the interior of $\mathcal{A}(x(0), T - \varepsilon)$, for some $\varepsilon > 0$, which is not possible, because that would violate the time optimality of (x, u) on the interval $[0, T]$. The argument is as follows:

Let $y_1, \ldots, y_n$ be the vertices of a simplex contained in $\mathcal{A}(x(0), T)$, with $x(T)$ in its center. Let (x_i, u_i) be any trajectories such that $x_i(0) = x(0)$ and $x_i(T) = y_i$. For small $\varepsilon > 0$, the simplex with vertices $x_i(T - \varepsilon)$, with $i = 1, 2, \ldots, n$, contains $x(T)$ in its interior. Each convex combination $\sum_{i=1}^{m} \lambda_i x_i(T - \varepsilon)$ is reachable from $x(0)$ by the control $\sum_{i=1}^{m} \lambda_i u_i$. Therefore, $x(T)$ belongs to $\mathcal{A}(x(0), T - \varepsilon)$. This argument proves part (a).

To prove (b), note that the sets reachable from the origin satisfy the following relations:

$$\mathcal{A}(0, \leq t_1) = \mathcal{A}(0, t_1) \subseteq \mathcal{A}(0, t_2) = \mathcal{A}(0, \leq t_2)$$

for any $t_1 \leq t_2$. Because linear systems are analytic, they are Lie-determined, and therefore $\mathcal{A}(a, \leq t) \subset \text{int}(\mathcal{A}(a, \leq t + \varepsilon))$ for any $a \in M, t > 0$, and any $\varepsilon > 0$ (Theorem 2, Chapter 3). Consequently, $\mathcal{A}(0, t) \subset \text{int}(\mathcal{A}(0, t + \varepsilon))$ for any $t > 0$ and $\varepsilon > 0$.

The preceding argument shows that for any $T > 0$, points on the boundary of $\mathcal{A}(0, T)$ cannot be reached in a time shorter than T. On the other hand, $\mathcal{A}(0, T)$ is compact for each $T > 0$ (a consequence of Theorem 11, Chapter 4). Therefore, for each b on $\partial \mathcal{A}(0, T)$ there exists a trajectory (x, u) defined on the interval $[0, T]$ such that $x(0) = 0$ and $x(T) = b$. It follows by the foregoing argument that (x, u) is time-optimal on $[0, T]$, and the proof is finished. ∎

Problem 1 Consider $dx_1/dt = x_2 + u$ and $dx_2/dt = -x_1$, with $|u(t)| \leq \varepsilon$. Show that for $a = (1, 0)$, there exist $\varepsilon > 0$ and $T > 0$ such that $a \in \partial \mathcal{A}(a, T)$, but that a is not a terminal point of any time-optimal trajectory (x, u) on $[0, T]$.

Theorem 2 (the maximum principle) *Any time-optimal trajectory* $(\bar{x}, \bar{u})$ *on an interval* $[0, T]$ *is the projection of an integral curve* $(\bar{x}, \bar{p}, \bar{u})$ *of the Hamiltonian vector field* $\overrightarrow{H}$ *associated with* $H(x, p, u) = -p_0 + p(Ax + Bu)$, *with* p_0 *equal to either* 0 *or* 1, *such that*

(a) $H(\bar{x}(t), \bar{p}(t), \bar{u}(t)) = \max_{u \in U_c} H(\bar{x}(t), \bar{p}(t), u)$ *for almost all t in* $[0, T]$,

(b) $H(\bar{x}(t), \bar{p}(t), \bar{u}(t)) = 0$ *a.e. in* $[0, T]$, *and*

(c) $\bar{p}(t) \neq 0$ *for any t, if* $p_0 = 0$.

Before going into the details of the proof, several explanatory remarks will be helpful:

1. H should be regarded as a function on $T^*M = M \times M^*$ parametrized by both the choice of a control function and the value of p_0.

2. Assume that $u(t)$ is a given measurable control function with values in U_c. Each integral curve $\sigma(t) = (x(t), p(t))$ of the Hamiltonian vector field $\vec{H}$ associated with $H(x, p, u(t)) = -p_0 + p(Ax + Bu(t))$, when expressed in canonical coordinates, satisfies the following pair of differential equations:

$$\frac{dx(t)}{dt} = Ax(t) + Bu(t), \qquad \frac{dp}{dt} = -A^*p(t). \tag{2}$$

3. The maximality condition (a) of Theorem 2 is equivalent to $\bar{p}(t)B\bar{u}(t) = \max_{u \in U_c} \bar{p}(t)Bu$ for almost all t in $[0, T]$.

The proof of Theorem 2 depends on the following auxiliary lemma:

Lemma *Let* $p(t)$ *denote an absolutely continuous curve in* M^* *defined on an interval* $[0, T]$. *Then* $m(t) = \max_{u \in U_c} p(t)Bu$ *is an absolutely continuous curve on* $[0, T]$. *Moreover, for each t for which* dm/dt *exists, there is an element* $u(t)$ *in* U_c *such that* $dm/dt = (dp(t)/dt)Bu(t)$, *and* $p(t)Bu(t) = m(t)$.

Proof For each t $(0 \leq t \leq T)$, denote by K_t the set

$$\left\{ v : p(t)Bv = \max_{u \in U_c} p(t)Bu \right\}.$$

Because U_c is a compact and convex set, it follows that K_t is also compact and convex. The collection $\{K_t : 0 \leq t \leq T\}$ of compact sets has the following continuity property:

$$\lim_{h \to 0} |K_{t+h} - K_t| = 0 \quad \text{for each } t.$$

For this notation, $|A - B| = \inf\{\|a - b\| : a \in A, \ b \in B\}$.

The argument is as follows: Let $\{h_n\}$ be any sequence such that $\lim h_n = 0$. For each n, let $u_n \in K_{t+h_n}$. Because $\{u_n\} \subset U_c$, which is compact, it contains a convergent subsequence. We shall assume that $\{u_n\}$ itself is convergent, and we denote

$$\lim_{n \to \infty} u_n = u_\infty.$$

It follows that for each n, and each u in K_t, $p(t + h_n)Bu_n \geq p(t + h_n)Bu$, and then, by passing to the limit, $p(t)Bu_\infty \geq p(t)Bu$. Therefore, $p(t)Bu_\infty = p(t)Bu$, and consequently $u_\infty \in K_t$. This shows that

$$\lim_{n \to \infty} \left| K_{t+h_n} - K_t \right| = 0.$$

The remaining part of the proof depends on the following estimates: For any $u(t + h)$ in K_{t+h} and any $u(t)$ in K_t,

$$m(t+h) - m(t) = p(t+h)Bu(t+h) - p(t)Bu(t) \leq (p(t+h) - p(t))Bu(t+h)$$

and

$$m(t+h) - m(t) = p(t+h)Bu(t+h) - p(t)Bu(t) \geq (p(t+h) - p(t))Bu(t).$$

Therefore,

$$(p(t + h) - p(t))Bu(t) \leq m(t + h) - m(t) \leq (p(t + h) - p(t))Bu(t + h).$$

It follows from the preceding argument concerning the continuity of $\{K_t\}$ that $u(t + h)$ and $u(t)$ can be so selected that $\lim_{h \to 0} u(t + h) = u(t)$. Therefore,

$$\frac{dm(t)}{dt} = \frac{dp(t)}{dt} Bu(t)$$

for each t for which $dp(t)/dt$ exists. The proof of the lemma is now finished. ∎

To prove Theorem 2 it will be necessary to use the following separation theorem for convex compact sets in a finite-demensional vector space M. Recall first that a hyperplane in M is any linear subspace of M having codimension 1. Any hyperplane is the zero set of a nonzero linear function. Thus, linear functions are in exact correspondence with hyperplanes in M. A linear function p is said to define a supporting hyperplane at a point y of a set K if $p(x - y) \leq 0$ for all x in K. We shall use the fact that a compact convex set has a supporting hyperplane at each of its boundary points. We now return to the proof of Theorem 2.

Proof Denote by $(\bar{x}(t), \bar{u}(t))$ a trajectory of (1) that is time-optimal on an interval $[0, T]$. Let a denote its initial point $x(0)$. According to Theorem 1, $\bar{x}(T)$ belongs to the boundary $\partial \mathcal{A}(a, T)$ of the set reachable from a at $t = T$. Because $\mathcal{A}(a, T)$ is a convex compact body, there is a supporting hyperplane at $\bar{x}(T)$. Let p_T denote the linear function that defines a supporting hyperplane at $\bar{x}(T)$. Then $p_T(x(T) - \bar{x}(T)) \leq 0$ for any trajectory (x, u) on $[0, T]$ with $x(0) = a$. That means that

$$ \int_0^T p_T \left(e^{A(T-t)} B(u(t) - \bar{u}(t)) \, dt \right) \leq 0. $$

Define $\bar{p}(t) = e^{A^*(T-t)} p_T$. The preceding inequality becomes

$$ \int_0^T \bar{p}(t) B(u(t) - \bar{u}(t)) \, dt \leq 0 $$

for any measurable control function $u(t)$ that takes values in U_c. Hence, $\bar{p}(t) B(u(t) - \bar{u}(t)) \leq 0$ for almost all t in $[0, T]$. If we write $m(t) = \max_{u \in U_c} \bar{p}(t) Bu$, then $\bar{p}(t) B \bar{u}(t) = m(t)$ almost everywhere in the interval $[0, T]$. It follows from the preceding lemma that

$$ \frac{dm}{dt} = \frac{d\bar{p}}{dt} B \hat{u}(t) $$

for some $\hat{u}(t)$ such that

$$ \bar{p}(t) B \hat{u}(t) = \max_{u \in U_c} \bar{p}(t) Bu. $$

Because $d\bar{p}/dt = -A^*(t) \bar{p}(t)$, it follows that $dm/dt = -A^* \bar{p}(t) B \hat{u}(t)$ for all t. But then $dm/dt = -A^* \bar{p}(t) B \bar{u}(t)$ for almost all t, because $\bar{p}(t) B \bar{u}(t) = \bar{p}(t) B \hat{u}(t)$ almost everywhere. Thus,

$$ \frac{d}{dt} \bar{p}(t)(A\bar{x}(t) + B\bar{u}(t)) = \frac{d}{dt} \bar{p}(t) A\bar{x}(t) + \frac{dm}{dt} $$
$$ = A^* \bar{p}(t) B \bar{u}(t) - A^* \bar{p}(t) B \hat{u}(t) = 0 $$

a.e. in $[0, T]$, and hence $\bar{p}(t)(A\bar{x}(t) + m(t)) = $ constant.

It remains to show that there exists a supporting hyperplane at $\bar{x}(T)$ such that the preceding constant is non-negative, for then the constant can be reduced to either 1 or 0 by suitably normalizing p_T.

A trajectory $(\bar{x}, \bar{u})$ that is time-optimal on $[0, T]$ must be time-optimal on every subinterval $[0, t]$. Hence $\bar{x}(t)$ belongs to the boundary $\partial \mathcal{A}(0, t)$ for each

t $(t \leq T)$. Moreover, it follows from time optimality that $\bar{x}(T)$ does not belong to $\mathcal{A}(a, t)$ for each t $(t < T)$.

Let p_t be any nonzero element in M^* that separates $\bar{x}(T)$ from $\mathcal{A}(a, t)$; that is, $p_t(x(t) - \bar{x}(t)) \leq 0$ for any trajectory (x, u) on $[0, t]$, with $x(0) = a$ and $p_t(\bar{x}(T) - \bar{x}(t)) > 0$. This implies that there is a set of positive measure contained in the interval $[t, T]$ such that $p_t(A\bar{x}(\tau) + B\bar{u}(\tau)) > 0$ for all τ in this set.

The remaining part of the proof is based on compactness arguments. For that reason, we shall assume that M^* is equipped with a Euclidean norm $\| \ \|$, in which case we shall assume that $\|p_t\| = 1$.

Let $\{t_n\} \subset [0, T]$ be a sequence such that $\lim_{n \to \infty} t_n = T$. Because $\{p_{t_n}\}$ is a bounded sequence, it contains a convergent subsequence. Therefore, we may as well assume that $\{p_{t_n}\}$ converges. Let $p_T = \lim_{n \to \infty} p_{t_n}$.

Let τ_n be any instant in $[t_n, T]$ such that $p_{t_n}(A\bar{x}(\tau_n) + B\bar{u}(\tau_n)) > 0$. The sequence $\{\bar{u}(\tau_n)\}$ contains a convergent subsequence. We shall assume that $\{\tau_n\}$ is so chosen that $\{\bar{u}(\tau_n)\}$ converges. Let $u_\infty = \lim_{n \to \infty} \bar{u}(\tau_n)$. It then follows from our construction that $p_T(x(T) - \bar{x}(T)) \leq 0$ for each trajectory (x, u) on $[0, T]$, with $x(0) = a$ and $p_T(A\bar{x}(T) + Bu_\infty) \geq 0$.

As in the first part of the theorem, define $\bar{p}(t) = e^{A^*(T-t)} p_T$. Then $\bar{p}(T) A\bar{x}(T) + m(T) \geq p_T(A\bar{x}(T) + Bu_\infty) \geq 0$. Because $\bar{p}(t) A\bar{x}(t) + m(t)$ is constant, this constant is non-negative. Let $p_0 = \bar{p}(t)(A\bar{x}(t) + m(t))$. If p_0 is zero, our proof is finished. Otherwise, $p_0 > 0$, in which case we multiply p_T by a suitable positive scalar such that $p_0 = 1$. The proof of the maximum principle is now finished. ∎

Definition 2 An absolutely continuous curve $\sigma(t) = (x(t), p(t))$ defined on an interval $[0, T]$ is called an *extremal curve* if there exists an integrable control function $u(t)$ on $[0, T]$ with values in U_c such that the following hold:

(a) $\sigma(t)$ is an integral curve of the Hamiltonian vector field $\vec{H}$ corresponding to $H(x, p, u(t)) = -p_0 + p(Ax + Bu(t))$, with p_0 equal to either 0 or 1, such that

(b) $H(x(t), p(t), u(t)) = \max_{v \in U_c} H(x(t), p(t), v)$ for almost all t in $[0, T]$,

(c) $H(x(t), p(t), u(t)) = 0$ for almost all t in $[0, T]$, and $p(t) \neq 0$ if $\lambda = 0$.

The extremal curves that correspond to $p_0 = 1$ will be called "regular." Following the classic terminology due to Bliss (1925), the extremals that correspond to $p_0 = 0$ will be called "abnormal." Carathéodory (1935), in his treatment of Zermelo's navigation problem, calls abnormal extremals "anomalous." As he notes in the same problem, abnormal directions have special geometric properties. For time-optimal problems, abnormal extremals appear for both the minimal-time problem and the maximal-time problem, and the maximum principle does not differentiate between the two problems in such a case. We shall illustrate some of these issues with the applications that follow.

As we shall see, there is ample evidence to suggest that abnormal extremal curves have particular geometric significance, reflecting the nature of the constraints that are present in the problem, and it is only with respect to the historical origins of the calculus of variations that they appear abnormal or anomalous. For general control systems, they are integral parts of the solutions and should be treated as such.

1.1 Time-optimal control of linear mechanical systems

The preceding theory yields easy solutions when applied to a second-order linear system controlled by an external force u restricted in magnitude to $|u| \leq \varepsilon$. The equations are

$$\frac{dy^2}{dt^2} + k\frac{dy}{dt} + gy = u(t), \tag{3}$$

for some non-negative constants k and g. The equivalent first-order system in $\mathbb{R}^2$ induced by the coordinates $x_1 = y$ and $x_2 = dy/dt$ is given by

$$\frac{dx_1}{dt} = x_2, \qquad \frac{dx_2}{dt} = -gx_1 - kx_2 + u.$$

In vector notation,

$$\frac{dx}{dt} = Ax + bu, \quad \text{with } A = \begin{pmatrix} 0 & 1 \\ -g & -k \end{pmatrix}, \ b = e_2, \ \text{and } x(t) = \begin{pmatrix} x_1(t) \\ x_2(t) \end{pmatrix}. \tag{4}$$

It follows from Theorem 6 in Chapter 5 that $k = 0$ is a necessary and sufficient condition for controllability of the foregoing system. So we shall consider only the cases for which $k = 0$.

Case 1 $g = 0$. In this case, $A^2 = 0$, and therefore any two states a and b in $\mathbb{R}^2$ can be reached from each other along a trajectory of the system. Hence, b can be reached from an initial point a time-optimally. Time-optimal trajectories are contained among the extremal curves that are solutions of

$$\frac{dx_1}{dt} = x_2, \qquad \frac{dx_2}{dt} = u(t), \qquad \frac{dp_1}{dt} = 0, \qquad \frac{dp_2}{dt} = -p_1, \tag{5}$$

with $u(t)$ satisfying $p_2(t)u(t) = \max_{|u| \leq \varepsilon} p_2(t)u$, a.e.

It follows that $u(t) = 1$ when $p_2(t) > 0$, and $u(t) = -1$ when $p_2 < 0$. The surface $p_2 = 0$ in $\mathbb{R}^4$ is called the *switching surface*. Because any solution curve for (5) satisfies $p_1 = $ constant and $p_2 = -p_1 t + c$, each extremal curve crosses the switching surface at most once. Hence, each time-optimal control is a piecewise-constant control with values $\pm\varepsilon$ having at most one discontinuity. This means that time-optimal paths require maximum possible acceleration/deceleration, with at most one switch from one to the other.

The projections (x_1, x_2) of the extremal curves are solutions of $dt_1/dt = x_2$, $dx_2/dt = \pm\varepsilon$. The solutions are semiparabolas in $\mathbb{R}^2$ traced by

$$x_2(t) = a_2 \pm \varepsilon t, \qquad x_1(t) = a_1 + a_2 t \pm \varepsilon \frac{t^2}{2}, \quad \text{for} \quad t > 0.$$

That is,

$$t = \frac{1}{\varepsilon} |x_2 - a_2|, \quad \text{and} \quad x_1 = a_1 + \frac{a_2(x_2 - a_2)}{\varepsilon} \pm \frac{(x_2 - a_2)^2}{2\varepsilon}.$$

Figure 10.1 depicts some typical extremal paths.

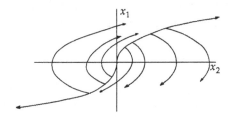

Fig. 10.1.

Abnormal extremal curves are contained to the surface $p_1 x_2 + \varepsilon(\operatorname{sgn} p_2)$ $p_2 = 0$. In particular, if $x(0) = 0$, then $\varepsilon(\operatorname{sgn} p_2) p_2(0) = 0$. So $p_2(0) = 0$, and therefore $p_2(t) = -p_1 t$. Since $p_1 \neq 0$, because $p_1^2 + p_2^2 \neq 0$, it follows that abnormal extremal curves that originate at $x_2(0)$ cannot contain a switch in the control. In general, at the switching time t, $x_2(t) = 0$ for any abnormal extremal curve, since $(p_1, p_2) \neq 0$.

Regular extremal curves are constrained to the surface $p_1 x_2 + \varepsilon(\operatorname{sgn} p_2)$ $p_2 = 1$. The problem of deciding which of these types of extremals corresponds to the correct optimal solution is a recurring problem in optimal control theory. In this problem, it is easy to see that the time-optimal solutions that either originate or terminate at the origin are projections of the regular extremals, because the terminal point that corresponds to a nonswitching extremal is a singular point on the boundary of the reachable set, admitting a cone of separating hyperplanes at that point (Figure 10.2).   ∎

Fig. 10.2.

Case 2 $k = 0$, $g \neq 0$. This case corresponds to the control of a linear harmonic oscillator through an external force u. The harmonic oscillator is the only remaining case in which the corresponding linear system is controllable. In this case,

$$\frac{dp}{dt} = -A^* p = \begin{pmatrix} 0 & g \\ -1 & 0 \end{pmatrix} \begin{pmatrix} p_1 \\ p_2 \end{pmatrix}.$$

The most general solution is of the form $p_2(t) = A \cos(\omega t - \varphi)$, with $\omega = \sqrt{g}$, and some fixed angle φ.

Because $p(t)Bu = p_2(t)u$, it follows that extremal control $u(t)$ is given by $u(t) = \varepsilon(\operatorname{sgn} p_2)(t)$, with $p_2 = 0$ the switching surface. In this case $p_2(t)$ crosses the switching surface whenever $A \cos(\omega t_n + \varphi) = 0$. Thus $p_2(t)$ crosses the switching surface at time intervals π/ω.

The trajectories that correspond to the extremal curves are solutions of the following differential equations:

$$\frac{dx_1}{dt} = x_2, \qquad \frac{dx_2}{dt} = -g x_1 \pm \varepsilon.$$

The solutions are ellipses centered at $x_1 = \pm\varepsilon/\omega^2$, $x_2 = 0$:

$$x_1(t) = B \sin(\omega t - \theta) \pm \frac{\varepsilon}{\omega^2}, \qquad x_2(t) = B\omega \cos(\omega t - \theta),$$

for constants B and θ. Time-optimal solutions reach the origin by jumping from one semi-arc of these ellipses to another, as shown in Figure 10.3.

Summarizing the information obtained earlier, it follows that time-optimal control functions are piecewise-constant functions $u(t)$ that take values $\pm\varepsilon$ and whose discontinuities occur at time intervals π/ω. The extremal

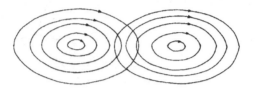

Fig. 10.3.

control functions also belong to this class and contain the time-optimal controls.

The extremal curve $(x(t), p(t))$ that accompanies an extremal control $u(t)$ is regular provided that

$$p_1(t)x_2(t) - \omega^2 p_2(t)x_1(t) + u(t)p_2(t) = 1$$

almost everywhere. Because $p_2(t) = A\cos(\omega t - \varphi)$, it follows that $p_1(t) = A\omega\sin(\omega t - \varphi)$, and therefore

$$
\begin{aligned}
p_1 x_2 - \omega^2 p_2 x_1 + up_2 &= AB\omega^2(\sin(\omega t - \varphi)\cos(\omega t - \theta) \\
&\quad - \cos(\omega t - \varphi)\sin(\omega t - \varphi)) \mp \varepsilon A\cos(\omega t - \varphi) \\
&\quad + \varepsilon(\operatorname{sgn} A)(\cos\omega t - \varphi) = AB\omega^2\sin(\theta - \varphi)
\end{aligned}
$$

almost everywhere.

Regular extremal curves correspond to $\sin(\theta - \varphi) \neq 0$, and abnormal extremals correspond to $\sin(\theta - \varphi) = 0$. For this problem, there are points b that can be reached time-optimally from the origin only by an abnormal extremal. For instance, let $b = (2\varepsilon/\omega^2, 0)$. Then $b = (2\varepsilon/\omega^2, 0)$ is reached time-optimally by the constant control $u = \varepsilon$ in $\omega T = \pi$ units of time. The corresponding adjoint variable p_2 must be of the form $p_2(t) = A\sin\omega t$, because $p_2(t) \neq 0$ in the interval $0 < \omega t < \pi$. Then $p_1(t) = A\omega\cos\omega t$, because $dp_2/dt = p_1(t)$. Then

$$p_1 x_2 - \omega^2 p_2 x_1 + \varepsilon p_2 = -(A\omega\cos\omega t)\left(\frac{\varepsilon}{\omega}\sin\omega t\right)$$

$$+ (A\sin\omega t)(-\varepsilon + \varepsilon\cos\omega t + \varepsilon) = 0.$$

This conclusion is evident from the diagram describing the shape of the set reachable from the origin: Among all the supporting hyperplanes at $(2\varepsilon/\omega^2, 0)$, the only hyperplane whose normal makes a sharp angle with the optimal trajectory is the vertical plane. Its normal gives rise to an abnormal extremal (Figure 10.4).   ∎

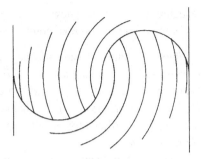

Fig. 10.4.

2 The brachistochrone problem and Zermelo's navigation problem

Time-optimal problems naturally extend to arbitrary control systems, and as we shall see in the next chapter, the maximum principle remains a valid necessary condition of optimality. In this section we shall illustrate its significance with two classic problems from the calculus of variations. To set the stage for these problems, it is necessary to replace a linear system with an arbitrary control affine system:

$$\frac{dx}{dt} = X_0(x) + \sum_{l=1}^{m} u_i(t) X_i(x) \tag{6}$$

on a manifold M.

The corresponding Hamiltonian is a function on T^*M further parametrized by a constant p_0 and the controls $u(t) = (u_1(t), \ldots, u_m(t))$. It is given by

$$H(\xi, u(t)) = -p_0 + \xi X_0(x) + \sum_{l=1}^{m} u_i(t)(\xi X_i(x)) \quad \text{for each } \xi \text{ in } T_x^*M.$$

As on vector spaces, so it is on manifolds, any function on T^*M defines a vector field, called the Hamiltonian vector field (to be defined in the next chapter). The Hamiltonian vector field that corresponds to $H(\xi, u(t))$ will be denoted by $\vec{H}(\xi, u(t))$. The integral curves of $H(\xi, u(t))$ project onto the trajectories of differential system (6).

The maximum principle for time-optimal problem for (6), with controls taking values in a set U, is as given by Theorem 2, with H replaced by the Hamiltonian of system (6), and the pair $(\bar{x}(t), \bar{p}(t))$ replaced by an integral curve $\xi(t)$ of H.

Leaving all technical details aside, we shall now consider the brachistochrone problem and Zermelo's problem as time-optimal control problems.

The brachistochrone problem Assume that a and b are any points in the physical space E^3. Among all the planar curves that connect a to b, find the one along which the time of travel, from a to b, of a mechanical bead is minimal. It is assumed that the gravitational force is the only force that acts on the bead, and it is also assumed that a and b are sufficiently close to the surface of the earth that the gravitational force is constant.

We shall regard this mechanical problem as a control problem, the controls being the curves along which the bead slides. Let e_1, e_2, e_3 denote a fixed orthonormal frame in E^3 centered at the initial point a. If m denotes the mass of the bead, then we shall assume that the orthonormal frame is so oriented that the gravitational force on the bead is equal to $\vec{F} = -mg e_2$ and that the terminal point b is contained in the plane spanned by e_1 and e_2.

A curve $x : [0, T] \to E^3$ is called "regular" if $(dx/dt)(t) \neq 0$ for each $t \in (0, T)$. We shall restrict the preceding problem to the regular curves that pass through a and are contained in the plane spanned by e_1 and e_2. Each such curve will be parametrized by its arc length from a. The coordinates of $x(s)$ relative to the frame e_1, e_2 are denoted by $x_1(s)$, $x_2(s)$. It follows that $(dx_1/ds)^2 + (dx_2/ds)^2 = 1$, and $dx/ds = (dx_1/ds)e_1 + (dx_2/ds)e_2$.

For each curve $x(s)$, ds/dt denotes the speed of the bead along $x(s)$. According to Newton's law of motion,

$$m\frac{d^2s}{dt^2} = F_T(s(t)),$$

where F_T denotes the tangential force on the bead.

The tangential force along a curve x is most naturally expressed in terms of a right-handed orthonormal frame $(T(s), N(s))$ along the curve. In fact,

$$T(s) = \frac{dx_1}{ds}e_1 + \frac{dx_2}{ds}e_2, \quad \text{and} \quad N(s) = -\frac{dx_2}{ds}e_1 + \frac{dx_1}{ds}e_2.$$

Then

$$\vec{F} = -mg e_2 = \alpha T(s) + \beta N(s)$$
$$= \alpha\left(\frac{dx_1}{ds}e_1 + \frac{dx_2}{ds}e_2\right) + \beta\left(-\frac{dx_2}{ds}e_1 + \frac{dx_1}{ds}e_2\right),$$

and

$$\alpha = -mg\frac{dx_2}{ds}, \quad \text{and} \quad \beta = -mg\frac{dx_1}{ds}.$$

Because $\alpha = F_T$, it follows that

$$m\frac{d^2s}{dt^2} = -mg\frac{dx_2}{ds}, \quad \text{or} \quad \frac{d^2s}{dt^2} = -g\frac{dx_2}{ds}.$$

After multiplying by ds/dt and integrating, we get that

$$\left(\frac{ds}{dt}\right)^2 = -gx_2(s), \quad \text{or} \quad \frac{ds}{dt} = \sqrt{-gx_2(s)}, \text{ with } x_2(s) \le 0.$$

Let $(x_1(t), x_2(t))$ denote the position of the bead at time t along a given curve x; that is, $x_1(t) = x_1(s(t))$, and $x_2(t) = x_2(s(t))$. The dynamics are given by

$$\frac{dx_1}{dt} = \frac{dx_1}{ds}\frac{ds}{dt} = \frac{dx_1}{ds}\sqrt{-gx_2} = u_1\sqrt{-gx_2},$$

$$\frac{dx_2}{dt} = \frac{dx_2}{ds}\sqrt{-gx_2} = u_2\sqrt{-gx_2}, \tag{7}$$

subject to $u_1^2 + u_2^2 = 1$ and $x_2 \le 0$. Our problem is to choose $u = (u_1, u_2)$ such that b is reached in minimal time.

There are two technical issues associated with this problem: The first is that the manifold (x_1, x_2), with $x_2 \le 0$, has a boundary and that the initial point a is in its boundary. For that reason we shall assume that the optimal trajectory does not intersect the boundary $x_2 = 0$ for any of its points, apart from the initial point. The second issue is that the sphere S^2 is not convex. Therefore, a priori there is no guarantee that there will be time-optimal solutions. For that reason we shall first convexify the space of controls, which means that instead of $u_1^2 + u_2^2 = 1$ we shall assume that $u_1^2 + u_2^2 \le 1$. The reader may verify that the system consists of complete vector fields. Then for any a and b, with $a_2 < 0$, $b_2 < 0$, there exists a time-optimal trajectory $(\bar{x}(t), \bar{u}(t))$ such that $\bar{x}(0) = a$ and $\bar{x}(T) = b$. T is the minimal time (Theorem 11 in Chapter 4).

Let p_1 and p_2 be the canonical coordinates associated with x_1 and x_2. The Hamiltonian for this problem, H_λ, is given by $H_\lambda = -\lambda + u_1 p_1\sqrt{-gx_2} + u_2 p_2\sqrt{-gx_2}$. (It is more convenient to write λ instead of p_0.) Each optimal trajectory is a projection of the following system:

$$\frac{d\bar{x}_1}{dt} = \bar{u}_1\sqrt{-g\bar{x}_2}, \qquad \frac{d\bar{x}_2}{dt} = \bar{u}_2\sqrt{-g\bar{x}_2}, \qquad \frac{d\bar{p}_1}{dt} = -\frac{\partial H}{\partial x_1} = 0,$$

$$\frac{d\bar{p}_2}{dt} = -\frac{\partial H_\lambda}{\partial x_2} = \frac{g\bar{p}_1}{2\sqrt{-g\bar{x}_2}}\bar{u}_1(t) + \frac{g\bar{p}_2}{2\sqrt{-g\bar{x}_2}}\bar{u}_2(t),$$

subject to

$$H_\lambda(\bar{x}(t), \bar{p}(t), \bar{u}(t)) = \sup_{\|u\|\le 1} (H_\lambda(\bar{x}(t), \bar{p}(t), u)) \quad \text{for almost all } t,$$

and

$$H_\lambda(\bar{x}(t), \bar{p}(t), \bar{u}(t)) = 0 \quad \text{for almost all } t.$$

If $\lambda = 0$, then $\bar{p}(t) \ne 0$, because of condition (c) of the maximum principle. Furthermore, $\bar{x}_2(t) \ne 0$, by our assumption, and therefore $\bar{u}_1(t) = \bar{u}_2(t) = 0$ almost everywhere, by condition (b) of the maximum principle. But then $\bar{x}(t)$ cannot reach b, because $d\bar{x}(t)/dt = 0$ almost everywhere.

This argument shows that it suffices to consider only the integral curves of H_λ with $\lambda = 1$. Let $p(t)$ and $x(t)$ denote any extremal curve of H_λ corresponding to an extremal control $u(t)$. It is evident that $p(t)$ is not equal to zero on any set of positive measure, because of condition (b). Then the maximality condition

$$\frac{u_1(t)p_1(t) + u_2(t)p_2(t)}{\sqrt{-gx_2(t)}} = \max_{\|v\|\le 1} \frac{v_1 p_1(t) + v_2 p_2(t)}{\sqrt{-gx_2(t)}}$$

immediately yields that $u(t) = p(t)/\|p(t)\|$. Hence $u(t)$ takes values in the unit circle. Therefore $u(t)$ can be parametrized by an angle θ defined by $u_1(t) = \cos\theta(t)$ and $u_2(t) = -\sin\theta(t)$. Angle θ also measures the rotation of the moving frame relative to the absolute frame, because

$$T = \frac{dx_1}{ds}\mathbf{e}_1 + \frac{dx_2}{ds}\mathbf{e}_2 = (\cos\theta)\mathbf{e}_1 - (\sin\theta)\mathbf{e}_2,$$

and

$$N = -\frac{dx_2}{ds}\mathbf{e}_1 + \frac{dx_1}{ds}\mathbf{e}_2 = (\sin\theta)\mathbf{e}_1 + (\cos\theta)\mathbf{e}_2.$$

The differential equation for θ is obtained by differentiating

$$\cos\theta(t) = \frac{p_1(t)}{\|p(t)\|}, \qquad -\sin\theta(t) = \frac{p_2(t)}{\|p(t)\|}.$$

It follows that

$$-\|p(t)\|\sin\theta\frac{d\theta}{dt} + \cos\theta\frac{d}{dt}\|p(t)\| = \frac{dp_1}{dt} = 0,$$

and

$$-\|p(t)\| \cos\theta \, \frac{d\theta}{dt} - \sin\theta \, \frac{d}{dt} \|p(t)\|$$

$$= \frac{dp_2}{dt} = \frac{g}{2\sqrt{-gx_2}} (p_1(t)u_1(t) + p_2(t)u_2(t))$$

$$= -\frac{1}{2} \frac{g}{\sqrt{-gx_2}} \|p(t)\|.$$

After multiplying the first equation by $\sin\theta$ and the second by $\cos\theta$ and adding the results, the foregoing equations yield

$$\frac{d\theta}{dt} = \frac{g}{2\sqrt{-gx_2}} \cos\theta.$$

This equation can be further simplified to

$$\frac{d\theta}{dt} = \frac{g}{2} p_1(t) \tag{8}$$

because $\sqrt{-gx_2} = 1/\|p\|$. Because p_1 is constant, it follows that $\theta(t) = Kt + \theta_0$ for some initial angle θ_0 with $K = (g/2)p_1$.

It is now convenient to parametrize the extremal curves in terms of θ rather than t. We have

$$\frac{dx_1}{d\theta} = \frac{dx_1}{dt} \Big/ \frac{d\theta}{dt} = u_1 \sqrt{-gx_2} \Big/ \frac{g}{2} p_1 = \frac{2\cos\theta}{g\|p\|p_1} = \frac{2\cos^2\theta}{gp_1^2}$$

$$= \frac{2}{gp_1^2} \left(\frac{1 + \cos 2\theta}{2} \right) = \frac{1}{p_1^2 g} (1 + \cos 2\theta),$$

and

$$\frac{dx_2}{d\theta} = \frac{dx_2}{dt} \Big/ \frac{d\theta}{dt} = \frac{2u_2}{gp_1} \sqrt{-gx_2} = \frac{-2\sin\theta}{gp_1\|p\|} = \frac{-2\sin\theta\cos\theta}{gp_1^2} = -\frac{1}{gp_1^2} \sin 2\theta.$$

Therefore, $x_1(\theta) = (1/2gp_1^2)(2\theta + \sin 2\theta)$ and $x_2(\theta) = (1/2gp_1^2)(\cos 2\theta - 1)$ are the extremal curves through the origin.

It follows that $(x_1(\theta), x_2(\theta))$ define a one-parameter family of cycloids all passing through the origin. The reader is invited to compare this solution with the ones obtained by the classic methods.

Problem 2 Show that for any $b = (b_1, b_2)$, with $b_2 < 0$, there exist unique p_1 and θ such that $x_1(\theta) = b_1$ and $x_2(\theta) = b_2$.

Zermelo's navigation problem This problem consists of finding the quickest nautical path for a ship at sea in the presence of currents. Assuming that the currents are stationary, in the sense that they do not change with time, they can be modeled by a planar vector field $X(x_1, x_2)$. We shall assume that X is at least continuously differentiable. In addition, it will be assumed that the speed of the ship cannot exceed a constant k and that the ship can be oriented in each planar direction. The overall dynamics are given by

$$\frac{dx}{dt} = X(x) + u_1(t)\,\mathbf{e}_1 + u_2(t)\,\mathbf{e}_2, \tag{9}$$

with $u_1^2(t) + u_2^2(t) \le k^2$, where $x(t)$ is the position of the ship at time t, and $\mathbf{e}_1$ and $\mathbf{e}_2$ are constant orthonormal vector fields in $\mathbb{R}^2$. The problem is to navigate the ship from a given departure point a in $\mathbb{R}^2$ to a given destination point b in $\mathbb{R}^2$ in the least amount of time. We denote by X_1 and X_2 the coordinates relative to $\mathbf{e}_1$ and $\mathbf{e}_2$. The foregoing differential system becomes

$$\frac{dx_1}{dt} = X_1(x_1, x_2) + u_1, \quad \text{and} \quad \frac{dx_2}{dt} = X_2(x_1, x_2) + u_2.$$

In contrast to the brachistochrone problem, the navigation problem has a drift term in the dynamics that creates several noteworthy distinctions.

The reachable sets $\mathcal{A}(a, \le T)$ associated with (9) are compact, and they grow with time T. Evidently, if the terminal point is reachable at any time, then it is reachable in the minimal time. However, in general there may be points that are not reachable at all. For instance, if the drift current is too strong, then the ship cannot return to its departure point. In the case of a linear drift, a partial controllability answer is provided by Theorem 5 in Chapter 5. Controllability questions for an arbitrary drift are subtle and require detailed analysis. We shall not go into those issues here.

For points that are reachable from a, the maximum principle is applicable, and hence optimal paths are projections of extremal curves. Let

$$H = -\lambda + p_1(X_1 + u_1) + p_2(X_2 + u_2) = -\lambda + p_1 X_1 + p_2 X_2 + u_1 p_1 + u_2 p_2.$$

The extremal curves $x_1(t)$, $x_2(t)$, $p_1(t)$, and $p_2(t)$ are defined by extremal controls $u(t)$, and the following must hold:

$$\frac{dp_1}{dt} = -\frac{\partial H}{\partial x_1}(x(t), p(t), u(t)) = -p_1 \frac{\partial X_1}{\partial x_1} - p_2 \frac{\partial X_2}{\partial x_1},$$

$$\frac{dp_2}{dt} = -\frac{\partial H}{\partial x_2}(x(t), p(t), u(t)) = -p_1 \frac{\partial X_1}{\partial x_2} - p_2 \frac{\partial X_2}{\partial x_2}. \tag{10}$$

It is easy to deduce that each extremal control $u(t)$ must satisfy $\|u(t)\| = k$ for almost all t. The argument is as follows: Equations (10) imply that $p(t)$ cannot be equal to zero at any time without being identically equal to zero. In the regular case, $p(t)$ cannot be identically zero because of condition (b) of the maximum principle, and in the abnormal case $p(t) = 0$ is ruled out by condition (c). So in either case, $p(t)$ is never equal to zero.

The maximality condition $\max_{\|v\| \leq k} H(x(t), p(t), v) = H(x(t), p(t), u(t))$ immediately yields that $u(t) = k(p(t))/\|p(t)\|$ for almost all t. It will be convenient to take $u(t) = k(p(t))/\|p(t)\|$ for all t and dispense with equalities defined only up to a set of measure zero.

Zermelo's navigation formula consists of a differential equation for $u(t)$ expressed in terms of only the drift vector and its first derivatives. It will be derived as follows: Let $\theta(t)$ be the angle given by $u_1(t) = k\cos\theta(t)$, $u_2(t) = k\sin\theta(t)$. Then

$$\cos\theta(t) = \frac{p_1(t)}{\|p(t)\|}, \qquad \sin\theta(t) = \frac{p_2(t)}{\|p(t)\|}.$$

Differentiating the relations $\|p\|\cos\theta = p_1$ and $\|p\|\sin\theta = p_2$ leads to the following system of equations:

$$\cos\theta \frac{d}{dt}\|p(t)\| - \|p(t)\|\sin\theta \frac{d\theta}{dt} = \frac{dp_1}{dt} = -p_1 \frac{\partial X_1}{\partial x_1} - p_2 \frac{\partial X_2}{\partial x_1},$$

$$\sin\theta \frac{d}{dt}\|p(t)\| + \|p(t)\|\cos\theta \frac{d\theta}{dt} = \frac{dp_2}{dt} = -p_1 \frac{\partial X_1}{\partial x_2} - p_2 \frac{\partial X_2}{\partial x_2}.$$

The foregoing equations imply that

$$\|p\| \frac{d\theta}{dt} = -p_1 \cos\theta \frac{\partial X_1}{\partial x_2} - p_2 \cos\theta \frac{\partial X_2}{\partial x_2} + p_1 \sin\theta \frac{\partial X_1}{\partial x_1} + p_2 \sin\theta \frac{\partial X_2}{\partial x_1},$$

as can be shown by multiplying the first equation by $-\sin\theta$ and the second by $\cos\theta$ and adding the resulting equations. The last equation can be written more simply as

$$\frac{d\theta}{dt} = -\cos^2\theta \frac{\partial X_1}{\partial x_2} + \sin\theta \cos\theta \left(\frac{\partial X_1}{\partial x_1} - \frac{\partial X_2}{\partial x_2} \right) + \sin^2\theta \frac{\partial X_2}{\partial x_1}. \qquad (11)$$

Equation (11) agrees with the equation of Carathéodory (1935), which Carathéodory calls the navigation formula of Zermelo. The same equation can be regarded as a generalized Euler-Lagrange equation for the navigation problem. Its solutions determine the extremal controls, which in turn determine the extremal ship paths through equations (8).

Let us now consider some distinctions between the two kinds of extremal curves:

$$0 = H(x(t), p(t), u(t)) = -1 + p_1 X_1 + p_2 X_2 + p_1 u_1 + p_2 u_2$$

$$= -1 + \|p\|(\cos\theta)X_1 + \|p\|(\sin\theta)X_2 + \frac{p_1^2 + p_2^2}{\|p\|}k$$

$$= -1 + \|p\|((\cos\theta)X_1 + (\sin\theta)X_2 + k).$$

The quantity $\Psi(t) = (\cos\theta(t))X_1(x(t)) + (\sin\theta(t))X_2(x(t)) + k$ is therefore always positive along a regular extremal curve. For abnormal extremal curves, the quantity $\Psi(t)$ is identically equal to zero. In his treatment of this problem, Carathéodory noticed that if instead of minimizing the time of travel, the problem was changed to that of finding the maximal time of travel, then all of the equations would remain the same, except that the quantity $\Psi(t)$ would be negative along its regular extremals (corresponding to the multiplier equal to -1). The abnormal extremals for the maximal-time problem would also live on the curve $\Psi(t) = 0$, and in his view, that fact accounted for the existence of two abnormal extremal curves, which he called "anomalous." He also noticed that $\Psi(t)$ is a solution of the following differential equation:

$$\frac{d\Psi}{dt} = \Psi(t)\left(\frac{\partial X_1(x(t))}{\partial x_1}\cos^2\theta(t) + \left(\frac{\partial X_1(x(t))}{\partial x_2} + \frac{\partial X_2(x(t))}{\partial x_1}\right)\right.$$

$$\left. \times \sin\theta(t)\cos\theta(t) + \frac{\partial X_2(x(t))}{\partial x_2}\sin^2\theta(t)\right). \tag{12}$$

Hence, if an initial point of $\Psi(0) = \Psi_0$ is chosen, $\Psi(t)$ is uniquely determined. That observation allowed him to conclude that the projection of an abnormal extremal curve on the state space was the limit of the projections of the regular extremal curves. His argument was as follows: If the initial point a is such that

$$X_1^2(a) + X_2^2(a) < k^2, \quad \text{then} \quad (\cos\theta)X_1(a) + (\sin\theta)X_2(a) + k > 0$$

for all angles θ, and therefore abnormal extremals do not exist through this point. This region consists of points where the speed of the current is less than the speed of the ship.

In the region $X_1^2(a) + X_2^2(a) \geq k^2$ there always exist directions θ_0 such that $(\cos\theta_0)X_1(a) + (\sin\theta_0)X_2(a) + k = 0$. Let $\theta(t)$ be the solution of the navigation equation (11) with $\theta(0) = \theta_0$. The corresponding solution of equation (10) defines an abnormal extremal curve, because $\Psi(t) = 0$ for all t. But θ_0 is the limit of a sequence of initial angles θ for which $(\cos\theta)X_1(a) + (\sin\theta)X_2(a) + k > 0$. The resulting sequence of regular extremals generated by these initial values of θ converges uniformly to the abnormal extremal curve.

In general, it is difficult to decide which of these extremal curves will produce the quickest path, so both need to be considered.

Problem 3 Derive equation (12).

Problem 4 Assume that $X_1(x_1, x_2) = x_2$ and that $X_2 = 0$. Solve equation (11), and determine all extremal curves.

Problem 5 Do the same for $X_1(x_1, x_2) = -x_2$, $X_2(x_1, x_2) = x_1$.

3 Linear quadratic problems with constraints, and Fuller's phenomenon

In the case of time-optimal solutions for linear systems with polyhedral control constraints, optimal controls jump abruptly from one face of the control set to another, with the discontinuities in the control occurring at discrete time points only.

Fuller (1960) was the first to discover an example demonstrating that the switching times in an optimal-control problem can have a finite limit point. That behavior later became known as the "Fuller phenomenon," and subsequently it was recognized in many other physical systems. A class of systems that naturally exhibit Fuller's phenomenon, of which Fuller's example is a particular case, is the class of singular linear quadratic problems with an extra assumption on the bounds on the controls.

For simplicity of proofs, we shall restrict our attention to convex singular problems of the following type: The dynamics will be given by a linear control system,

$$\frac{dx}{dt} = Ax + Bu, \tag{13}$$

defined in a real vector space M, with control functions taking values in a given set that is compact, convex, and symmetric about the origin U_c in the control vector space U.

We shall resume the notation from preceding chapters dealing with linear systems with quadratic cost and assume that $c(x, u) = \frac{1}{2}(Pu, u) + (Qx, u) + \frac{1}{2}(Rx, x)$, with P, Q, and R denoting the appropriate linear mappings. For further simplicity, we shall take $P \geq 0$ and assume that $c(x, u)$ is convex. The convexity assumption implies, in particular, that $c(x, u) \geq 0$ for all (x, u) in $M \times U$.

In order to take advantage of the material in Chapter 9, we shall assume that the singular quadratic system satisfies all the assumptions for the existence of the optimal space Ω in $M \times M^*$ that is symplectic. That is, it will be assumed that the linear system that defines the dynamics is of full-controllability rank and that the zero trajectory generated by the zero control is the only optimal trajectory that satisfies $x(0) = x(T) = 0$.

We shall consider the problem of minimizing $\int_0^T c(x(t), u(t)) \, dt$ among all trajectories $(x(t), u(t))$ of system (13) that satisfy the given boundary conditions $x(0) = a$ and $x(T) = b$, under the assumption that T is fixed. Because the cost extended system

$$\frac{dx_0}{dt} = c(x(t), u(t)), \qquad \frac{dx}{dt} = Ax(t) + Bu(t), \qquad u(t) \in U_c, \qquad (14)$$

is convex, it follows that its reachable set $\mathcal{A}_{\text{ext}}((0, a), T)$ is convex and compact for each $T > 0$. Therefore, if the terminal point b can be reached at all, then it can be reached optimally. As in the unconstrained case, the terminal data $(\int_0^T c(x(t), u(t)) \, dt, \, x(T))$ for an optimal trajectory (x, u) on an interval $[0, T]$ necessarily belong to the boundary of the extended reachable set $\mathcal{A}_{\text{ext}}((0, a), T)$. Moreover, we have the following:

Theorem 3 *For each a and b in M, with $T > 0$, there is at most one optimal trajectory relative to these boundary conditions.*

Proof Assume that both (x_1, u_1) and (x_2, u_2) are optimal relative to a, b, and T. Let $u_\lambda = \lambda u_1 + (1 - \lambda)u_2$, and $x_\lambda = \lambda x_1 + (1 - \lambda)x_2$. Because U_c is convex, $u_\lambda(t) \in U_c$ for all t in $[0, T]$ and each λ in $[0,1]$. Because $x_\lambda(0) = a$ and $x_\lambda(T) = b$ for each λ, it follows that

$$\int_0^T c(x_\lambda(t), u_\lambda(t)) \, dt \geq \int_0^T c(x_1(t), u_1(t)) \, dt = \int_0^T c(x_2(t), u_2(t)) \, dt.$$

As a consequence of convexity,

$$\int_0^T c(x_\lambda(t), u_\lambda(t)) \, dt \leq \lambda \int_0^T c((x_1(t), u_1(t)) \, dt + (1 - \lambda)) \int_0^T c(x_2(t), u_2(t)) \, dt$$
$$= \int_0^T c(x_2(t), u_2(t)) \, dt.$$

Therefore,

$$\int_0^T c(x_\lambda(t), u_\lambda(t)) \, dt = \int_0^T c(x_2(t), u_2(t)) \, dt, \qquad \text{for all } \lambda \text{ in } [0, 1].$$

On expanding,

$$\int_0^T c(x_\lambda, u_\lambda)\, dt = \int_0^T c(x_2, u_2)\, dt + \lambda^2 \int_0^T c(x_1 - x_2, u_1 - u_2)\, dt + \lambda C,$$

where C is equal to the integral of the appropriate products containing x_1, x_2, u_1, and u_2. Therefore,

$$\lambda^2 \int_0^T c(x_1 - x_2, u_1 - u_2)\, dt + \lambda C = 0,$$

which can hold only if both $C = 0$ and $\int_0^T c(x_1 - x_2, u_1 - u_2)\, dt = 0$.

The trajectory $(x_1 - x_2, u_1 - u_2)$ is a closed trajectory that originates and terminates at zero. Because its total cost is zero, it is optimal. Our assumption about uniqueness of optimal trajectories (for the unconstrained situation) implies that $x_1(t) = x_2(t)$ and $u_1(t) = u_2(t)$ for all t. End of the proof. ∎

In contrast to the unconstrained case, there may be only one trajectory (x, u) that satisfies the given boundary conditions $x(0) = a$ and $x(T) = b$. Such a trajectory is necessarily optimal, not only with respect to quadratic cost but also relative to other criteria, such as time optimality. This complication necessitates an additional multiplier p_0 in the maximum principle for this problem, as was the case for time-optimal problems. For each integrable control function $u(t)$, we define the corresponding Hamiltonian

$$H(x, p, u(t)) = -p_0 c(x, u(t)) + p\,(Ax + Bu(t)).$$

H is to be regarded as a function on $M \times M^*$, parametrized by the control u and a nonpositive parameter p_0. We shall normalize the multiplier to be either -1 or 0.

Definition 3 An absolutely continuous curve $\sigma(t) = (x(t), p(t))$ in $M \times M^*$ defined on an interval $[0, T]$ is called an extremal curve if there exists an integrable control function $u(t)$ taking values in U_c such that σ is an integral curve of the Hamiltonian vector field defined by $H(x, p, u(t))$, as defined earlier, and

$$H(\sigma(t), u(t)) = \max_{v \in U_c}(H(\sigma(t), v))$$

for almost all t in $[0, T]$. In addition, it is required that $p(t) \neq 0$ whenever $p_0 = 0$. The extremal curves that correspond to $p_0 = -1$ are called "regular," and others "abnormal" ($p_0 = 0$).

The maximum principle for the constrained linear quadratic problem is given by the next theorem.

Theorem 4

(a) *Any optimal trajectory* $(x(t), u(t))$ *is the projection of an extremal curve* $\sigma(t) = (x(t), p(t))$ *of* $H(x, p, u(t))$.

(b) *When the terminal point* $x(T) = b$ *belongs to the interior of the reachable set* $\mathcal{A}(a, T)$ *in* M, *then the extremal curve that projects onto* x *is regular.*

(c) *The projection* $x(t)$ *of a regular extremal curve* $\sigma(t) = (x(t), p(t))$ *defined by the control* $u(t)$ *is optimal provided that the terminal point* $x(T) = b$ *is contained in the interior of* $\mathcal{A}(a, T)$.

Proof Denote by $\alpha = \int_0^T c(\bar{x}(t), \bar{u}(t)) \, dt$ and $b = \bar{x}(T)$ the terminal data corresponding to an optimal trajectory $(\bar{x}, \bar{u})$ on an interval $[0, T]$. Because the reachable set $\mathcal{A}_{\text{ext}}((0, a), T)$ is compact and convex, there exists a nonzero linear function (q^0, q) on $\mathbb{R}^1 \times M$ that defines a supporting hyperplane to the reachable set at (α, b). That is, $q^0(x^0(t) - \alpha) + q(x(t) - b) \leq 0$ for all cost extended trajectories $(x^0(t), x(t))$ of (14) with $x^0(0) = 0$ and $x(0) = a$.

Consider first the case where b belongs to the interior of $\mathcal{A}(a, T)$. Then $q^0 < 0$, for the following reasons: To begin with, there is more than one trajectory $x(t)$ of (13) that satisfies $x(0) = a$ and $x(T) = b$. The total cost $x^0(T)$ of this trajectory is strictly larger than α, because optimal trajectories are unique. But then $q^0(x^0(T) - \alpha) \leq 0$, and therefore $q_0 \leq 0$. It then follows that $q_0 \neq 0$, because the terminal points $x(T)$ cover a neighborhood of b, and therefore $q(x(T) - b) \leq 0$ cannot hold without $q = 0$.

We shall now show that there exists a linear function (q^0, q) that defines a supporting hyperplane for the reachable set at (α, b) for which $q^0 \leq 0$, even when b belongs to the boundary of $\mathcal{A}(a, T)$. Because we have controllability of (13), the interior of $\mathcal{A}(a, T)$ is dense in $\mathcal{A}(a, T)$. Let $\{b_n\}$ be any sequence of states contained in the interior of $\mathcal{A}(a, T)$ that converges to b. Denote by (x_n, u_n) the corresponding sequence of optimal trajectories relative to the boundary conditions a, b_n, and T. If we denote $\alpha_n = \int_0^T c(x_n(t), u_n(t)) \, dt$, then each point (α_n, b_n) is on the boundary of the extended reachable set.

The sequence of control functions $\{u_n\}$ has a weakly convergent subsequence $\{u_{n_k}\}$ on $[0, T]$. Then the corresponding trajectories $\{x_{n_k}\}$ converge uniformly. Hence, there is no loss of generality in assuming that the original sequence of trajectories is such that $\{u_n\}$ is a weakly convergent sequence of functions on the interval $[0, T]$. Therefore, the corresponding solution curves $\{x_n\}$ converge uniformly, and consequently $\{(\alpha_n, b_n)\}$ converges to (α, b). Let (q_n^0, q_n) denote a linear function on $\mathbb{R}^1 \times M$ such that

$$q_n^0(x^0(T) - x_n^0(T)) + q_n(x(T) - x_n(T)) \leq 0$$

for all trajectories $(x(t), u(t))$ on $[0, T]$ that satisfy $x(0) = a$. We know that $q_n^0 < 0$ for each integer $n > 0$.

Let $\|q\|$ denote any norm on M^*, and normalize the quantities (q_n^0, q_n) such that $(q_n^0) + \|q_n\| = 1$. It follows from the usual compactness arguments that there is no loss of generality in assuming that the sequence $\{(q_n^0, q_n)\}$ converges to some point (q^0, q). Then $q^0 \leq 0$ and $(q^0)^2 + \|q\|^2 = 1$.

It follows that (q^0, q) defines a supporting hyperplane at (α, b), because the reachable sets $\mathcal{A}_{\text{ext}}((0, a), t)$ vary continuously with t. Let $\bar{p}(t)$ denote the solution curve of $dp/dt = q^0 \frac{\partial c}{\partial x}(\bar{x}(t), \bar{u}(t)) - A^* p(t)$, with $p(T) = q$. Assume that the preceding equation is expressed in some canonical coordinates. If $q^0 = 0$, the remaining part of the proof is as in the corresponding proof for the time-optimal problem and will not be repeated here.

In the remaining case, q_0 will be normalized to -1. Then

$$-\left(\int_0^T c(x(t), u(t)) \, dt - \int_0^T c(\bar{x}(t), \bar{u}(t)) \, dt \right) + p(T)(x(T) - \bar{x}(T)) \leq 0$$

for any trajectory (x, u) with $x(0) = a$ and $u(t) \in U_c$. Let $u_\lambda = \lambda u + (1 - \lambda)\bar{u}$, and $x_\lambda = \lambda x + (1 - \lambda)\bar{x}$. Then

$$c(x_\lambda, u_\lambda) = c(\bar{x}, \bar{u}) + \lambda \frac{\partial c}{\partial x}(\bar{x}, \bar{u})(x - \bar{x}) + \lambda \frac{\partial c}{\partial u}(\bar{x}, \bar{u})(u - \bar{u})$$
$$+ \frac{\lambda^2}{2} c(x - \bar{x}, u - \bar{u}),$$

and therefore

$$-\left(\int_0^T (c(x_\lambda(t), u_\lambda(t)) - c(\bar{x}(t), \bar{u}(t))) \, dt \right) + \bar{p}(T)(x_\lambda(T) - \bar{x}(T))$$
$$= -\lambda \int_0^T \left(\frac{\partial c}{\partial x}(\bar{x}, \bar{u})(x - \bar{x}) + \frac{\partial c}{\partial u}(\bar{x}, \bar{u})(u - \bar{u}) \, dt \right.$$
$$+ \frac{\lambda^2}{2} \int_0^T \left. c(x - \bar{x}, u - \bar{u}) \, dt \right) + p(T)(x_\lambda(T) - \bar{x}(T)).$$

On dividing by λ, and letting λ tend to zero, the preceding inequality becomes

$$-\int_0^T \left(\frac{\partial c}{\partial x}(\bar{x}, \bar{u})(x - \bar{x}) + \frac{\partial c}{\partial u}(\bar{x}, \bar{u})(u - \bar{u}) \right) dt + p(T)(x(T) - \bar{x}(T)) \leq 0.$$

The identity

$$p(T)(x(T) - \bar{x}(T)) = \int_0^T \left(\frac{d\bar{p}}{dt} (x - \bar{x}) + \bar{p}(A(x - \bar{x}) + B(u - \bar{u})) \right) dt$$

yields

$$\int_0^T \left(-\frac{\partial c}{\partial u}(\bar{x}(t), \bar{u}(t)) + B^* \bar{p}(t) \right)(u(t) - \bar{u}(t)) \, dt \leq 0.$$

The Hamiltonian $H(x, p, u) = -c(x, u) + p(Ax + Bu)$ is concave in u, because c is convex. Therefore,

$$H(\bar{x}(t), \bar{p}(t), u(t)) \leq H(\bar{x}(t), \bar{p}(t), \bar{u}(t)) + \frac{\partial H}{\partial u}(\bar{x}, \bar{p}, \bar{u})(u - \bar{u}).$$

But

$$\frac{\partial H}{\partial u}(\bar{x}, \bar{p}, \bar{u}) = -\frac{\partial c}{\partial u}(\bar{x}, \bar{u}) + B^* \bar{p},$$

and hence

$$\int_0^T H(\bar{x}(t), \bar{p}(t), u(t)) - H(\bar{x}(t), \bar{p}(t), \bar{u}(t)) \, dt \leq 0.$$

Because $u(t)$ can take any value in U_c, it follows that

$$H(\bar{x}(t), \bar{p}(t), \bar{u}(t)) \geq H(\bar{x}(t), \bar{p}(t), u(t))$$

almost everywhere in $[0, T]$. The proof of parts (a) and (b) is now finished. The proof of part (c) consists in reversing the steps in the preceding argument. The details are left to the reader. ∎

The optimal synthesis for a singular linear quadratic problem subject to constraints on the magnitude of controls incorporates the extremal curves arising from controls whose values are in the interior of U_c and those extremal curves that correspond to controls that take values on the boundary of U_c. The extremal curves generated by controls with values in the interior of U_c are independent of the constraints and therefore are contained in the optimal space Ω described in Chapter 9. In general, an extremal curve may be a concatenation of these two distinct types of curves, often exhibiting dramatic oscillations at the juncture points. Fuller's problem is the simplest such example demonstrating infinite chattering at the juncture points.

Fuller's problem Fuller's problem consists in minimizing $\frac{1}{2} \int_0^T x_1^2 \, dt$ over the trajectories of $dx_1/dt = x_2$ and $dx_2/dt = u(t)$, subject to the constraint that

$|u(t)| \leq 1$. In Fuller's original paper (1960), the terminal time T was not fixed. We shall, however, take T as fixed, because that problem fits more naturally with the exposition in Chapter 9.

In what follows, it will be assumed that the initial point a, the terminal point b, and the time of transfer T are such that b is in the interior of the reachable set $\mathcal{A}(a, T)$. According to Theorem 4, the optimal trajectories for these boundary conditions coincide with the regular extremal curves.

It is easy to verify that zero trajectory is the only optimal trajectory that satisfies $x(0) = x(T) = 0$. Hence the system satisfies the uniqueness assumption of Chapter 9. If there were no bounds on the size of controls, both e_1 and e_2 would be jump directions. Therefore, the optimal space Ω would reduce to the origin, and the only optimal trajectory is the zero trajectory generated by the zero control. This trajectory remains optimal even in the presence of constraints.

The extremal curves that correspond to the boundary controls are the integral curves of the following Hamiltonian:

$$H_+(x, p) = -\frac{1}{2}x_1^2 + p_1 x_2 + p_2 \quad \text{for} \quad p_2 > 0,$$

$$H_-(x, p) = -\frac{1}{2}x_1^2 + p_1 x_2 - p_2 \quad \text{for} \quad p_2 < 0.$$

The surface Σ in $\mathbb{R}^4$ defined by $p_2 = 0$ is the switching surface, which separates the half-space Σ_+, defined by $p_2 > 0$, from Σ_-, defined by $p_2 < 0$. Each extremal curve σ is an integral curve of $\vec{H}_+$ in the region Σ_+, and an integral curve of $\vec{H}_-$ in Σ_-. The only possibility of fusing such an extremal curve with the singular extremal curve is to go through the origin in $\mathbb{R}^4$. The integral curves of $H_\pm$ satisfy the following system of equations:

$$\frac{dx_1}{dt} = \frac{\partial}{\partial p_1}H_\pm = x_2, \qquad \frac{dx_2}{dt} = \frac{\partial}{\partial p_2}H_\pm = \pm 1,$$

$$\frac{dp_1}{dt} = -\frac{\partial}{\partial x_1}H_\pm = x_1, \qquad \frac{dp_2}{dt} = -\frac{\partial}{\partial x_2}H_\pm = -p_1.$$

The solutions that originate at $\sigma = (a_1, a_2, q_1, q_2)$ are given by

$$x_1(t) = a_1 + a_2 t \pm \frac{t^2}{2!}, \qquad x_2(t) = a_2 \pm t,$$

$$p_1(t) = q_1 + a_1 t + \frac{a_2 t^2}{2!} \pm \frac{t^3}{3!}, \tag{15}$$

$$p_2(t) = q_2 - \left(q_1 t + \frac{a_1 t^2}{2!} + \frac{a_2 t^3}{3!} \pm \frac{t^4}{4!}\right).$$

The first observation is that there is no extremal curve $\sigma(t)$ that passes through the origin and is a finite concatenation of the integral curves of $\vec{H}_+$ and $\vec{H}_-$. The argument is simple: In the last interval $[t_{n-1}, t_n]$, $\sigma(t_{n-1}) \in \Sigma$, $\sigma(t_n) = 0$, and σ is an integral curve of either $\vec{H}_+$ or $\vec{H}_-$. Denoting $(t_n - t_{n-1})$ by t, the foregoing condition means that

$$a_1 + a_2 t \pm \frac{t^2}{2} = 0, \qquad a_2 \pm t = 0, \qquad q_1 + a_1 t + \frac{a_2 t^2}{2!} \pm \frac{t^3}{3!} = 0,$$

$$q_1 t + \frac{a_1 t^2}{2} + \frac{a_2 t^3}{3!} \pm \frac{t^4}{4!} = 0.$$

Zero is the only solution of this system of equations for $t \geq 0$. Hence, any extremal curve that originates on Σ and passes through the origin must enter the origin with infinitely many switchings. The switching times must accumulate, since the total time interval is finite.

We shall use $(\exp t H_\pm)(\sigma)$ to denote the one-parameter groups of diffeomorphisms defined by equations (15). The second important observation is that each of these one-parameter groups is invariant under the group of dilations $T_\rho(\sigma) = (\rho^2 a_1, \rho a_2, \rho^3 q_1, \rho^4 q_2)$ for $\rho > 0$, in the following sense:

$$(\exp \rho t H_+) \circ T_\rho = T_\rho \circ (\exp t H_+) \quad \text{and} \quad (\exp \rho t H_-) \circ T_\rho = T_\rho \circ (\exp t H_-).$$

The proof is simple:

$$\rho^2 a_1 + (\rho a_2)(\rho t) \pm \frac{(\rho t)^2}{2!} = \rho^2 \left(a_1 + a_2 t \pm \frac{t^2}{2} \right), \qquad \rho a_2 \pm \rho t = \rho(a_2 \pm t),$$

$$\rho^3 q_1 + (\rho^2 a_1)(\rho t) + (\rho a_2)\frac{(\rho t)^2}{2!} \pm \frac{(\rho t)^3}{3!} = \rho^3 \left(q_1 + a_1 t + \frac{a_2 t^2}{2!} \pm \frac{t^3}{3!} \right),$$

and

$$\rho^4 q_2 - \left(\rho^3 q_1 (\rho t) + (\rho^2 a_1)\frac{(\rho t)^2}{2!} + (\rho^3 a_1)\frac{(\rho t)^3}{3!} \pm \frac{(\rho t)^4}{4!} \right)$$

$$= \rho^4 \left(q_2 - \left(q_1 t + \frac{a_1 t^2}{2!} + \frac{a_2 t^3}{3!} + \frac{t^4}{4!} \right) \right).$$

The group generated by $\{\exp t H_+ : t \in \mathbb{R}^1\}$ and $\{\exp t H_- : t \in \mathbb{R}^1\}$ is also invariant under the reflections $R(\sigma) = -\sigma$, because $(\exp t H_+) \circ R = R \circ (\exp t H_-)$, and $(\exp t H_-) \circ R = R \circ (\exp t H_+)$.

Suppose now that $(\exp t H_\pm)(\sigma_0) \in \Sigma$ for an arbitrary point $\sigma_0 = (a_1, a_2, q_1, q_2)$ in Σ, with $\sigma_0 \neq 0$, for some positive time t. Assume that t is the smallest such time. We shall now show (what is probably the most remarkable feature of this example) that there exist points σ_0 and a positive number ρ such that

$$(\exp t H_\pm)(\sigma_0) = -T_\rho(\sigma_0).$$

The calculations are as follows: Because $\sigma_0 \in \Sigma, q_2 = 0$. If for some t, $p_2(t) = 0$, then

$$-\left(q_1 t + \frac{a_1 t^2}{2!} + \frac{a_2 t^3}{3!} \pm \frac{t^4}{4!}\right) = 0, \quad \text{or} \quad q_1 = -\left(\frac{a_1 t}{2!} + \frac{a_2 t^2}{3!} \pm \frac{t^3}{4!}\right).$$

After substitutions,

$$p_1(t) = -\left(\frac{a_1 t}{2!} + \frac{a_2 t^2}{3!} \pm \frac{t^3}{4!}\right) + a_1 t + \frac{a_2 t^2}{2!} \pm \frac{t^3}{3!} = \frac{a_1 t}{2!} + \frac{2a_2 t^2}{3!} \pm \frac{3t^3}{4!}.$$

When $x_1(t) = -\rho^2 a_1, x_2(t) = -\rho a_2$, and $p_1(t) = -\rho^3 q_1$, then $a_2 \pm t = -\rho a_2$ and $a_1 + a_2 t \pm t^2/2 = -\rho^2 a_1$. Therefore,

$$\mp(1+\rho)a_2 = t \quad \text{and} \quad a_1 = \pm \frac{a_2^2}{2}\left(\frac{1-\rho^2}{1+\rho^2}\right).$$

$p_1(t) = -\rho^3 q_1$ is the same as

$$\frac{a_1 t}{2} + \frac{2a_2 t^2}{3!} \pm \frac{3t^3}{4!} = \rho^3\left(\frac{a_1 t}{2} + \frac{a_2 t^2}{3!} \pm \frac{t^3}{4!}\right).$$

Therefore,

$$(1-\rho^3)\frac{a_1 t}{2!} + (2-\rho^3)\frac{a_2 t^2}{3!} \pm (3-\rho^3)\frac{t^3}{4!} = 0.$$

After the substitutions $t = \mp(1+\rho)a_2$ and $a_1 = \pm(1-\rho^2)/(2(1+\rho^2))a_2^2$, the preceding equation becomes (after factoring out a_2^3)

$$-\frac{(1-\rho^3)(1-\rho^2)(1+\rho)}{4(1+\rho^2)} + \frac{(2-\rho^3)(1+\rho)^2}{3!} - \frac{(3-\rho^3)(1+\rho)^3}{4!} = 0.$$

Because we are looking for positive roots, there is no loss in factoring out $(1+\rho)^2/2$ to get

$$-\frac{(1-\rho^3)(1-\rho)}{2(1+\rho^2)} + \frac{2-\rho^3}{3} - \frac{(3-\rho^3)(1+\rho)}{12} = 0.$$

The foregoing equation simplifies to

$$\rho^6 - 3\rho^5 - 5\rho^4 + 5\rho^2 + 3\rho - 1 = 0. \tag{16}$$

Evidently both $\rho = +1$ and $\rho = -1$ are roots of equation (16). After factoring out the corresponding factors, equation (16) becomes

$$\rho^4 - 3\rho^3 - 4\rho^2 - 3\rho + 1 = 0. \tag{17}$$

Denoting $f(\rho) = \rho^4 - 3\rho^3 - 4\rho^2 - 3\rho + 1$, it follows that $f(0) = 1$ and $f(1) = -8$. Therefore, there is a positive root ρ_1 less than 1. Then $\rho_2 = 1/\rho_1$ is also a root because of the symmetry $f(1/\rho) = f(\rho)(1/\rho^4)$. It is easy to verify that $f'(\rho) < 0$ in the interval $[0,1]$, and therefore f has no other positive roots.

As in Fuller's original paper (1960), let h denote the quantity $(1 - \rho^2)/(2(1 + \rho^2))$. Then $\rho = \sqrt{1 - 2h}/(1 + 2h)$. This substitution in equation (17) leads to

$$(1 - 2h)^2 - 3(1 - 2h)^{3/2}(1 + 2h)^{1/2} - 4(1 - 2h)(1 + 2h)$$
$$- 3(1 - 2h)^{1/2}(1 + 2h)^{3/2} + (1 + 2h)^2 = 0.$$

The preceding equation can be written more simply as

$$3(2h)^2 - 1 = 3\sqrt{1 - (2h)^2}.$$

After squaring and collecting like terms, we get

$$h^4 + \frac{h^2}{12} - \frac{1}{18} = 0. \tag{18}$$

The root h_1 of (18) that is contained in the interval $(0,1)$ corresponds to ρ_1, and its negative $-h_1$ corresponds to the reciprocal $\rho_2 = 1/\rho_1$. For each root ρ of equation (17),

$$t = \mp(1 + \rho)a_2, \qquad a_1 = \pm\frac{1 - \rho^2}{2(1 + \rho^2)}a_2^2,$$

$$q_1 = -\left(\frac{a_1 t}{2} + \frac{a_2 t^2}{3!} \pm \frac{t^3}{4!}\right) = \frac{a_2^3}{24}\frac{(1 + \rho)^2}{(1 + \rho^2)}(\rho^3 - 3\rho^2 - 5\rho + 3).$$

Because

$$\frac{(1 + \rho)^2}{12(1 + \rho^2)}(\rho^3 - 3\rho^2 - 5\rho + 3) - \frac{(1 - \rho^2)^2}{4(1 + \rho^2)^2}$$

$$= \frac{(1 + \rho)^2}{12(1 + \rho^2)^2}((1 + \rho^2)(\rho^3 - 3\rho^2 - 5\rho + 3) - 3(1 - \rho)^2)$$

$$= \frac{(1 + \rho)^2}{12(1 + \rho^2)^2}(\rho^5 - 3\rho^4 - 4\rho^3 - 3\rho^2 + \rho) = 0$$

it follows that

$$q_1 = \frac{a_2^3}{2} \frac{(1-\rho^2)^2}{4(1+\rho^2)^2} = \frac{a_2^3}{2} h^2.$$

Of course, $q_2 = 0$.

The foregoing analysis shows that if σ_0 is on the curve $\Gamma = \{(-ha_2|a_2|, a_2, (a_2^3/2)h^2, 0) : a_2 \in \mathbb{R}^1\}$, then there are positive real numbers ρ and $(1/\rho)$ $(\rho \neq 1)$ such that $(\exp t H_\pm)(\sigma_0) = -T_\rho(\sigma_0)$. The meaning of this notation is that for points σ_0 with $a_2 > 0$, $t = (1+\rho)a_2$ and $(\exp t H_+)(\sigma_0) = -T_\rho(\sigma_0)$; but for σ_0 with $a_2 < 0$, $t = -(1+\rho)a_2$ and $(\exp t H_-)(\sigma_0) = -T_\rho(\sigma_0)$.

Suppose now that $0 < \rho < 1$. Then

$$(\exp \rho t H_\mp) \circ (\exp t H_\pm)(\sigma_0) = (\exp \rho t H_\mp)(R T_\rho(\sigma_0)) = R(\exp \rho t H_\pm)(T_\rho(\sigma))$$

$$= R T_\rho(\exp t H_\pm)(\sigma_0) = R T_\rho R T_\rho(\sigma_0) = T_\rho^2(\sigma_0).$$

Continuing,

$$(\exp \rho t H_\pm)(T_\rho^2(\sigma_0)) = T_\rho^2(\exp t H_\pm)(\sigma_0) = T_\rho^2 R T_\rho(\sigma_0) = R T_\rho^3(\sigma_0).$$

This procedure defines an extremal curve $\sigma(t)$ that originates at σ_0, with switching times $0 < t_1 < \cdots < t_n < \cdots$, with $t_n = t + \rho t + \cdots + \rho^{n-1}t = t(1 - \rho^n)/(1 - \rho)$. Because $\sigma(t_n) = R^n T_\rho^n(\sigma_0)$, it follows that $\lim_{n \to \infty} \sigma(t_n) = 0$. The total time required to reach the origin is given by

$$t \frac{1}{1-\rho} = |a_2| \frac{1+\rho}{1-\rho}.$$

By taking $t = -|a_2|(1+\rho)$, the foregoing argument would lead to a negative sequence of switching times $t_n = t + \rho t + \cdots + \rho^{n-1}t = t(1-\rho^n)/(1-\rho)$ such that the iterative sequence $\sigma(t_n)$ would tend to zero at $t/(1-\rho)$. Alternatively, by reversing the time, this information shows that any point b on Γ can be reached from the origin in $|b_2|(1+\rho)/(1-\rho)$ units of time.

The optimal synthesis resembles the turnpike solution for the unconstrained case and is as follows: Let a, b be given departure and destination points in $\mathbb{R}^2$. There is a unique point $\sigma_0 = (a_1, a_2, q_1, 0)$ on Σ and an unique time t_1 such that one of the two $(\exp t_1 H_\pm)(\sigma_0)$ belongs to Γ. Denote by $\sigma(t)$ the appropriate extremal curve such that $\sigma(0) = \sigma_0$ and $\sigma(t_1)$ belongs to Γ. Denote $\sigma(t_1) = \sigma_1$. In the interval $[t_1, t_2)$, define $\sigma(t)$ to be the chattering curve that connects σ_1 to the origin. The time t_2 required to bring σ_1 to the origin is given by $|\bar{a}_2|(1+\rho)/(1-\rho)$, with $\bar{a}_2$ equal to the x_2 coordinate of σ_1.

In the interval $[t_2, t_3]$, the extremal curve is the chattering curve that takes the origin to a unique point σ_3 on Γ determined by the destination b. Point σ_3 is equal to the point on Γ for which there is a positive time $(t_4 - t_3)$ such that either $(\exp(t_4 - t_3)H_+)(\sigma_3) = b$ or $(\exp(t_4 - t_3)H_-)(\sigma_3) = b$. (It is not possible for both to be true.) Time t_3 is equal to $|\bar{b}_2|(1 + \rho)/(1 - \rho)$, with $\bar{b}_2$ equal to the x_2 coordinate of σ_3.

Let T be any time greater than or equal to t_4. The optimal trajectory $\sigma(t)$ that reaches b in exactly T units of time is a concatenation of the extremal curve $\sigma(t)$ described earlier in the interval $[0, t_2]$, followed by the singular trajectory $\sigma(t) = 0$ corresponding to $u = 0$ for the duration $T - t_4$ units of time, followed by the chattering trajectory to σ_3, which is finally followed by the terminal piece to b in $t_4 - t_3$ units of time. Figure 10.5 illustrates this kind of optimal synthesis.

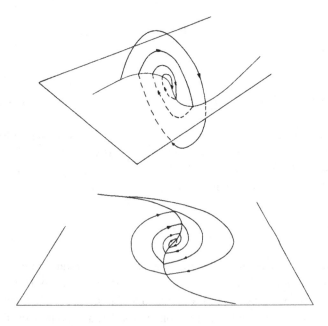

Fig. 10.5.

Problem 6 Show that the problem of minimizing $\frac{1}{2} \int_0^T x_2^2 \, dt$ over the trajectories of $dx_1/dt = x_2$ and $dx_2/dt = u$, with $|u(t)| \le 1$, does not exhibit Fuller's phenomenon. Find its optimal synthesis.

Problem 6 shows that not every singular linear quadratic problem exhibits Fuller-type behavior when constrained in regard to the size of control functions. The question of fusing the extremals defined by the constraints with the singular extremals has not been studied in sufficient detail. It would make an excellent research topic.

Notes and sources

Much of the material on linear time-optimal control problems is standard in the literature on control theory. The present exposition essentially follows Lee and Markus (1967), except for the additional focus on abnormal extremal curves, which are not specifically recognized in this literature. Abnormal extremal curves, in spite of their name, are integral parts of optimal control, although their limiting behavior (as the multiplier tends to zero) has not been properly appreciated in the existing literature.

The brachistochrone problem and Zermelo's navigation problem illustrate the classic origins of optimal control in the calculus of variations. Their control-theory solutions clarify the Hamiltonian methods used by Carathéodory (1935). The reader may also consult Bliss (1925) for his discussion of the brachistochrone problem.

The present treatment of Fuller's problem is in many respects similar to that of Kupka (1990). Some of the calculations used in Fuller's problem can be also found in the work of Marchal (1973).

11

The maximum principle

Our study thus far points to the maximum principle as the fundamental principle of optimality and identifies the symplectic structure and the associated Hamiltonian formalism as the main theoretical ingredients required for its proper understanding. In this chapter we shall take that direction to its natural end and ultimately arrive at a geometric formulation of the maximum principle for optimal problems on arbitrary differentiable manifolds, rather than solely on $\mathbb{R}^n$ as is customarily done in the literature on control theory. The geometric formulation of the maximum principle, essential for effective use of the principle for problems of mechanics and geometry, in a larger context illuminates the contribution of optimal control to the classic theory of Hamiltonian systems. Both of these points will become clearer in the next chapter, which deals with optimal problems on Lie groups.

This chapter begins with an initial formulation of the maximum principle for optimal problems defined on open subsets of $\mathbb{R}^n$. Rather than seeking the most general conditions under which the principle is valid, we shall follow the original presentation by Pontryagin et al. (1962). This level of generality is sufficient for the applications that follow and is at the same time relatively free of the technicalities that could obscure its geometric content.

The initial formulation is followed by a comparison between the maximum principle and other necessary conditions for optimality that emphasize the distinctions between strong and weak minima: The maximum principle, as an extension of the condition of Weierstrass, corresponds to the strong minimum, whereas the Euler-Lagrange equation corresponds to the weak minimum. Continuing with the minor theme that the maximum principle can illuminate even some of the most classic solutions of the calculus of variations, we shall then obtain solutions to the problem of the minimal surface of revolution through control-theory perspectives.

The emphasis then shifts to arbitrary manifolds, with a detailed discussion of the symplectic structure of their cotangent bundles. This material forms the theoretical foundation required for a proper understanding of the use of the maximum principle for variational problems on manifolds.

The chapter ends with applications to mechanics and a selection of geometric problems on the group of motions of a plane.

1 The maximum principle in $\mathbb{R}^n$

1.1 Background

Let $f_0, f_1, \ldots, f_m$ be given functions of $n + m$ variables $(x_1, \ldots, x_n, u_1, \ldots, u_m)$. It will be assumed that the variables $(x_1, \ldots, x_n)$ belong to a fixed open subset S of $\mathbb{R}^n$, and the variables $(u_1, \ldots, u_m)$ belong to an arbitrary subset U of $\mathbb{R}^m$. S is called the state space, and U is called the control set. It will be convenient to use single letters x and u for points $(x_1, \ldots, x_n)$ and $(u_1, \ldots, u_m)$.

We shall assume that for $u \in U$, each function f_i is a continuously differentiable function on S, and we shall further assume that each of the functions $f_i(x, u)$ and $(\partial f_i / \partial x_j)(x, u)$ is a continuous function on $S \times U$ in the relative topology induced by $\mathbb{R}^n \times \mathbb{R}^m$. By "a control function $u(t)$" we shall always mean a U-valued function defined on some compact interval $[t_0, t_1]$ that is essentially bounded and measurable on $[t_0, t_1]$.

With each control function $u(t)$, the preceding functions $f_1, \ldots, f_m$ define a differential system on S:

$$\frac{dx_i}{dt} = f_i(x_1(t), \ldots, x_n(t), u_1(t), \ldots, u_m(t)), \qquad i = 1, 2, \ldots, n. \quad (1)$$

The following discussion contains a summary of results (from the theory of ordinary differential equations) concerning the existence and uniqueness of solutions and their dependence on the initial data.

Let $\text{dom}(u)$ denote the interval on which the control u is defined. For each a in S and each τ in $\text{dom}(u)$, there exists an interval I that contains τ in its interior and is contained in $\text{dom}(u)$, and there exists an absolutely continuous curve $x(t)$ in S defined on I such that $x(\tau) = a$ and

$$\frac{dx_i}{dt} = f_i(x(t), u(t)), \qquad i = 1, 2, \ldots, n \quad (2)$$

for almost all t in I. (The existence of solutions.)

Solution curves can always be continued to the maximal interval of existence. I is said to be the maximal interval of existence if there exists an absolutely continuous curve $x(t)$ defined for all t in I that satisfies equation (2) almost everywhere and that satisfies the initial condition $x(\tau) = a$ and if I' is any interval in dom(u) with these properties then $I' \subset I$. Assuming that I is the maximal interval, then the solution curve is unique (up to a set of measure zero). We shall refer to it as the integral curve of (1), or a solution curve, or sometimes a trajectory of (1).

Then, according to the theory of differential equations, there exist a neighborhood O of a in S and a neighborhood N of τ in $\mathbb{R}^1$ such that for each initial condition $x(\overline{\tau}) = b$, with $b \in O$ and $\overline{\tau} \in N$, the corresponding integral curve $x(t)$ is defined on the entire interval I. It is convenient to denote the dependence of $x(t)$ on its initial data explicitly as $\Phi_u(a, t_0, t) = x(t)$. Paraphrasing the preceding statements in terms of this notation,

$$O \times N \times I \subset \text{dom}(\Phi_u).$$

It further follows from the results concerning the continuous dependence of solutions on the initial data that for any sequence $\{a_n\} \subset O$ such that $\lim_{n \to \infty} a_n = a$ and any sequence $\{\tau_n\} \subset N$ such that $\lim_{n \to \infty} \tau_n = t_0$, the corresponding sequence of curves $x_n(t) = \Phi_u(a_n, \tau_n, t)$ converges uniformly on I to the curve

$$x(t) = \Phi_u(a, t_0, t).$$

The variational equation For any integral curve $x(t) = \Phi_u(a, t_0, t)$, $A(t)$ denotes the matrix with entries

$$A_{ij}(t) = \frac{\partial f_i}{\partial x_j}(x(t),\, u(t)).$$

Each entry A_{ij} is an essentially bounded and measurable function on the interval I. The following linear system of differential equations is called the variational system along the curve $x(t)$, $u(t)$:

$$\frac{dv_i}{dt} = \sum_{j=1}^{n} A_{ij}(t)\, v_j(t), \qquad i = 1, 2, \ldots, n. \tag{3}$$

It follows from the theory of linear differential equations that for each v_0 in $\mathbb{R}^n$ there exists an absolutely continuous curve $v(t)$ defined on the entire interval I such that $v(t_0) = v_0$ and such that (3) is fulfilled for almost all t in I.

The adjoint system is given by

$$\frac{dp_i}{dt} = -\sum_{j=1}^{n} p_j A_{ji}(t), \qquad i = 1, 2, \ldots, n. \tag{4}$$

The solution curves for the adjoint system are also defined on the entire interval I for each initial value p_0 in $\mathbb{R}^n$.

The solutions $v(t)$ of the variational system and the solution curves $p(t)$ of the adjoint system always satisfy

$$\sum_{i=1}^n p_i(t)\, v_i(t) = \text{constant},$$

as can be verified by differentiating.

The pair of differential systems

$$\frac{dx_i}{dt} = f_i(x(t),\, u(t)),$$

and

$$\frac{dp_i}{dt} = \sum_{j=1}^n -p_j\, \frac{\partial f_j}{\partial x_i}\, (x(t),\, u(t)), \quad i = 1, 2, \ldots, n, \tag{5}$$

can be expressed in terms of a single function H, given by

$$H(x, p, u) = \sum_{i=1}^n p_i\, f_i(x, u),$$

by the formulas

$$\frac{dx_i}{dt} = \frac{\partial H}{\partial p_i}\, (x(t),\, p(t),\, u(t)) \quad \text{and} \quad \frac{dp_i}{dt} = -\frac{\partial H}{\partial x_i}\, (x(t),\, p(t),\, u(t)),$$

valid for any control function u.

1.2 The basic optimal problem and the maximum principle

We shall conform to the foregoing terminology and refer to $(x(t), u(t))$ as a trajectory of (1). Because f_0 is assumed to be continuous on $S \times U$, it follows that $t \to f_0(x(t), u(t))$ is an essentially bounded and measurable function on any interval $[t_0, t_1]$ contained in the domain of the trajectory $(x(t), u(t))$. Hence,

$$\int_{t_0}^{t_1} f_0(x(t), u(t))\, dt$$

is a well-defined number.

Definition 1 A trajectory $(x(t), u(t))$ is said to transfer a point a of S to another point b of S if there exists an interval $[t_0, t_1]$ contained in the domain of (x, u) such that $x(t_0) = a$ and $x(t_1) = b$. The cost of the transfer is given by

$$\int_{t_0}^{t_1} f_0(x(t), u(t)) \, dt.$$

Definition 2 A trajectory $(\bar{x}(t), \bar{u}(t))$ is optimal relative to the given points a and b if $\bar{x}(t_0) = a$ and $\bar{x}(t_1) = b$ and if $\int_{t_0}^{t_1} f_0(\bar{x}(t), \bar{u}(t)) \, dt$ is the minimal cost among all costs of trajectories that transfer a to b.

Remark Note that the length of time required to transfer a to b is not fixed in advance. On the other hand, if (x, u) transfers a to b in the interval $[t_0, t_1]$, then the time-shifted trajectory $y(t) = x(t + t_0)$, $v(t) = u(t + t_0)$ transfers a to b in the interval $[0, t_1 - t_0]$, and the cost of the transfer along (y, v) is the same as the cost of the transfer along (x, u). Hence, the initial time can always be taken equal to zero.vspace*2pt   ∎

The basic optimal problem is to find an optimal trajectory, assuming that the latter exists. An indirect approach to this problem consists in first determining some properties of optimal trajectories that will be sufficiently distinctive to narrow the class of candidates for optimal solutions to a small class of curves. The maximum principle provides a list of necessary conditions that an optimal trajectory must fulfill. It derives the required properties from the following geometric observations: Both the cost that is to be minimized and the dynamics of the control system can be viewed as an extended dynamical control system in $\mathbb{R}^1 \times S$, called the cost extended system, whose dynamics are given by

$$\frac{dx_0}{dt} = f_0(x(t), u(t)), \qquad \frac{dx_i(t)}{dt} = f_i(x(t), u(t)), \qquad i = 1, \ldots, n. \quad (6)$$

The trajectories of the cost extended system are curves $(x_0(t), x(t))$ in $\mathbb{R}^1 \times S$ parametrized by the control functions $u(t)$.

The basic optimal problem admits an equivalent formulation purely in terms of the trajectories of the cost extended system, as follows: Assume that a and b are given points in S. Let L denote the line $\{(x_0, b) : x_0 \in \mathbb{R}^1\}$. The basic optimal problem is to select a trajectory $(\bar{x}_0(t), \bar{x}(t))$, from among all trajectories of the cost extended system that transfer $(0, a)$ to the line L, for which the vertical coordinate $\bar{x}_0(T)$ on L is minimal.

Evidently, for any optimal solution curve $(\bar{x}_0(t), \bar{x}(t))$ that transfers $(0, a)$ to L in T units of time, the terminal point $(\bar{x}_0(T), b)$ cannot be an interior point in the intersection between L and the set of points of the cost extended system reachable from the initial point $(0, a)$, because $(\bar{x}_0(T) - \varepsilon, b)$ is not in that intersection for any $\varepsilon > 0$. Equivalently, the terminal point $(\bar{x}_0(T), b)$ of an optimal trajectory cannot be in the interior of the set of points of the cost extended system reachable from the initial point $(0, a)$, because $(\bar{x}_0(T) - \varepsilon, b)$ is not reachable for any $\varepsilon > 0$. Figure 11.1 depicts this geometric situation.

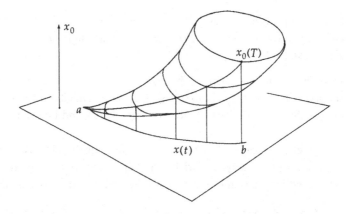

Fig. 11.1.

The maximum principle derives from the foregoing geometric descriptions of optimality. Its ultimate statement, however, requires a few more concepts: The cost extended system (6) defines its family of Hamiltonian functions $\mathcal{H}(x, p, u(t))$ in $\mathbb{R}^{n+1} \times \mathbb{R}^{n+1}$, parametrized by the control functions, given by

$$\mathcal{H}(x, p, u(t)) = \sum_{i=0}^{n} p_i f_i(x, u(t)).$$

For each control function $u(t)$, $\vec{\mathcal{H}}(x, p, u(t))$ denotes the Hamiltonian vector field that corresponds to $\mathcal{H}$. The integral curves of $\vec{\mathcal{H}}$ satisfy the following system of equations for almost all t:

$$\frac{dx_i}{dt} = \frac{\partial \mathcal{H}}{\partial p_i} = f_i(x(t), u(t)) \quad \text{and} \quad \frac{dp_i}{dt} = -\frac{\partial \mathcal{H}}{\partial x_i} = \sum_{j=0}^{n} p_j \frac{\partial f_j}{\partial x_i} (x(t), u(t))$$

$$\text{for all } i = 0, 1, \ldots, n. \quad (7)$$

Definition 3 The maximal Hamiltonian associated with each integral curve $(x(t), p(t), u(t))$ of (7) is defined by

$$\mathcal{M}(x(t), p(t)) = \sup_{u \in U} \mathcal{H}(x(t), p(t), u).$$

The maximum principle consists of the following necessary conditions of optimality. Suppose that $(\bar{x}(t), \bar{u}(t))$ transfers a to b optimally in an interval $[0, T]$. Then there exists a nonzero, absolutely continuous curve $p(t) = (p_0(t), \ldots, p_n(t))$ defined on the interval $[0, T]$ such that

1. $(x(t), p(t), u(t))$ is a solution curve of differential system (7),
2. $\mathcal{H}(x(t), p(t), u(t)) = \mathcal{M}(x(t), p(t))$ for almost all t in $[0, T]$, and
3. $p_0(T) \leq 0$ and $\mathcal{M}(x(T), p(T)) = 0$.

For proof of the maximum principle, the reader is referred to the original publication (Pontryagin et al., 1962) or to subsequent improvements and generalizations, such as the work of Clarke (1983), Warga (1972), and Lee and Markus (1967).

It is easy to see that the coordinate $p_0(t)$ associated with the adjoint curve $p(t)$ is constant, because x_0 is a cyclic variable, and therefore $dp_0/dt = 0$. It is then convenient to reduce the Hamiltonians to $S \times \mathbb{R}^n$ and regard p_0 as a parameter.

Definition 4 A curve $(x(t), p(t), u(t))$ with values in $S \times \mathbb{R}^n \times U$ is called an extremal triple if there exists a constant $p_0 \leq 0$ such that $x(t)$, $p(t)$, $u(t)$, and p_0 satisfy conditions 1–3 of the maximum principle and, in addition, satisfy $p(t) \neq 0$ whenever $p_0 = 0$. We shall also say that $x(t), p(t))$ is the extremal curve generated by $u(t)$. Sometimes we may also refer to $u(t)$ as the extremal control. It is known that $\mathcal{M}(x(t), p(t))$ is constant along each extremal curve $(x(t), p(t))$. Consequently, condition 3 of the maximum principle can be replaced by $\mathcal{M}(x(t), p(t)) = 0$ for all t in the interval $[0, T]$.

In the language of extremal curves, the maximum principle says that an optimal solution curve $(x(t), u(t))$ is the projection of an extremal curve. We have already seen (with time-optimal linear problems) that an optimal solution curve $(x(t), u(t))$ can be the projection of more than one extremal triple.

1.3 The maximum principle and the classic necessary conditions for optimality

In the original publication of the maximum principle, the authors considered a slightly more general optimal problem than the one presented in the preceding section, for which they obtained the necessary conditions of optimality that include the so-called transversality conditions. The optimal problem that leads to these conditions consists of the following: Instead of considering trajectories that transfer an initial point to a terminal point, consider trajectories that originate on an initial submanifold S_0 of S and terminate on a given submanifold S_1 of S. Associated with each such trajectory is a total cost incurred during the time of transfer. The trajectory whose total cost of transfer is minimal is called optimal.

Optimal trajectories relative to the preceding problem also satisfy the maximum principle, and each optimal trajectory, in addition to meeting conditions 1–3 of the maximum principle, must also satisfy an extra condition: If $x(t)$ denotes the optimal trajectory, and if $p(t)$ denotes the corresponding adjoint curve, then $p(0)$ must be orthogonal to each tangent vector of the initial manifold S_0 at $x(0)$, and $p(T)$ must be orthogonal to each tangent vector of the terminal manifold S_1 at the terminal point $x(T)$. These conditions are known as the transversality conditions.

The transversality conditions, important in their own right, also provide the rationale for the modifications of the maximum principle to optimal problems with a fixed time of transfer and the developments in earlier chapters involving the use of the maximum principle.

The passage to problems with a fixed time of transfer is as follows: Suppose that a, b, and T are fixed in advance and that $(x(t), u(t))$ is the solution curve that minimizes

$$\int_0^T f_0(x(t), u(t))\, dt$$

among all other solutions that transfer a to b in T units of time.

Let $x_{n+1}(t) = t$ be another coordinate attached to the solution curve $x(t) = (x_1(t), \ldots, x_n(t))$. Denote by $\tilde{a} = (a, 0)$, and $\tilde{b} = (b, T)$ the points in $\mathbb{R}^{n+1}$ defined by the boundary conditions $\tilde{a}$ and $\tilde{b}$. Then $\tilde{x}(t) = (x_1(t), \ldots, x_n(t), x_{n+1}(t))$, and $u(t)$ is the solution curve for the basic optimal problem for the extended system

$$\frac{dx_0}{dt} = f_0(x(t), u(t)), \qquad \frac{dx_i}{dt} = f_i(x(t), u(t)), \qquad i = 1, 2, \ldots, n,$$

$$\frac{dx_{n+1}}{dt} = 1. \tag{8}$$

An extended trajectory $(\tilde{x}(t), u(t))$ of the foregoing system can transfer $\tilde{a}$ to $\tilde{b}$ only in T units of time. Therefore, the adjoint curve defined by the maximum principle is defined on $[0, T]$. Let $p_{n+1}(t)$ denote the component of the adjoint curve that corresponds to the optimal solution $(\tilde{x}(t), u(t))$ of (8). Then $dp_{n+1}/dt = 0$, because x_{n+1} is a cyclic variable (the original system is autonomous). Therefore, $p_{n+1}(t) = \text{constant}$, in which case condition 3 becomes

$$\sum_{i=0}^{n} p_i(t) f_i(x(t), u(t)) + p_{n+1} = 0$$

for almost all t. Thus,

$$\sum_{i=1}^{n} p_i(t) f_i(x(t), u(t)) = \mathcal{H}(x(t), p(t), u(t)) = \text{constant}$$

for almost all t. The foregoing argument shows that the necessary conditions given by the maximum principle for the fixed time of transfer differ from the conditions corresponding to a variable time of transfer only by the constant defined by $\mathcal{M}(x(t), p(t))$. Finally, it should be remarked that nonautonomous problems can be reduced to autonomous problems by a similar trick, leading to the same necessary conditions, except that $\mathcal{M}(x(t), p(t))$ may no longer be constant along extremal curves.

In order to make explicit connections between the maximum principle and the classic necessary conditions from the calculus of variations, it will first be necessary to recall the distinctions between strong and weak extremals.

Definition 5 An absolutely continuous curve $x(t)$ in $\mathbb{R}^n$ defined on an interval $[t_0, t_1]$ is called *admissible* (Young, 1969) if there exists a number M, with $0 < M < \infty$, such that for each coordinate function $x_i(t)$, $|dx_i(t)/dt| \leq M$ for each t in $[t_0, t_1]$ for which dx_i/dt exists.

A strong δ neighborhood of an admissible curve $x(t)$ consists of all other admissible curves $y(t)$, defined on the same interval $[t_0, t_1]$ as $x(t)$, such that

$$\sum_{i=1}^{n} |x_i(t) - y_i(t)| < \delta \quad \text{for all } t \in [t_0, t_1].$$

Example 1 The piecewise-linear function y, whose graph is shown in Figure 11.2, is in the $\delta > 0$ neighborhood of $x(t) = 0$ for each division of $[0, 1]$.

Definition 6 A weak δ neighborhood of a continuously differentiable curve $x(t)$ on an interval $[t_0, t_1]$ consists of all continuously differentiable curves $y(t)$

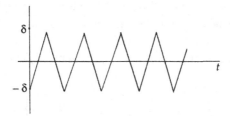

Fig. 11.2.

on $[t_0, t_1]$ such that

$$\sum_{i=1}^{n} |x_i(t) - y_i(t)| + \left(\frac{dx_i(t)}{dt} - \frac{dy_i(t)}{dt} \right) < \delta$$

for all t in $[t_0, t_1]$. The topology induced by the weak neighborhoods coincides with the $C^1[t_0, t_1]$ topology of $\mathbb{R}^n$-valued curves.

Suppose that $f_0 : O \times \mathbb{R}^n \to \mathbb{R}^1$ is a given continuously differentiable function defined on an open subset O of $\mathbb{R}^n$. Then we have the following:

Definition 7 An admissible curve $x(t)$ defined on an interval $[t_0, t_1]$ that takes values in O is said to be a strong local minimum for f_0 if there exists $\delta > 0$ such that

$$\int_{t_0}^{t_1} f_0 \left(x(t), \frac{dx}{dt} \right) dt \leq \int_{t_0}^{t_1} f_0 \left(y(t), \frac{dy}{dt} \right) dt$$

for any admissible curve $y(t)$ in the strong δ neighborhood of $x(t)$ that also satisfies $y(t_0) = x(t_0)$ and $y(t_1) = x(t_1)$.

An analogous definition will apply to weak local minima if we replace admissible functions with continuously differentiable functions and strong neighborhoods with weak neighborhoods.

Evidently any strong local minimum that is continuously differentiable is also a weak local minimum, but the converse is not always true.

Problem 1 Show that $x(t) = 0$ is the weak local minimum for the problem of minimizing

$$\int_0^1 \left(\frac{dx}{dt} \right)^2 \left(1 - \left(\frac{dx}{dt} \right)^2 \right) dt$$

subject to $x(0) = 0 = x(1)$, but that $x(t) = 0$ is not a strong local minimum. Show further that there is no strong minimum in this case.

Assume that a curve $x(t)$ is a strong local minimum for f_0 in some δ neighborhood of x. Let

$$
S = \left\{ y \in \mathbb{R}^n : \sum_{i=1}^n |y_i - x_i(t)| < \delta \right\}
$$

for all t in $[t_0, t_1]$. The corresponding cost extended control system is given by

$$
\frac{dx_0}{dt} = f_0(x(t), u(t)), \qquad \frac{dx_i(t)}{dt} = u_i(t), \qquad i = 1, 2, \ldots, n, \qquad (9)
$$

with U equal to $\mathbb{R}^n$. Evidently, $(x(t), u(t))$, with $u(t) = dx/dt$, is an optimal trajectory for system (9) relative to $a = x(t_0)$ and $b = x(t_1)$, and the maximum principle is applicable. The system Hamiltonian $\mathcal{H}$ is given by

$$
\mathcal{H}(x, p, u) = p_0 f_0(x, u) + \sum_{i=1}^n p_i u_i.
$$

Let $(p(t), x(t), u(t))$ denote an extremal triple associated with $(x(t), u(t))$. Then the corresponding constant p_0 cannot be equal to zero, for if $p_0 = 0$, then

$$
\mathcal{H}(x(t), p(t), u(t)) = \sum_{i=1}^n u(t)u_i(t).
$$

The maximality condition

$$
\mathcal{H}(x(t), p(t), u(t)) = \sup_{u \in \mathbb{R}^n} \mathcal{H}(x(t), p(t), u)
$$

implies that each $p_i(t) = 0$, which is ruled out by the maximum principle.

The conditions of the maximum principle determine the adjoint variable up to a positive scalar. Therefore, p_0 can be normalized to -1 whenever it is not equal to zero, and this choice of the value for p_0 leads to the usual expressions in the classic literature. Then the maximality condition of the maximum principle implies that

$$
\frac{\partial \mathcal{H}}{\partial u_i}(x(t), p(t), u(t)) = 0
$$

for each $i = 1, \ldots, n$ and almost all t. Hence, $p_i(t) = (\partial f_0/\partial u_i)(x(t), u(t))$ for $i = 1, 2, \ldots, n$ and almost all t in $[t_0, t_1]$, and therefore

$$\mathcal{H}(x(t), p(t), u(t)) = -f_0(x(t), u(t)) + \sum_{i=1}^{n} \frac{\partial f_0}{\partial u_i}(x(t), u(t)) u_i(t).$$

The maximality condition

$$\mathcal{H}(x(t), p(t), u(t)) - \mathcal{H}(x(t), p(t), \bar{u}) \geq 0,$$

for any $\bar{u} \in \mathbb{R}^m$ and almost all t in $[t_0, t_1]$, can be restated as

$$f_0(x(t), \bar{u}) - f_0(x(t), u(t)) + \sum_{i=1}^{n} \frac{\partial f_0}{\partial u_i}(x(t), u(t))(u(t) - \bar{u}) \geq 0.$$

According to the terminology of Weierstrass, the function

$$E(x, u, \bar{u}) = f_0(x, \bar{u}) - f_0(x, u) + \sum_{i=1}^{n} \frac{\partial f_0}{\partial u_i}(x, u)(u - \bar{u})$$

is called the excess function, or simply the E function. The foregoing argument shows that the maximality condition of the maximum principle is equivalent to the condition of Weierstrass that $E \geq 0$ along a strong minimum (Young, 1969).

Because $dp_i/dt = -(\partial \mathcal{H}/\partial x_i)(x(t), p(t), u(t))$, it follows that $dp_i/dt = (\partial f_0/\partial x_i)(x(t), u(t))$, and therefore

$$p_i(t) = \int_{t_0}^{t} \frac{\partial f_0}{\partial x_i}(x(\tau), u(\tau)) \, d\tau + \text{constant}.$$

Combining this formula with $p_i(t) = (\partial f_0/\partial u_i)(x(t), u(t))$ yields Euler's equation in integrated form:

$$\frac{\partial f_0}{\partial u_i}(x(t), u(t)) = \int_{t_0}^{t} \frac{\partial f_0}{\partial x_i}(x(\tau), u(\tau)) \, d\tau + \text{constant}. \tag{10}$$

Under suitable extra-differentiability assumptions, equation (10) can be differentiated to yield Euler's equation

$$\frac{d}{dt} \frac{\partial f_0}{\partial u_i}(x(t), u(t)) - \frac{\partial f_0}{\partial x_i}(x(t), u(t)) = 0. \tag{11}$$

(Remember that $dx/dt = u(t)$.) Under a further assumption that f_0 has continuous second derivatives relative to the variables $u_1, \ldots, u_n$, the maximality condition of the maximum principle implies that the Hessian of $\mathcal{H}$ satisfies

$$\left(\frac{\partial^2 \mathcal{H}}{\partial u_i \partial u_j} \ (x(t), p(t), u(t)) \right) \leq 0$$

or that

$$\sum_{j=1}^{n} \sum_{i=1}^{n} \frac{\partial^2 f_0}{\partial u_i \partial u_j} (x(t), u(t)) \, \bar{u}_i \bar{u}_j \geq 0 \quad \text{for all } \bar{u} \in \mathbb{R}^n \tag{12}$$

and for almost all t in $[t_0, t_1]$.

Condition (12) is called Legendre's necessary condition of optimality. The Legendre condition is said to be *strong* if the Hessian matrix

$$\left(\frac{\partial^2 f_0}{\partial u_i \partial u_j} \ (x(t), u(t)) \right)$$

is strictly positive-definite along the entire trajectory $(x(t), u(t))$. Then the implicit relation $p_i(t) = (\partial f_0/\partial x_i)(x(t), u(t))$ is locally uniquely solvable for $u(t)$ in terms of $p(t)$ and $x(t)$. The transformation $p \to u$ is called the Legendre transformation; in the literature on mechanics, the same transformation is called the momentum map.

If $u(t) = \varphi(x(t), p(t))$ denotes the solution of the implicit relation between p and u, then $(x(t), p(t))$ is an integral curve of a single Hamiltonian vector field $\vec{H}$, corresponding to $H(x(t), p(t)) = \mathcal{H}(x(t), p(t), \varphi(x(t), p(t)))$. When the Hessian $(\partial^2 f_0/\partial u_i \partial u_j)(x(t), p(t))$ has null directions, the one-to-one correspondence between $u(t)$ and $p(t)$ is lost, and instead of a single Hamiltonian there is a family of Hamiltonians associated with the extremal equations, similar to the singular linear quadratic problem discussed earlier in this text (Dirac, 1950).

As awkward as the maximum principle appears at first sight, it is a more direct test for optimality than are its classic counterparts, because of constraints that seem to be present in all applications, as, for instance, in the brachistochrone problem. We shall further illustrate this point by discussing the problem of finding the minimal surface of revolution, which, next to the brachistochrone problem, is the most classic variational problem. The subtlety of this problem and the beauty of its solutions are particularly well revealed by control-theory methods.

1.4 The minimal surface of revolution

Recall that the area of a surface of revolution obtained by revolving a curve $x(t)$ in the plane about the x_1 axis is given by

$$2\pi \int_0^T x_2(t) \sqrt{\left(\frac{dx_1}{dt}\right)^2 + \left(\frac{dx_2}{dt}\right)^2}\, dt,$$

assuming that $x_2(t) \geq 0$. When restricted to curves parametrized by arc length, the formula simplifies to

$$2\pi \int_0^T x_2(t)\, dt.$$

Assume now that each curve $x(t)$ is an absolutely continuous curve defined on some interval $[0, T]$, having a bounded derivative that is never equal to zero. Any such curve can be parametrized by its arc length measured from an initial point a, in which case

$$\frac{dx_1}{dt} = u_1(t), \quad \frac{dx_2}{dt} = u_2(t), \quad (u_1(t))^2 + (u_2(t))^2 = 1 \text{ a.e. on } [0, T]. \quad (13)$$

Consider the problem of finding a solution $x(t)$ of equation (13) that satisfies $x(0) = a$ and $x(T) = b$, with $x_2(t) \geq 0$ for all t in $[0, T]$, and that minimizes $\int_0^T x_2(t)\, dt$ among all other solutions satisfying the same boundary conditions. (We shall suppress the factor 2π.) The associated cost extended problem is given by

$$\frac{dx_0}{dt} = x_2(t), \quad \frac{dx_1}{dt} = u_1(t), \quad \frac{dx_2}{dt} = u_2(t), \quad (13a)$$

restricted to

$$\tilde{M} = \mathbb{R}^1 \times M, \quad \text{with} \quad M = \{(x_1, x_2) : x_2 \geq 0\} \quad \text{and} \quad u_1^2 + u_2^2 = 1.$$

$\tilde{M}$ is a three-dimensional manifold whose boundary is the plane $x_2 = 0$. Even though system (13a) is a linear system, the integral curves that originate in M may leave M in a finite time, and so the method of convexifying the space of controls does not a priori lead to the existence of optimal controls. However, the convexification procedure can be modified to yield the existence of solutions, as follows:

Let a and b denote given points in M such that $b_2 \geq a_2$. Let $\alpha = \inf\{\int_0^T x_2(t)\, dt : (x, u) \text{ is a solution curve of (13) in } M \text{ on an interval } [0, T] \text{ such that}$

$x(0) = a, x(T) = b$}. Let $(x^{(n)}, u^{(n)})$ be any sequence of solutions in M of (13) defined on intervals $[0, T_n]$ such that

$$\lim_{n \to \infty} \int_0^{T_n} x_2^{(n)}(t)\, dt = \alpha.$$

Bliss (1925) showed that if an arc $x(t)$ connecting a to b has an arc length greater than $a_2 + b_2$, then the broken line L shown in Figure 11.3 generates a smaller surface of revolution than $x(t)$. Because the arc length is equal to the time of travel from a to b, it follows that T_n cannot tend to ∞.

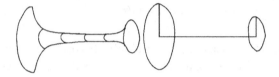

Fig. 11.3.

Let T be any number greater than T_n for $n = 1, \ldots$. Extend each control $u^{(n)}$ to the interval $[0, T]$. The choice of extension is unimportant for our purposes as long as the extended control function takes values in the unit disk in $\mathbb{R}^2$. The sequence $\{u^{(n)}(t) : t \in [0, T]\}$ is weakly sequentially compact in the space of measurable functions on the interval $[0, T]$, with values in the unit disk in $\mathbb{R}^2$. For notational simplicity, we shall assume that $\{u^{(n)}\}$ converges weakly to an element $u^{(\infty)}$. Let $x^{(\infty)}$ denote the solution $(dx^{(\infty)}/dt)(t) = u^{(\infty)}(t)$, with $x^{(\infty)}(0) = a$.

Because $\{x^{(n)}\}$ converge uniformly to $x^{(\infty)}$, it follows that $x^{(\infty)}(t) \in M$ for all t in $[0, T]$ and that

$$\lim_{n \to \infty} \int_0^{T_n} x_2^{(\infty)}(t)\, dt = \alpha.$$

This means that for some $\widehat{T}$ in the interval $[0, T]$,

$$\int_0^{\widehat{T}} x_2^{(\infty)}(t)\, dt = \alpha.$$

Therefore, for any a, b in M, there is an optimal solution curve $(x(t), u(t))$, with the control $u(t)$ taking values in the unit disk.

The following arguments based on the maximum principle will rule out the possibility that an optimal control can take values in the interior of the unit disk, and so the convexification procedure does not lead to extraneous solutions.

Each optimal solution $(x(t), u(t))$ that remains in the interior of M is the projection of an extremal triple $(x(t), p(t), u(t))$ subject to the following conditions:

$$\frac{dp_1}{dt} = -\frac{\partial \mathcal{H}}{\partial x_1} = 0, \qquad \frac{dp_2}{dt} = -\frac{\partial \mathcal{H}}{\partial x_2} = p_0,$$

where

$$\mathcal{H}(x, p, u) = -p_0 x_2 + p_1 u_1 + p_2 u_2, \qquad p_0 \leq 0,$$

and

$$\mathcal{H}(x(t), p(t), u(t)) = \max_{\|\bar{u}\| \leq 1} \mathcal{H}(x(t), p(t), \bar{u}) \text{ a.e.}$$

First note that $p_0 = 0$ is not possible, for if $p_0 = 0$, then $p(t) \neq 0$ for all t, and therefore $u(t) = p(t)/\|p(t)\|$, as can be easily verified through the Cauchy-Schwarz inequality. Hence, $\mathcal{H}(x(t), p(t), u(t)) = \|p(t)\|$, which violates condition 3 of the maximum principle that $\mathcal{H}(x(t), p(t), u(t)) = 0$ for almost all t.

Therefore, p_0 can be normalized to -1. At all times t such that $p(t) \neq 0$, $u(t) = p(t)/\|p(t)\|$, and consequently $x_2(t) = \|p(t)\|$, because

$$0 = -x_2(t) + p_1 u_1 + p_2 u_2.$$

It is easy to see from the foregoing equation that $p(t) \neq 0$ as long as $x_2(t) > 0$. Therefore, optimal solutions that are in the interior of M are generated by controls that are on the boundary of the unit disk.

Incorporating all this information into the extremal equations, we get

$$\frac{dx_1}{dt} = u_1(t) = \frac{p_1}{\sqrt{p_1^2 + p_2^2}}, \qquad \frac{dx_2}{dt} = u_2 = \frac{p_2}{\sqrt{p_1^2 + p_2^2}},$$

$$\frac{dp_1}{dt} = 0, \qquad \frac{dp_2}{dt} = 1.$$

When $p_1 = 0$, $x_1(t) = $ constant. Then $dx_2/dt = \pm 1$, depending on the sign of p_2. The graphs of these solutions are vertical lines.

For any other choice of p_1, $dx_1/dt \neq 0$, and so the extremal curves can be represented as graphs of functions of x_1. Then

$$\frac{dx_2}{dx_1} = \frac{dx_2}{dt} \bigg/ \frac{dx_1}{dt} = \frac{p_2}{p_1} = \pm \frac{\sqrt{x_2^2 - p_1^2}}{p_1}.$$

The last relation follows from $x_2 = \sqrt{p_1^2 + p_2^2}$. Hence,

$$\frac{dx_2}{\sqrt{x_2^2 - p_1^2}} = \pm \frac{dx_1}{p_1}, \quad \text{or} \quad \int_0^T \frac{dx_2}{\sqrt{x_2^2 - p_1^2}} = \pm \int_0^T \frac{dx_1}{p_1}.$$

Let $x_2 = p_1 \sec\theta$. Then $dx_2 = p_1 \sec\theta \tan\theta \, d\theta$, and

$$\int \frac{dx_2}{\sqrt{x_2^2 - p_1^2}} = \int \sec\theta \, d\theta = \ln|\sec\theta + \tan\theta| = \ln\left|\frac{x_2}{p_1} + \frac{\sqrt{x_2^2 - p_1^2}}{p_1}\right|.$$

Assuming that the terminal point b is to the right of the initial point a, then $p_1 > 0$, and

$$\ln\left(\frac{x_2}{p_1} + \frac{\sqrt{x_2^2 - p_1^2}}{p_1}\right) = \frac{x_1 - a_1}{p_1} + \text{constant.}$$

Upon simplifying, $x_2 = a_2 \cosh((x_1 - a_1)/p_1)$, which agrees with the classic finding that the solutions are catenaries.

It is easy to see that when b_1 is sufficiently far from a_1, there is no catenary that connects a to b. In such a case, the optimal trajectory is given by the broken line consisting of the vertical segments joined by the segment on the x_1 axis from a_1 to b_1. This solution is known as the *Goldschmidt solution*. These solutions are limits of smooth curves whose surface areas tend to the sum of the areas of circles formed by revolving the vertical segments from the x_1 axis to a and b, respectively. The surface area of the Goldschmidt solution is equal to the combined area of the two circles. The catenary solutions and the Goldschmidt solutions account for all possible types of optimal solutions. We shall not pursue this problem further. The interested reader can find a beautiful discussion (Bliss, 1925) concerning the interplay of these competing types of solutions and interpretation of the results in terms of the stability of the surfaces formed by soap films.

2 Extensions to differentiable manifolds

Variational problems in mechanics and geometry invariably have symmetries that determine the geometry of their solutions. Each of such problems usually has a preferred choice of coordinates that reveals this geometry, and such choices of coordinates are hardly ever canonical. We have already seen that for either the brachistochrone problem or the problem of the minimal surface of revolution, the final solutions are expressed in non-canonical coordinates. As we are about to see, the same is true for the motions of a rigid body fixed at a point, or, more generally, for optimal problems on Lie groups having either left- or right-invariant symmetries. In order to make the maximum principle directly applicable to such problems, it is necessary to further geometrize its content.

2.1 The symplectic structure of the cotangent bundle

The most immediate task is to become familiar with the symplectic structure of the cotangent bundle of a differentiable manifold M. To begin with, the cotangent bundle T^*M is equipped with a natural differential form θ that is defined through the projection $\pi : T^*M \to M$ carrying each covector in T^*M to its base point. Let π_* denote the tangent map defined by π. That is, for any curve $\xi(t)$ in T^*M that originates in T_x^*M, $\pi(\xi(t)) = \sigma(t)$ is a curve in M that originates at x, and

$$\pi_* \left(\frac{d\xi}{dt}(0) \right) = \frac{d\sigma}{dt}(0).$$

The tangent map π_* is a linear mapping from $T_{\xi(0)}(T^*M)$ onto $T_x M$.

The differential form θ is equal to the dual mapping π^* of the projection π_*. That is, $\theta : T^*M \to T^*(T^*M)$, and for each tangent vector $(d\xi/dt)(0)$ in the tangent space $T_\xi(T^*M)$,

$$\theta_\xi \left(\frac{d\xi}{dt}(0) \right) = \xi \left(\pi_* \left(\frac{d\xi}{dt}(0) \right) \right) \bigg| = \xi \left(\frac{d\sigma}{dt}(0) \right).$$

This abstract definition of θ takes on a familiar look when expressed in terms of local coordinates $x_1, \ldots, x_n$ on M. Let $v_1, \ldots, v_n$ denote the coordinates of a tangent vector at x, relative to the basis $\partial/\partial x_1, \ldots, \partial/\partial x_n$. We shall write $p_1, \ldots, p_n$ for the coordinates of a covector in T_x^*M relative to the dual basis $dx_1, \ldots, dx_n$. In these coordinates,

$$\theta_{(x,p)} = \sum_{i=1}^{n} p_i v_i.$$

Problem 2 Verify the preceding expression.

Definition 8 A differentiable 2-form ω on a manifold M is any mapping $x \to \omega_x$, $x \in M$, such that

(a) $\omega_x : T_x(M) \times T_x(M) \to \mathbb{R}^1$ is a bilinear, antisymmetric mapping, and
(b) the dependence $x \to \omega_x$ is smooth.

A 2-form ω is called nondegenerate, or regular, if ω_x is nondegenerate for each x; that is, $\omega_x(v, w) = 0$ for some v and for all w can obtain only if $v = 0$.

Definition 9 The exterior derivative $d\theta$ of a 1-form θ on a manifold M is a 2-form given by the following expression:

$$d\theta_x(X_1(x), X_2(x)) = X_1(\theta(X_2)) - X_2(\theta(X_1)) + \theta([X_1, X_2])$$

for any vector fields X_1 and X_2 on M.

In this notation, $\theta(X)$ is regarded as a function on M, and $Y(\theta(X))$ is the derivation in the direction of a vector field Y. The preceding formula for the exterior derivative of a 1-form will agree with that of Spivak (1970) if the definition for the Lie bracket is changed to its negative, which is the choice adopted by Spivak.

Definition 10 A symplectic form is any nondegenerate and closed 2-form on a manifold M.

We remind the reader that a form ω is closed if its exterior derivative $d\omega$ is equal to zero. Conversely any differential form that is an exterior derivative is closed. We shall not take time to define the exterior derivative of a 2-form. The reader can find these definitions in the books by Arnold (1976) and Spivak (1970).

Definition 11 The symplectic form ω that we shall use in the text is equal to the negative of $d\theta$, with θ equal to the canonical 1-form defined earlier.

Remark There is no agreement about the choice of sign for ω. The present definition of ω agrees with that of Abraham and Marsden (1985) and is the opposite of the one adopted by Arnold (1976).   ■

We shall now show that in terms of canonical coordinates on T^*M, ω agrees with the usual definition in $\mathbb{R}^{2n}$ used earlier, from which the nondegeneracy of ω will be more readily apparent.

Let $x_1, \ldots, x_n$ denote the coordinates of a point x on M, and let $p_1, \ldots, p_n$ denote the coordinates of a covector p in T_x^*M relative to the basis $dx_1, \ldots, dx_n$.

Any tangent vector to T^*M at p can be written as

$$\sum_{i=1}^{n} V_i \frac{\partial}{\partial x_i} + P_i \frac{\partial}{\partial p_i}$$

for some coordinates $V_1, \ldots, V_n, P_1, \ldots, P_n$ in $\mathbb{R}^{2n}$. Any choice of constant functions $V_1, \ldots, V_n, P_1, \ldots, P_n$ defines a vector field X in a neighborhood of a point p in T_x^*M through the preceding formula. Then

$$\theta(X) = \sum_{i=1}^{n} p_i V_i.$$

If

$$Y = \sum_{i=1}^{n} W_i \frac{\partial}{\partial x_i} + Q_i \frac{\partial}{\partial p_i}$$

denotes another vector field defined by constant functions $W_1, \ldots, W_n, Q_1, \ldots, Q_n$, then

$$Y(\theta(X)) = \sum_{i=1}^{n} Q_i V_i.$$

Because V_i, P_i, W_i, and Q_i are constant functions, $[X, Y] = 0$, and Definition 11 yields

$$\omega_{(x,p)}(X, Y) = \sum_{i=1}^{n} Q_i V_i - P_i W_i. \tag{14}$$

The choice of coordinates on T^*M that results in expression (14) will be called *canonical* or *symplectic*. This expression for ω coincides with the symplectic form defined in Chapter 7.

Evidently, expression (14) is nondegenerate, which implies that ω is nondegenerate. Being nondegenerate, ω sets up a one-to-one correspondence between forms and vectors on T^*M, which leads to the following:

Definition 12 The Hamiltonian vector field $\vec{H}$ associated with a function H on T^*M is the unique vector field with the property that

$$dH_\xi(Y(\xi)) = \omega_\xi(\vec{H}(\xi), Y(\xi)) \quad \text{for all } \xi \text{ in } T^*M$$

and all tangent vectors $Y(\xi)$.

This coordinate-free definition yields the familiar expressions when special-
ized to canonical coordinates, as demonstrated in Chapter 7. That is, the integral curves of $\vec{H}$, when expressed in canonical coordinates, are given by the
formulas:

$$\frac{dx_i}{dt} = \frac{\partial H}{\partial p_i}(x, p) \quad \text{and} \quad \frac{dp_i}{dt} = -\frac{\partial H}{\partial x_i}(x, p), \qquad i = 1, \ldots, n. \quad (15)$$

Definition 13 The Hamiltonian H_X associated with a vector field X on M is
a function on T^*M defined by $H_X(\xi) = \xi(X(x))$ for each $\xi \in T_x^*M$. The
Hamiltonian vector field $\vec{H}_X$ is called the Hamiltonian lift of X.

Problem 3 If

$$X(x) = \sum_{i=1}^{n} f_i(x_1, \ldots, x_n) \frac{\partial}{\partial x_i}$$

is the representation of X in some local coordinates, show that in symplectic
coordinates on T^*M the integral curves of the Hamiltonian lift of X satisfy the
following equations:

$$\frac{dx_i}{dt} = f_i(x_i, \ldots, x_n) \quad \text{and} \quad \frac{dp_i}{dt} = -\sum_{j=1}^{n} p_j \frac{\partial f_j}{\partial x_i}(x_1, \ldots, x_n).$$

Definition 14 The Poisson bracket of a function F with a function G, both
defined on T^*M, is a function H defined by $H(\xi) = (d/dt)F \circ (\exp t\vec{G})(\xi)$
for $t = 0$. We shall use $\{F, G\}$ to denote the function H.

Because

$$d(F_1 \cdot F_2)(\vec{G}(\xi)) = F_1(\xi)\, dF_2(\vec{G}(\xi)) + F_2(\xi)\, dF_1(\vec{G}(\xi)),$$

it follows that

$$\{F_1 F_2, G\} = F_1\{F_2, G\} + F_2\{F_1, G\}.$$

In particular, $\{F, G\} = 0$ if either F or G is a constant function.

Theorem 1

(a) $\{F, G\}(\xi) = \omega_\xi(\vec{F}(\xi), \vec{G}(\xi))$ *for all* $\xi \in T^*M$.
(b) $\{\{F, G\}, H\} + \{\{H, F\}, G\} + \{\{G, H\}, F\} = 0$ *(Jacobi identity).*
(c) $\vec{H} = [\vec{F}, \vec{G}]$ *whenever* $H = \{F, G\}$.

(d) $\{H_X, H_Y\} = H_{[X,Y]}$ *for the Hamiltonians corresponding to arbitrary vector fields X and Y in M.*

Proof For part (a),

$$H(\xi) = \frac{d}{dt} F \circ (\exp t\vec{G})(\xi)|_{t=0} = dF(\vec{G}(\xi)) = \omega_\xi(\vec{F}(\xi), \vec{G}(\xi)).$$

It follows from (a) that $\{F, G\} = -\{G, F\}$. Also note that $\{F, G\} = \vec{G}(F)$, regarding the Hamiltonian vector field $\vec{G}$ as the derivation on T^*M. We shall use this observation to prove (b):

Let F, G, and H be any functions on T^*M. Then

$$\{\{F, G\}, H\} + \{\{H, F\}, G\} + \{\{G, H\}, F\}$$
$$= \vec{H}(\vec{G}(F)) + \vec{G}(\vec{F}(H)) + \vec{F}(\vec{H}(G)).$$

Each of $\vec{H} \circ \vec{G}$, $\vec{G} \circ \vec{F}$, and $\vec{F} \circ \vec{H}$ is a second-degree differential operator; hence the preceding expression is a linear combination of second-order derivatives of the functions F, G, and H. The terms that involve second-degree partial derivatives of F are the terms $\vec{H}(\vec{G}(F))$ and $\vec{G}(\vec{F}(H))$, since $\vec{G}(\vec{F}(H)) = -\vec{G}(\vec{H}(F))$. But $\vec{H}(\vec{G}(F)) - \vec{G}(\vec{H}(F)) = [\vec{G}, \vec{H}](F)$.

The Lie bracket of vector fields is a vector field, and therefore $[\vec{G}, \vec{H}](F)$ contains only the first partial derivatives of F. Therefore, the preceding expression does not contain any partial derivatives of F. The same argument applies to H and G. The conclusion is that the right-hand side of part (b) is constant. But the only possible constant is zero. So part (b) is proved.

To prove (c), note that

$$[\vec{F}, \vec{G}](f) = \vec{G}(\vec{F}(f)) - \vec{F}(\vec{G}(f)) = \{\{f, F\}, G\} - \{\{f, G\}, F\}$$
$$= -\{\{G, f\}, F\} - \{\{F, G\}, f\} - \{\{f, G\}, F\}$$
$$= \{f, \{F, G\}\} = \overrightarrow{\{F, G\}}(f).$$

Hence,

$$[\vec{F}, \vec{G}] = \overrightarrow{\{F, G\}} = \vec{H}.$$

It remains to prove part (d). Let X and Y be arbitrary vector fields in M. Let θ denote the canonical 1-form on T^*M such that $\omega = -d\theta$. The canonical lifts $\vec{H}_X$ and $\vec{H}_Y$ satisfy $H_X = \theta(\vec{H}_X)$ and $H_Y = \theta(\vec{H}_Y)$. It then follows from

Definition 11 and part (a) of this theorem that

$$\{H_X, H_Y\}(\xi) = \omega_\xi (\vec{H}_X(\xi),\, \vec{H}_Y(\xi)) = \vec{H}_Y(H_X) - \vec{H}_X(H_Y) - \theta([\vec{H}_X, \vec{H}_Y])$$
$$= 2\{H_X, H_Y\} - \theta([\vec{H}_X, \vec{H}_Y]).$$

Hence,

$$\{H_X, H_Y\} = \theta([\vec{H}_X, \vec{H}_Y]).$$

Because

$$\theta([\vec{H}_X, \vec{H}_Y]) = H_{[X,Y]},$$

the theorem is proved.   ∎

Corollary *Suppose that* $F(x_1, \ldots, x_n, p_1, \ldots, p_n)$ *and* $G(x_1, \ldots, x_n, p_1,$ $\ldots, p_n)$ *are any functions on* T^*M *coordinatized by canonical coordinates* $x_1, \ldots, x_n,\ p_1, \ldots, p_n$. *The Poisson bracket* $\{F, G\}$ *in these coordinates is given by*

$$\sum_{i=1}^{n} \frac{\partial F}{\partial x_i}\frac{\partial G}{\partial p_i} - \frac{\partial F}{\partial p_i}\frac{\partial G}{\partial x_i}. \tag{16}$$

Problem 4 Let $X = \sum_{i=1}^{n} X_i(\partial/\partial x_i)$, and $Y = \sum_{i=1}^{n} Y_i(\partial/\partial x_i)$. Use expression (16) for the definition of $\{F, G\}$ to show directly that $\{H_X, H_Y\} = H_{[X,Y]}$ for

$$H_X = \sum_{i=1}^{n} p_i X_i\,(x_1, \ldots, x_n) \quad \text{and} \quad H_Y = \sum_{i=1}^{n} p_i Y_i\,(x_1, \ldots, x_n).$$

2.2 Variational problems on manifolds and the maximum principle

Typically, variational problems in mechanics and geometry are defined either on the entire tangent bundle of the underlying manifold M or on its subsets. Such variational problems are defined by a pair (f, C), with C a subset of the tangent bundle of M, and f a real-valued function defined on C.

A trajectory of C is any absolutely continuous curve $x(t)$ in M defined on some interval $[0, T]$ for which its tangent vector $(dx/dt)(t)$ belongs to C for almost all t in $[0, T]$. For any such curve $x(t)$, $f((dx/dt)(t))$ is a real-valued function defined for almost all t in the domain of x. Assuming further integrability properties, $\int_0^t f((dx/dt)(t))\, d\tau$ is a well-defined function on $[0, T]$. Depending on the application, the function f can be thought of as a Lagrangian, or a metric, or, more generally, any optimality criterion. In the literature on optimal problems, such a function is usually called an objective function. The integral $\int_0^T f((dx/dt)(t))\,dt$ will be called the cost of transfer from $x(0) = a$ to $x(T) = b$ along the trajectory $x(t)$.

The pair (f, C) defines an optimal-control problem of transferring a given initial point a in M to a given terminal point b in M along the trajectories of C that will minimize the cost of transfer. In order to make the maximum principle applicable to such problems, it will be necessary to assume that C has control-like structure. The most convenient way of introducing control-like structure is by means of "sections."

A (smooth) local section of a (smooth) mapping $\phi : X \rightarrow Y$ is a (smooth) mapping ψ from an open subset A of Y onto X, such that $\phi \circ \psi =$ identity on A. ψ is called a local section at y if A is a neighborhood of y, and ψ is called a section, or a global section, whenever $A = Y$.

Let π_c denote the restriction to C of the natural projection $\pi : TM \rightarrow M$. C is said to have control-like structure if π_c admits a local section at each point of M of the form

$$v = F(\pi_c(v), u_1, \ldots, u_m), \qquad v \in C,$$

for some family of smooth vector fields F parametrized by points $u = (u_1, \ldots, u_m)$ belonging to a subset U of $\mathbb{R}^m$.

In addition, it will be assumed that the number of parameters m does not vary with the choice of a section and that the passage from one local section to another is realized by a smooth change of parameters. Regarding each set of parameters as controls, then the passage from one control section to another is described by smooth feedback in the respective control spaces. It is further convenient to assume that points of C above any given point x in M are in exact correspondence with the controls $u = (u_1, \ldots, u_m)$.

Then each trajectory $x(t)$ of C is traced by a measurable control $u(t)$, with values in U defined by $(dx/dt)(t) = f(x(t), u(t))$ for almost all t for which $x(t)$ remains in a given control section, and the cost along $x(t)$ can be expressed in terms of $x(t)$ and $u(t)$ by $\int_0^T f_0(x(t), u(t))\,dt = \int_0^T f \circ F(x(t), u(t))\,dt$. Thus, in each control section of C, the original problem can be viewed as an optimal control problem, because optimal trajectories that emanate from a fixed initial point are in exact correspondence with optimal controls, as a consequence of the existence and uniqueness of solutions of differential equations.

The optimal-control problem defined by the pair (f, C) lends itself to the same geometric formulation in the extended space $\mathbb{R}^1 \times M$ that leads to the maximum principle as an optimal-control problem in $\mathbb{R}^n$. The extended trajectories of (f, C) in $\mathbb{R}^1 \times M$ are absolutely continuous curves $\tilde{x}(t) = (x_0(t), x(t))$ defined by $dx_0/dt = f((dx/dt)(t))$, $(dx/dt)(t) \in C$ for almost all t in the domain of x. The extended trajectories that emanate from an initial point $(0, a)$ define the set of points of (f, C) reachable from $(0, a)$. The reachable set intersects the line $\ell = \{(x_0, b) : x_0 \in \mathbb{R}^1\}$ in $\mathbb{R}^1 \times M$ if and only if there exists a trajectory $x(t)$ of C defined on an interval $[0, T]$ for which $x(T) = b$. Any such trajectory of C marks a line segment between $(0, b)$ and $(x_0(T), b)$, with $x_0(T) = \int_0^T f((dx/dt)(t)) \, dt$. The line segment marked by an optimal trajectory is minimal among all extended trajectories that emanate from $(0, a)$ and intersect ℓ. Any control section of (f, C) induces an extended control system

$$\frac{dx_0}{dt} = f_0(x(t), u(t)), \qquad \frac{dx}{dt}(t) = F(x(t), u(t)). \tag{17}$$

Each control u defines a vector field $X_u(x) = (f_0(x, u), F(x, u))$ on $\mathbb{R}^1 \times M$, and this vector field defines a Hamiltonian function H_u on the cotangent bundle of $\mathbb{R}^1 \times M$ according to Definition 13. The totality of Hamiltonians H_u defines a family of Hamiltonians $\mathcal{H}$ parametrized by the controls. $\mathcal{H}$ is called the Hamiltonian lift of (17).

Bounded and measurable control functions define a time-varying Hamiltonian $\mathcal{H}(\xi, u(t))$, with $\xi \in T^*(\mathbb{R}^1 \times M)$. $\vec{\mathcal{H}}(\xi, u(t))$ will denote the corresponding time varying Hamiltonian vector field. An integral curve $\xi(t)$ of this field, defined by $(d\xi/dt)(t) = \vec{\mathcal{H}}(\xi(t), u(t))$, projects onto a solution curve of (17); that is, $x(t) = \pi(\xi(t))$. Because the vector fields in (17) do not depend explicitly on the variable x_0, $\mathcal{H}$ can be reduced to the cotangent bundle of M by introducing another parameter p_0 defined as follows:

The cotangent bundle of $\mathbb{R}^1 \times M$ is equal to the product bundle $T^*\mathbb{R}^1 \times T^*M$. Furthermore, $T^*\mathbb{R}^1 = \mathbb{R}^1 \times \mathbb{R}^1$. Then each point of $T^*\mathbb{R}^1$ is represented by the coordinates (x_0, p_0) relative to the basis dx_0 in $T^*\mathbb{R}^1$. p_0 is a cyclic coordinate for each $\mathcal{H}(\cdot, u(t))$, in the sense that p_0 is constant along the integral curves of $\vec{\mathcal{H}}(\xi, u(t))$. Then

$$\mathcal{H}_{p_0}(\xi, u) = p_0 f_0(x, u) + \xi F(x, u) \quad \text{for all} \quad \xi \in T_x^* M.$$

Even though the Hamiltonian lifts depend on the choice of a control section, the supremal Hamiltonian $\mathcal{M}_{p_0}(\xi) = \sup_{u \in U} \mathcal{H}_{p_0}(\xi, u)$ depends only on f and C. The maximum principle for (f, C) is as follows:

The maximum principle Suppose that $x(t)$ is an optimal trajectory of (f, C) defined on an interval $[0, T]$. Let F be any control section of C that contains

$x(t)$, for $t \in [0, T]$, and let $u(t)$ be the control that generates $x(t)$. Then there exists a constant p_0 ($p_0 \leq 0$) and an integral curve $\xi(t)$ of $\vec{\mathcal{H}}_{p_0}(\cdot, u(t))$ defined on the entire interval $[0, T]$ such that the following hold:

1. $x(t)$ is the projection of $\xi(t)$ for all t in $[0, T]$. $\xi(t) \neq 0$ for any t when $p_0 = 0$.
2. $\mathcal{H}_{p_0}(\xi(t), u(t)) = \mathcal{M}_{p_0}(\xi(t))$ for almost all t in $[0, T]$.
3. $\mathcal{M}_{p_0}(\xi(t)) = 0$ for all t in $[0, T]$.

If, instead of the variable time of transfer from a to b, the problem is restricted to a fixed time interval $[0, T]$, then condition 3 of the maximum principle is changed to $\mathcal{M}_{p_0}(\xi(t)) = $ constant, and conditions 1 and 2 remain unchanged.

Evidently the statement of the maximum principle given here coincides with the version in Section 1.2 whenever the points ξ in T^*M are expressed in terms of the canonical coordinates $x_1, \ldots, x_n, p_1, \ldots, p_n$. The meaning of extremal curves remains unchanged in this more general setting. The extremal curves that correspond to the multipliers $p_0 = 0$ will be called abnormal, and others will be called regular, or normal.

Remark Strictly speaking, the statement of the maximum principle for arbitrary manifolds requires its own proof, because an extremal curve may not be contained in any coordinate chart corresponding to canonical coordinates. However, the proof of the maximum principle on manifolds is essentially a paraphrase of the original one in $\mathbb{R}^n$, modified only linguistically to conform to the language of the cotangent bundle of the manifold. ∎

For problems of Lagrangian mechanics or Riemannian geometry, $\mathcal{C}$ is the entire tangent bundle TM. The extremal curves that correspond to such problems can only be regular, by the following argument: Because every absolutely continuous curve is a trajectory, it follows that each basis of vector fields $X_1, \ldots, X_n$ in M defines a control system

$$\frac{dx(t)}{dt} = \sum_{i=1}^{n} u_i(t) X_i(x(t)).$$

The Hamiltonian of this system is $\sum_{i=1}^{n} u_i(t) H_i(\xi)$, with H_i denoting the Hamiltonian of X_i; that is, $H_i(\xi) = \xi(X_i(x))$ for $\xi \in T_x^*M$.

When the multiplier p_0 is equal to zero,

$$\mathcal{H}_{p_0}(\xi, u) = \sum_{i=1}^{n} u_i H_i(\xi).$$

Because $\mathcal{H}_{p_0}(\xi, u)$ is an affine function of the control variables, the maximality condition 2 implies that $H_i(\xi(t)) = 0$ for each $i = 1, \ldots, n$. This means that $\xi(t)$ annihilates each tangent vector $X_i(x(t))$ and therefore must be identically zero. That conclusion contradicts condition 1.

When p_0 is reduced to -1, and when $C = TM$, the maximum principle produces the classic Hamiltonian function, provided that the strong Legendre condition holds. For instance, for problems of Lagrangian mechanics, the maximum principle leads directly to the expression for the total energy of the system. The argument is as follows: According to the principle of least action, any conservative mechanical system minimizes $\int L \, dt$, where $L = T - V$, with T equal to the kinetic energy and V equal to the potential energy. T is a smooth function on the tangent bundle of M, which is a positive-definite quadratic function of the velocities dx/dt at each point x. Let $X_1, \ldots, X_n$ denote any basis of vector fields on M defined in a neighborhood of an arbitrary point. This choice of basis induces a control section

$$\frac{dx}{dt} = \sum_{i=1}^{n} u_i X_i(x),$$

with $u \in \mathbb{R}^n$. In terms of this choice of controls, $T = \sum_{i,j=1}^{n} P_{ij}(x) u_i u_j$, and $L(x, u_1, \ldots, u_n) = \sum_{i,j=1}^{n} P_{ij} u_i u_j - V(x)$. The matrix $P(x) = (P_{ij})$ is a positive-definite matrix. The Hamiltonian $\mathcal{H}$ associated with the control system $dx/dt = \sum_{i=1}^{n} u_i X_i(x)$ and the Lagrangian L is given by

$$\mathcal{H}(\xi, u) = -L(x, u) + \sum_{i=1}^{n} u_i H_i(\xi),$$

with $H_i(\xi) = \xi(X_i(x))$.

$$\mathcal{M}(\xi) = \sup_{u \in \mathbb{R}^n} \mathcal{H}(\xi, u)$$

occurs for

$$-2 \sum_{j=1}^{n} P_{ij} u_j + H_i = 0, \qquad i = 1, \ldots, n,$$

because $\mathcal{H}(\xi, u)$ is a concave function of u, and its maximum is given by $(\partial \mathcal{H}/\partial u_i)(\xi, u) = 0$, $i = 1, 2, \ldots, n$. Hence,

$$u = \frac{1}{2} P^{-1} \hat{H}, \quad \text{with} \quad \hat{H} = \begin{pmatrix} H_1 \\ \vdots \\ H_n \end{pmatrix},$$

and therefore

$$\mathcal{M}(\xi) = \sup_{u \in \mathbb{R}^n} \mathcal{H}(\xi, u)$$

$$= -\frac{1}{4}(P^{-1}\hat{H}(\xi), \hat{H}(\xi)) + V(x) + \frac{1}{2}(P^{-1}\hat{H}, \hat{H})$$

$$= \frac{1}{4}(P^{-1}\hat{H}(\xi), \hat{H}(\xi)) + V(x) \quad \text{for any} \quad \xi \text{ in } T_x^* M.$$

Each extremal curve $(x(t), u(t))$ is the projection of an integral curve $\xi(t)$ of the Hamiltonian vector field $\vec{\mathcal{M}}$, with $u(t) = \frac{1}{2}P(x(t))\hat{H}(\xi(t))$. Consequently, $\mathcal{M}(\xi(t)) = $ constant, and therefore the total energy is conserved.

Note that this derivation of the equations of motion does not make use of canonical coordinates, nor does it make any reference to the Euler-Lagrange equation. Of course, for those problems, the maximum principle does not provide any new information beyond the classic theory. For all those problems, a weak solution given by the Euler-Lagrange equation is in exact correspondence with the strong solution obtained by the maximum principle, because of the uniqueness of a control that maximizes $\mathcal{H}$ (a consequence of the convexity with respect to the velocities).

Let us now turn our attention to some of the typical systems controlled by a single control function and make some elementary observations concerning its Hamiltonians. Suppose that the system is given by

$$\frac{dx}{dt} = X_0(x) + uX_1(x),$$

and suppose further that we want to minimize $\frac{1}{2}\int_0^T u^2 \, dt$, assuming no bounds on the control. Let H_0 and H_1 denote respectively the Hamiltonian lifts of X_0 and X_1. The maximum principle says that each extremal curve $(\xi(t), u(t))$ is an integral curve of either of the following Hamiltonians:

$$\mathcal{H}_1 = -\frac{1}{2}u^2 + H_0(\xi) + u\,H_1(\xi), \qquad \mathcal{H}_0 = H_0 + uH_1,$$

subject to the maximality condition that

$$\mathcal{H}_i(\xi(t), u(t)) = \sup_{u \in \mathbb{R}^1} \mathcal{H}_i(\xi(t), u), \qquad i = 1, 2.$$

In the regular case, $u(t) = H_1(\xi(t))$, and therefore $\xi(t)$ is an integral curve of $H = H_0 + \frac{1}{2}H_1^2$. In the abnormal case, the maximality condition does not immediately yield an expression for $u(t)$. Instead, it produces a constraint $H_1(\xi(t)) = 0$. Then

$$0 = \frac{d}{dt} H_1(\xi(t)) = \{H_1, H_0 + u\, H_1\}\, (\xi(t)) = \{H_1, H_0\}\, (\xi(t)),$$

so there is always another constraint $\{H_0, H_1\}(\xi(t)) = 0$, provided that $\{H_0, H_1\}$ is not dependent on H_1. Differentiating $\{H_0, H_1\}(\xi(t)) = 0$ yields another relation

$$\{\{H_0, H_1\}, H_0\} + u(t)\{\{H_0, H_1\}, H_1\}\, (\xi(t)) = 0.$$

Assuming that $\{\{H_0, H_1\}, H_1\}\, (\xi(t)) \neq 0$, then the extremal control $u(t)$ that generates $\xi(t)$ is given by the following expression:

$$u(t) = -\frac{\{\{H_0, H_1\}, H_0\}\, (\xi(t))}{\{\{H_0, H_1\}, H_1\}\, (\xi(t))}.$$

This procedure defines two Hamiltonian systems:

$$H = H_0 + \frac{1}{2} H_1 \quad \text{and} \quad G = H_0 - \frac{(\operatorname{ad} H_0)^2(H_1)}{(\operatorname{ad} H_1)^2(H_0)}\, H_1.$$

The first Hamiltonian H is defined on T^*M, whereas the second is valid only on the set $H_1 = 0$ and $\operatorname{ad} H_0(H_1) = 0$. For a given set of boundary conditions, it may happen that the extremal curve that projects on an optimal solution is either an integral curve of $\vec{H}$ or an integral curve of $\vec{G}$ and in some situations both.

2.3 Euler's elastic problem and the problem of Dubins

Rather than pursuing this level of generality, we shall now return to some classic problems that were introduced in Chapter 1 and show how the maximum principle can be used to get to their solutions. The first of these problems goes back to Euler in 1744 and consists of the following:

Assume that a and b are arbitrary points in $\mathbb{R}^2$ and that $\dot{a}$ and $\dot{b}$ are fixed tangent vectors of unit length at a and b, respectively. The problem is to find a curve $\gamma(t)$ such that $\gamma(0) = a$, $(d\gamma/dt)(0) = \dot{a}$, $\gamma(T) = b$, and $(d\gamma/dt)(T) = \dot{b}$ and such that it minimizes

$$\frac{1}{2} \int_0^T k^2(t)\, dt$$

among all such curves, where $k(t)$ is the geodesic curvature along $\gamma(t)$. As stated in Chapter 1, the solutions to this problem were named *elastica* by Euler. We shall refer to this problem as the *elastic problem of Euler*. Recall that the elastic problem of Euler is a particular case of a problem of Radon, who considered the problem of minimizing an arbitrary function of curvature. That is, instead of minimizing $\frac{1}{2}\int_0^T k^2\,dt$, Radon considered the problem of minimizing $\int_0^T f(k(t))\,dt$ for some function f.

The second problem that we shall now consider is the problem of Dubins. Recall that this problem consists of finding the curves of minimal length that connect $(a, \dot{a})$ to $(b, \dot{b})$ and that satisfy an additional constraint that $|k(t)| \leq 1$ for all t.

All of these problems admit natural formulations on the group of motions E_2 of $\mathbb{R}^2$. Recall that E_2 is the semidirect product of $\mathbb{R}^2$ with $SO_2(R)$, which can also be regarded as the subgroup of $GL_3(R)$ consisting of 3×3 matrices

$$\begin{pmatrix} 1 & 0 & 0 \\ x_1 & \alpha & \beta \\ x_2 & -\beta & \alpha \end{pmatrix},$$

with $\alpha^2 + \beta^2 = 1$, and $(x_1, x_2) \in \mathbb{R}^2$. This group can also be viewed as the set of all pairs (x, F), with x a point in $\mathbb{R}^2$, and F a positively oriented orthonormal frame at x. Relative to a fixed frame (e_1, e_2), F can be represented by a rotation matrix

$$\begin{pmatrix} \alpha & \beta \\ -\beta & \alpha \end{pmatrix},$$

which explains the correspondence with the foregoing matrix.

As stated earlier, the Lie algebra of E_2 consists of all 3×3 matrices of the form

$$\begin{pmatrix} 0 & 0 & 0 \\ \alpha_1 & 0 & -\alpha_3 \\ \alpha_2 & \alpha_3 & 0 \end{pmatrix},$$

with $(\alpha_1, \alpha_2, \alpha_3)$ in $\mathbb{R}^3$. Let

$$A_1 = \begin{pmatrix} 0 & 0 & 0 \\ 1 & 0 & 0 \\ 0 & 0 & 0 \end{pmatrix}, \qquad A_2 = \begin{pmatrix} 0 & 0 & 0 \\ 0 & 0 & 0 \\ 1 & 0 & 0 \end{pmatrix}, \qquad A_3 = \begin{pmatrix} 0 & 0 & 0 \\ 0 & 0 & -1 \\ 0 & 1 & 0 \end{pmatrix}$$

denote the standard basis, so that any element in the Lie algebra is written $\alpha_1 A_1 + \alpha_2 A_2 + \alpha_3 A_3$. Corresponding to each element A in the Lie algebra, $\vec{A}$ denotes the left-invariant vector field $\vec{A}(g) = gA$ for $g \in E_2$. We shall consider the following control system on E_2:

$$\frac{dg}{dt} = \vec{A}_1(g) + u(t)\,\vec{A}_3(g), \tag{18}$$

with

$$g(t) = \begin{pmatrix} 1 & 0 \\ x(t) & R(t) \end{pmatrix}, \qquad x(t) = \begin{pmatrix} x_1(t) \\ x_2(t) \end{pmatrix}, \qquad \text{and } R(t) \in SO_2(R).$$

This control system is the classic Serret-Frenet system associated with a curve $x(t)$ parametrized by its arc length. That is, equation (18) can also be written as

$$\frac{dx}{dt} = R(t)e_1, \qquad \frac{dR}{dt} = R(t) \begin{pmatrix} 0 & -u(t) \\ u(t) & 0 \end{pmatrix}.$$

As demonstrated in Chapter 1, the rotation matrix $R(t)$, when parametrized by an angle θ, yields the following differential system in $\mathbb{R}^3$:

$$\frac{dx_1}{dt} = \cos\theta, \qquad \frac{dx_2}{dt} = \sin\theta, \qquad \frac{d\theta}{dt} = u(t). \tag{19}$$

Because $dx/dt = Re_1$, it follows that $d^2x/dt^2 = (dR/dt)e_1 = u(t)Re_2$. Therefore,

$$\left\| \frac{d^2x}{dt^2} \right\| = |u(t)|,$$

and the control u is equal to the geodesic curvature.

All three of the aforementioned problems are optimal-control problems concerning the trajectories of (18). Radon's problem consists in minimizing

$$\int_0^T f(u)\,dt$$

over the trajectories of (18) that satisfy the prescribed boundary conditions $g(0) = (a, R(0))$ and $g(T) = (b, R(T))$, with $R(0)$ and $R(T)$ equal to the elements in $SO_2(R)$ defined by $\dot{a}$ and $\dot{b}$, respectively.

The problem of Dubins is the same as finding the time-optimal trajectories of (18) with $|u(t)| \le 1$, because the parameter t is equal to the arc length. Denote by H_1, H_2, H_3 the Hamiltonians of the left-invariant vector fields $\vec{A}_1$, $\vec{A}_2$, $\vec{A}_3$, respectively. Because $[\vec{A}_1, \vec{A}_2] = 0$, $[\vec{A}_1, \vec{A}_3] = \vec{A}_2$, and $[\vec{A}_2, \vec{A}_3] = -\vec{A}_1$, it follows that the Poisson brackets of H_1, H_2, and H_3 satisfy the same relations.

We shall first determine the extremals for Euler's elastic problem: It follows from the preceding general discussion that the regular extremals are the integral curves of the Hamiltonian vector field $\vec{H}$, defined by $H = \frac{1}{2}H_3^2 + H_1$, and that along each extremal curve $\xi(t)$ the corresponding control $u(t)$ is equal to $H_3(\xi(t))$. For abnormal extremals, $H_3(\xi(t)) = 0$ and $\{H_3, H_1\}(\xi(t)) = 0$. Because $\{H_3, H_1\} = -H_2$, it follows that $H_2(\xi(t)) = 0$. Then

$$0 = \frac{d}{dt}H_2(\xi(t)) = \{H_2, H_1 + uH_3\}$$

$$= \{H_2, H_1\}(\xi) + u\{H_2, H_3\} = -u(t)H_1(\xi(t)).$$

$H_1(\xi(t))$ cannot be equal to zero, because $\xi(t) \neq 0$. Hence, abnormal extremals are produced by zero controls. Because u is equal to the curvature, the projections of these extremal curves on $\mathbb{R}^2$ are straight lines (geodesics).

It remains to examine the integral curves of $\vec{H}$. There is an important conservation law that comes from the left-invariant symmetries of the problem that will enable us to integrate the relevant differential equations. This conservation law is given by

$$H_1^2 + H_2^2 = M^2 = \text{constant.}$$

It is easy to verify that along an extremal curve $\xi(t)$,

$$\frac{d}{dt}\left(H_1^2 + H_2^2\right)(\xi(t)) = 2H_1\left\{H_1, \frac{1}{2}H_3^2 + H_1\right\}(\xi(t))$$

$$+ 2H_2\left\{H_2, \frac{1}{2}H_3^2 + H_1\right\}(\xi(t))$$

$$= H_1H_3\{H_1, H_3\}(\xi(t)) + H_2H_3\{H_2, H_3\}(\xi(t)) = 0,$$

but the explanation for the existence of this conserved quantity will be left for the next chapter.

Rather than write the extremal equations, we shall take a shortcut to the solutions and define an angle θ by $H_1(\xi(t)) = M\cos\theta(t)$ and $H_2(\xi(t)) = M\sin\theta(t)$. Then

$$M\cos\theta\frac{d\theta}{dt} = \frac{d}{dt}H_2(\xi(t)) = \left\{H_2, \frac{1}{2}H_3^2 + H_1\right\}(\xi(t))$$

$$= H_3\{H_2, H_3\}(\xi(t)) = H_3(\xi(t))H_1(\xi(t)).$$

After substituting $H_1(\xi(t)) = M \cos \theta$, we obtain

$$\frac{d\theta}{dt} = -H_3(\xi(t)) = -u(t). \tag{20}$$

Because the total energy $H = \frac{1}{2}H_3^2 + H_1$ is constant along the integral curves of $\vec{H}$, it follows that

$$H = \frac{1}{2}\left(\frac{d\theta}{dt}\right)^2 + M \cos \theta,$$

or that

$$\frac{d\theta}{dt} = \pm\sqrt{2H - 2M \cos \theta}.$$

Comparing expressions (19) and (20), we conclude that the two angles differ by a constant angle. We can then assume that the axes are so rotated that these angles coincide. The solution curves, when parametrized by the angle θ, are then given by

$$\frac{dx_1}{d\theta} = \frac{dx_1}{dt} \Big/ \frac{d\theta}{dt} = \frac{\cos \theta}{\pm\sqrt{2(H - M \cos \theta)}},$$

$$\frac{dx_2}{d\theta} = \frac{dx_2}{dt} \Big/ \frac{d\theta}{dt} = \frac{\sin \theta}{\pm\sqrt{2(H - M \cos \theta)}}.$$

The second equation immediately integrates to yield $x_2(\theta) = c + (1/M)$ $\sqrt{2(H - M \cos \theta)}$, and the solutions of the first equation are given by the following integral:

$$x_1(\theta) = c + \int \frac{\cos \theta \, d\theta}{\sqrt{2(H - M \cos \theta)}}.$$

The last integral is an elliptic integral, and it can be integrated by the means of elliptic functions.

These equations agree with Euler's solutions. The angle θ satisfies the differential equation for the mathematical pendulum, because

$$\frac{d^2\theta}{dt^2} = -\frac{d}{dt} H_3(\xi(t)) = \{H_3, H_1\}(\xi(t)) = -H_2(\xi(t)) = -M \sin \theta,$$

or

$$\frac{d^2\theta}{dt^2} + M\sin\theta = 0.$$

This equation provides a basis for Kirchhoff's analogy of the elastica with the motion of the mathematical pendulum, which is known in the theory of elastic rods as the *kinetic analogue of the elastica*. We defer discussion of this analogy to the next chapter, which contains additional ideas required for its understanding. The sketch of Euler's original solutions is included in Figure 11.4, which Euler produced in 1744, before the discovery of elliptic functions.

The extremal curves for the problem of Dubins are much simpler than the elastic curves. They are obtained as follows: The Hamiltonian for the problem of Dubins is given by $\mathcal{H}_{p_0} = -p_0 + H_1 + uH_3$. Because the time interval is not fixed, $\mathcal{H}_{p_0}(\xi(t), u(t)) = 0$ for almost all t along each extremal curve $\xi(t), u(t)$. The maximality condition implies that $u(t) = \operatorname{sgn} H_3(\xi(t))$ for all t for which $H_3(\xi(t)) \neq 0$.

The surface $H_3 = 0$ is called the switching surface. $H_3(\xi(t))$ is a continuously differentiable function for each extremal curve $\xi(t)$, as can be seen from the following differential equation:

$$\frac{d}{dt} H_3(\xi(t)) = \{H_3, -p_0 + H_1 + uH_3\}(\xi(t)) = -H_2(\xi(t)).$$

Therefore, if $H_3(\xi(t)) = 0$ on a set of positive measure, $H_2(\xi(t)) = 0$ in any subinterval, and consequently $H_3(\xi(t)) = 0$ for all t in this interval. This argument shows that the curve $H_3(\xi(t))$ either crosses the switching surface transversally or remains in it for some interval I. If it remains in the switching surface for all t in I, then $H_2(\xi(t)) = 0$ for all t in I, and therefore

$$\frac{d}{dt} H_2(\xi(t)) = \{H_2, -p_0 + H_1 + uH_3\}(\xi(t)) = -u(t)H_1(\xi(t)) = 0.$$

It is easy to verify that $H_1^2(\xi(t)) + H_2^2(\xi(t))$ is constant for this problem also. Therefore $H_1(\xi(t)) = $ constant for all t in I. $H_1(\xi(t))$ cannot be zero, since $\mathcal{H}_{p_0}(\xi(t), u(t)) = 0$ for almost all t, for that would imply that $p_0 = 0$, and $H_1(\xi(t)) = H_2(\xi(t)) = H_3(\xi(t)) = 0$ cannot hold because $\xi(t) \neq 0$. The conclusion is that $u(t) = 0$ for all t in I.

Our analysis shows that the solutions to the problem of Dubins are concatenations of circles and straight lines. In his original paper, Dubins (1957) showed that the optimal solutions cannot contain more than three concatenations. We

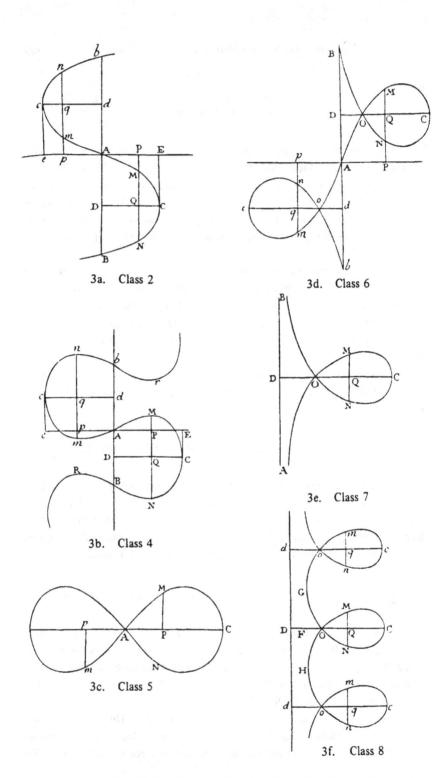

3a. Class 2

3d. Class 6

3b. Class 4

3e. Class 7

3c. Class 5

3f. Class 8

Fig. 11.4. Euler's sketches for various classes of elastica.

shall come back to the finer issues of the problem in the next chapter, aided by further theoretical insights.

Notes and sources

The basic optimality principle – that an optimal trajectory remains on the relative boundary of the reachable set – and its geometric realization in the form of a maximum principle are valid under very general conditions, and that recognition has led to large numbers of publications on optimal control ever since the work of Pontryagin et al. (1962). The works by Lee and Markus (1967), Young (1969), Warga (1972), Clarke (1983), and Albrecht (1968) are only a few of the major publications on the subject. With one exception (Albrecht, 1968), all of those works are confined to systems in $\mathbb{R}^n$. The book by Young (1969) provides a charming transition from the work of Carathéodory (1935) to more contemporary literature on optimal control. The books by Clarke (1983) and Warga (1972) obtain their necessary conditions of optimality in a more general framework (based on methods of non-smooth analysis) by viewing a control system as a multivalued differential equation.

12
Optimal problems on Lie groups

In approaching variational problems on Lie groups, optimal control theory finds itself on the same ground as Hamiltonian mechanics, facing an already developed theory of Hamiltonian systems that it needs to understand and absorb in order to arrive at the proper solutions of its own problems. This theory of Hamiltonian systems on Lie groups is based on a particular realization of the cotangent bundle of a Lie group G as the product of G and the dual of its Lie algebra $\mathcal{L}$. For vector spaces, which are also commutative Lie groups, that realization of the cotangent bundle coincides with the usual $M \times M^*$ representation. However, for non-commutative Lie groups, the representation $T^*G = G \times \mathcal{L}^*$ invariably leads to non-canonical coordinates and accounts for the somewhat mysterious appearance of the Hamiltonian equations in mechanics. For instance, the famous equation of Euler, $dp/dt = \Omega \times p$, describing the motion of a rigid body about its center of mass is the projection of the Hamiltonian system on the dual of the Lie algebra of $SO_3(R)$.

For variational problems with either left- or right-invariant symmetries, the projection of the Hamiltonian equations on $\mathcal{L}^*$ accounts for the conservation laws induced by the structure of the group. In the case of a rigid body, Euler's equation produces two integrals of motion: the total energy and the angular momentum. As we shall presently see, the same conservation laws apply to arbitrary Hamiltonian left-invariant systems on $SO_3(R)$, not just to mechanical systems.

This chapter begins with a self-contained treatment of the symplectic structure of T^*G realized as $G \times \mathcal{L}^*$. The development leads to an explicit description of Hamiltonian equations corresponding to an arbitrary function H on T^*G. We then illustrate the importance of this geometric setting with an easy derivation of the equations for the "heavy top," a rigid body fixed at some point on its body and moving around that point under the influence of gravity.

368

The Hamiltonian differential equations corresponding to left-invariant optimal problems are strongly rooted in the geometric properties of the group, and most of the material in this chapter is directed to a proper understanding of this geometry. In particular, we shall show that the projections of the extremal curves on $\mathcal{L}^*$ evolve on co-adjoint orbits of G in $\mathcal{L}^*$, which reveals the role of symmetry and also points to the Casimir elements as the universal conservation laws on G.

For semisimple Lie groups, the trace form sets up a natural correspondence between the adjoint and co-adjoint orbits, which in turn converts the extremal equations on $\mathcal{L}^*$ to their dual forms on $\mathcal{L}$. The dual forms of extremal equations corresponding to left-invariant problems are always of the so-called Lax pair form (i.e., of the form $dP/dt = [\Omega, P]$), like the equation for the Euler top.

We further illustrate the merits of this formalism with an easy derivation of the Hamiltonian equations that describe the equilibrium configurations for an elastic rod in $\mathbb{R}^3$, a result classically known as the kinetic analogue of Kirchhoff. The space of configurations for an elastic rod is the frame bundle of $\mathbb{R}^3$, which is then identified with the group of motions of $\mathbb{R}^3$ as its isometry group.

1 Hamiltonian vector fields

1.1 Realization of the cotangent bundle as the product $G \times \mathcal{L}^*$

Let G denote a real Lie group G, with its Lie algebra $\mathcal{L}$, and let e denote the group identity of G. For each g in G, let L_g denote the left-translation by g. The tangent map dL_g maps $T_x G$ onto $T_{gx} G$ for each x in G. Then dL_g^* denotes the dual mapping of dL_g. It follows that $dL_g^* : T_{gx}^* G \to T_x^* G$ for each x in G, and $dL_g^*(p)$ evaluated at a vector $X_x \in T_x G$ is equal to $p \circ dL_g(X_x)$ for each $p \in T_{gx}^* G$. In particular, at $x = g$, $dL_{g^{-1}}^*$ maps $T_e^* G$ onto $T_g^* G$. The correspondence $(g, p) \leftrightarrow dL_{g^{-1}}^*(p)$ realizes $T^* G$ as $G \times \mathcal{L}^*$.

In this representation of $T^* G$, the Hamiltonians of left-invariant vector fields are linear functions on $\mathcal{L}^*$, and the Hamiltonians of right-invariant vector fields are functions that depend on both factors G and $\mathcal{L}^*$. The explicit expressions are as follows:

The Hamiltonian H_X of a left-invariant vector field X is given by $H_X(g, p) = p(X(e))$, and the Hamiltonian of a right-invariant vector field X is given by

$$H_X(g, p) = p(dL_{g^{-1}} X(g)) = p(dL_{g^{-1}} dR_g(X(e))).$$

$T^* G$ could also have been realized as $G \times \mathcal{L}^*$ in terms of the right-multiplications $R_g(x) = xg$. Then the correspondence would be given by $(g, f) \to dR_{g^{-1}}^* f$, and therefore the Hamiltonians or right-invariant vector fields would become linear functions on $\mathcal{L}^*$. These representations are equally suitable for

applications. I have chosen the left-invariant realization because it is better
for the applications in mechanics and geometry that follow.

The tangent bundle of $G \times \mathcal{L}^*$ is naturally identified with $TG \times T\mathcal{L}^*$. We shall
further identify TG with $G \times \mathcal{L}$ via the correspondence $(g, L) \leftrightarrow (dL_g)(X)$
for each g in G and for X in $\mathcal{L}$. Then $T\mathcal{L}^* = \mathcal{L}^* \times \mathcal{L}^*$.

The preceding identifications realize $T(T^*G)$ as the product $(G \times \mathcal{L}) \times (\mathcal{L}^* \times \mathcal{L}^*)$. In this realization, each element $((g, X), (p, Y^*))$ is a tangent vector
(X, Y^*) based at (g, p) in T^*G. With these conventions, vector fields on T^*G
will be represented by pairs (X, Y^*), with X taking values in $\mathcal{L}$, and Y^* taking
values in $\mathcal{L}^*$.

Having identified T^*G with $G \times \mathcal{L}^*$, functions on T^*G become functions
on $G \times \mathcal{L}^*$. For each such function f, $\partial f/\partial g$ and $\partial f/\partial p$ will denote the partial
differentials of f; $\partial f/\partial g$ is the differential of the restriction of f to $p = $ constant,
and $\partial f/\partial p$ is the differential of the restriction of f to $g = $ constant. Then $\partial f/\partial g$
belongs to T_g^*G and $\partial f/\partial g$ belongs to $T_p^*\mathcal{L}^*$ at each point (g, p). It further
follows that $\partial f/\partial p$ belongs to $\mathcal{L}$, because $(\mathcal{L}^*)^* = \mathcal{L}$. In terms of these notations,

$$(Vf)(g, p) = \left(dL_g^* \frac{\partial f}{\partial g}(g, p) \right) (X(g, p)) + Y^*(g, p) \left(\frac{\partial f}{\partial p}(g, p) \right) \quad (1)$$

for any vector field $V = (X, Y^*)$ on T^*G and any function f on T^*G.

1.2 The symplectic form

Recall now the expression

$$\omega_\xi(V_1(\xi), V_2(\xi)) = V_2(\theta(V_1))(\xi) - V_1(\theta(V_2))(\xi) - \theta_\xi([V_1, V_2](\xi))$$

for the symplectic form ω defined in Chapter 11. It will be shown next that the
foregoing expression can be written as

$$\omega((X_1, Y_1^*), (X_2, Y_2^*)) = Y_2^*(X_1) - Y_1^*(X_2) - p([X_1, X_2]) \quad (2)$$

on the cotangent bundle of a Lie group G provided that the latter is realized as
$G \times \mathcal{L}^*$.

To begin with, the natural projection $\pi : T^*G \to G$ is given by $\pi(g, p) = g$.
Therefore, $\pi_*((g, X), (p, Y^*)) = (g, X)$ for any point $((g, X), (p, Y^*))$ of
$T_{(g,p)}(T^*G)$. Recall that the canonical 1-form θ is defined by

$$\theta_\xi \left(\frac{d\xi}{dt}(0) \right) = \pi_* \left(\frac{d\xi}{dt}(0) \right)$$

for each $\xi \in T^*G$ and each tangent vector $(d\xi/dt)(0)$ at ξ. In the realization $T^*G = G \times \mathcal{L}^*$, $\xi = (g, p)$, and $(d\xi/dt)(0) = ((g, X), (p, Y^*))$ for some elements X in $\mathcal{L}$ and Y^* in $\mathcal{L}^*$. Therefore, $\xi(d\pi((d\xi/dt)(0))) = p(X)$; that is,

$$\theta_{(g,p)}(X, Y^*) = p(X)$$

for any tangent vector (X, Y^*) at (g, p).

Each tangent vector (X, Y^*) in $\mathcal{L} \times \mathcal{L}^*$ defines a vector field V on $G \times \mathcal{L}^*$ equal to (X, Y^*) at each point (g, p) in $G \times \mathcal{L}^*$. It follows that the projection of V on G is the left-invariant vector field equal to X at the identity of G, and the projection of V on $\mathcal{L}^*$ is the translation in the Y^* direction. $\mathcal{L}^*$, being a vector space, is an abelian Lie group, and translations on $\mathcal{L}^*$ are left- and/or right-invariant vector fields on $\mathcal{L}^*$. Thus V could alternatively be described as a left-invariant vector field on the product $G \times \mathcal{L}^*$.

It then follows that $\theta_\xi(V(\xi)) = p(X)$ for any $\xi = (g, p)$ in $G \times \mathcal{L}^*$. Therefore, $\theta(V)$ is a linear function on $\mathcal{L}^*$. When $V_1 = (X_1, Y_1^*)$ and $V_2 = (X_2, Y_2^*)$ are arbitrary left-invariant fields on $G \times \mathcal{L}^*$, then

$$[V_1, V_2] = ([X_1, X_2], 0) \quad \text{and} \quad V_i(\theta(V_j)) = Y_i^*(X_j) \quad \text{for } i = 1, 2,$$

as can be seen from expression (1). Thus

$$\omega_\xi(V_1(\xi), V_2(\xi)) = Y_2^*(X_1) - Y_1^*(X_2) - p([X_1, X_2]),$$

and (2) is proved.

The foregoing expression can also be written as

$$\omega_\xi(V_1(\xi), V_2(\xi)) = Y_2^*(X_1) - Y_1^*(X_2) - ((\text{ad})^* X_1(p))(X_2),$$

with $(\text{ad})^* X$ denoting the dual mapping of ad X. The expression (2) will be called the left-invariant realization of ω.

As remarked earlier, functions on T^*G will be expressed as functions on $G \times \mathcal{L}^*$, and consequently the Hamiltonian vector fields on $T(T^*G)$ will take values in $\mathcal{L} \times \mathcal{L}^*$ according to the preceding formalism. The next theorem specifies the Hamiltonian vector fields in terms of the partial derivatives of the defining function.

Theorem 1 *Let $\vec{H}(g, p) = (X(g, p), Y^*(g, p))$ denote the Hamiltonian vector field corresponding to a function H on $G \times \mathcal{L}^*$. Then*

$$X(g, p) = \frac{\partial H}{\partial p}(g, p) \quad and \quad Y^*(g, p) = -dL_g^*\left(\frac{\partial H}{\partial g}(g, p)\right) - (\text{ad})^* X(p).$$

Proof Recall the formula

$$dH_\xi(V(\xi)) = \omega_\xi(\vec{H}(\xi), V(\xi))$$

for any vector field V and any point $\xi = (g, p)$. Denote V by the pair (Z, W^*), with Z in $\mathcal{L}$ and W^* in $\mathcal{L}^*$. It follows from expression (1) that

$$dH_{(g,p)}(Z, W^*) = \left(dL_g^*\left(\frac{\partial H}{\partial g}(g, p)\right)\right)(Z) + W^*\left(\frac{\partial H}{\partial p}(g, p)\right).$$

Equating that expression to $\omega_\xi(\vec{H}(\xi), V(\xi))$, in its left-invariant representation, leads to

$$dL_g^*\left(\frac{\partial H}{\partial g}\right)(Z) + W^*\left(\frac{\partial H}{\partial p}\right) = W^*(X) - Y^*(Z) - ((\mathrm{ad})^*X(p))(Z).$$

Because Z and W^* are arbitrary tangent vectors, $dL_g^*(\partial H/\partial g) = -Y^* - (\mathrm{ad})^*X(p)$, and $\partial H/\partial p = X$, and the proof is finished.   ∎

Corollary 1 *Each integral curve* $(g(t), p(t))$ *of* $\vec{H}$ *satisfies the following differential equations:*

$$\frac{dg}{dt} = dL_g\left(\frac{\partial H}{\partial p}\right),$$
$$\frac{dp}{dt} = -dL_{g(t)}^*\left(\frac{\partial H}{\partial g}(g(t), p(t))\right) - \left((\mathrm{ad})^*\frac{\partial H}{\partial p}(g(t), p(t))\right)(p(t)).$$

The foregoing formulas reduce to the usual formulas $dg/dt = dL_g(\partial H/\partial p)$ and $dp/dt = -dL_{g(t)}^*(\partial H/\partial g)$ when G is abelian, because $(\mathrm{ad})^*(\partial H/\partial p) = 0$.

The relevance of this formalism is best appreciated through the equations for the heavy top in mechanics, which we shall derive next through a control theory formalism. The derivation of this remarkable equation is indicative of possible uses of optimal control-theory in mechanics: Lagrange's principle of least action defines an optimal-control problem for which the appropriate Hamiltonian is obtained through an application of the maximum principle. The corresponding equations of motion are then obtained from Theorem 1 and Corollary 1.

2 The rigid body and the equations for the heavy top

Because the derivation of the equations for a rigid body makes use of several subtle geometric facts that are often obscured in the literature, we shall begin at the very beginning by deriving the expression for the appropriate Lagrangian associated with the heavy top.

We shall assume that the body is moving in a Euclidean space E^3. For any vectors $\mathbf{v}$ and $\mathbf{w}$ in E^3 we denote by $\langle \mathbf{v}, \mathbf{w} \rangle$ their inner product, and by $\|\mathbf{v}\|$ the associated norm. Let $\mathbf{e}_1$, $\mathbf{e}_2$, $\mathbf{e}_3$ denote an orthonormal frame fixed at the origin O_f of E^3, and $\mathbf{a}_1$, $\mathbf{a}_2$, $\mathbf{a}_3$ an orthonormal frame fixed on the body with its origin at a point O_b. We shall refer to $\mathbf{a}_1$, $\mathbf{a}_2$, $\mathbf{a}_3$ as the *moving frame*. Our notational conventions will follow Arnold (1976), whenever possible. Thus, capital letters will refer to the coordinates relative to the moving frame, and small letters will refer to coordinates relative to the fixed frame. This notation coincides with the notation used for deriving the kinematic equations for rolling spheres used in Chapters 4 and 6.

Let $r(t)$ denote the coordinates, relative to the fixed frame, of the vector $\overrightarrow{O_f O_b}$. For any point P on the body, let q denote the coordinate vector of $\overrightarrow{O_f P}$ relative to the fixed frame, and let Q denote the coordinate vector of $\overrightarrow{O_b P}$ relative to the moving frame. Finally, let R denote the rotation matrix defined by $q = r + RQ$. It follows that the ith column of R is equal to the coordinates of $\mathbf{a}_i$ relative to the fixed frame.

As the body moves through space, the velocity of the point P is given by $dq/dt = dr/dt + (dR/dt)Q$. Because $R(t)$ is a path in $SO_3(R)$, dR/dt is a tangent vector to $SO_3(R)$ at $R(t)$. The tangent space at R can be described by either left- or right-translation by R of the tangent space at the identity. Using the right-invariant representation of $SO_3(R)$, then $dR/dt = A(\omega(t))R$, with

$$A(\omega(t)) = \begin{pmatrix} 0 & -\omega_3(t) & \omega_2(t) \\ \omega_3(t) & 0 & -\omega_1(t) \\ -\omega_2(t) & \omega_1(t) & 0 \end{pmatrix}$$

for some functions $\omega_1(t)$, $\omega_2(t)$, $\omega_3(t)$.

The vector $\omega(t) = \omega_1(t)\mathbf{e}_1 + \omega_2(t)\mathbf{e}_2 + \omega_3(t)\mathbf{e}_3$ is called the angular velocity of the body (relative to the fixed frame); $\omega(t)$ will be used to denote its coordinate vector

$$\begin{pmatrix} \omega_1(t) \\ \omega_2(t) \\ \omega_3(t) \end{pmatrix}.$$

Then $A(\omega(t))q(t) = \omega(t) \times q(t)$, and therefore

$$\frac{dq}{dt} = \frac{dr}{dt} + A(\omega(t))R(t)Q = \frac{dr}{dt} + \omega(t) \times (q(t) - r(t)).$$

Denoting by m the point mass of a point P, its kinetic energy is given by $\frac{1}{2}m \, \|dq/dt\|^2$. The total kinetic energy T of the body is equal to the sum of the kinetic energies of the point masses in the body. T is usually given in terms of the mass density ρ as follows: We have

$$T = \lim\left(\sum \frac{1}{2}m_i \left\|\frac{dq_i}{dt}\right\|^2\right)$$

$$= \int_{\text{body}} \frac{1}{2}\rho\left(\left\|\frac{dr}{dt}\right\|^2 + 2\left\langle\frac{dr}{dt}, \omega \times (q - r)\right\rangle + \|\omega \times (q - r)\|^2\right) dQ.$$

Denote by $A(\Omega)$ the antisymmetric matrix

$$A(\Omega) = \begin{pmatrix} 0 & -\Omega_3 & \Omega_2 \\ \Omega_3 & 0 & -\Omega_1 \\ -\Omega_2 & \Omega_1 & 0 \end{pmatrix}$$

defined by $\Omega = R^{-1}\omega$. It follows that $A(\omega) = RA(\Omega)R^{-1}$, as can easily be verified through the following argument:

$$R^{-1}A(\omega)q = R^{-1}(\omega \times q) = R^{-1}\omega \times R^{-1}q = \Omega \times R^{-1}q = A(\Omega)R^{-1}q.$$

Therefore, $R^{-1}A(\omega) = A(\Omega)R^{-1}$. Ω is called the *angular velocity relative to the body coordinates.*

The total kinetic energy is conveniently expressed in terms of Ω, as follows: Denote by M the total mass of the body. Then $M = \int_{\text{body}} \rho \, dQ$. By substituting $q - r = RQ$ and $\omega \times RQ = R(\Omega \times Q)$, we get that

$$T = \frac{M}{2}\left\|\frac{dr}{dt}\right\|^2 + \left\langle\frac{dr}{dt}, \int_{\text{body}} \rho R(\Omega \times Q) \, dQ\right\rangle + \frac{1}{2}\int_{\text{body}} \rho\|\Omega \times Q\|^2 \, dQ.$$

In particular, if the center of the moving frame is situated at the center of mass of the body, then the middle integral is equal to zero, and

$$T = \frac{M}{2}\left\|\frac{dr}{dt}\right\|^2 + \frac{1}{2}\int_{\text{body}} \rho\|\Omega \times Q\|^2 \, dQ.$$

We shall now assume that the point O_b is fixed in absolute space, and therefore $dr/dt = 0$. Hence, $T = \frac{1}{2}\int_{\text{body}} \rho\|\Omega \times Q\|^2 \, dQ$. It now follows that the kinetic energy, considered as a function of Ω, can be written as $T(\Omega) = \frac{1}{2}\langle\Omega, P\Omega\rangle$ for some positive-definite matrix P, because the quadratic form $(\Omega_1, \Omega_2) \to \frac{1}{2}\int_{\text{body}}\langle\Omega_1 \times Q, \Omega_2 \times Q\rangle \, dQ$ is a positive-definite quadratic form.

Problem 1 Show that the preceding form is symmetric.

The eigenvalues of P are called the principal moments of inertia of the body. A rigid body fixed at some point on the body and moving under gravitational force is called a "heavy top." Let $\vec{F} = -C\mathbf{e}_3$ denote the gravitational force, with C the gravitational constant. Then the potential energy of a point mass m at q is given by $Cmq_3 = Cm\langle q, \mathbf{e}_3\rangle$. The total potential energy consists of the sum of the contributions due to point masses. It follows that

$$U = C \int_{\text{body}} \rho\langle RQ, \mathbf{e}_3\rangle \, dQ = CM\langle RQ_0, \mathbf{e}_3\rangle,$$

with Q_0 denoting the coordinates of the center of mass of the body. (We are not assuming that O_b is necessarily situated at the center of mass of the body.) Figure 12.1 describes this physical situation pictorially. The associated Lagrangian L is given by $L = T - U = \frac{1}{2}\langle \Omega, P\Omega\rangle - c\langle RQ_0, \mathbf{e}_3\rangle$, with $c = MC$.

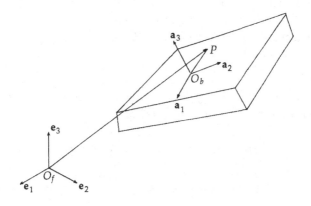

Fig. 12.1.

In order to formulate this problem as an optimal-control problem, Ω will be regarded as the control function. The dynamics of the associated control problem are given by $dR/dt = A(\omega)R = RA(\Omega)R^{-1}R = RA(\Omega)$. Let

$$A_1 = \begin{pmatrix} 0 & 0 & 0 \\ 0 & 0 & -1 \\ 0 & 1 & 0 \end{pmatrix}, \quad A_2 = \begin{pmatrix} 0 & 0 & 1 \\ 0 & 0 & 0 \\ -1 & 0 & 0 \end{pmatrix}, \quad A_3 = \begin{pmatrix} 0 & -1 & 0 \\ 1 & 0 & 0 \\ 0 & 0 & 0 \end{pmatrix}$$

be the standard basis for antisymmetric matrices. Then, letting $\vec{A}_i$ denote the left-invariant vector field on $SO_3(R)$ whose value at the identity is A_i, the differential equation for R becomes

$$\frac{dR}{dt} = \Omega_1 \vec{A}_1(R) + \Omega_2 \vec{A}_2(R) + \Omega_3 \vec{A}_3(R). \tag{3}$$

Lagrange's principle of least action leads to minimizing

$$\int_0^T \left(\frac{1}{2} \langle \Omega, P\Omega \rangle - c\langle RQ_0, \mathbf{e}_3 \rangle \right) dt$$

over the trajectories of the foregoing control system that satisfy the prescribed boundary conditions.

It is preferable to remain consistent with the earlier notation and use g to denote the rotation matrix R. This choice of notation is also consistent with the notation in the remaining parts of this chapter, and it further avoids possible confusion with right-translations. For simplicity, we shall assume that the matrix P is diagonal, with its principal moments of inertia I_1, I_2, I_3 equal to the diagonal entries. Alternatively, P can be diagonalized by a rotation matrix, which will produce new controls. Except for the renaming of controls, the problem remains the same. Then we have the following:

Theorem 2

(a) *Each extremal curve $\xi(t)$ of the heavy top is an integral curve of the Hamiltonian vector field $\vec{H}$ that corresponds to*

$$H = \frac{1}{2} \left(\frac{H_1^2}{I_1} + \frac{H_2^2}{I_2} + \frac{H_3^2}{I_3} \right) + c\langle gQ_0, \mathbf{e}_3 \rangle,$$

with each H_i equal to the Hamiltonian of the left-invariant vector field $\vec{A}_i$, with $i = 1, 2, 3$.

(b) *Along $\xi(t)$, $\Omega_i(t) = (1/I_i)H_i(\xi(t))$, with $i = 1, 2, 3$.*

(c) *Denote by $(g(t), p(t))$ the components of an extremal curve $\xi(t)$ in the realization $T^*G = G \times \mathcal{L}^*$. Then*

$$\frac{dg}{dt} = \sum_{i=1}^3 (1/I_i)H_i(p(t))\vec{A}_i(g(t)) \tag{4a}$$

and

$$\frac{dp}{dt} = -((\mathrm{ad})^*\Omega)(p(t)) - dL_g^*(dU), \tag{4b}$$

with $\Omega = \sum_{i=1}^3 (1/I_i)H_i(p)A_i$.

Proof Statements (a) and (b) are direct consequences of the maximum principle and follow from the discussion at the end of Chapter 11. Statement (c) follows from Theorem 1 and Corollary 1. End of the proof. ∎

Equation (4b) is the classic equation of Euler. In order to recognize it in its more familiar form, it is first necessary to identify $\mathcal{L}^*$ with $\mathcal{L}$ via the trace form.

Definition 1 K denotes the bilinear form on $\mathcal{L}$ defined by $K(A, B) = -\frac{1}{2}$ trace (AB) for any antisymmetric matrices A and B.

Definition 2 For each antisymmetric matrix

$$A = \begin{pmatrix} 0 & -a_3 & a_2 \\ a_3 & 0 & -a_1 \\ -a_2 & a_1 & 0 \end{pmatrix}, \quad \hat{A} \text{ will denote the column vector } \begin{pmatrix} a_1 \\ a_2 \\ a_3 \end{pmatrix}.$$

Theorem 3

(a) $K([A, B], C) = K(A, [B, C])$ *for all antisymmetric matrices A, B, and C.*
(b) $K(A, B) = \langle \hat{A}, \hat{B} \rangle$, *where $\langle \, , \, \rangle$ denotes the Euclidean inner product on $\mathbb{R}^3$.*
(c) $\widehat{[A, B]} = \hat{B} \times \hat{A}$ *for any antisymmetric matrices A and B.*
(d) $Ax = \hat{A} \times x$ *for any antisymmetric matrix A and any column vector x.*

Remark Recall that the cross-product of vectors is defined by

$$a \times b = \begin{pmatrix} a_2 b_3 - a_3 b_2 \\ a_3 b_1 - a_1 b_3 \\ a_1 b_2 - a_2 b_1 \end{pmatrix}. \qquad ∎$$

Problem 2 Prove Theorem 3. Keep in mind that $[A, B] = BA - AB$.

The trace form identifies each element p in $\mathcal{L}^*$ with a unique element P in $\mathcal{L}$ through the formula

$$K(P, A) = p(A) \quad \text{for all } A \text{ in } \mathcal{L}.$$

It then follows from part (b) of Theorem 3 that $p(A) = \langle \hat{P}, \hat{A} \rangle$, and therefore

$$\hat{P} = \begin{pmatrix} p(A_1) \\ p(A_2) \\ p(A_3) \end{pmatrix}.$$

Based on this identification, the extremal equations can be written in $\mathbb{R}^3$ according to the following theorem.

Theorem 4 *Let $(g(t), p(t))$ denote an extremal curve of the heavy top. Let $P(t)$ denote the antisymmetric matrix that is identified with $p(t)$. Denote by $\Omega(t)$ the antisymmetric matrix*

$$\sum_{i=1}^{3} \frac{1}{I_1} H_i(p(t)) A_i,$$

and by $Q(t)$ the vector $cg^{-1}(t)e_3$. Then

$$\frac{dg}{dt} = g(t)\Omega(t) \quad and \quad \frac{d\hat{P}}{dt} = \hat{P} \times \hat{\Omega} + Q(t) \times Q_0.$$

Proof The proof consists in identifying each term of equation (4b) with its equivalent in $\mathbb{R}^3$. To begin with, $U(g) = c\langle gQ_0, e_3\rangle$, and therefore

$$dL_g^*(dU)(A) = \frac{d}{d\varepsilon} U(ge^{A\varepsilon})|_{\varepsilon=0} = c\langle gAQ_0, e_3\rangle$$

is valid for any A in $\mathcal{L}$. In particular, along an extremal curve $g(t)$, $dL_{g(t)}^*(dU)$ $(A) = \langle AQ_0, Q(t)\rangle$. Combining further with part (d) of Theorem 3,

$$-\langle AQ_0, Q(t)\rangle = -\langle \hat{A} \times Q_0, Q(t)\rangle = -\langle Q_0 \times Q(t), \hat{A}\rangle = \langle Q(t) \times Q_0, \hat{A}\rangle.$$

Then $(dp/dt)(A) = \langle d\hat{P}/dt, \hat{A}\rangle$. Using the results in Theorem 3,

$$-p(t)([\Omega, A]) = -K(P(t), [\Omega, A]) = -K([P(t), \Omega], A)$$
$$= -\langle [\widehat{P(t)}, \Omega], \hat{A}\rangle = -\langle \hat{\Omega} \times \hat{P}, \hat{A}\rangle = \langle \hat{P} \times \hat{\Omega}, \hat{A}\rangle.$$

Therefore, equation (4b) becomes

$$\left\langle \frac{d\hat{P}}{dt}, \hat{A} \right\rangle = \langle \hat{P} \times \hat{\Omega}, \hat{A}\rangle + \langle Q(t) \times Q_0, \hat{A}\rangle.$$

Our proof is now finished, since $\hat{A}$ is an arbitrary element of $\mathbb{R}^3$. ∎

Recall that $g(t)$ is equal to the rotation matrix that transforms the coordinates in the moving frame into coordinates in the fixed frame. In particular, the moving vector $Q(t)$ represents the coordinates of ce_3 relative to the moving frame. $Q(t)$

is often expressed in the following form:

$$\frac{dQ}{dt} = \frac{d}{dt} R^{-1}(t)\, c e_3 = -A(\Omega)\, c R^{-1} e_3$$
$$= -A(\Omega)\, Q(t) = Q(t) \times \hat{\Omega}(t). \tag{5}$$

Remark Even though the equations for the heavy top evolve on $SO_3(R) \times \mathcal{L}(SO_3(R))$, equation (5) is often adjoined to the Hamiltonian equations in Theorem 4, with the overall system then being regarded as a differential system in $SO_3(R) \times \mathbb{R}^3 \times \mathbb{R}^3$. We shall discuss these equations again in the context of Kirchhoff's theorem. ∎

The quantity $m(t) = R(t)\hat{P}(t)$ is called the total angular momentum of the body. Then

$$\frac{dm}{dt} = \frac{d}{dt} R(t)\hat{P}(t) = R(t)A(\Omega)\hat{P}(t) + R(t)((\hat{P} \times \hat{\Omega}) + Q(t) \times Q_0)$$
$$= R(t)(\hat{\Omega} \times \hat{P}) + R(t)(\hat{P} \times \hat{\Omega}) + R(t)(Q(t) \times Q_0)$$
$$= R(t)(Q(t) \times Q_0) = R(t)Q(t) \times R(t)Q_0.$$

$R(t)Q(t) = c e_3$ describes the coordinates of $\vec{F}$ relative to the fixed frame. Recall that M denotes the total mass of the body, C denotes the gravitational constant, and $c = MC$. The quantity $q(t) = R(t)Q_0$ yields the coordinates of the center of gravity of the body relative to the fixed frame. In terms of these notations

$$\frac{dm}{dt} = q(t) \times \vec{F}(t), \quad \text{with } \vec{F}(t) = -MCe_3. \tag{6}$$

Equation (6) expresses a well-known law of mechanics: *The rate of change of the total angular momentum is equal to the external torque.*

3 Left-invariant control systems and co-adjoint orbits

An arbitrary control system $dg/dt = F(g, u)$, defined on a real Lie group G, with control functions $u(t) = (u_1(t), \dots, u_m(t))$ taking values in a subset U of $\mathbb{R}^m$, is said to be left-invariant if $dL_g F(h, u) = F(L_g(h), u)$ for each g and h in G. L_g denotes the left-translation by g, and dL_g denotes its tangent map at h. Each left-invariant system is defined by its values at the group identity, because $dL_g F(e, u) = F(g, u)$.

Definition 3 A function $f(g, u)$ is left-invariant if $f(L_g(h), u) = f(h, u)$ for all g and h in G. An optimal problem on G is said to be left-invariant if both the cost (objective) function f and the control system are left-invariant.

Example   The Lagrangian that corresponds to the heavy top is left-invariant on $SO_3(R)$ only when the center of gravity Q_0 coincides with the fixed point O_b. Then the total potential energy U is zero. (It would be correct to say that U is constant, because the potential energy is determined only up to a constant. In any case, the equations of motion are independent of this constant.) The left-invariant heavy top is called the *Euler top*.

The Hamiltonians of left-invariant systems are functions on $\mathcal{L}^*$ only. The reason is fairly evident: The Hamiltonian H_F of F is defined by $H_F(\xi) = \xi(F(g, u))$ for each ξ in $T_g^* G$. When $\xi = (g, p)$ with p in $\mathcal{L}^*$, then $\xi = dL_{g^{-1}}^*(p)$:

$$H_F(g, p) = p dL_{g^{-1}}(F(g, u)) = p(dL_{g^{-1}} dL_g F(e, u)) = p(F(e, u)).$$

Hence H_F is a linear function on $\mathcal{L}^*$ parametrized by the control values. The cost extended Hamiltonian of the left-invariant optimal problem is given by

$$\mathcal{H}_{p_0}(p, u) = -p_0 f(u) + p(F(e, u)), \qquad u = (u_1, \ldots, u_m),$$

with p_0 equal to zero in the abnormal case and equal to unity in the regular case. Because left-invariant functions are constant over G, f depends only on the controls.

Hamiltonian vector fields defined by Hamiltonians that are functions on $\mathcal{L}^*$ (i.e., constant over G) admit reductions caused by left-invariant symmetries. These reductions are most naturally expressed in the language of co-adjoint orbits of G, and for that reason we shall digress briefly into the basic definitions of co-adjoint action.

The diffeomorphism $R_{g^{-1}} \circ L_g : G \to G$ is given by $x \to gxg^{-1}$. Notice that the right-translations commute with left-translations; therefore, the order in which the foregoing composition is written is immaterial.

For each g in G, $R_{g^{-1}} \circ L_g$ leaves the group identity e fixed. Hence the tangent map of $R_{g^{-1}} \circ L_g$ at the identity is an automorphism of $\mathcal{L}$. This tangent map is denoted by Ad_g. Because $R_{(gh)^{-1}} \circ L_{gh} = R_{g^{-1}} \circ L_g \circ (R_{h^{-1}} \circ L_h)$, it follows that $\mathrm{Ad}_{gh} = \mathrm{Ad}_g \mathrm{Ad}_h$. Therefore $g \to \mathrm{Ad}_g$ is a group representation of G into the linear group of automorphisms of the Lie algebra of G. It is easy to check that

$$\mathrm{Ad}_g[L, M] = \mathrm{Ad}_g(M)\mathrm{Ad}_g(L) - \mathrm{Ad}_g(L)\mathrm{Ad}_g(M) = [\mathrm{Ad}_g(L), \ \mathrm{Ad}_g(M)]$$

for any L and M in $\mathcal{L}$.

The foregoing representation of G is called the *adjoint representation of* G. The co-adjoint representation of G is defined through duality, as follows: For each g in G, $\mathrm{Ad}_g^* : \mathcal{L}^* \to \mathcal{L}^*$ denotes the dual of Ad_g. Recall that it is defined by $(\mathrm{Ad}_g^*(f))(L) = f(\mathrm{Ad}_g(L))$ for each f in $\mathcal{L}^*$ and L in $\mathcal{L}$. The co-adjoint representation is given by $g \to \mathrm{Ad}_{g^{-1}}^*$ and represents G in the group of automorphisms of $\mathcal{L}^*$.

Problem 3 Verify that $gh \to \mathrm{Ad}_{g^{-1}}^* \mathrm{Ad}_{h^{-1}}^*$ for any g and h in G.

Definition 4 The set $\{\mathrm{Ad}_{g^{-1}}^*(p) : g \in G\}$ is called the co-adjoint orbit of G through p.

The set of co-adjoint orbits of G partitions $\mathcal{L}^*$. It is known that each co-adjoint orbit is a symplectic submanifold of $\mathcal{L}^*$, and consequently each orbit must be an even-dimensional submanifold of $\mathcal{L}^*$.

Problem 4 Let G be the subgroup of $\mathrm{GL}_2(R)$ of all matrices

$$\begin{pmatrix} a & b \\ 0 & 1 \end{pmatrix} \quad \text{with } a > 0.$$

This group is called the *affine group of the plane*. The Lie algebra of G consists of all matrices

$$\begin{pmatrix} \alpha & \beta \\ 0 & 0 \end{pmatrix}.$$

Let

$$L_1 = \begin{pmatrix} 1 & 0 \\ 0 & 0 \end{pmatrix} \quad \text{and} \quad L_2 = \begin{pmatrix} 0 & 1 \\ 0 & 0 \end{pmatrix}.$$

Denote by L_1^* and L_2^* the dual basis in $\mathcal{L}^*$ associated with L_1, L_2.

(a) Show that the co-adjoint orbit of any scalar multiple of L_1^* consists of a single point. Also show that the co-adjoint orbit of an element $p = \lambda_1 L_1^* + \lambda_2 L_2^*$, with $\lambda_2 \neq 0$, is an open subset of $\mathcal{L}^*$.

(b) Show that there exist adjoint orbits that are odd-dimensional.

With this terminology at our disposal, we now return to the initial development of left-invariant optimal problems on an arbitrary Lie group G, with the following theorem.

Theorem 5 *Suppose that $(g(t), p(t))$ is an integral curve of the Hamiltonian vector field $\vec{\mathcal{H}}_{p_0}(p, u(t))$ for some control function $u(t)$, with $\mathcal{H}_{p_0}(p, u) = -p_0 f(u) + pF(e, u)$. Then*

$$\frac{dg}{dt} = F(g(t), u(t)) \quad and \quad p(t) = \mathrm{Ad}^*_{g(t)}(p(0))$$

for some $p(0)$ in $\mathcal{L}^*$. Consequently, $p(t)$ is contained in the co-adjoint orbit of G through $p(0)$.

Proof Because $\mathcal{H}_{p_0}$ is constant on G, $\partial \mathcal{H}_{p_0}/\partial g = 0$. It then follows from Theorem 1 that $p(t)$ is a solution of

$$\frac{dp}{dt} = -\left((\mathrm{ad})^* \frac{\partial \mathcal{H}}{\partial p}\right) p(t).$$

But $\partial \mathcal{H}_{p_0}/\partial p = F(e, u(t))$, and therefore

$$\frac{dp}{dt} = -((\mathrm{ad})^* F(e, u))(p(t)).$$

But then $p(t) = \mathrm{Ad}^*_{g(t)}(p(0))$, as can be verified by differentiation. The proof is now finished. ∎

Corollary $\mathrm{Ad}^*_{g(t)^{-1}}(p(t)) = $ *constant for each integral curve $(g(t), p(t))$ of* $\vec{\mathcal{H}}_{p_0}(p, u(t))$.

Proof

$$\mathrm{Ad}^*_{g(t)^{-1}}(p(t)) = \mathrm{Ad}^*_{g(t)^{-1}} \mathrm{Ad}^*_{g(t)}(p(0)) = \mathrm{Ad}^*_e(p(0)) = p(0). ∎$$

In all the applications that follow, G is a subgroup of the general linear group. For such groups, $\mathrm{Ad}_g(A) = gAg^{-1}$. Therefore, adjoint orbits coincide with the conjugacy classes. The co-adjoint orbits through f coincide with all linear functions f_g, where $f_g(A) = f(g^{-1}Ag)$ for all A in $\mathcal{L}$ and g in G. Therefore, the preceding corollary can be paraphrased as

$$p(t)(g^{-1}(t) Ag(t)) = f(A) \quad \text{for all } A \text{ in } \mathcal{L}, \text{ with } f = p(0).$$

For many left-invariant optimal-control problems the extremal curves are integral curves of a single Hamiltonian vector field $\vec{H}$ defined by a function H on $\mathcal{L}^*$. In such a case, we have the following:

Theorem 6 *Let H denote any continuously differentiable function on $\mathcal{L}^*$, and let $(g(t), p(t))$ denote an integral curve of the corresponding Hamiltonian vector field $\vec{H}$. Then*

$$dL_{g^{-1}(t)}\left(\frac{dg}{dt}\right) = dH(p(t)) \quad \text{and} \quad \frac{dp}{dt} = -(\text{ad})^*(dH(p))(p(t)).$$

Consequently, $p(t) = \text{Ad}^*_{g(t)}(f)$ for some f in $\mathcal{L}^*$.

Euler's top is the most notable left-invariant problem on a Lie group. We saw earlier that its Hamiltonian is given by

$$H(p) = \frac{1}{2}\left(a_1 H_1^2(p) + a_2 H_2^2(p) + a_3 H_3^2(p)\right),$$

with $H_i(p) = p(A_i)$, and A_1, A_2, A_3 denoting the standard basis in the Lie algebra of $SO_3(R)$.

Theorem 6 easily leads to its Hamiltonian equations: Let $\sigma(\varepsilon) = p + \varepsilon M^*$ be a curve in $\mathcal{L}^*$. Then

$$\frac{d}{d\varepsilon}H(p + \varepsilon M^*)|_{\varepsilon=0}$$
$$= \frac{1}{2}\frac{d}{d\varepsilon}\left(a_1 H_1^2(p + \varepsilon M^*) + a_2 H_2^2(p + \varepsilon M^*) + a_3 H_3^2(p + \varepsilon M^*)\right)$$
$$= a_1 H_1(p)M^*(A_1) + a_2 H_2(p)M^*(A_2) + a_3 H_3(p)M^*(A_3)$$
$$= M^*(a_1 H_1(p)A_1 + a_2 H_2(p)A_2 + a_3 H_3(p)A_3).$$

Regarding $dH(p)$ as an element in $\mathcal{L}$, it follows that

$$dH(p) = a_1 H_1(p)A_1 + a_2 H_2(p)A_2 + a_3 H_3(p)A_3.$$

According to Theorem 6, the integral curves $(g(t), p(t))$ of $\vec{H}$ satisfy

$$g^{-1}(t)\frac{dg}{dt} = dH(p(t)) \quad \text{and} \quad \frac{dp}{dt} = -(\text{ad})^*(dH(p))(p(t)).$$

Hence

$$\frac{dg}{dt} = g(t)(a_1 H_1(p(t))A_1 + a_2 H_2(p(t))A_2 + a_3 H_3(p(t))A_3) = g(t)\Omega(t).$$

$\Omega(t)$ is the antisymmetric matrix defined by

$$\Omega_1(t) = a_1 H_1(p(t)), \qquad \Omega_2(t) = a_2 H_2(p(t)), \qquad \Omega_3(t) = a_3 H_3(p(t)).$$

Then

$$\frac{dp(t)}{dt} = -(\mathrm{ad})^*\Omega(t)(p(t)), \quad \text{or } \frac{dp}{dt}(A) = -p(t)([\Omega(t), A]),$$

for any A in $\mathcal{L}(SO_3(R))$.

As we saw in the section on the heavy top, $p(t)$ can also be identified with $P(t) = H_1(p(t))A_1 + H_2(p(t))A_2 + H_3(p(t))A_3$ via the trace form, in which case

$$\frac{dP}{dt} = [\Omega(t), P(t)]. \tag{7}$$

With this identification of $\mathcal{L}$ with $\mathcal{L}^*$, the relation $p(t) = \mathrm{Ad}_g^*(f)$ becomes

$$P(t) = g(t)^{-1}P(0)g(t). \tag{8}$$

Equation (8) is the integrated form of equation (7), as can be verified by differentiating. Each of them implies the other.

Any differential equation of the same form as equation (7) is said to be in Lax pair form. Evidently the solutions of such differential systems are isospectral; that is, the spectrum of $P(t)$ must be constant. The reader can verify that the nonzero eigenvalue of $P(t)$ for the Euler top is given by $H_1^2 + H_2^2 + H_3^2$. Each eigenvalue determines the sphere $H_1^2 + H_2^2 + H_3^2 = M$, called the momentum sphere. The particular value of M is determined by the initial condition $p(0)$. Together with the total energy $H = \frac{1}{2}(a_1H_1^2 + a_2H_2^2 + a_3H_3^2)$, M is another integral of motion. This information recovers the classic result, which states that the momentum curves $p(t)$ are contained in the intersection of the momentum sphere with the energy ellipsoid (see Figure 4.7).

It will be shown later that much of the underlying geometry of the solutions for Euler's top extends to arbitrary semisimple Lie groups and that the extra integrals of motion, such as the momentum integral, always exist and are known as Casimir elements. Casimir elements may even exist on groups that are not semisimple, such as the group of motions of $\mathbb{R}^n$.

4 The elastic problem in $\mathbb{R}^3$ and the kinetic analogue of Kirchhoff

Euler's planar elastic problem has several interesting extensions to curves in $\mathbb{R}^3$. We shall begin with one such extension, motivated by the equilibrium theory for a thin, inextensible elastic rod in $\mathbb{R}^3$ subjected to bending and twisting at its ends. According to a classic treatise on elasticity (Love, 1927), an elastic rod is described by its central line and an orthonormal frame along the central line that measures the amounts of bending and twisting relative to some fixed frame in the unstressed state of the rod. Each equilibrium configuration minimizes the total elastic energy among all other configurations that satisfy the same boundary conditions. This problem admits a natural description as an optimal, left-invariant problem on the group of motions of the Euclidean space E^3, as follows:

The space of all configurations is the totality of all curves $x(t)$ in $\mathbb{R}^3$ along with the orthonormal frames $\mathbf{a}_1(t)$, $\mathbf{a}_2(t)$, $\mathbf{a}_3(t)$ defined along each curve $x(t)$. The curve $x(t)$ is to be regarded as the central line of the rod. We shall assume that the frame is adapted to the curve $x(t)$ by requiring that $(dx/dt)(t) = \mathbf{a}_1(t)$. This requirement reflects the physical assumption that the rod is nonsheared and is inextensible. (In a more general situation, we can imagine that dx/dt is a linear combination of the frame vectors, with the coefficients conforming to some symmetry conditions.)

Let $\mathbf{e}_1$, $\mathbf{e}_2$, $\mathbf{e}_3$ denote a fixed right-handed orthonormal frame in E^3. Suppose that the coordinates of each vector $\mathbf{a}_i(t)$ relative to the fixed, or absolute, frame are given by

$$\mathbf{a}_i(t) = \sum_{j=1}^{3} a_i^j(t)\,\mathbf{e}_i.$$

Denote by $R(t)$ the matrix whose ith column consists of the coordinates of $\mathbf{a}_i$ relative to the fixed frame. Within the context of the theory of elastic rods, $R(t)$ measures the amounts of twisting and bending along the central line $x(t)$.

Having identified adapted frames with rotations, we shall think of the configuration space as the set of all pairs $(x(t), R(t))$, with $x(t)$ a curve in E^3 and $R(t)$ a curve in $SO_3(R)$ such that $dx/dt = R(t)\mathbf{e}_1$. For each configuration $(x(t), R(t))$, the derivative $(dR/dt)(t)$ is an element of the tangent space of $SO_3(R)$ at $R(t)$. In the left-invariant representation of the tangent bundle of $SO_3(R)$,

$$\frac{dR}{dt} = R(t)\begin{pmatrix} 0 & -u_3(t) & u_2(t) \\ u_3(t) & 0 & -u_1(t) \\ -u_2(t) & u_1(t) & 0 \end{pmatrix}$$

for some functions $u_1(t)$, $u_2(t)$, and $u_3(t)$. The preceding equation can also be written as

$$\frac{dR}{dt} = u_1(t)\vec{A}_1(R) + u_2(t)\vec{A}_2(R(t)) + u_3(t)\vec{A}_3(R(t)), \tag{9}$$

with $\vec{A}_i$ equal to the left-invariant vector field whose value at the identity is equal to A_i. (As in the case of the heavy top, A_1, A_2, A_3 is the standard basis in $\mathcal{L}(SO_3(R))$.)

The functions $u_1(t), u_2(t), u_3(t)$ are called strains. If T is equal to the total length of the curve $x(t)$, then

$$\frac{1}{2} \int_0^T \left(c_1 u_1^2(t) + c_2 u_2^2(t) + c_3 u_3^2(t) \right) dt$$

is called the total elastic energy associated with the configuration $(x(t), R(t))$ in the interval $[0, T]$. The constants c_1, c_2, and c_3 reflect the physical characteristics of the bar related to the geometric shape of its cross section.

The constraint $dx(t)/dt = \mathbf{a}_1(t)$ and equation (9) define a differential system on E_3, the group of motions of E^3. The reader should recall that E_3 is the semidirect product of $SO_3(R)$ with $\mathbb{R}^3$. For our purposes, it is convenient to regard E_3 as the closed subgroup of $GL_4(R)$ consisting of all matrices

$$\begin{pmatrix} 1 & 0 \\ x & R \end{pmatrix} \quad \text{with } x \text{ in } \mathbb{R}^3,$$

written as a column vector, and R an element of $SO_3(R)$.

Elements on $GL_4(R)$ act linearly on points in $\mathbb{R}^4$ (regarded as column vectors). In particular,

$$\begin{pmatrix} 1 & 0 \\ x & R \end{pmatrix} \begin{pmatrix} 1 \\ y \end{pmatrix} = Ry + x$$

for each y in $\mathbb{R}^3$. Thus x can be identified with a translation in $\mathbb{R}^3$ given by $y \to y+x$. The set of all translations in E_3 is a three-dimensional vector space, and its tangent space at the identity is a linear span of

$$B_1 = \begin{pmatrix} 0 & 0 & 0 & 0 \\ 1 & 0 & 0 & 0 \\ 0 & 0 & 0 & 0 \\ 0 & 0 & 0 & 0 \end{pmatrix}, \quad B_2 = \begin{pmatrix} 0 & 0 & 0 & 0 \\ 0 & 0 & 0 & 0 \\ 1 & 0 & 0 & 0 \\ 0 & 0 & 0 & 0 \end{pmatrix}, \quad B_3 = \begin{pmatrix} 0 & 0 & 0 & 0 \\ 0 & 0 & 0 & 0 \\ 0 & 0 & 0 & 0 \\ 1 & 0 & 0 & 0 \end{pmatrix}.$$

The matrices

$$A_1 = \begin{pmatrix} 0 & 0 & 0 & 0 \\ 0 & 0 & 0 & 0 \\ 0 & 0 & 0 & -1 \\ 0 & 0 & 1 & 0 \end{pmatrix}, \quad A_2 = \begin{pmatrix} 0 & 0 & 0 & 0 \\ 0 & 0 & 0 & 1 \\ 0 & 0 & 0 & 0 \\ 0 & -1 & 0 & 0 \end{pmatrix}, \quad A_3 = \begin{pmatrix} 0 & 0 & 0 & 0 \\ 0 & 0 & -1 & 0 \\ 0 & 1 & 0 & 0 \\ 0 & 0 & 0 & 0 \end{pmatrix},$$

along with the matrices B_1, B_2, and B_3, form a basis for the Lie algebra of E_3. The Lie-bracket table for the corresponding left-invariant vector fields is given by Table 12.1. Using these notations, the configuration space can be regarded as the solution space of the following differential system in E_3:

$$\frac{dg}{dt} = \vec{B}_1(g(t)) + \sum_{i=1}^{3} u_i(t)\, \vec{A}_i(g(t)). \tag{10}$$

The total elastic energy of each solution curve $(g(t), u(t))$ in a fixed interval $[0, T]$ is given by

$$\frac{1}{2} \int_0^T \left(c_1 u_1^2(t) + c_2 u_2^2(t) + c_3 u_3^2(t) \right) dt.$$

Table 12.1.

$[\,,\,]$	$\vec{A}_1$	$\vec{A}_2$	$\vec{A}_3$	$\vec{B}_1$	$\vec{B}_2$	$\vec{B}_3$
$\vec{A}_1$	0	$-\vec{A}_3$	$\vec{A}_2$	0	$-\vec{B}_3$	$\vec{B}_2$
$\vec{A}_2$	$\vec{A}_3$	0	$-\vec{A}_1$	$\vec{B}_3$	0	$-\vec{B}_1$
$\vec{A}_3$	$-\vec{A}_2$	$\vec{A}_1$	0	$-\vec{B}_2$	$\vec{B}_1$	0
$\vec{B}_1$	0	$-\vec{B}_3$	$\vec{B}_2$	0	0	0
$\vec{B}_2$	$\vec{B}_3$	0	$-\vec{B}_1$	0	0	0
$\vec{B}_3$	$-\vec{B}_2$	$\vec{B}_1$	0	0	0	0

By "the elastic problem in $\mathbb{R}^3$" we shall mean the problem of finding a solution curve $(g(t), u(t))$ in a given interval $[0, T]$ that will satisfy the prescribed boundary conditions $g(0) = g_0$ and $g(T) = g_1$ and that will minimize the total elastic energy among all other solution curves with the same boundary conditions. The elastic problem in $\mathbb{R}^3$ is a left-invariant optimal-control problem on the group of motions of $\mathbb{R}^3$. It differs from the heavy-top problem in several respects. First, the group of motions is not a semisimple Lie group, because the trace form is degenerate. Indeed, it is easy to verify that trace$(AB) = 0$ for all matrices A in the Lie algebra of E_3, whenever B is in the linear span of B_1, B_2, B_3.

Second, equation (10) defines a proper subset of the tangent bundle of E_3, because for each g in E_3 the right-hand side of equation (10) is a three-dimensional affine subspace of the six-dimensional tangent space at g. In particular, abnormal extremals may occur in the elastic problem, whereas they do not occur in the heavy-top problem. These remarks will be of relevance for a theorem of Kirchhoff that relates the equations of the elastic problem to the equations of the heavy top.

We shall continue with the notation used in the section on the rigid body and let H_i denote the Hamiltonians $H_i(p) = p(A_i)$, $i = 1, 2, 3$. The Hamiltonians that correspond to the translations will be denoted by small letters. In particular, h_1, h_2, h_3 will denote the Hamiltonians $h_i(p) = p(B_i)$, $i = 1, 2, 3$. The first theorem describes the structure of the abnormal extremals.

Theorem 7 *An absolutely continuous curve* $(g(t), p(t))$ *is an abnormal extremal curve of the elastic problem if and only if* $(g(t), p(t))$ *is an integral curve of* $\vec{H} = \vec{h}_1 + u_1(t)\vec{H}_1$ *restricted to the set* $h_2 = h_3 = 0$, *and* $H_1 = H_2 = H_3 = 0$. *The function* $u_1(t)$ *is an arbitrary integrable function, and* $u_2(t) = u_3(t) = 0$ *for all* t. *For abnormal extremals that project onto optimal trajectories,* $u_1(t)$ *must be constant.*

Proof Each abnormal extremal curve $(g(t), p(t))$ is an integral curve of $H = h_1 + u_1(t)H_1 + u_2(t)H_2 + u_3(t)H_3$ for some integrable functions $u_1(t)$, $u_2(t)$, $u_3(t)$. The maximality condition of the maximum principle can hold only when

$$H_i(p(t)) = 0 \quad \text{for all } t \text{ and each } i = 1, 2, 3.$$

The Poisson brackets of the Hamiltonians $H_1, H_2, H_3, h_1, h_2, h_3$ satisfy the same relations as the Lie brackets of the left-invariant vector fields that define them (Table 12.1). Therefore,

$$\begin{aligned}
0 &= \{H_2, H\}(p(t)) \\
&= \{H_2, h_1\} + u_1(t)\{H_2, H_1\} + u_2(t)\{H_2, H_2\} + u_3(t)\{H_2, H_3\} \\
&= h_3(p(t)).
\end{aligned}$$

Repeating this argument with $\{H_3, H\}(p) = 0$ yields $h_2(p(t)) = 0$. Therefore, all abnormal extremals are contained in the zero set of $h_2 = 0, h_3 = 0$, $H_1 = H_2 = H_3 = 0$. Then

$$\begin{aligned}
0 &= \frac{d}{dt}h_3(p(t)) = \{h_3, H\}(p(t)) \\
&= u_1(t)\{h_3, H_1\} + u_2(t)\{h_3, H_2\} + u_3(t)\{h_3, H_3\} \\
&= u_2(t)h_1(p(t)).
\end{aligned}$$

Thus $h_1(p(t))$ cannot be equal to zero, because $p(t) \neq 0$. Therefore, $u_2(t) = 0$ for all t.

Differentiating $h_2(p(t)) = 0$ yields $u_3(t) = 0$ by a completely analogous argument. Hence, $(g(t), p(t))$ and $u(t)$ satisfy the conditions of the theorem.

Conversely, let $(g(t), p(t))$ be any integral curve of $\vec{H} = \vec{h}_1 + u_1(t)\vec{H}_1$ that originates in the manifold $h_2 = h_3 = 0$, $H_1 = H_2 = H_3 = 0$. Such a curve remains in the manifold, because all the Poisson brackets $\{h_i, H\}$, $i = 2, 3$, and $\{H_i, H\}$, $i = 1, 2, 3$, vanish along the curve. Moreover, along such a curve, $H(p(t), v_1, v_2, v_3) = h_1(p(t))$ for any strain functions v_1, v_2, v_3. Hence the maximality condition is satisfied, and therefore $(g(t), p(t))$ is an abnormal extremal curve generated by $u_1(t)$.

It remains to show that if $g(t)$ is an optimal curve in E_3, then $u_1(t) = $ constant. Because $u_2(t) = u_3(t) = 0$, the rotational part $R(t)$ of $g(t)$ satisfies

$$\frac{dR}{dt} = u_1(t)\vec{A}_1(R(t)).$$

Hence

$$R(t) = R_0\exp\left(\int_0^t u_1(\tau)\,d\tau\right)A_1,$$

with R_0 denoting the initial state $R(0)$. Let $T > 0$ denote the length of the rod, and let R_1 denote the terminal state of $R(t)$. The solution curves of $dR/dt = \varphi(t)\vec{A}_1(R(t))$ for which $R(T) = R_1$ are constrained by

$$\int_0^T \varphi(t)\,dt = \text{constant}.$$

The total elastic energy of such a solution curve is given by

$$\frac{1}{2}\int_0^T c_1\varphi^2(t)\,dt.$$

The minimum value of $\int_0^T \varphi^2(t)\,dt$, restricted to $\int_0^T \varphi(t)\,dt = c$, occurs for $\varphi(t) = c$. The argument is based on the Cauchy-Schwarz inequality:

$$c = \left|\int_0^T \varphi(t)\,dt\right| \leq \left(\int_0^T \varphi^2(t)\,dt\right)^{1/2} T^{1/2}.$$

Hence

$$\int_0^T \varphi^2(t)\,dt \geq \frac{(c)^2}{T},$$

with equality only for $\varphi(t) = c$. End of the proof. ∎

Corollary *The equilibrium configurations that are the projections of abnormal extremal curves consist of straight lines in E_3 whose frames rotate with constant angular velocity around the $\mathbf{e}_1$ direction (Figure 12.2).*

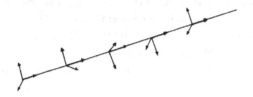

Fig. 12.2.

The essential properties of the regular extremal curves are described by the following theorem.

Theorem 8 *Each regular extremal curve $(g(t), p(t))$ defined by a strain function $u_1(t), u_2(t), u_3(t)$ satisfies the following properties:*

(a) *$p(t)$ is an integral curve of the Hamiltonian vector field $\vec{H}$ that corresponds to*

$$H = \frac{1}{2}\left(\frac{H_1^2}{c_1} + \frac{H_2^2}{c_2} + \frac{H_3^2}{c_3}\right) + h_1.$$

(b) *$u_1(t) = (1/c_1)H_1(p(t))$, $u_2(t) = (1/c_2)H_2(p(t))$, and $u_3(t) = (1/c_3)H_3(p(t))$ for all t.*

Proof The system Hamiltonian

$$\mathcal{H}(p(t), u(t)) = -\frac{1}{2}\left(c_1 u_1^2(t) + c_2 u_2^2(t) + c_3 u_3^2(t)\right) + h_1(p(t))$$
$$+ u_1(t)H_1(p(t)) + u_2(t)H_2(p(t)) + u_3(t)H_3(p(t))$$

attains its maximum over all other strain functions only for

$$u_1(t) = \frac{1}{c_1}H_1(p(t)), \qquad u_2(t) = \frac{1}{c_2}H_2(p(t)), \qquad u_3(t) = \frac{1}{c_3}H_3(p(t)).$$

Hence $p(t)$ is an integral curve of

$$H = \frac{1}{2}\left(\frac{H_1^2}{c_1} + \frac{H_2^2}{c_2} + \frac{H_3^2}{c_3}\right) + h_1.$$

Conversely, any integral curve of $\vec{H}$ is a regular extremal curve. We leave it to the reader to verify the required steps. ∎

Theorem 8 will be called Kirchhoff's theorem. According to Love (1927), Kirchhoff was, in 1859, the first to discover the Hamiltonian equations for the elastic problem, which he then attempted to solve by analogy with the heavy-top problem. In this analogy, the term h_1 corresponds to the potential energy, and the physical constants c_1, c_2, c_3 correspond to the principal moments of inertia.

According to Theorem 6, each integral curve of the Hamiltonian vector field $\vec{H}$ satisfies $p(t) = \text{Ad}^*_{g(t)}(f)$ for some f in $\mathcal{L}^*$. That is, $p(t)$ is a curve in the co-adjoint orbit of G through f. In order to extract the geometric significance of this relation, it is necessary to identify $\mathcal{L}$ with $\mathcal{L}^*$. This identification will be done via the following bilinear form on $\mathcal{L}$:

$$\langle (A \oplus B), \ (A' \oplus B') \rangle = -\frac{1}{2}\text{trace}(A' \cdot A) + B \cdot B'. \tag{11}$$

A and A' are elements of $\mathcal{L}$ in the linear span of A_1, A_2, A_3, and B and B' are elements of $\mathcal{L}$ in the linear span of B_1, B_2, B_3. $B \cdot B'$ is the standard Euclidean product in $\mathbb{R}^3$ given by $\sum_{i=1}^{3} b_i b_i'$ whenever $B = \sum_{i=1}^{3} b_i B_i$ and $B' = \sum_{i=1}^{3} b_i' B_i$. In terms of the earlier notations that identify antisymmetric matrices A with elements $\hat{A}$ in $\mathbb{R}^3$, the foregoing bilinear form can also be written as

$$\langle (A \oplus B, (A' \oplus B') \rangle = \hat{A} \cdot \hat{A}' + B \cdot B'.$$

Each element p of $\mathcal{L}^*$ will be identified with an element M_p in $\mathcal{L}$ via the formula $p(L) = \langle M_p, L \rangle$ for all L in $\mathcal{L}$. In particular, $\langle M_p, A_i \rangle = p(A_i) = H_i$, and $\langle M_p, B_i \rangle = p(B_i) = h_i$, for each $i = 1, 2, 3$. Therefore,

$$M_p = \begin{pmatrix} 0 & 0 & 0 & 0 \\ h_1 & 0 & -H_3 & H_2 \\ h_2 & H_3 & 0 & -H_1 \\ h_3 & -H_2 & H_1 & 0 \end{pmatrix} = \sum_{i=1}^{3} h_i B_i + H_i A_i.$$

Theorem 9 *Let* $(g(t), p(t))$ *be a regular extremal curve of the elastic problem. Denote* $g(t)$ *by*

$$\begin{pmatrix} 1 & 0 \\ x(t) & R(t) \end{pmatrix}$$

and $p(t)$ *by* $M_{p(t)} = \sum_{i=1}^{3} h_i(t) B_i + H_i(t) A_i$. *Then*

$$R(t)h(t) = a \quad \text{and} \quad R(t)\hat{H}(t) + a \times x(t) = b$$

for some constant vectors a *and* b *in* $\mathbb{R}^3$.

Proof For each extremal curve $(g(t), p(t))$, $p(t) = \text{Ad}^*_{g(t)}(f)$. Therefore, $\text{Ad}^*_{g(t)^{-1}}(p(t))$ is a fixed linear function on $\mathcal{L}$. Because E_3 is a linear group, $\text{Ad}^*_{g^{-1}}(p(t))(L) = p(t)(g(t)^{-1}Lg(t))$ for each L in $\mathcal{L}$. Denote $L = \sum_{i=1}^{3} v_i B_i + V_i A_i$ by

$$\begin{pmatrix} 0 & 0 \\ v & A \end{pmatrix}.$$

Then

$$g^{-1}Lg = \begin{pmatrix} 1 & 0 \\ -R^{-1}x & R^{-1} \end{pmatrix} \begin{pmatrix} 0 & 0 \\ v & V \end{pmatrix} \begin{pmatrix} 1 & 0 \\ x & R \end{pmatrix}$$

$$= \begin{pmatrix} 0 & 0 \\ R^{-1}(v + Vx) & R^{-1}VR \end{pmatrix}.$$

Hence

$$p(t)(g^{-1}Lg) = \langle M_p, g^{-1}Lg \rangle = h(t) \cdot (R^{-1}(v + Vx(t)) + \hat{H}(t) \cdot R^{-1}\hat{V}.$$

(The reader should recall that the antisymmetric matrix $R^{-1}VR$ is identified with $R^{-1}\hat{V}$ in $\mathbb{R}^3$.) It then follows that

$$h(t) \cdot (R^{-1}(v + Vx(t)) + \hat{H}(t) \cdot R^{-1}\hat{V} = \text{constant}$$

for each v and each $\hat{V}$ in $\mathbb{R}^3$. In particular, $\hat{V} = 0$ yields $h(t) \cdot R^{-1}v = \text{constant}$. Because V is arbitrary, $R(t)h(t) = a$ for some constant vector a in $\mathbb{R}^3$.

The other equality follows from holding $v = 0$ and letting $\hat{V}$ vary over $\mathbb{R}^3$. It follows that $h(t) \cdot (R^{-1}Vx(t)) + \hat{H} \cdot R^{-1}\hat{V} = a \cdot Vx(t) + \hat{H} \cdot R^{-1}\hat{V} = \text{constant}$. Combining $a \cdot Vx(t) = a \cdot (\hat{V} \times x(t)) = \hat{V} \cdot (x(t) \times a)$ and $\hat{H}(t) \cdot R^{-1}(t)\hat{V} = R(t)\hat{H}(t) \cdot \hat{V}$ into the foregoing equation gives $(R(t)\hat{H}(t) + x(t) \times a) \cdot \hat{V} = \text{constant}$ for all $\hat{V}$ in $\mathbb{R}^3$. Consequently, $R(t)\hat{H}(t) + x(t) \times a = b$ for some constant b in $\mathbb{R}^3$. End of the proof. ∎

Theorem 10 *Let $g(t)$ and $M_{p(t)}$ have the same meanings as in Theorem 9. Denote by $\Omega(t)$ the antisymmetric matrix*

$$\sum_{i=1}^{3} \frac{H_i(t)}{c_i} A_i.$$

Then the conditions of Theorem 9 are equivalent to the following differential system in $\mathbb{R}^3 \times \mathbb{R}^3$:

$$\frac{d\hat{H}}{dt} = \hat{H}(t) \times \hat{\Omega}(t) + h(t) \times e_1 \quad and \quad \frac{dh}{dt} = h(t) \times \hat{\Omega}(t).$$

Proof Recall that $dx/dt = R(t)e_1$, and $dR/dt = R(t)\Omega(t)$. Differentiating $R(t)h(t) = a$ yields

$$\frac{dR}{dt}h + R\frac{dh}{dt} = 0, \quad \text{or} \quad \frac{dh}{dt} = -R^{-1}\frac{dR}{dt}h = -\Omega h.$$

Thus $dh/dt = h(t) \times \hat{\Omega}(t)$. Differentiating $R(t)\hat{H}(t) + x(t) \times a = b$ yields

$$\frac{dR}{dt}\hat{H} + R\frac{d\hat{H}}{dt} + \frac{dx}{dt} \times a = 0.$$

Hence

$$\frac{d\hat{H}}{dt} = -R^{-1}(t)\frac{dR}{dt}\hat{H} - R^{-1}\left(\frac{dx}{dt} \times a\right) = -\Omega\hat{H} - R^{-1}(Re_1 \times a)$$

$$= \hat{H} \times \hat{\Omega} - e_1 \times h(t) = \hat{H} \times \hat{\Omega} + h(t) \times e_1.$$

Conversely, assume that $h(t)$ and $\hat{H}(t)$ satisfy the following differential equations:

$$\frac{d\hat{H}}{dt} = \hat{H} \times \Omega + h \times e_1, \quad \frac{dh}{dt} = h \times \hat{\Omega},$$

$$\frac{dx}{dt} = R(t)e_1, \quad \frac{dR}{dt} = R\Omega.$$

Then

$$\frac{d}{dt}R(t)h(t) = \frac{dR}{dt}h + R\frac{dh}{dt} = R\Omega h + R(h \times \hat{\Omega})$$

$$= R(\hat{\Omega} \times h) + R(h \times \hat{\Omega}) = 0,$$

and

$$\frac{d}{dt}R\hat{H} + x(t) \times a = \frac{dR}{dt}\hat{H} + R\frac{d\hat{H}}{dt} + \frac{dx}{dt} \times a$$

$$= R\Omega\hat{H} + R(\hat{H} \times \hat{\Omega} + h \times e_1) + Re_1 \times a$$

$$= R(\hat{\Omega} \times \hat{H}) + R(\hat{H} \times \Omega) + R(h \times e_1) + R(e_1 \times a)$$

$$= 0.$$

Hence

$$Rh = a \quad \text{and} \quad R\hat{H} + x(t) \times a = b$$

for some constants a and b. End of the proof.   ■

Corollary *Each projection of an abnormal extremal curve that is optimal is also a projection of a regular extremal curve.*

Proof Abnormal extremal curves that correspond to optimal trajectories are generated by the controls $u_1 = $ constant, $u_2 = u_3 = 0$ (Theorem 7). The curve $H_1 = H_2 = H_3 = 0$, $h_1 = $ constant, $h_2 = h_3 = 0$, satisfies both $u_1 = H_1/c_1$, $u_2 = H_2/c_2$, $u_3 = H_3/c_3$ and

$$\frac{d\hat{H}}{dt} = \hat{H} \times \hat{\Omega} + h \times e_1 \quad \text{and} \quad \frac{dh}{dt} = h \times \hat{\Omega}.$$

Therefore, it is a regular extremal curve. End of the proof. ∎

Therefore, all elastic configurations shown in Figure 12.2 are projections of both the regular and abnormal extremal curves.

The analogy of the elastic problem with the heavy top is indeed remarkable: The rigid body that corresponds to the elastic bar must have its center of gravity Q_0 along the $\mathbf{a}_1$ axis one unit from the fixed point O_b, and $h(t)$ is identified with $Q(t) = R^{-1}(t)ce_3$. Conversely, any heavy top whose center of gravity Q_0 is situated on the $\mathbf{a}_1$ axis one unit from the origin O_b has a "central line" $x(t)$ defined by $x(t) = R(t)e_1$.

It is important to note that the conditions $R(t)h(t) = a$ and $R(t)\hat{H}(t) + x(t) \times a = b$ imply the existence of two extra integrals of motion:

$$\|h(t)\|^2 = \|a\|^2 \quad \text{and} \quad h(t) \cdot \hat{H}(t) = a \cdot b.$$

It will be shown in the next section that $h_1^2 + h_2^2 + h_3^2$ and $h_1 H_1 + h_2 H_2 + h_3 H_3$ are integrals of motion for any left-invariant optimal problem on the group of motions of $\mathbb{R}^3$, not just for the elastic problems. Such conserved quantities depend solely on the structure of the group, not on the form of the Hamiltonian.

5 Casimir functions and the conservation laws

Let us now return to an arbitrary Lie group G, with its Lie algebra $\mathcal{L}$.

Definition 5 A Casimir function is any Ad_G^*-invariant real analytic function on $\mathcal{L}^*$.

A function f is Ad_G^*-invariant if $f(p) = f(\text{Ad}_g^*(p))$ for all $g \in G$ and $p \in \mathcal{L}^*$. It follows immediately that a Casimir function is an integral of motion for any left-invariant Hamiltonian control system H. The argument is as follows: If

$(g(t), p(t))$ denotes an integral curve of $\vec{H}(\cdot, u(t))$, then $p(t)$ is contained in some co-adjoint orbit of G (Theorem 5). Therefore,

$$f(p(t)) = f(\mathrm{Ad}^*_{g(t)}(p_0)) = f(p_0)$$

for some p_0 in $\mathcal{L}^*$.

Casimir functions can also be defined through the properties of the Poisson algebra of functions on $\mathcal{L}^*$. In order to get to these properties, it will be necessary to return briefly to the Poisson bracket.

Theorem 11 *Let f and g be any functions on $\mathcal{L}^*$. Then*

$$\{f, g\}(p) = p[df(p), dg(p)]$$

for all p in $\mathcal{L}^*$.

Proof A function f on $\mathcal{L}^*$ defines its Hamiltonian vector field $\vec{f}$ through the equations

$$dL_{g^{-1}} \frac{dg}{dt} = df(p(t)) \quad \text{and} \quad \frac{dp}{dt} = -(\mathrm{ad})^*(df(p(t))).$$

Then

$$\{f, g\}(p) = w_p(\vec{f}(p), \vec{g}(p)) = -(\mathrm{ad})^* dg(p)(df(p))$$
$$+ (\mathrm{ad})^* df(p)(dg(p)) - p[df(p), dg(p)] = -p[dg(p), df(p)]$$
$$+ p[df(p), dg(p)] - p[df(p), dg(p)] = p[df(p), dg(p)].$$

End of the proof. ∎

Remark In the theory of Hamiltonian systems, and in many papers in mechanics, the foregoing expression for the Poisson bracket is called the Poisson-Lie bracket. The reader should note that $df(p)$ is a linear function on $\mathcal{L}^*$ and hence is an element of $\mathcal{L}$. Therefore, the Lie bracket $[df(p), dg(p)]$ makes sense. ∎

It follows from Theorem 11 that functions on $\mathcal{L}^*$ form an algebra under the Poisson bracket. Generally this algebra is infinite-dimensional.

Problem 5 Verify the Jacobi identity

$$\{\{f, g\}, h\} + \{\{h, f\}, g\} + \{\{g, h\}, f\} = 0.$$

Definition 6 Let H be any function on T^*G. A function G on T^*G is said to be an integral of motion for H if $\{G, H\} = 0$.

It follows that G is an integral of motion for H if and only if $G(\xi(t)) = $ constant along any integral curve $\xi(t)$ of the Hamiltonian vector field $\vec{H}$. Evidently H is an integral of motion for G whenever G is an integral of motion for H.

Problem 6 Prove that $\{G_1, G_2\}$ is an integral of motion for H for any integrals of motion G_1 and G_2 of H.

In the literature on mechanics, Casimir functions are often referred to as the conservation laws, because of the following property.

Theorem 12 *A Casimir function f is an integral of motion for any function H on $\mathcal{L}^*$.*

Proof Let $\xi(t) = (g(t), p(t))$ denote an integral curve of the Hamiltonian vector field $\vec{H}$ corresponding to the function H. Because f is a function on $\mathcal{L}^*$, $f(\xi(t)) = f(p(t))$. The curve $p(t)$ is contained in a co-adjoint orbit of G, and f is constant in each co-adjoint orbit. Therefore, $f(p(t)) = $ constant. But then

$$0 = \frac{d}{dt} f(p(t)) = \frac{d}{dt} f(\xi(t)) = \{f, H\}(\xi(t)).$$

End of the proof. ■

Hamiltonian equations for Euler's top revealed $H_1^2 + H_2^2 + H_3^2$ as its integral of motion. It may already be clear from the preceding derivations that the foregoing function is a Casimir function on $SO_3(R)$ and therefore is a conservation law for any left-invariant optimal problem. This fact can be verified in several ways:

Recall first that each H_i is the Hamiltonian of the left-invariant vector field $\vec{A}_i(t)$, with A_1, A_2, A_3 denoting the standard basis in the space of 3×3 skew symmetric matrices. Also recall that each element p in $\mathcal{L}^*$ is identified with an element

$$P = \begin{pmatrix} 0 & -H_3(p) & H_2(p) \\ H_3(p) & 0 & -H_1(p) \\ -H_2(p) & H_1(p) & 0 \end{pmatrix}$$

in the Lie algebra $\mathcal{L}$ of $SO_3(R)$ via the trace form. The trace form identifies co-adjoint orbits with adjoint orbits as follows:

$$(\mathrm{Ad}^*_{g^{-1}}(p))A = p \; \mathrm{Ad}_{g^{-1}}(A) = p \, (g^{-1}Ag) = \hat{P} \cdot g^{-1}\hat{A} = g\hat{P} \cdot \hat{A}.$$

Hence, $\mathrm{Ad}^*_{g^{-1}}(p)$ is identified with gPg^{-1}. As g varies, gPg^{-1} describes the conjugacy class of P. Hence the eigenvalues of P are constant in each orbit. The characteristic polynomial of P is equal to

$$-\lambda \left(\lambda^2 + H_1^2 + H_2^2 + H_3^2\right).$$

The nonzero eigenvalue λ satisfies $-\lambda^2 = M$. Hence M is an invariant function on each orbit and therefore is a Casimir function for $SO_3(R)$.

There is an alternative way of verifying that $H_1^2 + H_2^2 + H_3^2$ is a Casimir function on $SO_3(R)$. Let A_j^* denote the dual basis in $\mathcal{L}^*$ defined by $A_j^*(A_i) = \delta_{ij}$. Then each p in $\mathcal{L}^*$ can be written as

$$p = \sum_{i=1}^{3} p_i A_i^*.$$

Because $H_j(p) = p(A_j) = \sum p_i A_i^*(A_j) = p_j$, it follows that

$$p = \sum_{i=1}^{3} H_i(p) A_i^*.$$

Any function f on $\mathcal{L}^*$, when expressed in the dual coordinates of p, becomes a function of H_1, H_2, H_3.

The Poisson brackets of H_1, H_2, H_3 satisfy

$$\{H_1, H_2\} = -H_3, \qquad \{H_3, H_1\} = -H_2, \qquad \{H_2, H_3\} = -H_1.$$

It is now easy to verify that $M = H_1^2 + H_2^2 + H_3^2$ Poisson-commutes with each H_1, H_2, H_3 and therefore commutes with any function of $H_1, H_2,$ and H_3.

The reader may want to verify that there are no other Casimir functions on $SO_3(R)$ that are functionally independent of M.

Problem 7 Let $(g(t), p(t))$ be an integral curve of the Hamiltonian vector field $\vec{M}$ that corresponds to $M = H_1^2 + H_2^2 + H_3^2$ on $SO_3(R)$. Prove that $p(t) =$ constant. Prove that the restriction of $\vec{M}$ to each co-adjoint orbit is null.

The properties of a Casimir function described in $SO_3(R)$ extend to any Lie group. Its key property is described in the next theorem.

Theorem 13 *A function f on $\mathcal{L}^*$ is a Casimir function if and only if f Poisson-commutes with any analytic function of $H_1, \ldots, H_n$, with each H_i equal to the Hamiltonian of a left-invariant vector field $\vec{A}_i$ on G corresponding to a basis $A_1, \ldots, A_n$ in $\mathcal{L}$.*

We leave the proof of this theorem to the reader.

5.1 Left-invariant optimal problems on the group of motions of a plane

The group of motions of a plane also has a Casimir function (even though it is not semisimple). Therefore its co-adjoint orbits are at most two-dimensional and account for much of the geometry of the solutions for its left-invariant optimal-control problems. We shall illustrate the basic geometry of the solutions in terms of the elastic problem of Euler and the problem of Dubins, introduced at the end of Chapter 11.

Continuing with the notation from Chapter 11, let

$$A_1 = \begin{pmatrix} 0 & 0 & 0 \\ 1 & 0 & 0 \\ 0 & 0 & 0 \end{pmatrix}, \qquad A_2 = \begin{pmatrix} 0 & 0 & 0 \\ 0 & 0 & 0 \\ 1 & 0 & 0 \end{pmatrix}, \qquad A_3 = \begin{pmatrix} 0 & 0 & 0 \\ 0 & 0 & -1 \\ 0 & 1 & 0 \end{pmatrix}$$

denote the standard basis elements for the Lie algebra of E_2. The corresponding Hamiltonians H_1, H_2, H_3 satisfy the following Poisson-bracket table:

$$\{H_1, H_2\} = 0, \qquad \{H_1, H_3\} = H_2, \qquad \{H_2, H_3\} = -H_1. \qquad (12)$$

It follows that $M = H_1^2 + H_2^2$ is a Casimir function on E_2, because $\{M, H_i\} = 0$ for each $i = 1, 2, 3$.

Consider, first, the geometry of the solutions of $H = \frac{1}{2}(a_1 H_1^2 + a_2 H_2^2 + a_3 H_3^2)$, with a_1, a_2, and a_3 positive constants. This Hamiltonian corresponds to a left-invariant Riemannian metric defined by

$$\frac{1}{2} \int \left(\frac{1}{a_1} u_1^2 + \frac{1}{a_2} u_2^2 + \frac{1}{a_3} u_3^2 \right) dt$$

over the trajectories of

$$\frac{dg}{dt} = u_1(t)\, \vec{A}_1(g) + u_2 \vec{A}_2(g) + u_3 \vec{A}_3(g).$$

Analogous to Euler's top, the geodesics $(g(t), p(t))$ are the integral curves of $\vec{H}$ for which

$$u_1(t) = a_1 H_1(p(t)), \qquad u_2(t) = a_2 H_2(p(t)), \qquad u_3(t) = a_3 H_3(p(t)).$$

The curves $p(t)$ are at the intersection of H = constant and M = constant. Regarded as surfaces in H_1, H_2, H_3 space, H = constant is an ellipsoid, and M = constant is a cylinder. The behavior of curves $p(t)$ is described by Figure 12.3. The vertices of the ellipsoid correspond to critical points $p(t)$ = constant. The two critical points on the H_3 axis are always stable. The longer axis in the horizontal plane contains the other stable equilibrium points, and the remaining axis contains hyperbolic equilibrium points. Apart from the separatrix emanating from the unstable equilibrium points, all other trajectories are closed.

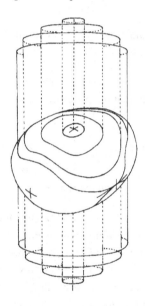

Fig. 12.3.

The Hamiltonian for Euler's elastic problem is given by $H = \frac{1}{2} H_3^2 + H_1$. H = constant corresponds to a cylindrical paraboloid in H_1, H_2, H_3 space. The intersections of these surfaces with M = constant are shown in Figure 12.4. In contrast to a left-invariant Riemannian problem, the elastic problem shows a greater diversity of solutions because of its non-holonomic character. Figure 12.4 shows geometric types of solutions. The middle figure (b) corresponds to the critical situation near which the solutions bifurcate. The mathematical pendulum, introduced in Chapter 11, has just the right energy H to reach the top in the critical case. Case (c) corresponds to an oscillating pendulum with a cutoff angle θ_c.

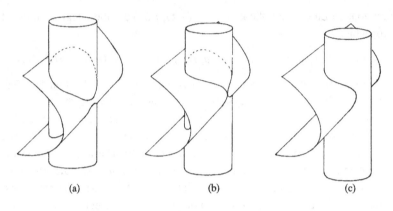

(a)   (b)   (c)

Fig. 12.4.

Case (a) corresponds to the pendulum with sufficient energy to clear the top and execute full revolutions.

The geometry of the solutions for the problem of Dubins can be seen through similar illustrations. Recall that the problem of Dubins generates the following Hamiltonians:

$$H_+ = -p_0 + H_1 + H_3, \qquad \text{valid in the region } H_3(p) > 0,$$
$$H_- = -p_0 + H_1 - H_3, \qquad \text{valid in the region } H_3(p) < 0,$$
$$H_0 = -p_0 + H_1, \qquad \text{in the region where } H_3(p) = 0.$$

Also recall that in the last case, $H_2 = 0$ contains the geodesic arcs. In all of these cases, the value of the Hamiltonian is equal to zero along its extremal curves. Note that $p_0 = 0$ is excluded from the last case, because that would imply that $H_1(p) = 0$, which in turn would imply that $p(t) = 0$.

Both the regular and the abnormal extremals are contained in the intersections of planes with the Casimir cylinder in H_1, H_2, H_3 space. Figure 12.5 depicts all the cases.

Let $\theta(t)$ be the angle defined along an extremal curve $p(t)$ by the formula $\sqrt{M} \sin\theta(t) = H_1(p(t))$ and $\sqrt{M} \cos\theta(t) = H_2(p(t))$. Then

$$\sqrt{M}\cos\theta(t)\frac{d\theta}{dt} = \frac{dH_1}{dt} = \{H_1, -p_0 + H_1 + u(t)H_3\} = u(t)\{H_1, H_3\}(p(t))$$
$$= u(t)H_2(p(t)) = (\sqrt{M}\cos\theta(t))u(t).$$

Hence

$$\frac{d\theta}{dt} = u(t) = \pm1, \quad \text{or} \quad \theta(t) = \pm t + \theta_0.$$

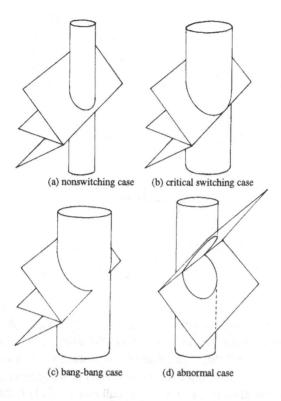

(a) nonswitching case   (b) critical switching case

(c) bang-bang case   (d) abnormal case

Fig. 12.5.

Along an abnormal extremal, $0 = H_1 \pm H_3$. The switching occurs when $H_3 = 0$, and therefore the switching times occur when $\sin(\pm t + \theta_0) = 0$. This means that the successive switchings occur at intervals all of length π. The projections of these extremals on the plane consists of concatenations of semicircles of radius 1, as shown in Figure 12.6. Figure 12.7 shows that these curves cannot have minimal length.

Problem 8 Give an analytical proof showing that concatenations of semicircles are not optimal.

5.2 Left-invariant optimal problems on $SO_3(R)$ and $SO(2, 1)$

$SO_3(R)$ and $SO(2, 1)$ are the isometry groups for a sphere and a hyperbolic plane and provide a natural setting for geometric variational problems on these non-Euclidean surfaces. Both the elastic problem of Euler and the problem of

Fig. 12.6.

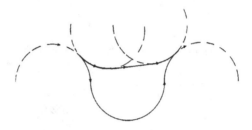

Fig. 12.7.

Dubins admit natural formulations on these spaces, and their solutions can be effectively analyzed by the theoretical framework established earlier.

We shall first recall the basic geometric facts about these groups. Because $SO_3(R)$ has already been discussed in sufficient detail in connection with the heavy top, we shall focus on $SO(2, 1)$. Recall that $SO(2, 1)$ is the group that leaves the following bilinear form in R^3 invariant:

$$\langle x, y \rangle = -x_1 y_1 + x_2 y_2 + x_3 y_3. \tag{13}$$

The Lie algebra $\mathcal{L}$ of $SO(2, 1)$ consists of all 3×3 matrices A such that

$$\langle Ax, y \rangle + \langle x, Ay \rangle = 0.$$

The reader can verify that $\mathcal{L}$ is equal to the space of matrices

$$A = \begin{pmatrix} 0 & a_1 & a_2 \\ a_1 & 0 & -a_3 \\ a_2 & a_3 & 0 \end{pmatrix}$$

for (a_1, a_2, a_3) in $\mathbb{R}^3$. Let

$$A_1 = \begin{pmatrix} 0 & 1 & 0 \\ 1 & 0 & 0 \\ 0 & 0 & 0 \end{pmatrix}, \qquad A_2 = \begin{pmatrix} 0 & 0 & 1 \\ 0 & 0 & 0 \\ 1 & 0 & 0 \end{pmatrix}, \qquad A_3 = \begin{pmatrix} 0 & 0 & 0 \\ 0 & 0 & -1 \\ 0 & 1 & 0 \end{pmatrix}.$$

Table 12.2.

[,]	A_1	A_2	A_3
A_1	0	$-\varepsilon A_3$	A_2
A_2	εA_3	0	$-A_1$
A_3	$-A_2$	A_1	0

Then

$$[A_1, A_2] = A_3, \qquad [A_3, A_1] = -A_2, \qquad [A_2, A_3] = -A_1.$$

Comparing with the corresponding Lie-bracket tables for E_2 and $SO_3(R)$, it follows that they can be expressed in terms of a single parameter ε. Table 12.2 shows the Lie brackets for all three groups, with $\varepsilon = 0$ for the group of motions of $\mathbb{R}^2$, $\varepsilon = -1$ for $SO(2, 1)$, and $\varepsilon = 1$ for $SO_3(R)$.

Let H^2 denote the hyperboloid $x_1^2 - (x_2^2 + x_3^2) = 1$, $x_1 > 0$. Because $SO(2, 1)$ leaves $\langle , \rangle$ invariant, it follows that $SO(2, 1)$ acts on H^2 by the matrix multiplication on the left; that is,

$$g \begin{pmatrix} x_1 \\ x_2 \\ x_3 \end{pmatrix} = gx.$$

Problem 9 Prove that $SO(2, 1)$ acts transitively on H^2. Also prove that $ge_1 = e_1$ if and only if

$$g = \begin{pmatrix} 1 & 0 & 0 \\ 0 & \alpha & \beta \\ 0 & -\beta & \alpha \end{pmatrix}, \quad \text{with} \quad \alpha^2 + \beta^2 = 1.$$

It follows from Problem 9 that $H^2 = SO(2, 1)/SO_2$, with SO_2 equal to the group

$$\begin{pmatrix} 1 & 0 & 0 \\ 0 & \alpha & \beta \\ 0 & -\beta & \alpha \end{pmatrix}, \qquad \alpha^2 + \beta^2 = 1.$$

$SO_3(R)$ acts on S^2 in an obvious way, and one can show that $S^2 = SO_3(R)/SO_2$ (R). Both S^2 and H^2 admit a left-invariant metric defined by

$$\varepsilon(dx_1)^2 + (dx_2)^2 + (dx_3)^2, \qquad \varepsilon = \pm 1,$$

where ε is equal to the curvature of each of these surfaces: $\varepsilon = 1$ on S^2, and $\varepsilon = -1$ on H^2. These symmetric spaces are respectively called elliptic and hyperbolic.

The Serret-Frenet system for each of these symmetric spaces, including the Euclidean case, can be expressed by the single equation

$$\frac{dg}{dt} = \vec{A}_1(g) + u(t)\vec{A}_3(g),\qquad(14)$$

with the notational understanding that

$$A_1 = \begin{pmatrix} 0 & -\varepsilon & 0 \\ 1 & 0 & 0 \\ 0 & 0 & 0 \end{pmatrix} \quad\text{and}\quad A_3 = \begin{pmatrix} 0 & 0 & 0 \\ 0 & 0 & -1 \\ 0 & 1 & 0 \end{pmatrix},$$

and also that (14) is an equation in $SO(2,1)$ for $\varepsilon = -1$, and correspondingly in $SO_3(R)$ for $\varepsilon = 1$.

The elastic problem for each of these groups consists in minimizing $\frac{1}{2}\int u^2 dt$ over the trajectories of (14), and the problem of Dubins becomes the time-optimal problem for (14) under an additional constraint that $|u(t)| \le 1$.

Denote by H_1, H_2, H_3 the Hamiltonians that correspond to $\vec{A}_1$, $\vec{A}_2$, and $\vec{A}_3$. It is easy to verify from Table 12.2 that $M = H_1^2 + H_2^2 + \varepsilon H_3^2$ is a Casimir function for the group that corresponds to ε. This finding agrees with the information discussed earlier for $SO_3(R)$ and E_2. For $SO(2,1)$, M corresponds to several geometric types. For $M > 0$, $M = H_1^2 + H_2^2 - H_3^2$ is a hyperboloid of one sheet. For $M = 0$, $0 = H_1^2 + H_2^2 - H_3^2$ is a double cone. For $M < 0$, $M = H_1^2 + H_2^2 - H_3^2$ is a hyperboloid of two sheets. The geometric types of solutions for the elastic problem and the problem of Dubins on $SO_3(R)$ correspond to the intersections between the momentum sphere and the appropriate Hamiltonian, as shown in Figures 12.8 and 12.9.

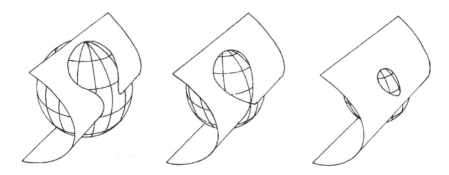

Fig. 12.8.

However, on $SO(2,1)$ the situation is a bit more complex, because of different shapes of the Casimir functions. Figure 12.10 shows some typical cases for the

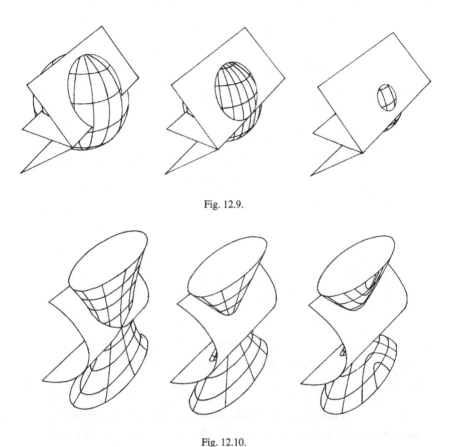

Fig. 12.9.

Fig. 12.10.

elastic problem, and Figure 12.11 shows some typical cases for the problem of Dubins.

Finally, note that the trace form is nondegenerate on each $SO_3(R)$ and $SO(2, 1)$, and therefore the co-adjoint orbits can be identified with the adjoint orbits to yield $g(t)P(t)g^{-1}(t) = $ constant for each integral curve $(g(t), P(t))$ in $G \times \mathcal{L}(G)$ of a left-invariant Hamiltonian vector field.

There are further symmetry considerations that complement and clarify the material in this chapter and that ultimately lead to natural integration procedures for left-invariant optimal problems on Lie groups. We shall return to the integration of the foregoing Hamiltonian systems in the next chapter, after first establishing the theoretical prerequisites.

Problem 10 Show that the control system (14) is not controllable on $SO(2, 1)$ when $|u(t)| \le 1$.

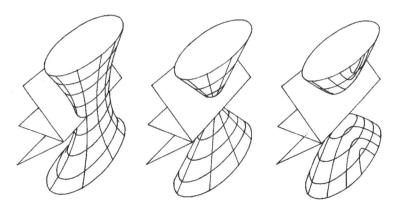

Fig. 12.11.

Problem 11 Show that (14) becomes controllable on $SO(2, 1)$ whenever the bound on the control is relaxed to $|u(t)| \leq 1 + \delta$, with $\delta > 0$.

Notes and sources

This chapter is a synthesis of classic results from the theory of Hamiltonian systems put into focus through the maximum principle, the left-invariant realization of the symplectic form, and the Hamiltonian equations in Theorem 1.

The theory of left-invariant Hamiltonian systems has a long history, originating with a paper by Poincaré (1901). The literature on this subject is vast. The interested reader can consult Arnold (1976, Appendix 2) for the literature on left-invariant metrics and their generalizations to hydrodynamics. The book by Abraham and Marsden (1985) has an extensive treatment of mechanical systems having either left or right invariance (systems with symmetries). In particular, their book contains the left-invariant realization of the symplectic form. The papers of Bogoyavlenski (1984) and Bobenko et al. (1989) treated left-invariant Hamiltonian systems on semisimple Lie groups through the formalism adopted in this chapter. Those papers also used the structure of semisimple Lie groups to express their Hamiltonian equations in the Lax pair form, and then extracted the integrals of motion through the isospectral properties of that representation.

Most of the applications considered in this chapter, particularly the elastic problem of Kirchhoff, were essentially motivated by my recent papers on left-invariant optimal-control problems on Lie groups (Jurdjevic, 1993a,b) originating with the paper on non-Euclidean elastica (Jurdjevic, 1995a).

13
Symmetry, integrability, and the Hamilton-Jacobi theory

Problems of optimal control, like the problems of its classic predecessor, mathematical physics, rely on the integration of Hamiltonian differential equations for their resolution. That remarkable discovery goes back to the work of R. W. Hamilton and C. G. Jacobi concerning the problems of classic mechanics in the 1830s. The content of their publications, subsequently known as the Hamilton-Jacobi theory, had profound impact on subsequent developments in mathematical physics and was the principal source of inspiration for the present theory of Hamiltonian systems. The maximum principle and contemporary optimal control theory are also anchored in the Hamilton-Jacobi theory, and the main issue before us is to understand those classic developments in modern geometric terms and make them accessible for problems of optimal control.

We shall begin this chapter with a related topic: The connection between symmetry and optimality. We shall first arrive at the appropriate definition for "symmetry," which extends the classic theorem of E. Noether concerning the existence of extra integrals of motion. An extension of that theorem implies, in particular, that a right-invariant vector field is a symmetry for any left-invariant control problem, and consequently the Hamiltonian of the right-invariant vector field is an integral of motion for the extremal system induced by the optimal problem.

The existence of extra integrals of motion for a given Hamiltonian system makes a link with another classic topic: the theory of integrable Hamiltonian systems. That theory, which was discussed extensively by Poincaré (1892) in his treatise on celestial mechanics and later by Carathéodory (1935) in his book on the calculus of variations, is most naturally expressed through the geometry of Lagrangian submanifolds of cotangent bundles.

Any Hamiltonian flow induces a flow on the space of Lagrangian manifolds, in much the same manner as in the case of linear Lagrangians and quadratic Hamiltonians discussed in Chapter 8 on the Riccati equation. Similar to linear

Lagrangians, which are graphs of symmetric matrices, Lagrangian manifolds that project diffeomorphically onto M are graphs of closed differential forms. Because closed differential forms are defined locally by functions on M, it follows that each Lagrangian manifold that projects onto M is locally represented by a function S. S is unique up to a constant.

As in the case of the Riccati equation, any Lagrangian manifold L_0 flows to a Lagrangian manifold L_t under the Hamiltonian flow induced by a function H on T^*M. If L_0 projects onto M, so does L_t for t sufficiently small. Let S_t denote a function on M that locally represents L_t in a neighborhood of some point. It will be shown that S_t is a solution of the Hamilton-Jacobi partial differential equation

$$\frac{\partial S_t}{\partial t}(x) + H\left(x, \frac{\partial S_t}{\partial x}\right) = 0, \tag{1}$$

with $x = (x_1, \ldots, x_n)$ and $p = (p_1, \ldots, p_n)$ denoting any choice of canonical coordinates, and

$$p_i = \frac{\partial S_t}{\partial x_i}(x), \qquad i = 1, \ldots, n.$$

Poincaré referred to the foregoing statement as the first theorem of Jacobi. In the literature on mechanics, S_t is called the "action."

Any $n - 1$ independent integrals of motion $h_1, \ldots, h_{n-1}$ of a given Hamiltonian H that also Poisson-commute with each other define a Lagrangian manifold: $H = $ constant, $h_1 = $ constant, $\ldots, h_{n-1} = $ constant. This Lagrangian manifold is invariant not only under $\vec{H}$ but also under the Hamiltonian flow of each h_i, because all these functions are symmetries of each other. Denoting by S the function that locally represents L in a neighborhood of a given point, it then follows, by the first theorem of Jacobi, that

$$H\left(x, \frac{\partial S}{\partial x}\right) = \text{constant}, \qquad h_i\left(x, \frac{\partial S}{\partial x}\right) = \text{constant} \quad \text{for } i = 1, \ldots, n-1.$$

The function S is a function of x and of the constants $y_1, \ldots, y_n$ that define the Lagrangian $H = y_n, h_i = y_i, i = 1, \ldots, n - 1$. Jacobi also showed that S defines a new set of canonical coordinates $y_1, \ldots, y_n, q_1, \ldots, q_n$ through the relations

$$p_i = \frac{\partial S}{\partial x_i}(x, y) \quad \text{and} \quad -q_i = \frac{\partial S}{\partial y_i}(x, y), \qquad i = 1, \ldots, n.$$

In terms of these coordinates, $H = y_n$, and therefore the integral curves of H can be expressed as

$$\frac{dy_i}{dt} = \frac{\partial H}{\partial q_i} = 0 \quad \text{and} \quad \frac{dq_i}{dt} = -\frac{\partial H}{\partial y_i} = -\delta_{in}.$$

A Hamiltonian system that admits $n - 1$ integrals of motion that Poisson-commute with each other is called completely integrable. The preceding formulas show that there exist canonical coordinates in terms of which the Hamiltonian equations of a completely integrable system can be integrated by a trivial procedure. Unfortunately, the theory does not provide a constructive route to those coordinates. In this chapter we shall also examine a geometric procedure for integrating the Hamiltonian equations of completely integrable systems on three-dimensional Lie groups.

1 Symmetry, Noether's theorem, and the maximum principle

Symmetry, a fundamental issue in any mathematical study, has a particular meaning in Lagrangian and Hamiltonian mechanics that naturally extends to optimal-control problems and leads to extra integrals of motion. The definitions are extensions of the following simple situation: A vector field X is said to be *a symmetry* of another vector field Y if for any integral curve $\sigma(t)$ of Y, $(\exp \lambda X)(\sigma(t))$ is also an integral curve of X for each value of the parameter λ. Stated in slightly different language, a symmetry preserves the solution curves of Y, or the flow of X permutes the integral curves of Y.

Expressing the conditions of symmetry in terms of the exponential maps yields

$$(\exp \lambda X)\,((\exp tY)(x)) = (\exp tY)((\exp \lambda X)(x)).$$

Therefore, X is a symmetry for Y if and only if X and Y commute, which in turn implies that Y is also a symmetry for X; therefore, symmetry is a reflexive relation on the pairs of vector fields.

There are several ways in which the notion of symmetry could be extended to families of vector fields. The most restrictive definition would require that a symmetry X commute with each vector field in the family. A much weaker notion of symmetry would require that the flow of a symmetry permute the reachable sets of the family. One could also consider the closure of the reachable sets as the criterion of symmetry, in which case the earlier notion of a normalizer would also be seen as a symmetry. For the time being, we shall adopt a definition of intermediate generality that is adequate for the applications that follow:

Definition 1 A complete vector field X is called a symmetry for a family of vector fields $\mathcal{F}$ if

$$(\exp \lambda X)_*(\mathcal{F}((\exp -\lambda X)(x))) \subset \mathcal{F}(x)$$

for all real numbers λ and each x in M.

It follows from this definition that for any trajectory σ of $\mathcal{F}$, $\xi_\lambda = (\exp \lambda X)(\sigma)$ is also a trajectory of $\mathcal{F}$. (A trajectory of a family of vector fields $\mathcal{F}$ is any absolutely continuous curve in M such that $(d\sigma/dt)(t) \in \mathcal{F}(\sigma(t))$ for almost all t in the domain of σ.) The argument is as follow:

$$\frac{d\xi_\lambda}{dt}(t) = (\exp \lambda X)_*\left(\frac{d\sigma}{dt}(t)\right) \in (\exp \lambda X)_*(\mathcal{F}(\sigma(t)))$$

$$= (\exp \lambda X)_*(\mathcal{F}((\exp -\lambda X)(\xi_\lambda))) \subset \mathcal{F}(\xi_\lambda(t)).$$

Therefore, ξ_λ is a trajectory of $\mathcal{F}$.

In order to extend the notion of symmetry to optimal-control problems, it is convenient to assume that the family $\mathcal{F}$ is parametrized by controls that take values in a set U in $\mathbb{R}^m$; that is,

$$\mathcal{F}(x) = \{F(x, u) : u \in U\}.$$

Assuming that X is a symmetry for $\mathcal{F}$, then

$$(\exp \lambda X)_*(F((\exp -\lambda X)(x), u)) \subset \{F(x, v) : v \in U\},$$

and therefore there exists a control u_λ for which

$$(\exp \lambda X)_*(F((\exp -\lambda X)(x), u)) = F(x, u_\lambda). \tag{2}$$

We shall now extend the notion of symmetry to the optimal problem of minimizing $\int_0^T f(x, u)\, dt$ over the trajectories of $dx/dt = F(x, u)$, with the controls u taking values in a fixed set U. It will be convenient to refer to this problem as the optimal problem defined by (f, F, U).

Definition 2 A complete vector field X is a symmetry for the optimal problem defined by (f, F, U) if X is a symmetry for F and if $f((\exp \lambda X)(x), u_\lambda) = f(x, u)$ for all x in M and u in U.

It follows that $f(\xi_\lambda(t), u_\lambda(t)) = f(\sigma(t), u(t))$ for the trajectories (σ, u) and (ξ_λ, u_λ), related by $\xi_\lambda = (\exp \lambda X)(\sigma)$.

We have already encountered systems with symmetries without making any explicit remarks about them. The first such occurrence of symmetry is described by

Theorem 1 *Each right-invariant vector field on a Lie group G is a symmetry for any left-invariant optimal-control problem on G.*

Proof Let $F(\cdot, u)$ be any family of left-invariant vector fields on a Lie group G. Recall that this means that $F(g, u) = dL_g F(e, u)$ for each $g \in G$, where dL_g denotes the tangent map of the left-translation $L_g(x) = gx$. Let $\vec{X}$ denote a right-invariant vector field whose value at the group identity is equal to X. Then

$$(\exp \lambda \vec{X})(g) = (\exp \lambda X)(g) = L_{(\exp \lambda X)}(g),$$

and therefore

$$
\begin{aligned}
(\exp \lambda \vec{X})_* &(F((\exp -\lambda \vec{X})(g), u)) \\
&= dL_{(\exp \lambda X)}(F((\exp -\lambda X)(g), u)) \\
&= F(g, u).
\end{aligned}
$$

Thus $u_\lambda(t) = u(t)$. Because the cost function is constant over G, it follows that $f(\xi_\lambda(t), u_\lambda) = f(\sigma, u)$. The proof is now finished. ∎

   In addition to right-invariant symmetries, the elastic problem in $\mathbb{R}^3$ (or any of its extensions to non-Euclidean spaces) has another symmetry when the constants c_2 and c_3 are equal. This case will be called the *symmetric elastic problem*. Recall that the elastic problem is defined on the group of motions of $\mathbb{R}^3$ by

$$\frac{dg}{dt} = \vec{B}_1(g) + \sum_{i=1}^{3} u_i \vec{A}_i(g),$$

with $\vec{B}_1$ and $\vec{A}_i$ denoting the left-invariant vector fields corresponding to B_1 and A_i, with

$$
B_1 = \begin{pmatrix} 0 & 0 & 0 & 0 \\ 1 & 0 & 0 & 0 \\ 0 & 0 & 0 & 0 \\ 0 & 0 & 0 & 0 \end{pmatrix}, \qquad
A_1 = \begin{pmatrix} 0 & 0 & 0 & 0 \\ 0 & 0 & 0 & 0 \\ 0 & 0 & 0 & -1 \\ 0 & 0 & 1 & 0 \end{pmatrix},
$$

$$
A_2 = \begin{pmatrix} 0 & 0 & 0 & 0 \\ 0 & 0 & 0 & 1 \\ 0 & 0 & 0 & 0 \\ 0 & -1 & 0 & 0 \end{pmatrix}, \qquad
A_3 = \begin{pmatrix} 0 & 0 & 0 & 0 \\ 0 & 0 & 1 & 0 \\ 0 & 1 & 0 & 0 \\ 0 & 0 & 0 & 0 \end{pmatrix}.
$$

The cost function for this problem is the elastic energy $f = \frac{1}{2}(c_1 u_1^2 + c_2 u_2^2 + c_3 u_3^2)$. We shall now show that the left-invariant vector field $\vec{A}_1(t)$ is a symmetry for this variational problem. Note that

$$(\exp \lambda \vec{A}_1)_* \left(\vec{B}_1 + \sum_{i=1}^{3} u_i \vec{A}_i \right) (\exp -\lambda \vec{A}_1)$$

is a left-invariant vector field whose value at the group identity is given by the following matrices:

$$e^{\lambda A_1} B_1 e^{-\lambda A_1} + \sum_{i=1}^{3} u_i e^{\lambda A_1} A_i e^{-\lambda A_1}.$$

Because A_1 and B_1 commute, $e^{\lambda A_1} B_1 e^{-\lambda A_1} = B_1$,

$$e^{\lambda A_1} = \begin{pmatrix} 1 & 0 & 0 & 0 \\ 0 & 1 & 0 & 0 \\ 0 & 0 & \cos\lambda & -\sin\lambda \\ 0 & 0 & \sin\lambda & \cos\lambda \end{pmatrix},$$

and therefore

$$\sum_{i=1}^{3} u_i e^{\lambda A_1} A_i e^{-\lambda A_1} = (u_2\cos\lambda - u_3\sin\lambda)A_2 + (u_3\cos\lambda + u_2\sin\lambda)A_3 + A_1.$$

The preceding relations define a new control function $u(\lambda) = (u_1, u_2\cos\lambda - u_3\sin\lambda, u_3\cos\lambda + u_2\sin\lambda)$. Therefore, $\exp \lambda \vec{A}_1$ is a symmetry for the control system. It is also a symmetry for the optimal problem, because

$$c_1 u_1^2 + c_2 u_2^2 + c_3 u_3^2 = c_1 u_1^2(\lambda) + c_2 u_2^2(\lambda) + c_3 u_3^2(\lambda)$$

when $c_2 = c_3$.

According to a famous theorem of E. Noether, each symmetry leads to an integral of motion. The next theorem shows that Noether's result extends to any optimal control system (f, F, U).

Denote the corresponding Hamiltonian $p_0 f(x, u) + \xi F(x, u)$ by $H_u(\xi)$ for each $\xi \in T_x^* M$. Recall that an absolutely continuous curve $\xi(t)$ is an extremal curve for this problem if

(i) there exists a control function $u(t)$ such that $\xi(t)$ is an integral curve of the Hamiltonian vector field $\vec{H}_{u(t)}$ corresponding to $H_{u(t)}$ and

(ii) $H_{u(t)}(\xi(t)) \geq p_0 f(x(t), v) + \xi(t)F(x(t), v)$ for almost all t and all v in U, where $x(t)$ denotes the projection of $\xi(t)$ on M.

With these notations in place, Noether's control theorem is as follows:

Theorem 2 *Let H_X denote the Hamiltonian of a symmetry X for the optimal problem (f, F, U). Then for any extremal curve $\xi(t)$ of the optimal problem (f, F, U), $\{H_X, H_{u(t)}\}(\xi(t)) = 0$. Consequently, $H_X(\xi(t)) = constant$.*

Proof Let $u(t)$ denote the extremal control that defines $\xi(t)$, and let $u_\lambda(t)$ denote the control that satisfies $(\exp \lambda X)_*(F((\exp -\lambda X)(x(t)), u(t))) = F(x(t), u_\lambda(t))$. Then

$$H_{u(t)}(\xi(t)) = p_0 f(x(t), u(t)) + \xi(t)F(x(t), u(t))$$
$$\geq p_0 f(x(t), u_\lambda(t)) + \xi(t)F(x(t), u_\lambda(t))$$

for all λ and almost all t. It follows that the expression

$$\alpha_t(\lambda) = p_0 f(x(t), u_\lambda(t)) + \xi(t)F(x(t), u_\lambda(t))$$

attains its maximum for $\lambda = 0$. Therefore, $(d/d\lambda)\alpha_t(\lambda)$ is equal to zero at $\lambda = 0$. We shall presently find that $(d/d\lambda)\alpha_t(\lambda)|_{\lambda=0} = \{H_X, H_{u(t)}\}(\xi(t))$.

Because $f((\exp \lambda X)(x(t)), u_\lambda(t)) = f(x(t), u(t))$, it follows that $(d/d\lambda)$ $f((\exp \lambda X)(x(t)), u_\lambda(t)) = 0$. Therefore, $(Xf)(x(t), u(t)) + (\partial f/\partial u)(x(t), u(t))(du_\lambda/d\lambda)(t) = 0$. Hence

$$\frac{d}{d\lambda} f(x(t), u_\lambda(t))|_{\lambda=0} = -(Xf)(x(t), u(t)).$$

Differentiating $F(x(t), u_\lambda(t)) = (\exp \lambda X)_*(F(\exp -\lambda X)(x(t)), u(t))$ with respect to λ gives

$$\frac{d}{d\lambda} F(x(t), u_\lambda(t))|_{\lambda=0} = [X, F](x(t), u(t)).$$

Therefore,

$$\frac{d\alpha_t}{d\lambda}(\lambda)|_{\lambda=0} = -p_0(Xf)(x(t), u(t)) + \xi(t)[X, F](x(t), u(t)).$$

But $-p_0(Xf)(x(t), u(t)) + \xi(t)[X, F](x(t), u(t)) = \{H_X, H_{u(t)}\}(\xi(t))$, and therefore the proof is finished. ∎

Each regular extremal curve of the elastic problem is an integral curve of the Hamiltonian vector field corresponding to $H = \frac{1}{2}(H_1^2/c_1 + H_2^2/c_2 + H_3^2/c_3) + h_1$. Then

$$\{H_1, H\} = \frac{H_2}{c_2}\{H_1, H_2\} + \frac{H_3}{c_3}\{H_1, H_3\} + \{H_1, h_1\} = H_2 H_3\left(\frac{1}{c_3} - \frac{1}{c_2}\right) = 0$$

when $c_2 = c_3$. This calculation confirms the results of Theorem 2, because $\vec{A}_1$ is a symmetry for the elastic problem.

For left-invariant Hamiltonian systems on a Lie group G, Theorem 2 provides explanations, complementary to those presented in Chapter 12, concerning the evolution of its solution curves along the co-adjoint orbits of G, as can be seen through the following arguments:

Let $(g(t), p(t))$ denote an extremal curve corresponding to a left-invariant optimal problem on G in the representation $T^*G = G \times \mathcal{L}^*$. According to Theorem 1, each right-invariant vector field X is a symmetry for this problem, and therefore $H_X(\xi(t)) = $ constant. Then

$$H_X(g(t), p(t)) = p(t)((dL_{g(t)})^{-1} R_{g(t)} X(e)) = p(t)(\mathrm{Ad}_{g(t)^{-1}}(X(e)))$$
$$= \left(\mathrm{Ad}^*_{g(t)^{-1}}(p(t))(X(e))\right).$$

Thus, $H_X(g(t), p(t)) = $ constant can hold only if $\mathrm{Ad}^*_{g(t)^{-1}}(p(t)) = f$ for some linear function f on $\mathcal{L}$. But then $p(t) = \mathrm{Ad}^*_{g(t)}(f)$.

The foregoing argument shows that the expression $p(t) = \mathrm{Ad}^*_{g(t)}(f)$, obtained in Chapter 12, can also be seen as a consequence of the left-invariant symmetries.

2 The geometry of Lagrangian manifolds and the Hamilton-Jacobi theory

The integrability theory of Hamiltonian systems is another classic topic inseparably linked with symmetry and the existence of extra integrals of motion. As we saw in the preceding chapters, problems of optimal control through the maximum principle lead to their own Hamiltonian systems and make new theoretical demands. In order to deal effectively with these demands, it becomes essential to understand the classic theory first.

The classic theory of integration of Hamiltonian systems is based on the Hamilton-Jacobi theory, which from a contemporary perspective is most easily described through the geometry of Lagrangian manifolds. The developments that follow will contain the essential ingredients of this theory.

Definition 3 A family Φ of functions on T^*M (or any symplectic manifold) is said to be in involution if $\{f, g\} = 0$ for any f and g in Φ.

As before, we shall use $\vec{f}$ to denote the Hamiltonian vector field corresponding to a function f on T^*M. It will then be convenient to denote by $\vec{\Phi}$ the family of Hamiltonian vector fields generated by the elements of Φ. That is, $\vec{\Phi} = \{\vec{f} : f \in \Phi\}$. If Φ is in involution, then $\vec{\Phi}$ consists of commuting vector fields, because $[\vec{f}, \vec{g}] = \overrightarrow{\{f, g\}} = 0$ for any f and g in Φ.

Definition 4 The functions $f_1, \ldots, f_m$ are said to be independent at ξ if their differentials are linearly independent at ξ.

Theorem 3 *Let Φ denote any family of smooth functions on T^*M in involution. Then the orbit of the corresponding family of Hamiltonian vector fields $\vec{\Phi}$ through a point ξ is an m-dimensional submanifold of T^*M, with m equal to the maximal number of independent functions Φ at ξ.*

Proof Because the elements of Φ are in involution, the corresponding Hamiltonian vector fields in $\vec{\Phi}$ form a set of commuting vector fields. Hence, the tangent space of the orbit through ξ is equal to the linear span of $\{\vec{f}(\xi) : f \in \Phi\}$.

Let $\vec{f}_1, \ldots, \vec{f}_m$ be the maximal number of vector fields of $\vec{\Phi}$ that are linearly independent at ξ. Then $f_1, \ldots, f_m$ are linearly independent at ξ, because $(df_i(\xi))(v) = w_\xi(\vec{f}_i(\xi), v)$ for each i and each v in $T_\xi(T^*M)$. End of the proof. ∎

Corollary *The maximal number of independent functions at ξ that are also in involution cannot exceed the dimension of M.*

Proof Each orbit of a family of Hamiltonian vector fields that correspond to functions in involution is an isotropic submanifold of T^*M; that is, the symplectic form restricted to each orbit is identically zero for any pair of tangent vectors. The maximal dimension of an isotropic submanifold is equal to the dimension of M. End of the proof. ∎

Definition 5 A submanifold N of T^*M is called Lagrangian if the restriction of the symplectic form to TN is identically zero, that is, if N is isotropic, and if the dimension of N is equal to the dimension of M. Alternatively, an isotropic submanifold of T^*M of maximal dimension is called Lagrangian.

The next theorem summarizes the most relevant property of functions in involution in terms of Lagrangian submanifolds.

Theorem 4 Let $f_1, \ldots, f_m$ be any functions on the cotangent bundle of M that are in involution. Assume that m is equal to the dimension of M. Then each point ξ in the level set Ω defined by $f_1 = \text{constant}, \ldots, f_m = \text{constant}$, for which $df_1(\xi), \ldots, df_m(\xi)$ are linearly independent, is contained in a Lagrangian submanifold of T^*M contained in Ω.

Proof The proof is left to the reader.   ∎

Lagrangian submanifolds of T^*M that project diffeomorphically onto open subsets of M are of particular interest in the integrability theory of Hamiltonian systems.

Definition 6 Let H be any function on T^*M. Suppose that H admits $n - 1$ independent integrals of motion $h_1, \ldots, h_{n-1}$ that are also in involution with each other. Then H is said to be completely integrable.

Example 1 Each left-invariant Hamiltonian on a three-dimensional Lie group G is completely integrable, for the following reasons: Each semisimple group G has a Casimir function h that is in the center of the Poisson-Lie algebra of functions defined on the dual $\mathcal{L}^*$ of the Lie algebra of G. Any left-invariant Hamiltonian H on G is a function on $\mathcal{L}^*$ and therefore Poisson-commutes with h. The remaining integral is the Hamiltonian h_r corresponding to any right-invariant vector field X_r on G. Even though h_r is not a function on $\mathcal{L}^*$, it Poisson-commutes with all functions on $\mathcal{L}^*$. Therefore the three integrals H, h, and h_r are all in involution, and hence H is completely integrable.

Example 2 A symmetric elastic problem in the group of motions of $\mathbb{R}^3$ is completely integrable.

Proof We have already shown that H_1 is its integral of motion for

$$H = \frac{1}{2}\left(\frac{H_1^2}{c_1} + \frac{H_2^2}{c_2} + \frac{H_3^2}{c_3}\right) + h_1.$$

The group of motions of $\mathbb{R}^3$ has two Casimir elements: $M_1 = H_1^2 + H_2^2 + H_3^2$ and $M_2 = H_1 h_1 + H_2 h_2 + H_3 h_3$. The functions H, H_1, M_1, and M_2 are all in involution.

The remaining two integrals of motion, which are in involution with each other, are the Hamiltonians of any two commuting right-invariant vector fields.

For instance, the right-invariant vector fields corresponding to the matrices

$$
B_1 = \begin{pmatrix} 0 & 0 & 0 & 0 \\ 1 & 0 & 0 & 0 \\ 0 & 0 & 0 & 0 \\ 0 & 0 & 0 & 0 \end{pmatrix} \quad \text{and} \quad A_1 = \begin{pmatrix} 0 & 0 & 0 & 0 \\ 0 & 0 & 0 & 0 \\ 0 & 0 & 0 & -1 \\ 0 & 0 & 1 & 0 \end{pmatrix}
$$

commute. Remember that left-invariant vector fields commute with right-invariant vector fields, and therefore their Hamiltonians Poisson-commute. ■

"Complete integrability" is a term that originated with Hamilton and Jacobi. The next theorem provides a geometric base for the Hamilton-Jacobi theory.

Theorem 5 *Any Lagrangian submanifold of T^*M that projects diffeomorphically onto M is the graph of a closed differential form on M. Conversely, the graph of any closed differential form on M is a Lagrangian submanifold of T^*M.*

For the proof of this theorem, the reader is referred to Abraham and Marsden (1985, p. 410).

Because closed differential forms are defined locally by functions, it follows that any point x in M has a neighborhood and has a function S defined on that neighborhood such that a given Lagrangian L is equal to the graph of dS in a neighborhood of $\pi^{-1}(x) \cap L$. S is called a *generating function* for the Lagrangian L. Strictly speaking, S is generating L only locally in a given neighborhood of $\xi = \pi^{-1}(x)$, and therefore any use of generating functions leads to local results only.

The reader should recall the earlier discussion of Lagrangian subspaces in connection with the Riccati equation and can view the present development as a nonlinear extension of the corresponding formalism. The space of all Lagrangian submanifolds of T^*M forms a vector bundle over M, and M itself is a Lagrangian submanifold, equal to the zero section of T^*M.

Theorem 6 *Let H be a smooth function on T^*M, and let L_0 be a given Lagrangian submanifold of T^*M that projects diffeomorphically onto M.*

(a) *$L_t = \{(\exp t\vec{H})(\xi) : \xi \in L_0\}$ is a Lagrangian submanifold of T^*M for each t.*

(b) *There exists an interval $[0, T]$, with $T > 0$, such that L_t projects diffeomorphically onto M for each t in $[0, T]$.*

(c) *For each x_0 in M there exists a generating function S_t for L_t defined in a neighborhood of x_0 that is a solution of the Hamilton-Jacobi equation*

$$\frac{\partial}{\partial t} S_t(x) + H(x, dS_t(x)) = 0.$$

Remark 1 The generating functions for Lagrangian manifolds that vary with time are determined up to an arbitrary function of time. Part (c) says that this function can be suitably chosen so that S_t will satisfy the Hamilton-Jacobi equation. ∎

Remark 2 Because H does not depend explicitly on time, it is enough to prove (c) for small time. But then the arguments are local and thus can be carried out in terms of canonical coordinates. In particular, if $p_1, \ldots, p_n$ denote the coordinates relative to the basis $dx_1, \ldots, dx_n$ induced by the coordinates $x_1, \ldots, x_n$ in M, then $dS_t = \sum_{i=1}^{n}(\partial S_t/\partial x_i)\, dx_i$. On L_t, $p_i = (\partial S_t/\partial x_i)(x_1, \ldots, x_n)$, $i = 1, \ldots, n$. In these coordinates, the Hamilton-Jacobi equation becomes

$$\frac{\partial}{\partial t} S_t(x_1, \ldots, x_n) + H\left(x_1, \ldots, x_n, \frac{\partial S_t}{\partial x_1}, \ldots, \frac{\partial S_t}{\partial x_n}\right) = 0. \quad ∎$$

Proof The symplectic form is invariant under the flow of $\vec{H}$; that is,

$$\omega_{(\exp t \vec{H})(\xi)}((\exp t \vec{H})_*(V_1), (\exp t \vec{H})_*(V_2)) = \omega_\xi(V_1, V_2)$$

for any vectors V_1 and V_2 in $T_\xi(T^*M)$. In particular, when V_1 and V_2 are tangent vectors to L_0 at ξ, then $\omega_\xi(V_1, V_2) = 0$. Because $\exp t \vec{H}$ is a diffeomorphism, it follows that L_t is isotropic and of the same dimension as L_0. Therefore, L_t is a Lagrangian submanifold for each t.

Property (b) is a consequence of transversality. If L_t projects diffeomorphically on M, then L_τ projects diffeomorphically on M for all τ with $|t - \tau| < \epsilon$.

It remains to prove part (c), which we shall prove locally in terms of canonical coordinates. So assume that $M = \mathbb{R}^n$ and that H is a function of $x_1, \ldots, x_n$, $p_1, \ldots, p_n$.

Let us denote by $x(\bar{x}, \bar{p}, t)$, $p(\bar{x}, \bar{p}, t)$ the general solution of $\vec{H}$; that is,

$$\frac{\partial x}{\partial t} = \frac{\partial H}{\partial p}(x, p), \qquad \frac{\partial p}{\partial t} = -\frac{\partial H}{\partial x}(x, p) \quad \text{for all } t,$$

and $x(\bar{x}, \bar{p}, 0) = \bar{x}$, $p(\bar{x}, \bar{p}, 0) = \bar{p}$.

Let S_t be any choice of generating functions for each Lagrangian submanifold L_t in $\mathbb{R}^n \times \mathbb{R}^n$. For notational simplicity we shall assume that each function S_t is defined on the entire $\mathbb{R}^n$, rather than just an open subset of $\mathbb{R}^n$.

Let x_0 denote an arbitrary point of $\mathbb{R}^n$, which will be held fixed, and consider arbitrary curves $y(\tau)$ in $\mathbb{R}^n$ based at x_0, that is, $y(0) = x_0$. It will be assumed that the parameter τ varies in $[0, 1]$, and we shall designate the terminal point $y(1)$ by x.

Each curve y can be uniquely lifted to a curve $\gamma(\tau) = (y(\tau), (\partial S_t/\partial x)(y(t)))$ in L_t. But then there exists a unique curve γ_t in L_0 such that $(\exp t\vec{H})(\gamma_t(\tau)) = \gamma(\tau)$.

Let $\gamma_t(\tau) = (\bar{x}_t(\tau), \bar{p}_t(\tau))$, with $\bar{p}_t(\tau) = (\partial S_0/\partial x)(\bar{x}_t(\tau))$ for all τ in $[0, 1]$. It then follows that

$$x(\bar{x}_t(\tau), \bar{p}_t(\tau), t) = y(\tau) \quad \text{and} \quad p(\bar{x}_t(\tau), \bar{p}_t(\tau), t) = \frac{\partial S_t}{\partial x}(y(\tau)) \quad (3)$$

for all τ in $[0, 1]$. Figure 13.1 shows these correspondences. Then

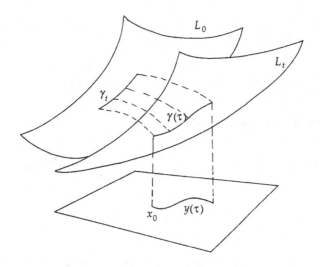

Fig. 13.1.

$$\int_0^1 \frac{\partial S_t}{\partial x}(y(\tau))\frac{dy}{d\tau}(\tau)\, d\tau = \int_0^1 \frac{d}{dt}S_t(y(\tau))\, d\tau = S_t(x) - S_t(x_0).$$

Combining the foregoing formula with (3) leads to $S_t(x) - S_t(x_0) = \int_0^1 p(\bar{x}_t(\tau),$

$\bar{p}_t(\tau), t)(dy/d\tau)(\tau)\, d\tau$. Differentiating this relation with respect to t gives

$$\frac{\partial}{\partial t} S_t(x) - \frac{\partial S_t}{\partial x}(x_0) = \int_0^1 \frac{d}{dt} p(\bar{x}_t(\tau), \bar{p}_t(\tau), t)\frac{dy}{d\tau}(\tau)\, d\tau. \qquad (4)$$

In order to differentiate $p(\bar{x}_t(\tau), \bar{p}_t(\tau), t)$ relative to t, it is convenient to begin with

$$\frac{d}{dt}(\exp t\vec{H})(\bar{x}_t(\tau), \bar{p}_t(\tau)) = \vec{H}(\exp t\vec{H})(\bar{x}_t(\tau), \bar{p}_t(\tau))$$

$$+ (\exp t\vec{H})_* \left(\frac{d\bar{x}_t}{dt}(\tau), \frac{d\bar{p}_t}{dt}(\tau) \right).$$

Let $(V_x(t,\tau), V_p(t,\tau))$ denote the tangent vector equal to $(\exp t\vec{H})_*((d\bar{x}_t/dt)(\tau),$ $(d\bar{p}_t/dt)(\tau))$. With these notations,

$$0 = \frac{\partial x}{\partial t}(\bar{x}_t(\tau), \bar{p}_t(\tau), t) = \frac{\partial H}{\partial p}\left(y(\tau), \frac{\partial S_t}{\partial x}(y(\tau)) \right) + V_x(t, \tau), \qquad (5a)$$

because $x(\bar{x}_t(\tau), \bar{p}_t(\tau), t) = y(\tau)$, and

$$\frac{\partial}{\partial t} p(\bar{x}_t(\tau), \bar{p}_t(\tau), t) = -\frac{\partial H}{\partial p}\left(y(t), \frac{\partial S_t}{\partial x}(y(\tau)) \right) + V_p(t, \tau). \qquad (5b)$$

Substitution of (5b) into (4) leads to

$$\frac{\partial}{\partial t} S_t(x) - \frac{\partial}{\partial t} S_t(x_0) = -\int_0^1 \frac{\partial H}{\partial x}\left(y(\tau), \frac{\partial S}{\partial t}(y(\tau)) \right)\frac{dy}{d\tau}(\tau)\, d\tau$$

$$+ \int_0^1 V_p(t, \tau)\frac{dy}{d\tau}(\tau)\, d\tau. \qquad (6)$$

The fact that L_t is a Lagrangian submanifold implies that $V_p(dy/d\tau) - V_x(d/d_\tau)$ $(\partial S_t/\partial x)(y(\tau)) = 0$. Hence,

$$\int_0^1 V_p(t, \tau)\frac{dy}{d\tau}(\tau)\, d\tau = \int_0^1 V_x(t, \tau)\frac{d}{d\tau}\frac{\partial S_t}{\partial x}(y(\tau))\, dt$$

$$= -\int_0^1 \frac{\partial H}{\partial p}\left(y(\tau), \frac{\partial S_t}{\partial x}(y(\tau)) \right)\frac{d}{d\tau}\left(\frac{\partial S_t}{\partial x}(y(\tau)) \right)\, d\tau.$$

Combining this expression with (6) leads to

$$\frac{\partial S_t}{\partial t}(x) - \frac{\partial S_t}{\partial t}(x_0)$$

$$= -\int_0^1 \left(\frac{\partial H}{\partial x}\left(y(\tau), \frac{\partial S_t}{\partial x}(y(\tau))\right) + \frac{\partial H}{\partial p}\left(y(\tau), \frac{\partial S_t}{\partial x}(y(\tau))\right) \frac{d}{d\tau} \frac{\partial S_t}{\partial x}(y(\tau)) \right) d\tau$$

$$= -\int_0^1 \frac{d}{d\tau} H\left(y(\tau), \frac{\partial S_t}{\partial x}(y(\tau))\right) d\tau$$

$$= -H\left(x, \frac{\partial S_t}{\partial x}(x)\right) + H\left(x_0, \frac{\partial S_t}{\partial x}(x_0)\right).$$

The foregoing relation shows that

$$\frac{\partial S_t}{\partial t}(x) + H\left(x, \frac{\partial S_t}{\partial x}(x)\right) = \frac{\partial S_t}{\partial t}(x_0) + H\left(x_0, \frac{\partial S_t}{\partial x}(x_0)\right).$$

The right-hand side of this equation is a function of time only; therefore, by the preceding Remark 1, S_t can be modified so that the right-hand side is equal to zero. The proof is now finished. ∎

In the literature on mechanics, the generating function $S_t(x)$ is called the *action*. The classic use of the action and its connection to the wave fronts and to the optimal principle of Huygens were discussed by Arnold (1976). Theorem 6 can be viewed as a geometric extension of that theory.

As mentioned in the Introduction, the Hamilton-Jacobi equation is a generalization of the Riccati equation. Here is the proof:

When M is a vector space, $T^*M = M \times M^*$. If H is a quadratic form on $M \times M^*$, then $\vec{H}$ is a linear vector field. Because $\exp t\vec{H}$ acts on the sub-bundle of linear Lagrangians, each Lagrangian in Theorem 6 is a linear Lagrangian. The generating function S_t for L_t is given by $S_t(x) = \sum_{i,j=1}^n S_{ij}(t)x_i x_j$, with $S_{ij}(t) = S_{ji}(t)$ for all i and j. Suppose that

$$H = \frac{1}{2}\sum_{i,j=1}^n P_{ij}p_i p_j + \sum_{i,j=1}^n Q_{ij}x_i p_j + \frac{1}{2}\sum_{i,j=1}^n R_{ij}x_i x_j.$$

In order to avoid cumbersome notation with the sums, we use Einstein's convention that any repeated index implies the sum relative to that index. Then

$$\frac{\partial}{\partial t}S_t(x) + H\left(x, \frac{\partial S_t}{\partial x}\right) = 0$$

is the same as

$$\frac{d}{dt} S_{ij}(t) x_i x_j + \frac{1}{2} P_{ij} S_{ik} x_k S_{j\ell} x_\ell + Q_{ij} x_i S_{jk} x_k + \frac{1}{2} R_{ij} x_i x_j = 0.$$

The foregoing equation can hold for all x only if

$$\frac{dS}{dt}(t) + S(t) P S(t) + Q S(t) + S(t) Q^* + R = 0,$$

and therefore the matrix that corresponds to the generating function S_t is a solution of the Riccati equation.

Problem 1 Let $H(x, p) = x$, $(x, p) \in \mathbb{R}^2$. Let L_0 be the linear Lagrangian $x = y$. Find the generating function $S_t(x)$ that satisfies the Hamilton-Jacobi equation.

Problem 2 Change the Hamiltonian in Problem 1 to $H(x, p) = p$. Repeat problem 1 with the new Hamiltonian.

Problem 3 How are the foregoing problems affected if H is either a function of x only or a function of p only?

The solutions of the Hamilton-Jacobi equation can be described in terms of the characteristics according to the general theory of first-order partial differential equations. It has been known since the work of Jacobi that the characteristic equations for the Hamilton-Jacobi equation recover Hamilton's equations, an important fact that complements the results in Theorem 6. We shall now derive the characteristic equations for the Hamilton-Jacobi equation by quoting the characteristic equations from John (1982) for a general first-order partial differential equation $F(x, y, S(x, y), (\partial S/\partial y)(x, y)) = 0$.

Let $z = S(x, y)$, $p = (\partial S/\partial x)(x, y)$, and $q = (\partial S/\partial y)(x, y)$. According to John (1982, p. 24), the differential equations for the characteristics of the foregoing equation are given by

$$\frac{dx}{dt} = F_p, \qquad \frac{dy}{dt} = F_q, \qquad \frac{dz}{dt} = p F_p + q F_q,$$
$$\frac{dp}{dt} = -F_x - F_z p, \qquad \frac{dq}{dt} = -F_y - F_z q.$$

For

$$F(x, y, z, p, q) = q + H(x, p) = 0,$$

the characteristics are given by

$$\frac{dx}{dt} = F_p = \frac{\partial H}{\partial p}(x(t), p(t)), \qquad \frac{dy}{dt} = F_q = 1,$$

$$\frac{dz}{dt} = p(t)\frac{\partial H}{\partial p} + q(t), \qquad \frac{dp}{dt} = -\frac{\partial H}{\partial x}(x(t), p(t)), \qquad \frac{dq}{dt} = 0.$$

It follows that $q(t) = $ constant. Because $z(t) = S(x(t), y(t))$, it follows that $dz/dt = (d/dt)S(x(t), y(t)) = p(t)(\partial H/\partial p)(x(t), p(t)) + q(t) = p(t)$ $(dx/dt)(t)+q$. But $q+H(x(t), p(t)) = 0$, and therefore $(d/dt)S(x(t), y(t)) = p(t)(dx/dt)-H(x(t), p(t))$. Moreover, $dy/dt = 1$ implies that y can be taken equal to t. Then

$$\frac{d}{dt}S(x(t), t) = p(t)\frac{dx}{dt} - H(x(t), p(t)).$$

Summarizing, the characteristic equations for the Hamilton-Jacobi equation corresponding to an arbitrary function H are given by

$$\frac{dx}{dt} = \frac{\partial H}{\partial p}(x(t), p(t)), \tag{7a}$$

$$\frac{dp}{dt} = -\frac{\partial H}{\partial x}(x(t), p(t)), \tag{7b}$$

$$\frac{d}{dt}S_t(x(t)) = p(t)\frac{dx}{dt} - H(x(t), p(t)). \tag{7c}$$

The extension of the Hamilton-Jacobi theory for problems of optimal control is known as the method of dynamic programming, a term coined by R. Bellman. Along with the maximum principle, the method of dynamic programming represents another significant contribution of control theory to the classic theory of the calculus of variations. However, we shall not go further into its theory, because its mathematical development would take us too far from our present objectives.

Instead, we shall now return to the integrability theory. Suppose that $h_1, \ldots,$ h_{n-1} are $n-1$ independent integrals of motion for a given Hamiltonian $H = h_n$. Assume that $\{h_1, \ldots, h_n\}$ are all in involution. Then each level set $h_1 = y_1, \ldots, h_n = y_n$ is a Lagrangian submanifold, that is invariant for each Hamiltonian flow $\exp t\vec{h}_i$, $i = 1, \ldots, n$. Let $S(x, y)$ denote the generating function for the level set $h_i = y_i$, $i = 1, \ldots, n$, defined by n numbers $y_1, \ldots, y_n$.

Consider a new choice of coordinates $y_1, \ldots, y_n, q_1, \ldots, q_n$, defined by S through the following relations:

$$p_i = \frac{\partial S}{\partial x_i}(x, y), \qquad -q_i = \frac{\partial S}{\partial y_i}(x, y), \qquad i = 1, \ldots, n.$$

Jacobi showed that $y_1, \ldots, y_n, q_1, \ldots, q_n$ are also canonical coordinates. Poincaré (1892) referred to this result as the second theorem of Jacobi. In terms of these coordinates, $H = y_n$, and the integral curves of H are given by the trivial equations

$$\frac{dy_i}{dt} = \frac{\partial H}{\partial q_i} = 0, \qquad i = 1, \ldots, n, \qquad \frac{dq_i}{dt} = -\frac{\partial H}{\partial y_i} = -\delta_{in}. \qquad (8)$$

It is very likely that equation (8) accounts for the term "completely integrable," although I have not been able to find the exact reference.

The reader should also note that S satisfies $h_i(x, \partial S/\partial x) = \text{constant}$, for $i = 1, \ldots, n-1$, as a consequence of the Hamilton-Jacobi equation. These relations, together with $H = y_n$, can be used to obtain $\partial S/\partial x$, which can be further integrated to determine S. The interested reader can find further applications of Jacobi's theory in the work of Poincaré (1892) and can refer to Carathéodory (1935) for different integration methods due to Lagrange, Meyer, and Lie.

As deep and beautiful as this theory is, it is nevertheless too confined to canonical coordinates, and for that reason its methodology does not readily reveal the geometry of solutions. For problems with symmetries, it is better to follow different integration methods, which will be described next through solutions on three-dimensional Lie groups.

3 Integrability

3.1 Integrable systems on the Heisenberg group

The three-dimensional Heisenberg group G is equal to $\mathbb{R}^3$, with the group operation $(x_1, y_1, z_1)(x_2, y_2, z_2) = (x_1 + x_2, y_1 + y_2, z_1 + z_2 + y_1 x_2 - x_1 y_2)$. The group identity is $e = (0, 0, 0)$, and the group inverse is $(x, y, z)^{-1} = -(x, y, z)$. G can also be seen as $\mathbb{C} \times \mathbb{R}^1$, with $(u, z_1)(v, z_2) = (u + v, z_1 + z_2 + \text{Im}(u\bar{v}))$ for $u = x_1 + iy_1$ and $v = x_2 + iy_2$.

It follows that left-invariant vector fields on G are of the form

$$X(x, y, z) = a\frac{\partial}{\partial x} + b\frac{\partial}{\partial y} + (c + ya - xb)\frac{\partial}{\partial z}$$

$$= a\left(\frac{\partial}{\partial x} + y\frac{\partial}{\partial z}\right) + b\left(\frac{\partial}{\partial y} - x\frac{\partial}{\partial z}\right) + c\frac{\partial}{\partial z}$$

for arbitrary constants a, b, c. Right-invariant vector fields are of the form $a(\partial/\partial x - y\partial/\partial z) + b(\partial/\partial y + x\partial/\partial z) + c\partial/\partial z$.

Let $X_1 = \partial/\partial x + y\partial/\partial z$, $X_2 = \partial/\partial y - x\partial/\partial z$, and $X_3 = \partial/\partial z$. X_1, X_2, X_3 is a standard basis for the Lie algebra of left-invariant vector fields. The Lie bracket $[X, Y]$ of $X = a_1X_1 + b_1X_2 + c_1X_3$ and $Y = a_2X_1 + b_2X_2 + c_2X_3$ is given by $[X, Y] = 2(b_2a_1 - b_1a_2)X_3$. From this formula it follows that $[X_1, X_2] = 2X_3$ and that $[X_1, X_3] = [X_2, X_3] = 0$.

Let h_1, h_2, h_3 denote the Hamiltonians that correspond to X_1, X_2, X_3. Because h_3 commutes with h_1 and h_2, it is a Casimir function on G. This fact can also be seen as a consequence of the extra symmetry of X_3 – it is the only vector field that is both right- and left-invariant.

Consider now a Hamiltonian function $H = \frac{1}{2}(a_1h_1^2 + a_2h_2^2 + a_3h_3^2)$ with $a_i > 0$, $i = 1, 2, 3$. This Hamiltonian corresponds to a left-invariant Riemannian metric on G defined by minimizing $\frac{1}{2}\int_0^T ((1/a_1)u_1^2 + (1/a_2)u_2^2 + (1/a_3)u_3^2)\, dt$ over the left-invariant system $dg/dt = u_1X_1(g) + u_2X_2(g) + u_3X_3(g)$. Each trajectory $g(t) = (x(t), y(t), z(t))$ of the foregoing differential system can be expressed more explicitly as

$$\frac{dx}{dt} = u_1(t), \qquad \frac{dy}{dt} = u_2(t), \qquad \frac{dz}{dt} = yu_1(t) - xu_2(t) + u_3(t).$$

The extremal equations satisfy

$$\frac{dx}{dt} = a_1h_1(p(t)), \qquad \frac{dy}{dt} = a_2h_2(p(t)),$$
$$\frac{dz}{dt} = a_1yh_1(p(t)) - a_2xh_2(p(t)) + a_3h_3(p(t)),$$

and $p(t) = \mathrm{Ad}^*_{g(t)}(p(0))$.

The differential equations for $p(t)$ can also be written explicitly as

$$\frac{dh_1}{dt} = 2a_2h_2h_3, \qquad \frac{dh_2}{dt} = -2a_1h_1h_3, \qquad \frac{dh_3}{dt} = 0,$$

with h_1, h_2, h_3 equal to the coordinates of p relative to the dual basis X_1^*, X_2^*, X_3^*.

The reader will notice a number of intriguing-looking first integrals. For instance, $h_1 - 2yh_3$ is constant along the solutions, and so is $h_2 + 2xh_3$. It is easy to check that $h_1 - 2yh_3$ is the Hamiltonian of the right-invariant vector field $\partial/\partial x - y(\partial/\partial z)$ expressed in noncanonical representation $T^*G = G \times \mathcal{L}^*$. Similarly, $h_1 - 2yh_3$ is the Hamiltonian of $\partial/\partial y + x(\partial/\partial z)$.

The integrals H, h_3, and $h_1 - 2yh_3$ are 3-integrals of motion in involution, and so are h, h_3, and $h_2 + 2xh_3$. The reader may want to find the generating function for the Lagrangian submanifold $H = $ constant, $h_3 = $ constant, $h_1 - 2yh_3 = $ constant, and thus provide the extremals for this problem using the Hamilton-Jacobi theory. We shall instead follow the ideas inherited from mechanics.

The intersection of $H = $ constant with $h_3 = $ constant is an ellipse $2H - a_3h_3^2 = a_1h_1^2 + a_2h_2^2$. Let $\theta(t)$ be the angle defined by $h_1 = (M/\sqrt{a_1})\cos\theta(t)$, $h_2 = (M/\sqrt{a_2})\sin\theta$, with $M = (2H - a_3h_3^2)^{1/2}$. Then

$$-\frac{M}{\sqrt{a_1}}\sin\theta\frac{d\theta}{dt} = \frac{dh_1}{dt} = 2a_2h_2h_3 = 2a_2h_3\frac{M}{\sqrt{a_2}}\sin\theta.$$

Hence

$$\frac{d\theta}{dt} = -2a_2h_3\sqrt{\frac{a_1}{a_2}} = \omega.$$

Because h_3 is a constant of motion, $\theta(t) = \omega t + \theta_0$. Then

$$\frac{dx}{dt} = \sqrt{a_1}M\cos(\omega t + \theta_0) \quad \text{and} \quad \frac{dy}{dt} = \sqrt{a_2}M\sin(\omega t + \theta_0),$$

and therefore

$$x(t) = a + K_1\sin(\omega t + \theta_0), \qquad y(t) = b - K_2\cos(\omega t + \theta_0),$$

with $\omega K_1 = \sqrt{a_1}M$ and $\omega K_2 = \sqrt{a_2}M$. In particular, the extremals that emanate from the origin are given by

$$x(t) = K_1(\sin(\omega t + \theta_0) - \sin\theta_0), \qquad y(t) = K_2(-\cos(\omega t + \theta_0) + \cos\theta_0).$$

The remaining coordinate $z(t)$ is given by

$$
\begin{aligned}
z(t) &= a_3h_3t + \int_0^t \left(y\frac{dx}{d\tau} - x\frac{dy}{d\tau}\right)d\tau \\
&= \int_0^t (K_2(-\cos(\omega\tau + \theta_0) + \cos\theta_0)K_1\omega\cos(\omega\tau + \theta_0) \\
&\quad - K_1(\sin(\omega\tau + \theta_0) - \sin\theta_0)K_2\omega\sin(\omega\tau + \theta_0))\,d\tau + a_3h_3t \\
&= K_1K_2\omega\int_0^t (\cos\omega\tau - 1)\,d\tau + a_3h_3t \\
&= K_1K_2\omega\left(\frac{1}{\omega}\sin\omega t - t\right) + a_3h_3t.
\end{aligned}
$$

Recall that $\omega = -2a_2h_3\sqrt{a_1/a_2} = -2h_3\sqrt{a_1a_2}$. Then $h_3^2 = \omega^2/4a_1a_2$, and therefore

$$K_i = \frac{1}{\omega}\sqrt{a_i\left(2H - a_3h_3^2\right)} = \frac{1}{\omega}\sqrt{a_i\left(2H - a_3\frac{\omega^2}{4a_1a_2}\right)} \quad \text{for} \quad i = 1, 2.$$

On substituting these values in the expression for $z(t)$,

$$z(t) = -a_3\frac{\omega}{2\sqrt{a_1a_2}}t + \frac{a_1a_2}{\omega}\left(2H - \frac{a_3\omega^2}{4a_1a_2}\right)\left(\frac{1}{\omega}\sin\omega t - t\right).$$

At each instant of time t, the coordinates $x(t)$ and $y(t)$ are on the ellipse $(1/K_1^2)x^2(t) + (1/K_2^2)y^2(t) = 2(1 - \cos\omega t)$. These ellipses become circles when $a_1 = a_2$.

Consider now the symmetrical case given by $a_1 = a_2 = 1$. Points along the z axis can be reached in two ways: When $2H - (a_3\omega^2/4) = 0$, then $x(t) = y(t) = 0$ for all t, and $z(t) = -\frac{1}{2}a_3\omega t$. Assuming now that $2H - a_3(\omega^2/4) > 0$, $x^2(T) + y^2(T) = 0$ when $\omega T = \pm 2\pi$, in which case

$$z(T) = -\frac{1}{2}a_3\omega T - \frac{1}{\omega}\left(2H - \frac{a_3\omega^2}{4}\right)T = \mp a_3\pi \mp \frac{2\pi}{\omega^2}\left(2H - a_3\frac{\omega^2}{4}\right).$$

Note that all the extremal curves that originate at the origin pass through $x = y = 0, z = z(T)$ at $T = \pm 2\pi$. Therefore, this point is a conjugate, or a focal point (Figure 13.2).

Problem 4 Show that every extremal curve is no longer optimal for $t > T$, with $\omega T = \pm 2\pi$.

Consider now a natural sub-Riemannian problem on G consisting of minimizing $\frac{1}{2}\int_0^T (u_1^2 + u_2^2)\, dt$ over the trajectories of

$$\frac{dg}{dt} = u_1(t)X_1(g(t)) + u_2(t)X_2(g(t)), \qquad g(t) \in G,$$

with X_1 and X_2 having the same meaning as in the Riemannian problem. The regular extremals for this problem are integral curves of the Hamiltonian $H = \frac{1}{2}(h_1^2 + h_2^2)$. The abnormal extremals are integral curves of the time-varying Hamiltonian $H_0 = u_1(t)h_1 + u_2(t)h_2$ subject to the conditions that $h_1(p(t)) = h_2(p(t)) = 0$ along each extremal curve $p(t)$. But then

$$0 = \frac{d}{dt}h_1(p(t)) = \{h_1, u_1(t)h_1 + u_2(t)h_2\}(p(t)) = u_2(t)\{h_1, h_2\}(p(t))$$

$$= 2u_2(t)h_3(p(t))$$

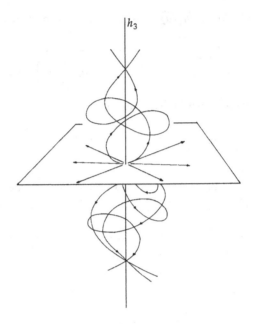

Fig. 13.2.

$h_3(p(t)) = 0$ would imply that $p(t) = 0$, which is ruled out by the maximum principle. Therefore $u_2(t) = 0$. An analogous argument with $h_2 = 0$ yields $u_1(t) = 0$. Therefore, abnormal extremal curves are generated by $u_1 = 0$, $u_2 = 0$. The corresponding integral curves $g(t)$ remain stationary.

The regular extremal equations are given by

$$\frac{dx}{dt} = h_1(p(t)), \qquad \frac{dy}{dt} = h_2(p(t)),$$

$$\frac{dz}{dt} = y(t)h_1(p(t)) - x(t)h_2(p(t)),$$

$$\frac{dh_1}{dt}(p(t)) = \{h_1, H\}(p(t)) = h_2(p(t))\{h_1, h_2\} = 2h_2(p(t))h_3(p(t)),$$

$$\frac{dh_2}{dt}(p(t)) = \{h_2, H\}(p(t)) = -2h_1(p(t))h_3(p(t)), \qquad \frac{dh_3}{dt}(p(t)) = 0.$$

The geometry of the solutions is shown in Figure 13.3. The energy ellipsoid is replaced by the energy cylinder $\frac{1}{2}(h_1^2 + h_2^2) = $ constant. The extremal curves are contained in the intersections of the plane $h_3 = $ constant with the energy cylinder $\frac{1}{2}(h_1^2 + h_2^2) = $ constant. The reader should notice that the geometry of the

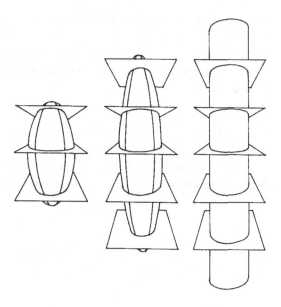

Fig. 13.3.

sub-Riemannian problem can be regarded as a deformation of the Riemannian problem with the parameter a_3 tending to zero. The limit contains the equations for the sub-Riemannian problem, because the abnormal extremals in this case are also regular (they correspond to $H = 0$).

The equations are integrated, as in the Riemannian case, to yield

$$x(t) = K(\sin(\omega t + \theta) - \sin\theta), \qquad y(t) = K(-\cos(\omega t + \theta) + \cos\theta),$$

$$z(t) = K^2\omega\left(\frac{1}{\omega}\sin\omega t - t\right), \qquad \text{with} \quad K = \frac{1}{\omega}\sqrt{2H}.$$

In contrast with the Riemannian case, where ω was restricted to the interval $a_3(\omega^2/4) \le 2H$, ω is unrestricted for the sub-Riemannian problem. It follows that on any energy level H,

$$x^2(t) + y^2(t) = 2K^2(1 - \cos\omega t) = \frac{4H}{\omega^2}(1 - \cos\omega t),$$

and $z(t) = (2H/\omega)((1/\omega)\sin\omega t - t)$.

As ω tends to ∞, the foregoing curves tend uniformly to $x(t) = y(t) = z(t) = 0$. That is, the projections of the abnormal extremals on G are uniform

limits of the projections of the regular extremals on G, all of which are contained
in an arbitrary energy level H.

Evidently, for any H and $T > 0$, $\omega = \pm 2\pi/T$ leads to the conjugate points
$x = y = 0, z = \mp HT^2$. These curves arrive at the conjugate points with the
following tangent vectors

$$\frac{dx}{d\omega} = \sqrt{2H} \cos\theta, \qquad \frac{dy}{d\omega} = \sqrt{2H} \sin\theta, \qquad \frac{dz}{d\omega} = \pm\frac{1}{2}\frac{HT^3}{\pi^2}.$$

In the interval $|\omega| \leq 2\pi/T$, with $0 \leq \theta \leq 2\pi$, these curves describe a closed
surface resembling the shape of an apple, as shown in Figure 13.4.

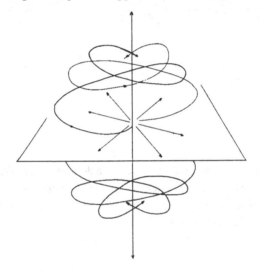

Fig. 13.4.

3.2 Integrable systems on the group of motions of a plane

Recall that the Lie algebra $\mathcal{L}$ of the group of motions E_2 is equal to the linear
span of the matrices

$$A_1 = \begin{pmatrix} 0 & 0 & 0 \\ 1 & 0 & 0 \\ 0 & 0 & 0 \end{pmatrix}, \qquad A_2 = \begin{pmatrix} 0 & 0 & 0 \\ 0 & 0 & 0 \\ 1 & 0 & 0 \end{pmatrix}, \qquad A_3 = \begin{pmatrix} 0 & 0 & 0 \\ 0 & 0 & -1 \\ 0 & 1 & 0 \end{pmatrix}.$$

We shall use the earlier convention denoting $\vec{A}_i$ the left-invariant vector field
whose value at the identity is equal to A_i. We have $[\vec{A}_1, \vec{A}_2] = 0$, $[\vec{A}_1, \vec{A}_3] =$
$\vec{A}_2$, and $[\vec{A}_2, \vec{A}_3] = -\vec{A}_1$.

Let h_1, h_2, h_3 denote the Hamiltonians of $\vec{A}_1, \vec{A}_2, \vec{A}_3$. Recall that both $M = h_1^2 + h_2^2$ and the Hamiltonian H_r of any right-invariant vector field are integrals of motion for any left-invariant problem.

Let us first obtain an explicit expression for the Hamiltonians corresponding to a right-invariant vector field in terms of g and h_1, h_2, h_3. Let

$$A = \begin{pmatrix} 0 & 0 & 0 \\ \alpha_1 & 0 & -\alpha_3 \\ \alpha_2 & \alpha_3 & 0 \end{pmatrix}$$

be an arbitrary element of the Lie algebra. The corresponding right-invariant vector field is given by $g \to Ag$ for any $g \in G$. Its Hamiltonian H_r is given by $H_r = p(g^{-1}Ag)$ (in the representation of T^*G as $G \times \mathcal{L}^*$).

Each h_i is a linear function on $\mathcal{L}^*$ and therefore is an element of $\mathcal{L}$. So in order to write H_r as a function of h_1, h_2, h_3, it is necessary to identify elements of $\mathcal{L}^*$ with those of $\mathcal{L}$.

Because E_2 is not a semisimple Lie group, the trace form is degenerate. As in Chapter 12, we use the following bilinear form to identify $\mathcal{L}$ with $\mathcal{L}^*$:

$$\langle A, B \rangle = a_1 b_1 + a_2 b_2 + a_3 b_3$$

for any

$$A = \begin{pmatrix} 0 & 0 & 0 \\ a_1 & 0 & -a_3 \\ a_2 & a_3 & 0 \end{pmatrix} \quad \text{and} \quad B = \begin{pmatrix} 0 & 0 & 0 \\ b_1 & 0 & -b_3 \\ b_2 & b_3 & 0 \end{pmatrix} \quad \text{in} \quad \mathcal{L}.$$

For each p in L^*, denote by P the element in $\mathcal{L}$ for which $p(A) = \langle P, A \rangle$ for all A in $\mathcal{L}$. The foregoing correspondence yields

$$P = \begin{pmatrix} 0 & 0 & 0 \\ h_1 & 0 & -h_3 \\ h_2 & h_3 & 0 \end{pmatrix}.$$

Hence, $H_r(p, g) = \langle P, g^{-1}Ag \rangle$. Each element g in G is described by the translation part x in $\mathbb{R}^2$ and a rotation R in $SO_2(R)$, with

$$g = \begin{pmatrix} 1 & 0 \\ x & R \end{pmatrix}.$$

Because

$$g^{-1} = \begin{pmatrix} 1 & 0 \\ -R^{-1}x & R^{-1} \end{pmatrix},$$

it follows that

$$g^{-1}Ag = \begin{pmatrix} 1 & 0 \\ -R^{-1}x & R^{-1} \end{pmatrix} \begin{pmatrix} 0 & 0 \\ \alpha & A_0 \end{pmatrix} \begin{pmatrix} 1 & 0 \\ x & R \end{pmatrix} = \begin{pmatrix} 0 & 0 \\ R^{-1}(A_0 x + \alpha) & A_0 \end{pmatrix},$$

with A_0 denoting the matrix

$$\begin{pmatrix} 0 & -\alpha_3 \\ \alpha_3 & 0 \end{pmatrix}.$$

Therefore,

$$H_r = R \begin{pmatrix} h_1 \\ h_2 \end{pmatrix} \cdot (A_0 x + \alpha) + h_3 \alpha_3.$$

Note that H_r depends explicitly on both g and p for each element A in $\mathcal{L}$. In particular, when $A_0 = 0$,

$$H_r = R \begin{pmatrix} h_1 \\ h_2 \end{pmatrix} \cdot \alpha.$$

Given any function H on $\mathcal{L}^*$, then H, $M = h_1^2 + h_2^2$, and H_r constitute three integrals in involution.

The cylinder $M = h_1^2 + h_2^2$ can be used to introduce an angle θ, along the extremal curves, defined as follows:

$$h_1(p(t)) = -\sqrt{M} \cos \theta(t), \qquad h_2(p(t)) = \sqrt{M} \sin \theta(t).$$

$H_r = $ constant implies an easy relation between θ and any choice of parametrization of the rotation matrix R, for if

$$R = \begin{pmatrix} \cos \phi & \sin \phi \\ -\sin \phi & \cos \phi \end{pmatrix},$$

then

$$R \begin{pmatrix} h_1 \\ h_2 \end{pmatrix} \cdot \alpha = \text{constant}$$

implies that

$$\sqrt{M}((-\cos\phi\cos\theta+\sin\phi\sin\theta)\alpha_1+(\sin\phi\cos\theta+\cos\phi\sin\theta)\alpha_2 = \text{constant}.$$

This means that

$$-\alpha_1\cos(\theta+\phi)+\alpha_2\sin(\phi+\theta) = \text{constant},$$

or that $\theta(t)+\phi(t)$ is constant along an extremal curve.

The time evolution of $\theta(t)$ depends on the given Hamiltonian. Among the quadratic Hamiltonians there are three natural choices:

1. $H = \frac{1}{2}(a_1h_1^2+a_2h_2^2+a_3h_3^2)$, with $a_i > 0$, $i = 1, 2, 3$, which corresponds to a left-invariant metric defined by minimizing $\frac{1}{2}\int(u_1^2/a_1+u_2^2/a_2+u_3^2/a_3)\,dt$ over an arbitrary left-invariant path

$$\frac{d}{dt}g(t) = u_1(t)\vec{A}_1(g(t)) + u_2(t)\vec{A}_2(g(t)) + u_3(t)\vec{A}_3(g(t)).$$

2. $H = \frac{1}{2}(a_1h_1^2 + a_3h_3^2)$, with $a_1 > 0$, $a_3 > 0$. This Hamiltonian corresponds to the left-invariant sub-Riemannian problem of minimizing $\frac{1}{2}\int(u_1^2/a_1 + u_3^2/a_3)\,dt$ of the trajectories of $dg/dt = u_1(t)\vec{A}_1(g(t)) + u_3(t)\vec{A}_3(g(t))$.

3. $H = \frac{1}{2}h_3^2 + h_1$ is the Hamiltonian corresponding to the elastic problem of Euler, as shown in Chapter 11.

The Hamiltonian $H = \frac{1}{2}(a_1h_1^2+a_2h_2^2+a_3h_3^2)+bh_1$ covers all three cases. For this choice of H, $\theta(t)$ satisfies the following differential equation:

$$\sqrt{M}\sin\theta\frac{d\theta}{dt} = \frac{dh_1}{dt}(p(t)) = \{h_1, H\}(p(t)) = a_3h_3\{h_1, h_3\}(p(t))$$

$$= a_3h_3h_2 = a_3h_3\sqrt{M}\sin\theta.$$

Therefore

$$\frac{d\theta(t)}{dt} = a_3h_3(p(t)).$$

But then

$$\left(\frac{d\theta}{dt}\right)^2 = a_3^2h_3^2 = 2Ha_3 - a_2a_3h_2^2 - a_1a_3h_1^2 + -2ba_3h_1$$

$$= 2Ha_3 - a_2a_3M\sin^2\theta - a_1a_3M\cos^2\theta - 2ba_3\sqrt{M}\cos\theta.$$

In general, this equation can be integrated in terms of the elliptic integrals. For instance, take the Riemannian case and the sub-Riemannian case ($b = 0$). Then

$\cos^2\theta = (1 + \cos 2\theta)/2$, and the foregoing equation becomes

$$\left(\frac{d\theta}{dt}\right)^2 = 2Ha_3 - a_2a_3M\frac{1-\cos 2\theta}{2} - a_1a_3M\frac{1+\cos\theta}{2} = a_3(K_1 - K_2\cos 2\theta),$$

with $K_1 = 2H - (M/2)(a_1 + a_2)$, and $K_2 = M/2(a_1 - a_2)$.

In each of these cases the differential equation for θ is essentially an equation of the mathematical pendulum. The argument is as follows: Because $d\theta/dt = a_3h_3$, it follows that

$$\frac{d^2\theta}{dt^2} = a_3\{h_3, H\} = a_2h_2\{h_3, h_2\} + a_3a_1h_1\{h_3, h_1\}$$
$$= a_3a_2h_2h_1 - a_3a_1h_1h_2 = a_3(a_1 - a_2)M\sin\theta\cos\theta.$$

Therefore

$$\frac{d^2}{dt^2}(2\theta) = a_3(a_1 - a_2)\sin 2\theta,$$

and this equation is the equation of the mathematical pendulum. The total energy of this pendulum is given by $E = 4a_3K_1$.

There are three qualitative cases to consider: (1) the case in which E is sufficiently large that the pendulum has enough energy to go around the top; (2) the critical case in which the pendulum reaches the top, but in infinite time; (3) the case in which there is a cutoff angle θ_c, and the pendulum oscillates in the interval $-\theta_c \le \theta \le \theta_c$.

As remarked earlier, the existence of the mathematical pendulum was observed by Kirchhoff in 1859 in the context of the elastic problem. For the elastic problem of Euler, $H = \frac{1}{2}h_3^2 + h_1$, the differential equation for θ, obtained in Section 2.3 of Chapter 11, is given by

$$\frac{d\theta}{dt} = h_3 = \pm\sqrt{2(H + M\cos\theta)},$$

where $d\theta/dt$ is equal to the geodesic curvature along the solution curve $x(t)$. In the case where $H > M$, the curvature never changes its sign along the solution curve. Such a case is called *noninflectional*. The remaining cases are called *critical* and *inflectional*.

Geometrically, all of these cases correspond to the type of intersection between the surfaces

$$H = \frac{1}{2}\left(a_1h_1^2 + a_2h_2^2 + a_3h_3^2\right) + bh_1 \quad \text{and} \quad M = h_1^2 + h_2^2,$$

as shown in Figure 13.5 (see also Figure 12.4).

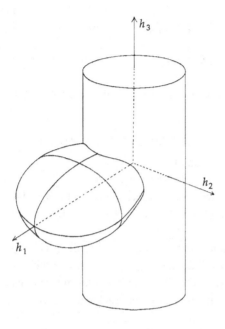

Fig. 13.5.

Problem 5 Show that the solution curves of the Riemannian problem can be parametrized by the angle θ to give

$$\frac{dx_1}{d\theta} = \pm \frac{\sqrt{M}}{2} \frac{(a_1 + a_2) + (a_1 - a_2)\cos 2\theta}{\sqrt{K_1 - K_2 \cos 2\theta}},$$

$$\frac{dx_2}{d\theta} = \pm \frac{\sqrt{M}(a_1 - a_2)}{2} \frac{\sin 2\theta}{\sqrt{K_1 - K_2 \cos 2\theta}},$$

$$K_1 = 2H - \frac{M}{2}(a_1 + a_2) \quad \text{and} \quad K_2 = \frac{M(a_1 - a_2)}{2}.$$

Problem 6 Show that in the sub-Riemannian problem, with $a_1 = a_3 = 1$,

$$\frac{dx_1}{d\theta} = \pm \frac{\sqrt{M}}{2} \frac{1 + \cos 2\theta}{\sqrt{K_1 - K_2 \cos 2\theta}}, \qquad \frac{dx_2}{d\theta} = \pm \frac{\sqrt{M}}{2} \frac{\sin 2\theta}{\sqrt{K_1 - K_2 \cos 2\theta}}.$$

Problem 7 Show that in the elastic problem of Euler,

$$\frac{dx_1}{d\theta} = \pm\frac{\cos\theta}{\sqrt{2(H + M\cos\theta)}}, \qquad \frac{dx_2}{d\theta} = \pm\frac{\sin\theta}{\sqrt{2(H + M\cos\theta)}}.$$

Problem 8 Find the expression for $x_2(\theta)$ in the preceding problems.

Problem 9 Find the conditions for the solution curve $x(\theta)$ to be a closed curve.

3.3 Integrability on $SO_3(\mathbb{R})$

We shall first summarize the basic facts learned in Chapter 12 through the discussion of the heavy top. The Lie algebra $\mathcal{L}$ of $SO_3(\mathbb{R})$ consists of all 3×3 antisymmetric matrices. Each element A in $\mathcal{L}$ is identified with an element $\hat{A}$ in $\mathbb{R}^3$. In this correspondence, the Lie bracket in $\mathcal{L}$ corresponds to the cross-product in $\mathbb{R}^3$.

$\vec{A}_1, \vec{A}_2, \vec{A}_3$ denote the left-invariant vector fields corresponding to the standard basis in $\mathcal{L}$. We shall use h_1, h_2, h_3 to denote their Hamiltonians. In the representation $T^*G = G \times \mathcal{L}^*$, each h_i is a linear function on $\mathcal{L}^*$. Then h_1, h_2, and h_3 satisfy the following Poisson-bracket relations:

$$\{h_1, h_3\} = h_2, \qquad \{h_2, h_1\} = h_3, \qquad \{h_3, h_2\} = h_1.$$

The trace form identifies each element p in L^* with $P(p)$, or simply P in $\mathcal{L}$, given by

$$P = \begin{pmatrix} 0 & -h_3 & h_2 \\ h_3 & 0 & -h_1 \\ -h_2 & h_1 & 0 \end{pmatrix}.$$

Any function H on $\mathcal{L}^*$ can be regarded as a function of h_1, h_2, h_3. Then its differential dH is identified with

$$\begin{pmatrix} 0 & -\frac{\partial H}{\partial h_3} & \frac{\partial H}{\partial h_2} \\ \frac{\partial H}{\partial h_3} & 0 & -\frac{\partial H}{\partial h_1} \\ -\frac{\partial H}{\partial h_2} & \frac{\partial H}{\partial h_1} & 0 \end{pmatrix} \quad \text{in } \mathcal{L}.$$

For any such function H, $M = h_1^2 + h_2^2 + h_3^2$, and the Hamiltoninan H_r of any right-invariant vector field are the integrals in involution. Each level set $H = \text{constant}$, $M = \text{constant}$, $H_r = \text{constant}$ is a Lagrangian submanifold of $G \times \mathcal{L}^*$, as explained in the preceding section. These level sets are in exact correspondence with the adjoint orbits of G in $\mathcal{L}$ because of the relation $P(t) = g(t)\Lambda g^{-1}(t)$, $P(0) = \Lambda$. Each matrix

$$\Lambda = \begin{pmatrix} 0 & -\lambda_3 & \lambda_2 \\ \lambda_3 & 0 & -\lambda_1 \\ -\lambda_2 & \lambda_1 & 0 \end{pmatrix}$$

is determined by the level sets $H = c_1$, $M = c_2$, and $H_r = c_3$, as follows: Let

$$\hat{A} = \begin{pmatrix} a_1 \\ a_2 \\ a_3 \end{pmatrix}$$

denote the entries of the matrix that defines H_r. Then

$$c_3 = H_r = -\frac{1}{2}\text{trace}(P(t)(g^{-1}(t)Ag(t))) = -\frac{1}{2}\text{trace}(g(t)P(t)g^{-1}(t)A)$$
$$= -\frac{1}{2}\text{trace}(\Lambda A) = \lambda_1 a_1 + \lambda_2 a_2 + \lambda_3 a_3.$$

The remaining relations consist of $H(P(t)) = c_1$ and $c_2 = \lambda_1^2 + \lambda_2^2 + \lambda_3^2$.

SO_3 acts transitively on any sphere. That is, for a given

$$\hat{\Lambda} = \begin{pmatrix} \lambda_1 \\ \lambda_2 \\ \lambda_3 \end{pmatrix},$$

there always exists an element g_0 in SO_3 such that $g_0 \hat{\Lambda} = (\lambda_1^2 + \lambda_2^2 + \lambda_3^2)^{1/2} e_3$. This fact implies that any orbit $g(t)P(t)g^{-1}(t) = \Lambda$ is conjugate to $\Lambda = \sqrt{M} A_3$. So it suffices to integrate the particular orbit $gPg^{-1} = \sqrt{M} A_3$.

Then the Euler angles ϕ_1, ϕ_2, ϕ_3 defined by $g = (\exp \phi_1 A_3)(\exp \phi_2 A_2) (\exp \phi_3 A_3)$ are particularly suitable for integration, because

$$\hat{P}(t) = \sqrt{M} g^{-1}(t) e_3 = \sqrt{M}(\exp -\phi_3 A_3)(\exp -\phi_2 A_2)(\exp -\phi_1 A_3)e_3$$
$$= \sqrt{M}(\exp -\phi_3 A_3)(\exp -\phi_2 A_2)e_3,$$

and P depends only on ϕ_3 and ϕ_2. More explicitly,

$$\exp \phi_3 A_3 = \begin{pmatrix} \cos \phi_3 & -\sin \phi_3 & 0 \\ \sin \phi_3 & \cos \phi_3 & 0 \\ 0 & 0 & 1 \end{pmatrix}, \quad \exp \phi_2 A_2 = \begin{pmatrix} \cos \phi_2 & 0 & \sin \phi_2 \\ 0 & 1 & 0 \\ -\sin \phi_2 & 0 & \cos \phi_2 \end{pmatrix},$$

and consequently

$$(\exp -\phi_3 A_3)(\exp -\phi_2 A_2)e_3 = \begin{pmatrix} -\sin\phi_2\cos\phi_3 \\ \sin\phi_2\sin\phi_3 \\ \cos\phi_2 \end{pmatrix}.$$

Therefore,

$$h_1 = -\sqrt{M}\sin\phi_2\cos\phi_3, \qquad h_2 = \sqrt{M}\sin\phi_2\sin\phi_3, \qquad h_3 = \sqrt{M}\cos\phi_2.$$
(9)

These relations can be used to calculate $\exp\phi_2 A_2$ and $\exp\phi_3 A_3$ in terms of h_1, h_2, h_3. It follows that $\cos\phi_2 = (1/\sqrt{M})h_3$, and therefore $\sin\phi_2 = \pm(1/\sqrt{M})(1 - h_3^2)^{1/2} = \pm 1/\sqrt{M}(h_1^2 + h_2^2)^{1/2}$. It also follows that $\tan\phi_3 = -h_1/h_2$, and therefore

$$\sin\phi_3 = \pm\frac{h_2}{\sqrt{h_1^2 + h_2^2}}, \qquad \cos\phi_3 = \mp\frac{h_1}{\sqrt{h_1^2 + h_2^2}}.$$

Hence

$$\exp\phi_2 A_2 = \frac{1}{\sqrt{M}}\begin{pmatrix} h_3 & 0 & \pm\sqrt{h_1^2 + h_2^2} \\ 0 & 1 & 0 \\ \mp\sqrt{h_1^2 + h_2^2} & 0 & h_3 \end{pmatrix},$$

$$\exp\phi_3 A_3 = \begin{pmatrix} \mp\frac{h_1}{\sqrt{h_1^2 + h_2^2}} & \mp\frac{h_2}{\sqrt{h_1^2 + h_2^2}} & 0 \\ \pm\frac{h_2}{\sqrt{h_1^2 + h_2^2}} & \mp\frac{h_1}{\sqrt{h_1^2 + h_2^2}} & 0 \\ 0 & 0 & 1 \end{pmatrix}.$$

The appropriate sign is determined from the initial conditions.

The remaining angle ϕ_1 is computed from the differential equation $g^{-1}(t)$ $(dg/dt) = dH$, as follows:

$$\frac{dg}{dt} = \frac{d}{dt}((\exp\phi_1 A_3)(\exp\phi_2 A_2)(\exp\phi_3 A_3))$$

$$= \frac{d\phi_1}{dt}A_3 g(t) + \frac{d\phi_2}{dt}(\exp\phi_1 A_3)A_2(\exp\phi_2 A_2)(\exp\phi_3 A_3) + \frac{d\phi_3}{dt}g(t)A_3.$$

Hence,

$$g^{-1}\frac{dg}{dt} = \frac{d\phi_1}{dt}g^{-1}(t)A_3 g(t) + \frac{d\phi_2}{dt}(\exp -\phi_3 A_3)A_2(\exp\phi_3 A_3) + \frac{d\phi_3}{dt}A_3.$$

The equation $g^{-1}(dg/dt) = dH$, when expressed as an equation in $\mathbb{R}^3$ (via the formula $\widehat{g^{-1}dg/dt} = \widehat{dH}$), and after the substitution $\sqrt{M}g^{-1}A_3g = P$, becomes

$$\frac{1}{\sqrt{M}}\frac{d\phi_1}{dt}\hat{P} + \frac{d\phi_2}{dt}(\exp -\phi_3 A_3)e_2 + \frac{d\phi_3}{dt}e_3 = \frac{\partial H}{\partial h_1}e_1 + \frac{\partial H}{\partial h_2}e_2 + \frac{\partial H}{\partial h_3}e_3.$$

Substitution of $(\exp -\phi_3 A_3)e_2 = (\sin \phi_3)e_1 + (\cos \phi_3)e_2$ in the foregoing differential equation leads to the following relations:

$$\frac{1}{\sqrt{M}}\frac{d\phi_1}{dt}h_1 + \frac{d\phi_2}{dt}(\sin \phi_3) = \sqrt{M}\frac{\partial H}{\partial h_1},$$

$$\frac{1}{\sqrt{M}}\frac{d\phi_1}{dt}h_2 + \frac{d\phi_2}{dt}(\cos \phi_3) = \frac{\partial H}{\partial h_2},$$

and $((d\phi_1/dt)/\sqrt{M})h_3 + (d\phi_3/dt) = \partial H/\partial h_3$.

The foregoing equations can be explicitly solved for $d\phi_1/dt, d\phi_2/dt$, and $d\phi_3/dt$ in terms of the variables h_1, h_2, h_3. The resulting equations are

$$\frac{d\phi_1}{dt} = \frac{\sqrt{M}\left(\sin \phi_3 \frac{\partial H}{\partial h_2} - \cos \phi_3 \frac{\partial H}{\partial h_1}\right)}{h_2\sin\phi_3 - h_1\cos\phi_3},$$

$$\frac{d\phi_2}{dt} = \frac{h_2 \frac{\partial H}{\partial h_1} - h_1 \frac{\partial H}{\partial h_2}}{h_2\sin\phi_3 - h_1\cos\phi_3},$$

$$\frac{d\phi_3}{dt} = \frac{\partial H}{\partial h_3} - h_3\frac{\sin \phi_3 \frac{\partial H}{\partial h_2} - \cos \phi_3 \frac{\partial H}{\partial h_1}}{h_2\sin\phi_3 - h_1\cos\phi_3}.$$

Combined with the earlier formulas $\cos \phi_2 = (1/\sqrt{M})h_3$ and $\tan \phi_3 = -h_1/h_2$,

$$h_2\sin\phi_3 - h_1\cos\phi_3 = \pm\frac{h_2^2}{\sqrt{h_1^2 + h_2^2}} \pm \frac{h_1^2}{\sqrt{h_1^2 + h_2^2}} = \pm\sqrt{h_1^2 + h_2^2},$$

and

$$\sin \phi_3\frac{\partial H}{\partial h_2} - \cos \phi_3\frac{\partial H}{\partial h_1} = \pm\frac{h_2}{\sqrt{h_1^2 + h_2^2}}\frac{\partial H}{\partial h_2} \pm \frac{h_1}{\sqrt{h_1^2 + h_2^2}}\frac{\partial H}{\partial h_1}.$$

Therefore

$$\frac{d\phi_1}{dt} = \frac{\sqrt{M}\left(h_1\frac{\partial H}{\partial h_1} + h_2\frac{\partial H}{\partial h_2}\right)}{h_1^2 + h_2^2}$$

$$\frac{d\phi_2}{dt} = \pm\frac{h_2\frac{\partial H}{\partial h_1} - h_1\frac{\partial H}{\partial h_2}}{\sqrt{h_1^2 + h_2^2}} \tag{10}$$

$$\frac{d\phi_3}{dt} = \frac{\partial H}{\partial h_3} - \frac{\sqrt{M}h_3\left(h_1\frac{\partial H}{\partial h_1} + h_2\frac{\partial H}{\partial h_2}\right)}{h_1^2 + h_2^2}.$$

Aside from their practical computational value, the foregoing equations demonstrate that the essential part of the integration procedure for an invariant Hamiltonian system is confined to the variables in the dual of the Lie algebra. We shall further illustrate this point by discussing the integration of $H = \frac{1}{2}h_3^2 + h_1$, which corresponds to the Hamiltonian of the elastic problem on $SO_3(R)$ (as the isometry group of the sphere).

We shall now outline two methods of integration, each based on the method of quadrature. The first method is based on expressing ϕ_1, ϕ_2, and ϕ_3 in terms of h_3, which in turn is determined through its own differential equation. This equation has the nicest form when expressed in terms of $v(t) = h_3^2(p(t))$. Then

$$\frac{dv}{dt} = 2h_3\{h_3, H\} = 2h_3\{h_3, h_1\} = -2h_3h_2.$$

Hence

$$\frac{1}{4}\left(\frac{dv}{dt}\right)^2 = h_2^2h_3^2 = v\left(M - h_1^2 - h_3^2\right)$$

$$= v\left(M - \left(H - \frac{1}{2}h_3^2\right)^2 - h_3^2\right) = v\left(M - \left(H - \frac{1}{2}v\right)^2 - v\right).$$

Therefore

$$\left(\frac{dv}{dt}\right)^2 + v^3 - 4v^2(H + 1) + 4v(H^2 - M) = 0.$$

The foregoing differential equation can be integrated by means of elliptic functions for various values of H and M. This equation can also be viewed as the Euler-Lagrange equation for the elastic problem, because its solutions determine the extremal control $u(t) = h_3(p(t))$. Then

$$h_1 = H - \frac{1}{2}v, \qquad h_2 = \pm\sqrt{M - \left(H - \left(H - \frac{1}{2}v\right)^2 - v\right)}.$$

The corresponding Euler angles ϕ_2 and ϕ_3 satisfy

$$\sin\phi_2 = \pm\frac{1}{\sqrt{M}}\sqrt{1-v}, \qquad \cos\phi_2 = \frac{\sqrt{v}}{\sqrt{M}},$$

$$\sin\phi_3 = \pm\frac{\sqrt{M - \left(H - \frac{1}{2}v\right)^2 - v}}{\sqrt{M-v}}, \qquad \cos\phi_3 = \mp\frac{H - \frac{1}{2}v}{\sqrt{M-v}}.$$

The remaining angle ϕ_1 is obtained from equations (10) by integrating

$$\frac{d\phi_1}{dt} = \frac{\sqrt{M}h_1}{h_1^2 + h_2^2} = \frac{\sqrt{M}\left(H - \frac{1}{2}v\right)}{M-v}.$$

The second method of integrating the extremal equations uses another "natural" coordinate θ defined by

$$h_1(t) - 1 = -J\cos\theta(t) \quad \text{and} \quad h_2(t) = J\sin\theta(t).$$

Note that $(h_1(t)-1)^2 - h_2^2(t) = h_1^2 - 2h_1 + 1 + h_2^2 = M - h_3^2 - (2H - h_3^2) + 1 = M - 2H + 1 = J^2$, and the foregoing angle is well defined. Then

$$J\sin\theta\frac{d\theta}{dt} = \frac{dh_1}{dt} = \{h_1, H\}(p(t)) = h_3\{h_1, h_3\} = h_3 h_2 = h_3 J\sin\theta.$$

Hence

$$\frac{d\theta(t)}{dt} = h_3(t).$$

Because $H = \frac{1}{2}h_3^2 + h_1 = \frac{1}{2}(d\theta/dt)^2 + 1 - J\cos\theta$, it follows that $H - 1$ is "the energy" of the mathematical pendulum corresponding to the movements of $\theta(t)$. (Note that $d^2\theta/dt^2 = dh_3/dt = \{h_3, H\} = \{h_3, h_1\} = -h_2 = -J\sin\theta$.) Therefore, $\theta(t)$ is the solution of

$$\frac{d\theta}{dt} = \pm\sqrt{2(H - 1 + J\cos\theta)}.$$

The sign of $h_3(t)$ determines the sign of $d\theta/dt$: $d\theta/dt > 0$ when $h_3 > 0$, and $d\theta/dt < 0$ when $h_3 < 0$.

Keeping the correct sign in mind, then all the variables can be parametrized by the angle θ, as in the planar case, to yield

$$h_1 = 1 - J\cos\theta, \qquad h_2 = J\sin\theta, \quad h_3 = \pm\sqrt{2(H - 1 + J\cos\theta)}.$$

Then $\exp\phi_2 A_2$ and $\exp\phi_3 A_3$ are determined through (9). The remaining angle $\phi_1(\theta)$ is obtained by integrating the equation

$$\frac{d\phi_1}{d\theta} = \frac{d\phi_1}{dt} \Big/ \frac{d\theta}{dt} = \pm\frac{\sqrt{M}h_1}{(M^2 - h_3^2)\sqrt{2(H - 1 + J\cos\theta)}},$$

obtained from (10). After further substitutions,

$$\frac{d\phi_1}{d\theta} = \pm\frac{\sqrt{M}(1 - J\cos\theta)}{(M^2 - (2(H - 1 + J\cos\theta)))\sqrt{2(H - 1 + J\cos\theta)}}.$$

Figure 13.6 shows the elastic curves on S^2 obtained by projecting the preceding solution curves down to S^2.

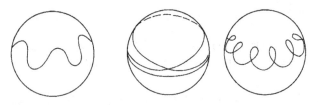

Fig. 13.6.

Problem 10 Let

$$A_1 = \frac{1}{2}\begin{pmatrix} 1 & 0 \\ 0 & -1 \end{pmatrix}, \qquad A_2 = \frac{1}{2}\begin{pmatrix} 0 & 1 \\ 1 & 0 \end{pmatrix}, \qquad A_3 = \frac{1}{2}\begin{pmatrix} 0 & -1 \\ 1 & 0 \end{pmatrix}.$$

A_1, A_2, A_3 is a basis for the matrices with trace equal to zero, which is the Lie algebra for $SL_2(R)$.

Find the Poisson table for the Hamiltonians h_1, h_2, h_3 of the corresponding left-invariant vector fields $\vec{A}_1, \vec{A}_2, \vec{A}_3$, and show that $M = h_1^2 + h_2^2 - h_3^2$ is the Casimir element for the Poisson algebra defined by $h_1, h_2,$ and h_3.

Notes and sources

Noether's theorem and its use in mechanics were described by Arnold (1976). The extension of Noether's theorem to control systems (Theorem 2) and the applications to left-invariant systems are original. These results have much in common with the Marsden-Weinstein reduction procedure for systems with symmetries (Arnold, 1976, appendix 5).

It seems that the use of Lagrangian manifolds in the Hamilton-Jacobi theory has been known for some time, although I have not been able to find a suitable reference. The extension of Hamilton-Jacobi theory to problems of optimal control is known as dynamic programming, or as the Bellman equation in some engineering texts. The reader interested in these extensions should consult Lions (1982) for further information.

This presentation of the integrability theory of three-dimensional Lie groups is essentially based on "Non-Euclidean Elastica" (Jurdjevic, 1995a). The sub-Riemannian problem on the Heisenberg group, also known as the Carnot-Carathéodory metric, has been studied extensively in connection with the fundamental solutions of certain hypoelliptic operators. The discovery of the singularity of the sub-Riemannian unit sphere along the z axis (the apple-like shape) goes back to Brockett (1981).

14

Integrable Hamiltonian systems on Lie groups: the elastic problem, its non-Euclidean analogues, and the rolling-sphere problem

Having incorporated into optimal control theory the contributions made by the calculus of variations, mechanics, and geometry, we can begin to appreciate the exciting and challenging possibilities that this enriched subject offers back to geometry and applied mathematics. We have seen that the most remarkable equations of classic applied mathematics appear as only natural steps in the context of geometric control theory. In this chapter we shall again turn to the classic past to find inspiration for the future. We shall reconsider Kirchhoff's elastic problem in the Euclidean space E^3 and his famous observation of 1859 that the elastic problem is "like" the heavy-top problem, known since then as the mechanical or kinetic analogue of the elastic problem.

Brilliant, but baffling, Kirchhoff's observation has had lasting influence on subsequent generations of mathematicians interested in the elastic problem who have sought to understand its solution through the analogy with the heavy-top problem. That view of the elastic problem partly explains why it has always been in the shadow of the heavy-top problem and is also partly responsible for its relative obscurity outside of the literature on elasticity.

Our treatment of variational problems on Lie groups in Chapters 12 and 13 shows clear connections between these two problems and provides natural interpretations for the famous observation of Kirchhoff: The heavy-top problem is "like" the elastic problem, and the analogy with the elastic problem illuminates its solutions, rather than the other way around as previously understood. As a left-invariant problem on the group of motions of E^3, the elastic problem has its own integrals of motion that explain much of the geometry of the solutions for the heavy top and, in particular, justify extension of its Hamiltonian equations from $SO_3(R)$ to the group of motions of E^3. The control-theory formulation of this problem also makes it clear why the classic calculus of variations could not adequately deal with this problem: The elastic assumptions of nonextensibility and nonshearing of the rod introduce non-holonomic constraints on the space of the possible configurations of the rod and turn the elastic problem into a

444

genuine control problem with three controls in a six-dimensional space, as we saw in Chapter 12.

We shall obtain the solutions to the elastic problem in a self-contained manner without any reference to the heavy top. The methods extend to other symmetric spaces of constant curvature, such as the sphere S^3 and the Lorentz space H^3. In contrast to the heavy-top problem, the elastic problem has a natural geometric formulation in terms of the moving frame that allows generalizations to arbitrary dimensions. The analogy with the heavy-top problem, however, reveals the integrable cases for the elastic problem, and we shall see an extra integral of motion under precisely the same conditions as in the heavy-top problem.

This chapter begins with the solutions of the symmetric elastic problem in E^3. This material is followed by extensions of the elastic problem to other three-dimensional symmetric spaces, the sphere S^3 and the Lorentz space H^3. In each of these cases, the frame bundle is identified with the isometry group, and therefore the configuration spaces of non-Euclidean elastic rods are $SO_4(R)$ and $SO(1, 3)$, the isometry groups of S^3 and H^3, respectively. Together with their Euclidean counterparts in E^3, all these problems define a natural left-invariant control system on each isometry group G of the form

$$\frac{dg}{dt} = \vec{B}_1(g) + \sum_{i=1}^{3} u_i(t)\vec{A}_i(g).$$

The controlled vector fields $\vec{A}_1$, $\vec{A}_2$, $\vec{A}_3$ provide an orthonormal basis for left-invariant vector fields in the maximal compact subgroup K of G, and in all three cases K is equal to $SO_3(R)$, with the drift vector field a left-invariant vector field defined by an element B_1 in the Cartan subspace of the Lie algebra of G.

The quadratic cost that is to be minimized is equal to the total elastic energy $E = \frac{1}{2} \int_0^T (c_1 u_1^2 + c_2 u_2^2 + c_3 u_3^2) \, dt$. As shown in Chapter 13, the symmetric problem, which occurs when $c_2 = c_3$, is completely integrable, because $\vec{A}_1$ is a symmetry for the problem, and therefore its Hamiltonian H_1 is an integral of motion. However, what is much more remarkable, and still puzzling, is that the elastic problem in all three cases is also completely integrable under precisely the same conditions as the Kowalewski top when either $c_1 = c_2 = 2c_3$ or $c_1 = c_3 = 2c_2$.

In this chapter we shall see that the integration techniques for the symmetric elastic problem can be handled by a uniform procedure in the Lie algebra of the group, in which the curvature of the underlying space appears as a parameter ε ($\varepsilon = 0$ in the Euclidean case, $\varepsilon = -1$ in the Lorentz case, and $\varepsilon = 1$ in the spherical case).

Apart from the elastic problem, we shall also investigate solutions of the sub-Riemannian problem associated with a rolling sphere on a horizontal plane in E^3. As was shown in Chapter 4, the space of configurations is a five-dimensional

Lie group $G = \mathbb{R}^2 \times SO_3(R)$, and all possible paths are generated by the controls u_1, u_2 corresponding to the angular velocities of the sphere. (The vertical component of the angular velocity is taken to be zero.)

The sub-Riemannian problem of minimizing $\int_0^T (u_1^2 + u_2^2)^{1/2} dt$ leads to a completely integrable Hamiltonian system that is then integrated in terms of elliptic integrals defined by the coordinates induced by the integrals of motion.

The solutions show a remarkable and also puzzling connection to the elastic problem of Euler: The point of contact of the sphere with the stationary plane traces the elastic curves of Euler. Aside from the movements along straight lines, all movements of the sphere fall into two geometric types: "rolls" and "wobbles" of the sphere, corresponding exactly to the noninflectional and inflectional elastica traced by the point of contact.

1 The symmetric elastic problem in $\mathbb{R}^3$

We shall continue with the notation developed in Chapter 12, with $H = \frac{1}{2}(H_1^2/c_1 + H_2^2/c_2 + H_3^2/c_3) + h_1$. Recall that along extremal curves $(g(t), p(t))$, the control functions $u_i(t)$ are equal to $(1/c_i)H_i(p(t))$ for $i = 1, 2, 3$. Denoting

$$g(t) = \begin{pmatrix} 1 & 0 \\ x(t) & R(t) \end{pmatrix},$$

recall the earlier results that

$$R(t)h(t) = a, \quad \text{and} \quad R(t)\vec{H}(t) + a \times x(t) = b$$

for constants a and b in $\mathbb{R}^3$. Aside from $H = $ constant along the extremal curves, it follows from preceding equations that

(a) $h_1^2(t) + h_2^2(t) + h_3^2(t) = \|a\|^2$ and that
(b) $h_1(t)H_1(t) + h_2(t)H_2(t) + h_3(t)H_3(t) = a \cdot b$ are its integrals of motion.

In the symmetric elastic problem, $c_2 = c_3$, and there is another integral of motion:

(c) $H_1(t) = $ constant. Recall the observation from Chapter 13 that the symmetric elastic problem is completely integrable. We shall now show how to integrate the Hamiltonian equations as a and b vary over $\mathbb{R}^3$ (for they correspond to different co-adjoint orbits).

For simplicity of notation, we shall take $c_2 = c_3 = 1$.

Definition 1 The projections of extremal curves on $\mathbb{R}^3$ are called the elastic curves.

In order to draw geometric conclusions about the elastic curves, it becomes necessary to relate the extremal curves to the geodesic curvature $k(t)$ and the torsion $\tau(t)$ along the central line. We shall use $\vec{T}(t)$, $\vec{N}(t)$, $\vec{B}(t)$ to denote the Serret-Frenet frame along an extremal curve $x(t)$: $\vec{T}(t)$ is the tangent vector, $\vec{N}(t)$ is the normal vector, and $\vec{B}(t)$ is the bi-normal vector at $x(t)$. Then

$$\frac{dx}{dt} = \vec{T}(t), \qquad \frac{d\vec{T}}{dt} = k(t)\vec{N}(t), \qquad \frac{d\vec{N}}{dt} = -k(t)\vec{T}(t) + \tau(t)\vec{B}(t),$$

$$\frac{d\vec{B}}{dt} = -\tau(t)\vec{N}(t).$$

The Serret-Frenet frame is related to the moving frame $\vec{a}_1, \vec{a}_2, \vec{a}_3$ through an angle β defined as follows:

Because $dx/dt = \vec{a}_1(t) = \vec{T}(t)$, it follows that $\vec{a}_2$ and $\vec{a}_3$ are in the normal plane spanned by $\vec{N}$ and $\vec{B}$. Let $\beta(t)$ denote the angle that the normal $\vec{N}(t)$ makes with $\vec{a}_2(t)$. Then

$$\vec{N}(t) = (\cos\beta)\vec{a}_2 + (\sin\beta)\vec{a}_3, \qquad \vec{B}(t) = (-\sin\beta)\vec{a}_2 + (\cos\beta)\vec{a}_3.$$

Differentiating $\vec{N}(t)$ produces

$$\frac{d\vec{N}}{dt} = \frac{d\beta}{dt}((-\sin\beta)\vec{a}_2 + (\cos\beta)\vec{a}_2) + (\cos\beta)\frac{d\vec{a}_2}{dt} + (\sin\beta)\frac{d\vec{a}_3}{dt}.$$

Recall now that $R(t)$ is the rotation matrix that relates the moving frame $\vec{a}_1, \vec{a}_2, \vec{a}_3$ to the fixed frame $\vec{e}_1, \vec{e}_2, \vec{e}_3$ along $x(t)$. Consequently, $d\vec{a}_2/dt = -u_3\vec{a}_1 + u_1\vec{a}_3$, and $d\vec{a}_3/dt = u_2\vec{a}_1 - u_1\vec{a}_2$. Incorporating these expressions into the preceding equation gives

$$\frac{d\vec{N}}{dt} = ((-\cos\beta)u_3 + (\sin\beta)u_2)\,\vec{a}_1 - \left(\frac{d\beta}{dt}\sin\beta + u_1\sin\beta\right)\vec{a}_2$$

$$+ \left(\frac{d\beta}{dt}\cos\beta + u_1\cos\beta\right)\vec{a}_3.$$

But $d\vec{N}/dt$ is also equal to $-k\vec{T} + \tau\vec{B}$, which in turn is equal to $-k\vec{a}_1 + \tau(-\sin\beta)\vec{a}_2 + (\cos\beta)\vec{a}_3$. Equating the two expressions for $d\vec{N}/dt$ yields

$$k = u_3\cos\beta - u_2\sin\beta \quad \text{and} \quad \frac{d\beta}{dt} + u_1(t) = \tau(t). \tag{1}$$

Performing similar calculations with $d\vec{B}/dt$ leads to

$$u_2\cos\beta + u_3\sin\beta = 0. \tag{2}$$

Thus

$$\tan \beta(t) = -\frac{u_2(t)}{u_3(t)} \quad \text{and} \quad k^2(t) = u_2^2(t) + u_3^2(t). \tag{3}$$

Having assumed that $c_2 = c_3 = 1$ implies that the extremal controls are given by $u_2(t) = H_2(t)$ and $u_3(t) = H_3(t)$. Therefore, the system Hamiltonian H can also be written as

$$H = \frac{1}{2}\frac{H_1^2}{c_1} + \frac{1}{2}k^2 + h_1.$$

It then becomes convenient to replace H with the reduced Hamiltonian H_0:

$$H_0 = H - \frac{1}{2}\frac{H_1^2}{c_1} = \frac{1}{2}k^2 + h_1.$$

Theorem 1 *Let $v(t) = k^2(t)$. Then $v(t)$ is a solution of the following differential equation:*

$$\frac{1}{4}\left(\frac{dv}{dt}\right)^2 = v\left(\|a\|^2 - \left(H_0 - \frac{1}{2}v\right)\right)^2 - \left(a \cdot b - H_1\left(H_0 - \frac{1}{2}v\right)\right)^2. \tag{4}$$

Also,

$$k^2(t)\tau(t) = v(t)\tau(t) = H_1\left(H_0 + \frac{v}{2}\right) - a \cdot b.$$

Proof

$$\frac{1}{2}\frac{dv}{dt} = \frac{1}{2}\frac{d}{dt}(H_2^2 + H_3^2) = H_2\frac{dH_2}{dt} + H_3\frac{dH_3}{dt} = H_2\{H_2, H\} + H_3\{H_3, H\}$$
$$= H_2(H_3\{H_2, H_3\} + \{H_2, h_1\}) + H_3(H_2\{H_3, H_2\} + \{H_3, h_1\})$$
$$= -H_2h_3 + H_3h_2$$

Therefore

$$\frac{1}{4}\left(\frac{dv}{dt}\right)^2 = (H_3h_2 - H_2h_3)^2$$
$$= H_3^2h_2^2 + H_2^2h_3^2 - 2H_2H_3h_2h_3$$
$$= H_3^2h_2^2 + H_2^2h_3^2 + H_2^2h_2^2 + H_3^2h_3^2 - (H_2h_2 + H_3h_3)^2$$
$$= (h_2^2 + h_3^2)(H_2^2 + H_3^2) - (a \cdot b - H_1h_1)^2$$
$$= (\|a\|^2 - h_1^2)(H_2^2 + H_3^2) - (a \cdot b - H_1h_1)^2.$$

Because $h_1 = H_0 - \frac{1}{2}v$, it follows that

$$\frac{1}{4}\left(\frac{dv}{dt}\right)^2 = v\left(\|a\|^2 - \left(H_0 - \frac{1}{2}v\right)^2\right) - \left(a \cdot b - H_1\left(H_0 - \frac{1}{2}v\right)\right)^2.$$

It remains to prove that $v\tau = H_1(H_0 - \frac{v}{2}) - a \cdot b$. It follows from (3) that $\tan \beta = -H_2/H_3$. Therefore

$$\sec^2 \beta \frac{d\beta}{dt} = \frac{H_2 \frac{dH_3}{dt} - H_3 \frac{dH_2}{dt}}{H_3^2}.$$

Then

$$H_3 \frac{dH_2}{dt} - H_2 \frac{dH_3}{dt} = H_3\{H_2, H\} - H_2\{H_2, H\}$$

$$= H_3 \left(h_3 + H_1 H_3 \left(\frac{1}{c_1} - 1\right)\right)$$

$$- H_2 \left(-h_2 + H_1 H_2 \left(1 - \frac{1}{c_1}\right)\right)$$

$$= H_3 h_3 + H_2 h_2 + H_1 \left(\frac{1}{c_1} - 1\right)(H_2^2 + H_3^2).$$

Moreover, $\sec^2 \beta = (H_2^2 + H_3^2)/H_3^2$, because $\tan \beta = -H_2/H_3$. Therefore

$$\frac{d\beta}{dt} = -\frac{H_3 h_3 + H_2 h_2}{k^2} - H_1 \left(\frac{1}{c_1} - 1\right).$$

Substitution of the foregoing expression into $\tau = d\beta/dt + u_1$ gives

$$\tau = -\frac{H_2 h_2 + H_3 h_3}{k^2} + H_1. \tag{5}$$

Therefore

$$v\tau = -(a \cdot b - H_1 h_1) + H_1 v = -a \cdot b + H_1(h_1 + v)$$

$$= -a \cdot b + H_1 \left(H_0 - \frac{1}{2}v + v\right)$$

$$= -a \cdot b + H_1 \left(H_0 + \frac{1}{2}v\right),$$

and the proof is finished.   ∎

Corollary 1 *The elastic curves that correspond to $a = 0$ are either straight lines or helices. These helices reduce to circles when $H_1 = 0$.*

Proof Because $\|h\|^2 = \|a\|^2$, $h(t) = 0$ when $\|a\| = 0$. Therefore $H_0 - \frac{1}{2}v = 0$, and $\frac{1}{4}(dv/dt)^2 = 0$. So $v(t) = $ constant. Either $v(t) = 0$, in which case the elastic curve is a straight line, or $v = v_0 \neq 0$. In the latter case, $v_0\tau = H_1(H_0 - \frac{1}{2}v_0 + v_0) = H_1 v_0$, and therefore $\tau = H_1$. Circles are the only curves with constant curvature and zero torsion, and our proof is finished. ∎

Remark Recall that the straight-line solutions are also the projections of the abnormal extremal curves. ∎

The qualitative behavior of an arbitrary elastic curve is determined by the differential equation

$$\frac{1}{4}\left(\frac{dv}{dt}\right)^2 = f(v) = v\left(\|a\|^2 - \left(H_0 - \frac{1}{2}v\right)^2\right) - \left(a \cdot b - H_1\left(H_0 - \frac{1}{2}v\right)\right)^2$$

$f = 0$ only for $\|a\| = 0$. Otherwise, f is a cubic polynomial whose graph tends to $-\infty$ as v tends to ∞. It follows that

$$f'(v) = \|a\|^2 - \left(H_0 - \frac{1}{2}v\right)^2 + v\left(H_0 - \frac{1}{2}v\right)$$
$$- H_1\left(a \cdot b - H_1\left(H_0 - \frac{1}{2}v\right)\right)$$
$$= -\frac{3}{4}v^2 + \left(2H_0 - \frac{H_1^2}{2}\right)v + \|a\|^2 - H_0^2 - H_1(a \cdot b - H_1 H_0),$$

and therefore the first critical point of $f'(v)$ corresponds to the minimum of f, and the second critical point corresponds to the maximum of f (Figure 14.1). The zeros of f correspond to the equilibrium solutions of the equation $dv/dt = 2\sqrt{f(v)}$. The corresponding elastic curves are helices, because then the torsion is also constant. The helix that corresponds to the first critical point v_1 degenerates into a straight line when $v_1 = 0$.

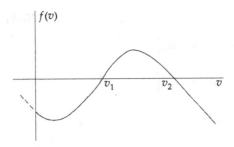

Fig. 14.1.

The remaining elastic curves correspond to $a \neq 0$. For $a \neq 0$ there is a natural choice of spherical coordinates determined by the sphere $\|h\|^2 = \|a\|^2$. Let θ and ψ be the angles defined by

$$h_1 = \|a\| \cos\theta, \qquad h_2 = \|a\| \sin\theta \sin\psi, \qquad h_3 = \|a\| \sin\theta \cos\psi.$$

Along each extremal curve, $\cos\theta(t) = h_1/\|a\| = (1/\|a\|)(H_0 - \frac{1}{2}k^2) = (1/\|a\|)(H_0 - \frac{1}{2}v(t))$. Therefore $(d/dt)(\cos\theta(t)) = -(1/2\|a\|)(dv/dt)$, and hence $-\frac{d\theta}{dt}\sin\theta = -(1/2\|a\|)(dv/dt)$. Thus

$$\sin^2\theta \left(\frac{d\theta}{dt}\right)^2 = \frac{1}{\|a\|^2}\frac{1}{4}\left(\frac{dv}{dt}\right)^2.$$

Consequently,

$$\sin^2\theta \left(\frac{d\theta}{dt}\right)^2 = \frac{1}{\|a\|^2}\frac{1}{4}\left(\frac{dv}{dt}\right)^2 = \frac{v}{\|a\|^2}(\|a\|^2 - h_1^2) - \frac{(a \cdot b - H_1 h_1)^2}{\|a\|^2},$$

or

$$\left(\frac{d\theta}{dt}\right)^2 = 2(H_0 - \cos\theta) - \frac{(a \cdot b - H_1\|a\|\cos\theta)^2}{\|a\|^2 \sin^2\theta}. \qquad (6)$$

In the literature of the heavy top, the angle θ is called the *nutation angle*. There is another angle in this literature called the *precession angle*, which is denoted by ϕ and is defined as the solution of the following differential equation:

$$\frac{d\phi}{dt} = \|a\| \frac{a \cdot b - H_1 h_1(t)}{\|a\|^2 - h_1^2(t)}. \qquad (7)$$

The significance of ϕ will be clear from subsequent developments.

In the meantime, we shall see that the extremal control functions $u_1(t) = H_1$, $u_2(t) = H_2(t)$, $u_3(t) = H_3(t)$ can be integrated by quadrature in terms of $\theta(t)$. It will first be necessary to obtain the differential equation for the angle ψ introduced earlier. The required differential equations will be extracted from the extremal equations as follows:

$$\frac{dh_2}{dt} = \{h_2, H\} = \frac{H_1}{c_1}\{h_2, H_1\} + H_2\{h_2, H_2\} + H_3\{h_2, H_3\} + \{h_2, h_1\}.$$

It follows from the Lie-bracket table (Table 12.1) that $dh_2/dt = (H_1/c_1)h_3 - H_3 h_1$. Similarly, $dh_3/dt = -(H_1/c_1)h_2 + H_2 h_1$. Then

$$\dot{\theta}\|a\|\cos\theta\sin\psi + \dot{\psi}\|a\|\sin\theta\cos\psi = \frac{dh_2}{dt} = \frac{H_1}{c_1}h_3 - H_3 h_1,$$

$$\dot{\theta}\|a\|\cos\theta\cos\psi - \dot{\psi}\|a\|\sin\theta\sin\psi = \frac{dh_3}{dt} = -\frac{H_1}{c_1}h_2 + H_2 h_1.$$

It then follows that

$$\begin{pmatrix} \dot{\theta} \\ \dot{\psi} \end{pmatrix} = \frac{1}{\|a\| \sin\theta \cos\theta} \begin{pmatrix} -\sin\theta \sin\psi & -\sin\theta \cos\psi \\ -\cos\theta \cos\psi & \cos\theta \sin\psi \end{pmatrix} \begin{pmatrix} -\frac{H_1}{c_1} h_3 + H_3 h_1 \\ \frac{H_1}{c_1} h_2 - H_2 h_1 \end{pmatrix},$$

and therefore

$$\dot{\theta} = \frac{H_1}{c_1 \|a\| \cos\theta}(h_3 \sin\psi - h_2 \cos\psi) - \frac{h_1}{\|a\| \cos\theta}(H_3 \sin\psi - H_2 \cos\psi)$$

$$\dot{\psi} = \frac{H_1}{c_1 \|a\| \sin\theta}(h_3 \cos\psi + h_2 \sin\psi) - \frac{h_1}{\|a\| \sin\theta}(H_3 \cos\psi + H_2 \sin\psi).$$

The quantity $h_3 \sin\psi - h_2 \cos\psi$ is equal to zero, because $h_2 = \|a\| \sin\theta \sin\psi$ and $h_3 = \|a\| \sin\theta \cos\psi$. Moreover, $h_3 \cos\psi + h_2 \sin\psi = \|a\| \sin\theta$. It then follows from $h_1 H_1 + h_2 H_2 + H_3 h_3 = a \cdot b$ that $H_3 \cos\psi + H_2 \sin\psi = (a \cdot b - h_1 H_1)/\|a\| \sin\theta$. After the substitutions,

$$\dot{\theta} = (H_2 \cos\psi - H_3 \sin\psi) \quad \text{and} \quad \dot{\psi} = \frac{H_1}{c_1} - \frac{h_1(a \cdot b - H_1 h_1)}{\|a\|^2 \sin^2\theta}. \tag{8}$$

The last differential equation can also be expressed in terms of the precession angle ϕ, as follows: Because

$$\frac{d\phi}{dt} = \|a\| \frac{a \cdot b - H_1 h_1}{\|a\|^2 - h_1^2} = \frac{\|a\|(a \cdot b - H_1 h_1)}{\|a\|^2 \sin^2\theta},$$

it follows that

$$\frac{h_1(a \cdot b - H_1 h_1)}{\|a\|^2 \sin^2\theta} = \cos\theta \frac{d\phi}{dt}.$$

Therefore

$$\dot{\psi} = \frac{H_1}{c_1} - (\cos\theta)\dot{\phi},$$

and consequently ψ and ϕ are not independent; ϕ gets preferential treatment in the literature on the heavy top because of an accustomed choice of coordinates on $SO_3(R)$ in terms of the Euler angles, as will be seen in the next section.

Continuing with the main issue of integration by quadrature, note that

$$H_2(t) = \dot{\phi} \sin\theta \sin\psi + \dot{\theta} \cos\psi \quad \text{and} \quad H_3(t) = \dot{\phi} \sin\theta \cos\psi - \dot{\theta} \sin\psi, \tag{9}$$

as can easily be verified from the following relations:

$$H_3 \cos\psi + H_2 \sin\psi = \frac{a \cdot b - h_1 H_1}{\|a\| \sin\theta} = \dot{\phi} \sin\theta,$$

$$H_3 \sin\psi - H_2 \cos\psi = \dot{\theta}.$$

1.1 Euler angles and elastic curves

Assuming that $a \neq 0$, then a can be rotated so that it is in the direction of e_1. This rotation changes the initial orientation of the rotation matrix $R(t)$, but otherwise does not affect the extremal curves. Then $R(t) = e^{\alpha_1(t)A_1}e^{\alpha_2(t)A_2}e^{\alpha_3(t)A_1}$ provides a natural system of coordinates adapted to this choice of a. It follows that

$$\frac{dR}{dt} = R(t)\left(\dot{\alpha}_1 R^{-1}A_1 R + \dot{\alpha}_2 e^{-\alpha_3 A_1}A_2 e^{\alpha_3 A_1} + \dot{\alpha}_3 A_1\right).$$

The foregoing differential equations for α_1, α_2, α_3 are easily calculated with the aid of relations used before that identify antisymmetric matrices with vectors in $\mathbb{R}^3$. Recall the notations that

$$\hat{A} = \begin{pmatrix} a_1 \\ a_2 \\ a_3 \end{pmatrix} \quad \text{corresponds to} \quad A = \begin{pmatrix} 0 & -a_3 & a_2 \\ a_3 & 0 & -a_1 \\ -a_2 & a_1 & 0 \end{pmatrix},$$

and also recall the basic fact that $\widehat{R^{-1}AR} = R^{-1}\hat{A}$ (Chapter 12). In particular, $R^{-1}e_1 = \widehat{R^{-1}A_1 R}$. Then

$$R^{-1}e_1 = e^{-\alpha_3 A_1}e^{-\alpha_2 A_2}e_1$$
$$= \begin{pmatrix} 1 & 0 & 0 \\ 0 & \cos\alpha_3 & \sin\alpha_3 \\ 0 & -\sin\alpha_3 & \cos\alpha_3 \end{pmatrix}\begin{pmatrix} \cos\alpha_2 & 0 & -\sin\alpha_2 \\ 0 & 1 & 0 \\ \sin\alpha_2 & 0 & \cos\alpha_2 \end{pmatrix}\begin{pmatrix} 1 \\ 0 \\ 0 \end{pmatrix}$$
$$= (\cos\alpha_2)e_1 + (\sin\alpha_2 \sin\alpha_3)e_2 + (\cos\alpha_3)e_3.$$

Similarly,

$$e^{-\alpha_3 A_1}e_2 = (\cos\alpha_3)e_2 - (\sin\alpha_3)e_3.$$

Therefore

$$\left(\widehat{R^{-1}\frac{dR}{dt}}\right) = (\dot{\alpha}_1 \cos\alpha_2 + \dot{\alpha}_3)e_1 + (\dot{\alpha}_1 \sin\alpha_2 \sin\alpha_3 + \dot{\alpha}_2 \cos\alpha_3)e_2$$
$$+ (\dot{\alpha}_1 \sin\alpha_3 \cos\alpha_3 - \dot{\alpha}_2 \sin\alpha_3)e_3.$$

It follows that

$$\dot{\alpha}_1 \cos\alpha_2 + \dot{\alpha}_3 = u_1(t),$$
$$\dot{\alpha}_1 \sin\alpha_2 \sin\alpha_3 + \dot{\alpha}_2 \cos\alpha_3 = u_2(t),$$
$$\dot{\alpha}_1 \sin\alpha_3 \cos\alpha_3 - \dot{\alpha}_2 \sin\alpha_3 = u_3(t),$$

because $\overbrace{R^{-1}(dR/dt)} = u_1e_1 + u_2e_2 + u_3e_3$. These equations can be inverted in a neighborhood of $\sin\alpha_2 \neq 0$ to yield

$$\dot{\alpha}_1 = \frac{u_2 \sin\alpha_3 + u_3 \cos\alpha_3}{\sin\alpha_2}, \qquad \dot{\alpha}_2 = u_2 \cos\alpha_3 - u_3 \sin\alpha_3,$$

$$\dot{\alpha}_3 = u_1 - \frac{\cos\alpha_2}{\sin\alpha_2}(u_2 \sin\alpha_3 + u_3 \cos\alpha_3).$$

It remains to relate these angles to the angles θ and ψ introduced earlier. Because $R(t)h(t) = a = \|a\|e_1$, it follows that

$$h(t) = \|a\|R^{-1}(t)e_1 = \|a\|e^{-\alpha_3 A_1}e^{-\alpha_2 A_2}e_1$$

$$= \|a\|((\cos\alpha_1)e_1 + (\sin\alpha_1 \sin\alpha_3)e_3 + (\sin\alpha_1 \cos\alpha_3)e_3).$$

Therefore

$$h_1(t) = \|a\| \cos\alpha_2, \quad h_2(t) = \|a\| \sin\alpha_2 \sin\alpha_3, \quad h_3(t) = \|a\| \sin\alpha_2 \cos\alpha_3,$$

and consequently α_2 and α_3 agree with the angles θ and ψ. But then

$$\dot{\alpha}_1 = \frac{u_2 \sin\alpha_3 + u_3 \cos\alpha_3}{\sin\alpha_2} = \frac{\cos\alpha_2}{\cos\alpha_2} \frac{(u_2 \sin\alpha_3 + u_3 \cos\alpha_3)}{\sin\alpha_2} = \frac{u_1 - \dot{\alpha}_3}{\cos\theta},$$

and therefore $\dot{\alpha}_1 = \dot{\psi}$.

Because each elastic curve $x(t)$ is related to the rotation matrix by $dx/dt = R(t)e_1$, it follows that $x(t)$ is independent of the angle ψ, which finally justifies our choice of coordinates on $SO_3(R)$, and at the same time illuminates the significance of the precession angle ϕ.

The qualitative behavior of the extremal curves is determined by the sign of $a \cdot b - H_1 h_1(t)$. If $a \cdot b - H_1 h_1(t)$ is always of the same sign along an extremal curve, then the precession angle is either increasing or decreasing, depending on the sign. If $a \cdot b - H_1 h_1(t)$ is equal to zero for some values of t, then $d\phi/dt$ changes its sign and causes $\phi(t)$ to oscillate. In either case, the nutation angle θ oscillates between its limits. The resulting motion, represented on the sphere $\|h\|^2 = \|a\|^2$, is shown in Figure 14.2.

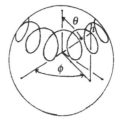

Fig. 14.2.

Problem 1 Show that the central line $x(t)$ is given by the following equations:

$$x_1(t) = \frac{1}{\|a\|} \int_0^t h_1(s)\, ds,$$

$$x_2(t) = \frac{1}{\|a\|}(b_3 - (\sin\phi)A - (\cos\phi)B),$$

$$x_3(t) = \frac{1}{\|a\|}(-b_2 - (\sin\phi)B + (\cos\phi)A),$$

with

$$A = \dot\theta = \sqrt{2(H_0 - \|a\|\cos\theta) - \frac{(a\cdot b - H_1\|a\|\cos\theta)^2}{\|a\|^2 \sin^2\theta}}$$

and

$$B = \frac{-H_1\|a\|^2 + a\cdot b\|a\|\cos\theta}{\|a\|^2 \sin\theta}.$$

Hint: Use $R\hat{H} + a\times x = b$.

The foregoing relations, or a more direct calculation based on $R\hat{H} + a\times x = b$, will show that

$$(\|a\|x_2 - b_3)^2 + (\|a\|x_3 - b_2)^2 + b_1^2 = H_1^2 + k^2 = H_1^2 + v.$$

Because v is constrained to the compact interval $[v_1, v_2]$ (Figure 14.1), it follows that

$$H_1^2 - b_1^2 + v_1 \le (\|a\|x_2 - b_3)^2 + (\|a\|x_3 - b_2)^2 \le H_1^2 - b_1^2 + v_2.$$

Therefore the elastic curves reside on a cylinder of varying radius. The cylinder can cross itself when $x_1(T) = x_1(0)$ for some value of T. Such crossings can be of several types, as shown in Figure 14.3.

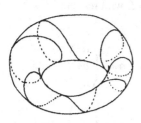
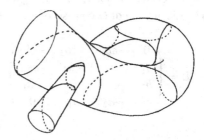

Fig. 14.3.

2 Non-Euclidean symmetric elastic problems

Kirchhoff's model of an elastic rod in E^3 admits natural extensions to other Riemannian spaces. Here we shall consider only the extensions to three-dimensional simply connected spaces of constant curvature: the sphere S^3 and the Lorentz space H^3. For these spaces, the frame bundles can be naturally identified with the isometry groups, and consequently the elastic configurations can be regarded as curves in each isometry group, $SO_4(R)$ for S^3, and $SO(1, 3)$ for H^3. In order to get to the equations as quickly as possible, it will be necessary to suppress much of the intermediate discussion concerning the geometric properties of symmetric spaces. For those details, the interested reader should consult Wolf (1984).

2.1 Algebraic preliminaries

Let G denote the group that leaves the quadratic form $x_1y_1 + \epsilon(x_2y_2 + x_3y_3 + x_4y_4)$ invariant, with $\epsilon = \pm 1$. It is well known that $G = SO_4(R)$ for $\epsilon = 1$ and that $G = SO(1, 3)$ for $\epsilon = -1$.

Because G leaves the quadratic form invariant, G acts on each surface $M = \{x \in \mathbb{R}^4 : x_1^2 + \epsilon(x_2^2 + x_3^2 + x_4^2) = \text{constant}\}$. We shall use H^3 to denote the hyperbolic sphere $x_1^2 - (x_2^2 + x_3^2 + x_4^2) = 1$, $x_1 > 0$, and S^3 denotes the standard unit sphere in $\mathbb{R}^4$. Both of these spaces are conveniently described by $x_1^2 + \epsilon(x_2^2 + x_3^2 + x_4^2) = 1$, with the understanding that $x_1 > 0$ when $\epsilon = -1$.

Readers can prove for themselves that $SO_4(R)$ acts transitively on S^3 and that $SO(1, 2)$ acts transitively on H^3. Taking $x = e_1$ (written as a column vector), the group K that leaves x fixed is equal to

$$\begin{pmatrix} 1 & 0 \\ 0 & R \end{pmatrix}, \quad \text{with } R \in SO_3(R).$$

Therefore each of S^3 and H^3 can be regarded as the homogeneous spaces G/K.

The Lie algebra of G consists of all matrices L with real entries such that

$$L = \begin{pmatrix} 0 & -\epsilon b_1 & -\epsilon b_2 & -\epsilon b_3 \\ b_1 & 0 & -a_3 & a_2 \\ b_2 & a_3 & 0 & -a_1 \\ b_3 & -a_2 & a_1 & 0 \end{pmatrix} \quad \text{for} \quad a = \begin{pmatrix} a_1 \\ a_2 \\ a_3 \end{pmatrix} \quad \text{and}$$

$$b = \begin{pmatrix} b_1 \\ b_2 \\ b_3 \end{pmatrix} \quad \text{in} \quad \mathbb{R}^3.$$

Table 14.1.

[,]	$\vec{A}_1$	$\vec{A}_2$	$\vec{A}_3$	$\vec{B}_1$	$\vec{B}_2$	$\vec{B}_3$
$\vec{A}_1$	0	$-\vec{A}_3$	$\vec{A}_2$	0	$-\vec{B}_3$	$\vec{B}_2$
$\vec{A}_2$	$\vec{A}_3$	0	$-\vec{A}_1$	$\vec{B}_3$	0	$-\vec{B}_1$
$\vec{A}_3$	$-\vec{A}_2$	$\vec{A}_1$	0	$-\vec{B}_2$	$\vec{B}_1$	0
$\vec{B}_1$	0	$-\vec{B}_3$	$\vec{B}_2$	0	$-\epsilon\vec{A}_3$	$\epsilon\vec{A}_2$
$\vec{B}_2$	$\vec{B}_3$	0	$-\vec{B}_1$	$\epsilon\vec{A}_3$	0	$-\epsilon\vec{A}_1$
$\vec{B}_3$	$-\vec{B}_2$	$\vec{B}_1$	0	$-\epsilon\vec{A}_2$	$\epsilon\vec{A}_1$	0

We shall use A_1, A_2, A_3 to denote the matrices L for which $b = 0$ and $a = e_i$, $i = 1, 2, 3$. B_1, B_2, B_3 will denote the matrices L for which $a = 0$ and $b = e_i$, $i = 1, 2, 3$, respectively. The reader should note that when $\epsilon = 0$, this choice of basis agrees with the basis for the Lie algebra for the group of motions of E^3 used in the preceding section. Following our established convention, for any L in $\mathcal{L}(G)$, $\vec{L}$ denotes the left-invariant vector field on G whose value at the identity is equal to L.

We leave it to the reader to verify that $\vec{A}_i$ and $\vec{B}_i$ satisfy Table 14.1 for Lie brackets. The reader may notice that Table 14.1 coincides with Table 12.1 in Chapter 12 for $\epsilon = 0$, and therefore it can be used to describe the structures of each Lie group, E_3, $SO(1, 3)$, and $SO_4(R)$.

We shall denote by $\mathcal{P}$ the linear span of B_1, B_2, B_3, and by $\mathcal{K}$ the linear span of A_1, A_2, A_3. The decomposition $\mathcal{L}(G) = \mathcal{P} \oplus \mathcal{K}$ is called the Cartan decomposition of $\mathcal{L}(G)$. It follows that $[\mathcal{P}, \mathcal{P}] = \mathcal{K}$ for $\epsilon \neq 0$, and $[\mathcal{P}, \mathcal{P}] = 0$ for $\epsilon = 0$. Furthermore, $[\mathcal{P}, \mathcal{K}] = \mathcal{P}$ and $[\mathcal{K}, \mathcal{K}] = \mathcal{K}$.

For any elements L and $\bar{L}$ in $\mathcal{L}(G)$, trace$(L\bar{L})$ denotes the trace of $L\bar{L}$. It follows that trace$(L\bar{L}) = -2 \sum_{i=1}^{3} a_i \bar{a}_i + \epsilon b_i \bar{b}_i$, with $L = \sum_{i=1}^{3} a_i A_i + b_i B_i$ and $\bar{L} = \sum_{i=1}^{3} \bar{a}_i A_i + \bar{b}_i B_i$. Therefore, trace$(L\bar{L})$ is negative-definite on SO_4, degenerate on E_3, and indefinite on $SO(1, 3)$.

With these algebraic considerations in mind, "the configuration space of an elastic rod in G/K" is equal to the space of trajectories of the left-invariant differential system

$$\frac{dg}{dt} = \vec{B}_1(g) + \sum_{i=1}^{3} u_i(t) \vec{A}_i(g) \tag{10}$$

as $u(t) = (u_1(t), u_2(t), u_3(t))$ varies over all bounded and measurable functions on the interval $[0, \infty)$.

Associated with each solution curve $(g(t), u(t))$, its total elastic energy in the interval $[0, T]$ is defined to be

$$\frac{1}{2} \int_0^T \left(c_1 u_1^2(t) + c_2 u_2^2(t) + c_3 u_3^2(t) \right) dt \tag{11}$$

for fixed positive constants c_1, c_2, c_3.

Each solution curve of (10) defines a curve $x(t)$ in $M = G/K$ and an orthonormal frame $\vec{a}_1$, $\vec{a}_2$, $\vec{a}_3$ along $x(t)$ that is adapted to the curve $x(t)$ in the sense that $(dx/dt)(t) = \vec{a}_1(t)$. The identification is as follows:

$$x(t) = g(t)e_1 \quad \text{and} \quad \vec{a}_i(t) = g(t)e_{i+1}, \quad i = 1, 2, 3.$$

Conversely, any curve $(x(t), \vec{a}_1(t), \vec{a}_2(t), \vec{a}_3(t))$ in the frame bundle that satisfies the constraint that $(dx/dt)(t) = \vec{a}_1(t)$ defines a unique trajectory of equation (10) for some control functions $u_1(t)$, $u_2(t)$, $u_3(t)$.

The non-Euclidean elastic problem consists of minimizing $\frac{1}{2} \int_0^T (c_1 u_1^2 + c_2 u_2^2 + c_3 u_3^2) \, dt$ over all solution curves $(g(t), u(t))$ of (10) for $\epsilon \neq 0$ that satisfy the fixed boundary conditions $g(0) = g_0$ and $g(T) = g_1$.

2.2 The structure of extremal curves

As in the Euclidean case $T^*G = G \times \mathcal{L}^*$, h_1, h_2, h_3 denote the Hamiltonians corresponding to $\vec{B}_1$, $\vec{B}_2$, $\vec{B}_3$, and the Hamiltonians that correspond to $\vec{A}_1$, $\vec{A}_2$, $\vec{A}_3$ will be denoted by H_1, H_2, H_3. The regular extremal curves are the integral curves of the Hamiltonian vector field defined by

$$H = \frac{1}{2} \left(\frac{H_1^2}{c_1} + \frac{H_2^2}{c_2} + \frac{H_3^2}{c_3} \right) + h_1. \tag{12}$$

The abnormal extremals $(g(t), u(t))$ are the integral curves of the time-varying Hamiltonian $H = h_1(t) + u_1(t)H_1 + u_2(t)H_2 + u_3(t)H_3$, subject to the constraints $H_i = 0, i = 1, 2, 3$. The structure of the abnormal extremals is similar to that in the Euclidean case, with the straight-line solutions replaced by geodesic curves. Because the arguments are the same in all cases, they will not be repeated here. Instead, we shall concentrate on the integrability properties of the Hamiltonian function (12).

As in the Euclidean case, there are two quadratic Casimir functions M_1 and M_2 given explicitly by

$$M_1 = h_1 H_1 + h_2 H_2 + h_3 H_3, \quad \text{and}$$
$$M_2 = \left(h_1^2 + h_2^2 + h_3^2 \right) + \epsilon \left(H_1^2 + H_2^2 + H_3^2 \right). \tag{13}$$

There are several ways to prove that M_1 and M_2 commute with each element in $\mathcal{L}^*$, as will become clear from the subsequent developments. For the time being, the reader may check directly that M_1 and M_2 commute with each element of the basis $h_1, h_2, h_3, H_1, H_2, H_3$, through Table 14.1.

As in the Euclidean case, each maximal commutative subalgebra in $\mathcal{L}(G)$ is of dimension 2. Hence, there are two linearly independent elements in $\mathcal{L}(G)$ that commute (e.g., B_1 and A_1). The Hamiltonians M_3 and M_4 of the corresponding right-invariant vector field form integrals of motion in involution with each other that are also in involution with any function on $\mathcal{L}^*(G)$ (Theorem 1, Chapter 13). It then follows that the Hamiltonian H itself is the fifth integral of motion, which is in involution with the remaining four integrals, M_1, M_2, M_3, M_4. Therefore the corresponding Hamiltonian system is completely integrable whenever there is an extra integral of motion.

Theorem 2 *Suppose that the constants c_2 and c_3 are equal to each other. Then $H_1 = $ constant along each extremal curve, and therefore the Hamiltonian system corresponding to $H = \frac{1}{2}(H_1^2/c_1 + H_2^2/c_2 + H_3^2/c_2) + h_1$ is completely integrable.*

Proof There are two ways to prove this result. The first way is to show that system (10) is invariant under the rotation around the e_1 axis and then use the extended Noether theorem described in Chapter 13 to conclude that H_1 is an integral of motion for H. The second way is to prove it directly, as follows: Let $p(t)$ be any integral curve of $\vec{H}$. Then

$$\frac{d}{dt} H_1(p(t)) = \{H_1, H\}(p(t))$$

$$= \frac{H_1}{c_1}\{H_1, H_1\} + \frac{H_2}{c_2}\{H_1, H_2\} + \frac{H_3}{c_3}\{H_1, H_3\} + \{H_1, h_1\}$$

$$= -\frac{H_2 H_3}{c_2} + \frac{H_3 H_2}{c_3} = 0. \qquad \blacksquare$$

As in the Euclidean case, the elastic problem is called *symmetric* whenever $c_2 = c_3$. For the symmetric problem, the constants c_2 and c_3 will be taken to be equal to unity, and H_0 will be used to denote the reduced Hamiltonian $H - H_1^2/2c_1$. Then we have the following:

Theorem 3 *Let $v = H_2^2 + H_3^2$. Then, along each extremal curve,*

$$\frac{1}{4}\left(\frac{dv}{dt}\right)^2 = v\left(M_2 - \left(H_0 - \frac{1}{2}v\right)^2 - \epsilon(v + H_1^2)\right)$$

$$- \left(M_1 - \left(H_0 - \frac{1}{2}v\right)H_1\right)^2. \qquad (14)$$

Proof

$$\frac{1}{2}\frac{d}{dt}v(t) = \frac{1}{2}\frac{d}{dt}\left(H_2^2(t) + H_3^2(t)\right) = H_2\frac{dH_2}{dt} + H_3\frac{dH_3}{dt}$$

$$= H_2\{H_2, H\} + H_3\{H_3, H\}$$

$$= H_2\left(\frac{H_1}{c_1}\{H_2, H_1\} + H_3\{H_2, H_3\} + \{H_2, h_1\}\right)$$

$$+ H_3\left(\frac{H_1}{c_1}\{H_3, H_1\} + H_2\{H_3, H_2\} + \{H_3, h_1\}\right)$$

$$= H_2h_3 - H_3h_2.$$

Therefore

$$\frac{1}{4}\left(\frac{dv}{dt}\right)^2 = (H_2h_3 - H_3h_2)^2 = H_2^2h_3^2 + H_3^2h_2^2 - 2H_2H_3h_2h_3$$

$$= H_2^2h_3^2 + H_3^2h_2^2 + H_2^2h_2^2 + H_3^2h_3^2 - (H_2h_2 + H_3h_3)^2$$

$$= \left(H_2^2 + H_3^2\right)\left(h_2^2 + h_3^2\right) - (M_1 - h_1H_1)^2$$

$$= v\left(M_2 - h_1^2 - \epsilon\left(H_1^2 + H_2^2 + H_3^2\right)\right) - (M_1 - h_1H_1)^2.$$

Because $H_0 = \frac{1}{2}v + h_1$, it follows by substitution that

$$\frac{1}{4}\left(\frac{dv}{dt}\right)^2 = v\left(M_2 - \left(H_0 - \frac{1}{2}v\right)^2 - \epsilon(v + H_1^2)\right)$$

$$- \left(M_1 - \left(H_0 - \frac{1}{2}v\right)H_1\right)^2,$$

and therefore the proof is finished.   ∎

The quantity $k = (H_2^2 + H_3^2)^{1/2}$ is equal to the geodesic curvature along the "central line" $x(t) = g(t)e_1$. This can be seen through the following argument:

Let $x(t) = g(t)e_1$, with $(g(t), p(t))$ equal to an extremal curve of H. The Serret-Frenet equation associated with the curve $x(t)$ is another differential equation in G:

$$\frac{d\bar{g}}{dt} = \vec{B}_1(\bar{g}) + k(t)\vec{A}_3(\bar{g}) + \tau(t)\vec{A}_1(\bar{g}),$$

with $\bar{g}(t)e_1 = x(t)$. Let $\vec{T}(t) = \bar{g}(t)e_2$, $\vec{N}(t) = \bar{g}(t)e_3$, and $\vec{B}(t) = \bar{g}(t)e_4$. Then

$$\frac{d\vec{T}}{dt} = \frac{d\bar{g}}{dt}e_2 = \bar{g}(B_1 + kA_3 + \tau A_1)e_2 = \bar{g}(-\epsilon e_1 + ke_3) = -\epsilon x(t) + k\vec{N}(t),$$

$$\frac{d\vec{N}}{dt} = \bar{g}(B_1 + kA_3 + \tau A_1)e_3 = \bar{g}(-ke_2 + \tau e_4) = -k\vec{T}(t) + \tau \vec{B}(t),$$

$$\frac{d\vec{B}}{dt} = \bar{g}(B_1 + kA_3 + \tau A_1)e_4 = \bar{g}(-\tau e_3) = -\tau \vec{N}(t).$$

The moving frame $\vec{a}_1(t), \vec{a}_2(t), \vec{a}_3(t)$ along $g(t)$ is given by $\vec{a}_i(t) = g(t)e_{i+1}$, $i = 1, 2, 3$. The same calculations as in the Euclidean case show that the angle $\beta(t)$ subtended by $\vec{N}(t)$ and $\vec{a}_2(t)$ satisfies the following relations:

$$k = u_3(t)\cos\beta(t) - u_2(t)\sin\beta(t), \qquad \frac{d\beta}{dt} + u_1(t) = \tau(t),$$

$$u_2(t)\cos\beta(t) + u_3(t)\sin\beta(t) = 0.$$

Along an extremal curve, $u_1(t) = (1/c_1)H_1(t)$, $u_2(t) = H_2(t)$, and $u_3(t) = H_3(t)$; therefore $k^2(t) = H_2^2 + H_3^2$, and $\tan\beta = -H_2/H_3$. It follows that $(\sec^2\beta)(d\beta/dt) = (H_2(dH_3/dt) - H_3(dH_2/dt))/H_3^2$. Then

$$H_2\frac{dH_3}{dt} - H_3\frac{dH_2}{dt} = H_2\{H_3, H\} - H_3\{H_2, H\}$$

$$= H_2\left(\frac{H_1}{c_1}\{H_3, H_1\} + H_2\{H_3, H_2\} + \{H_3, h_1\}\right)$$

$$- H_3\left(\frac{H_1}{c_1}\{H_2, H_1\} + H_3\{H_2, H_3\} + \{H_2, h_1\}\right)$$

$$= H_2\left(-\frac{1}{c_1}H_1H_2 + H_2H_1 - h_2\right)$$

$$- H_3\left(\frac{H_1H_3}{c_1} - H_3H_1 + h_3\right)$$

$$= H_1(H_2^2 + H_3^2)\left(1 - \frac{1}{c_1}\right) - H_2h_2 - H_3h_3,$$

and $\sec^2\beta = (H_2^2 + H_3^2)/H_3^2$ lead to $d\beta/dt = H_1(1 - 1/c_1) - (H_2h_2 + H_3h_3)/(H_2^2 + H_3^2)$. Hence $\tau = -(H_2h_2 + H_3h_3)/(H_2^2 + H_3^2) + H_1$, which in turn results in $k^2\tau = H_1(k^2 + h_1) - M_1 = H_1(H_0 + \frac{1}{2}k^2) - M_1$. Therefore the relationship between the curvature and the torsion along an elastic curve remains the same in all three cases ($\epsilon = 0$, $\epsilon = \pm 1$).

The integration of the extremal curves follows much the same technique described in the Euclidean case. There is a natural choice of spherical coordinates induced by the constant M_2, H_1, and the reduced Hamiltonian H_0. Recall that $M_2 = h_1^2 + h_2^2 + h_3^2 + \epsilon(H_1^2 + H_2^2 + H_3^2)$. Because $H_2^2 + H_3^2 = 2(H_0 - h_1)$, it follows that $(h_1 - \epsilon)^2 + h_2^2 + h_3^2 = M_2 - \epsilon(H_1^2 + 2H_0) - \epsilon^2$.

Let M^2 denote the quantity $M_2 - \epsilon(H_1^2 + 2H_0) - \epsilon^2$. Then each extremal curve $h(t)$ resides on the sphere $(h_1 - \epsilon)^2 + h_2^2 + h_3^2 = M^2$. Let θ and ψ denote the spherical coordinates defined by

$$h_1 - \epsilon = M \cos\theta, \qquad h_2 = M \sin\theta \sin\psi, \qquad h_3 = M \sin\theta \cos\psi.$$

The time evolution of these angles is determined from the differential equations for h_1, h_2, and h_3. It is easy to obtain the differential equation for $\dot{\theta}$ from equation (14): Because $2h_1 = 2H_0 - (H_1^2 + H_2^2) = 2H_0 - v$, it follows that equation (14) can be rewritten as

$$\left(\frac{dh_1}{dt}\right)^2 = 2(H_0 - h_1)(M^2 - (h_1 - \epsilon)^2) - (M_1 - h_1 H_1)^2. \qquad (15)$$

Then $dh_1/dt = -M \sin\theta(d\theta/dt)$, and hence $(dh_1/dt)^2 = M^2 \sin^2\theta(d\theta/dt)^2$. Therefore

$$\left(\frac{d\theta}{dt}\right)^2 = 2(H_0 - (\epsilon + M\cos\theta)) - \frac{(M_1 - H_1(\epsilon + M\cos\theta))^2}{M^2 \sin^2\theta}. \qquad (16)$$

The remaining differential equations are obtained from the differential equations for $h_2(t)$ and $h_3(t)$. It follows that

$$\frac{dh_2}{dt} = \{h_2, H\} = \frac{H_1}{c_1}h_3 - H_3 h_1 + \epsilon H_3,$$

$$\frac{dh_3}{dt} = \{h_3, H\} = -\frac{H_1}{c_1}h_2 - H_2 h_1 - \epsilon H_2.$$

Therefore

$$\dot{\theta} M \cos\theta \sin\psi + \dot{\psi} M \sin\theta \cos\psi = \frac{dh_2}{dt} = \frac{H_1}{c_1}h_3 - H_3 h_1 + \epsilon H_3,$$

$$\dot{\theta} M \cos\theta \cos\psi - \dot{\psi} M \sin\theta \sin\psi = \frac{dh_3}{dt} = -\frac{H_1}{c_1}h_2 + H_2 h_1 - \epsilon H_2.$$

Then

$$\begin{pmatrix} \dot{\theta} \\ \dot{\psi} \end{pmatrix} = \frac{1}{M \cos \theta \sin \theta} \begin{pmatrix} -\sin \theta \sin \psi & -\sin \theta \cos \psi \\ -\cos \theta \cos \psi & \cos \theta \sin \psi \end{pmatrix} \\ \times \begin{pmatrix} -\frac{H_1}{c_1} h_3 + H_3 (h_1 - \epsilon) \\ \frac{H_1}{c_1} h_2 - H_2 (h_1 - \epsilon) \end{pmatrix}.$$

Hence,

$$\dot{\theta} = \frac{H_1}{c_1 M \cos \theta} (h_3 \sin \psi - h_2 \cos \psi) + \frac{h_1 - \epsilon}{M \cos \theta} (H_3 \sin \psi - H_2 \cos \psi),$$

$$\dot{\psi} = \frac{H_1}{c_1 M \sin \theta} (h_3 \cos \psi + h_2 \sin \psi) + \frac{h_1 - \epsilon}{M \sin \theta} (H_3 \cos \psi + H_2 \sin \psi).$$

Consequently,

$$\dot{\theta} = -H_3 \sin \psi + H_2 \cos \psi, \qquad (17a)$$

because $h_3 \sin \psi - h_2 \cos \psi = 0$. The differential equation for ψ simplifies with the aid of the following expressions:

$$H_3 \cos \psi + H_2 \sin \psi = (H_3 h_3 + H_2 h_2) \frac{1}{M \sin \theta} = \frac{M_1 - H_1 h_1}{M \sin \theta},$$

and $h_3 \cos \psi + h_2 \sin \psi = M \sin \theta$. Substituting these expressions in the differential equation for ψ leads to

$$\dot{\psi} = \frac{H_1}{c_1} + \frac{(h_1 - \epsilon)}{M^2 \sin^2 \theta} (M_1 - H_1 h_1). \qquad (17b)$$

Equation (17a) and the relation $H_3 \cos \psi + H_2 \sin \psi = (M_1 - H_1 h_1)/(M \sin \theta)$ provide the expressions for $H_2(t)$ and $H_3(t)$ in terms of θ. It follows that

$$H_2(t) = \dot{\phi} \sin \theta \sin \psi + \dot{\theta} \cos \psi \quad \text{and} \quad H_3(t) = \dot{\phi} \sin \theta \cos \psi - \dot{\theta} \sin \psi, \quad (18)$$

with ϕ denoting the angle defined by

$$\frac{d\phi}{dt} = \frac{M(M_1 - H_1 h_1)}{M^2 - (h_1(t) - \epsilon)^2}. \qquad (19)$$

The foregoing calculations demonstrate that the integration procedure is essentially the same in all three symmetric spaces, $\epsilon = 0$ and $\epsilon = \pm 1$. It may be helpful to summarize the main points of our integration procedure.

Equation (14) is the fundamental equation. Its solutions are obtained by means of elliptic functions, because $dv/dt = \pm\sqrt{f_\varepsilon(v)}$, with f_ε equal to a third-degree polynomial in v. The extremal curves $h_1(t)$, $h_2(t)$, $h_3(t)$ are confined to the sphere

$$(h_1 - \epsilon)^2 + h_2^2 + h_3^2 = M^2, \quad \text{with} \quad M^2 = M_2 - \epsilon(H_1^2 + 2H_0) - \epsilon^2.$$

The corresponding spherical angles $\theta(t)$ and $\psi(t)$ are determined by $h_1(t)$ through the formulas $M \cos\theta(t) = h_1(t) - \epsilon = H_0 - \frac{1}{2}v^2 - \epsilon$ and

$$\psi(t) = \frac{H_1}{c_1}t + \int_0^t \frac{(h_1(s) - \epsilon)(M_1 - H_1 h_1(s))}{M^2 - (h_1(s) - \epsilon)^2} \, ds + \psi(0).$$

Finally, the extremal controls u_2, u_3 are then determined through formulas (18). Of course, $u_1(t) = H_1$ is constant.

The projections of extremal curves in G are given by the integral curves $g(t)$ of equation (10), with $u_1(t)$, $u_2(t)$, and $u_3(t)$ equal to the extremal controls. The elastic curves are the projections of the integral curves $g(t)$ on G/K. We shall not go into the details required to integrate the time-varying differential equation (10) that describes the extremal curves in G. That procedure is somewhat lengthy, but conceptually simple.

Instead, we shall reemphasize the important theoretical differences separating the non-Euclidean elastic problem from the Euclidean elastic problem. The essential difference is that both $SO(1, 3)$ and $SO_4(R)$ are semisimple Lie groups, whereas the group of motions of E^3 is not. That means that in the non-Euclidean cases the co-adjoint orbits are in one-to-one correspondence with the adjoint orbits, because of the nondegeneracy of the trace form.

Recalling the results from Chapter 12 concerning the properties of left-invariant Hamiltonians, it follows that $g(t)P(t)g^{-1}(t) = $ constant for any extremal curve $(g(t), p(t))$ after the identification of $p(t)$ with

$$P(t) = \begin{pmatrix} 0 & -h_1 & -h_2 & -h_3 \\ \epsilon h_1 & 0 & -H_3 & H_2 \\ \epsilon h_2 & H_3 & 0 & -H_1 \\ \epsilon h_3 & -H_2 & H_1 & 0 \end{pmatrix}.$$

Then

$$\frac{dP}{dt} = [P, \Omega], \quad \text{with} \quad \Omega = B_1 + \sum_{i=1}^{3} \frac{H_i(t)}{c_i} A_i,$$

as can be verified by differentiating $g(t)P(t)g^{-1}(t)$.

Because $g(t)P(t)g^{-1}(t)$ is constant, it follows that the spectrum of $P(t)$ must be constant along each extremal curve. It is left to the reader to show that the characteristic polynomial of $P(t)$ is equal to

$$C(\lambda) = \lambda^4 + \lambda^2 \epsilon (\|h\|^2 + \epsilon \|\hat{H}\|^2) + \epsilon (h \cdot \hat{H})^2,$$

with $\hat{H} = (H_1, H_2, H_3)$. This isospectral information furnishes an independent proof that $h_1^2 + h_2^2 + h_3^2 + \epsilon(H_1^2 + H_2^2 + H_3^2)$ and $h_1 H_1 + h_2 H_2 + h_3 H_3$ are integrals of motion for each left-invariant Hamiltonian H. Then the integrals of motion for the Euclidean group E_3 can be seen as the limits of the foregoing expressions as ϵ tends to zero (regarding it as a continuous parameter).

Problem 2 Show that the elastic curves that correspond to the equilibrium solutions of equations (14) are given by $x(t) = g_0 \exp(t(B_1 + \tau A_1 + k A_3)e_1)$ for constant k and τ and for an arbitrary g_0 in G.

2.3 The Kowalewski elastic problem

The classic literature on the heavy-top problem is restricted exclusively to the integrable cases. The first integrable case occurs when the center of gravity of the body coincides with the fixed point about which the body is free to move (Euler's top). The second integrable case is known as the top of Lagrange, and it occurs when the body is symmetrical about the axis that connects the fixed point of the body to the center of gravity of the body. In both of these cases the role of symmetry and its connections to the integrals of motion have been understood for a long time, as described in preceding chapters.

There is, however, another integrable case that occurs when two moments of inertia are equal and are each twice the remaining moment of inertia, while the fixed point of the body is in the plane spanned by the principal moments corresponding to the equal moments of inertia. This case was discovered by S. Kowalewski in 1899 and has since been known as the Kowalewski top.

Ever since its discovery, the Kowalewski top has held a somewhat special place in the literature on the heavy top, partly because of its integration procedure, but mostly because of the mysterious appearance of the extra integral of motion.

In analogy with the heavy top, the symmetric elastic problem corresponds to the top of Lagrange. Euler's top appears only in the particular orbit corresponding to $h_1^2 + h_2^2 + h_3^2 = M_2 = 0$, for then $H_1^2 + H_2^2 + H_3^2$ is constant, which accounts for the remaining integral. In the elliptic elastic problem, the orbit $M_2 = 0$ reduces to the zero orbit in $\mathcal{L}^*$, because all variables $h_1, h_2, h_3, H_1, H_2, H_3$

are zero. The resulting elastic curves are geodesics on S^3. For the hyperbolic problem, the orbit $M_2 = 0$ results in the extremals being constrained to the light cone $h_1^2 + h_2^2 + h_3^2 - (H_1^2 + H_2^2 + H_3^2) = 0$. I do not know if this orbit also admits an extra integral of motion.

It is remarkable, however, that the Kowalewski integral appears in all elastic problems under precisely the same conditions discovered by Kowalewski, namely, when either $c_1 = c_2 = 2c_3$ or $c_1 = c_3 = 2c_2$. We shall demonstrate the existence of the extra integral of motion using essentially the same argument used by Kowalewski (1899).

Assume that $c_1 = c_2 = 2$ and that $c_3 = 1$. The Hamiltonian H of the corresponding elastic problem is given by $H = \frac{1}{4}(H_1^2 + H_2^2) + \frac{1}{2}H_3^2 + h_1$, and each extremal curve of $\vec{H}$ defines the following differential equations:

$$\frac{dH_1}{dt} = \{H_1, H\} = \frac{1}{2}H_2 H_3, \qquad \frac{dH_2}{dt} = \{H_2, H\} = -\frac{1}{2}H_1 H_3 + h_3,$$

$$\frac{dh_1}{dt} = \{h_1, H\} = H_3 h_2 - \frac{1}{2}H_2 h_3, \qquad \frac{dh_2}{dt} = \{h_2, H\}$$

$$= \frac{1}{2}H_1 h_3 - H_3 h_1 + \epsilon H_3.$$

These relations are easily verifiable by reading off the appropriate Poisson brackets from Table 14.1.

Let $z(t)$ and $w(t)$ denote the complex-valued curves defined by $z(t) = \frac{1}{2}(H_1(t) + i H_2(t))$ and $w(t) = h_1(t) + i h_2(t) - \epsilon$. Then

$$2\frac{dz}{dt} = \frac{dH_1}{dt} + i\frac{dH_2}{dt} = \frac{1}{2}H_2 H_3 + i\left(-\frac{1}{2}H_1 H_3 + h_3\right)$$
$$= -i H_3(t) z(t) + i h_3(t).$$

Therefore

$$2z\frac{dz}{dt} = -i H_3 z^2 + i h_3 z.$$

Furthermore,

$$\frac{dw}{dt} = \frac{dh_1}{dt} + i\frac{dh_2}{dt} = \left(H_3 h_2 - \frac{1}{2}H_2 h_3\right) + i\left(\frac{1}{2}H_2 h_3 - H_3 h_1 + \epsilon H_3\right)$$
$$= i h_3 z - i H_3 w.$$

Then $(d/dt)(z^2 - w) = -i H_3(z^2 - w)$. But then the complex conjugate $\overline{z^2 - w}$ satisfies $(d/dt)\overline{z^2 - w} = i H_3(\overline{z^2 - w})$. Therefore, $|z^2 - w|^2$ is another integral of motion, because

$$\frac{d}{dt}|z^2 - w|^2 = \frac{d}{dt}(z^2 - w)\overline{(z^2 - w)}$$
$$= -iH_3(z^2 - w)\overline{(z^2 - w)} + iH_3\overline{(z^2 - w)}(z^2 - w)$$
$$= 0.$$

This extra integral of motion is a polynomial of degree 4 equal to

$$\left|\frac{1}{4}(H_1 + iH_2)^2 - (h_1 + ih_2 - \epsilon)\right|^2 = \left(\frac{1}{4}(H_1^2 - H_2^2) - h_1 + \epsilon\right)^2$$
$$+ \left(\frac{1}{2}H_1H_2 - h_2\right)^2.$$

Therefore we have the following:

Theorem 4 *The Hamiltonian system corresponding to $H = \frac{1}{4}(H_1^2 + H_2^2) + \frac{1}{2}H_3^2 + h_1$ is completely integrable on G. In addition to M_1, M_2, M_3, M_4,*

$$M_5 = \left(\frac{1}{4}(H_1^2 - H_2^2) - h_1 + \epsilon\right)^2 + \left(\frac{1}{2}H_1H_2 - h_2\right)^2$$

is also an integral of motion.

3 Rolling-sphere problems

The configuration space for an n-dimensional sphere rolling, without slipping, on an n-dimensional Euclidean space E^n is equal to $G = E^n \times SO_{n+1}(R)$, with the kinematic paths of the sphere given by the solution curves of the following differential system:

$$\frac{dx_1}{dt} = u_1, \ldots, \frac{dx_n}{dt} = u_n \quad \text{and} \quad \frac{dR}{dt} = R \begin{pmatrix} 0 & \cdots & 0 & -u_1 \\ & & & -u_2 \\ \vdots & & \vdots & \vdots \\ 0 & \cdots & 0 & -u_n \\ u_1 u_2 & \cdots & u_n & 0 \end{pmatrix}, \quad (20)$$

where

$$x = \begin{pmatrix} x_1 \\ \vdots \\ x_n \\ 1 \end{pmatrix}$$

are the coordinates of the center of the sphere expressed in a fixed orthonormal frame $\vec{e}_1, \ldots, \vec{e}_{n+1}$. In deriving equations (20), it is assumed that E^n is the projection of E^{n+1} corresponding to $x_{n+1} = 0$, and it is also implicitly assumed that the sphere is not allowed to spin in any direction perpendicular to the velocity vector of the point of contact between the sphere and the stationary space E^n.

Differential system (20) is spanned by n left-invariant vector fields $\vec{X}_1, \ldots, \vec{X}_n$ in G defined by $\vec{X}_i(x, R) = \vec{e}_i \oplus R A_i$, with A_i equal to the antisymmetric matrix $(e_i \otimes e_{n+1}) - (e_{n+1} \otimes e_i)$. The value of each vector field $\vec{X}_i$ at the group identity is equal to $X_i(0, I) = e_i \oplus A_i$.

Problem 3 Show that (20) is controllable on $E^n \times SO_{n+1}(R)$ for $n \geq 2$.

Each path in the configuration space projects onto a path $x(t)$ traced by the point of contact of the sphere with the horizontal space E^n. The Euclidean length of $x(t)$ is given by the integral

$$\int \left(\sum_{i=1}^{n} \frac{dx_i^2}{dt} \right)^{1/2} dt = \int \|u(t)\| \, dt,$$

where $\|u(t)\|^2 = \sum_{i=1}^{n} u_i^2(t)$.

The problem of finding the curve $x(t)$ in E^n of minimal length along which the sphere should be rolled from a prescribed initial configuration to a prescribed terminal configuration will be called the *rolling-sphere problem on* E^n. The rolling-sphere problem lifts to a left-invariant optimal-control problem on the configuration space $G = E^n \times SO_{n+1}(R)$ via equations (20): The solutions of the lifted problem project onto the solutions of the rolling-sphere problem. The rolling-sphere problem can also be regarded as a time-optimal problem, provided that the sphere is never stationary, that is, that the velocity of the center of the sphere is never equal to zero. All such paths can be parametrized by their arc lengths, in which case the Euclidean length of the curve is equal to the time required to bring the initial configuration into the terminal configuration.

3.1 The extremals for the rolling sphere in E^2

As tempting as it is to consider the problem in its full generality, we shall consider only the planar case, leaving the higher-dimensional investigations for further research.

For $n = 2$, $G = E^2 \times SO_3(R)$, and the distribution (20) is spanned by two left-invariant vector fields $\vec{X}_1$ and $\vec{X}_2$. Let a_1, a_2 denote the Hamiltonians of the constant vector fields $\vec{e}_1 \oplus 0$ and $\vec{e}_2 \oplus 0$ in G, and let h_1, h_2 denote the Hamiltonians of the left-invariant vector fields $0 \oplus \vec{A}_1$ and $0 \oplus \vec{A}_2$. In terms of this notation, the Hamiltonians of X_1 and X_2 are given by $a_1 + h_1$ and $a_2 + h_2$. Then h_3 will denote the Poisson bracket $\{h_1, h_2\}$. It follows that h_3 is the Hamiltonian of the left-invariant vector field X_3 on G whose value at the identity is equal to $0 \oplus A_3$, with

$$
A_3 = \begin{pmatrix} 0 & 1 & 0 \\ -1 & 0 & 0 \\ 0 & 0 & 0 \end{pmatrix}.
$$

It then follows from the Lie-algebra structure of $SO_3(R)$ that $\{h_3, h_1\} = h_2$, and $\{h_3, h_2\} = -h_1$.

E^2 regarded as a subgroup of G defines an abelian subalgebra. The right- and left-invariant vector fields on E^2 coincide, which further implies that a_1, a_2 are constants of motion for any left- or right-invariant Hamiltonian flow on G. Stated differently, a_1 and a_2 are Casimir elements on the dual $\mathcal{L}^*$ of the Lie algebra of G. As usual, $T^*G = G \times \mathcal{L}^*$. Then we have the following:

Theorem 5 *Abnormal extremal curves $\xi(t)$ are integral curves of the time-varying Hamiltonian vector field $\vec{H}(u_1(t), u_2(t))$ corresponding to*

$$
H(u_1(t), u_2(t)) = u_1(t)(a_1 + h_1) + u_2(t)(a_2 + h_2)
$$

and satisfying the constraints $(a_i + h_i)(\xi(t)) = 0$ for $i = 1, 2$, and $h_3(\xi(t)) = $ constant. Furthermore, $u_1(t) = f(t)h_1(\xi(t))$, and $u_2(t) = f(t)h_2(\xi(t))$ for some function $f(t)$ that is arbitrary. For abnormal extremal curves that project onto nonstationary optimal curves in G, f is constant, and $h_3(\xi(t)) = 0$.

Proof Suppose that $\xi(t)$ is an abnormal extremal curve generated by the control functions $u_1(t)$, $u_2(t)$. $\xi(t)$ is an integral curve of $\vec{H}$, with $H = u_1(t)(a_1 + h_1) + u_2(t)(a_2 + h_2)$, subject to the constraints that $(a_i + h_i)(\xi(t)) = 0$, with $i = 1, 2$, for almost all t in the domain of ξ. Because $\xi(t)$ is an absolutely continuous curve, it follows that $(a_i + h_i)(\xi(t)) = 0$ for all t. Differentiation of $(a_i + h_i)(\xi(t))$ leads to

$$
0 = \frac{d}{dt}(a_i + h_i)(\xi(t)) = \{a_i + h_i, \; u_1(a_1 + h_1) + u_2(a_2 + h_2)\}(\xi(t))
$$
$$
= u_1(t)\{h_i, h_1\} + u_2(t)\{h_i, h_2\}.
$$

Therefore $u_1(t)h_3(\xi(t)) = u_2(t)h_3(\xi(t)) = 0$.

Assume now that at least one of the control functions is not identically equal to zero. Then $h_3(\xi(t)) = 0$, and therefore

$$0 = \frac{d}{dt} h_3(\xi(t)) = \{h_3, u_1(t)(a_1 + h_1) + u_2(t)(a_2 + h_2)\}(\xi(t))$$

$$= u_1(t)\{h_3, h_1\} + u_2(t)\{h_3, h_2\} = u_1(t)h_2(\xi(t)) - u_2(t)h_1(\xi(t)).$$

Hence $u_1(t) = f(t)h_1(\xi(t))$ and $u_2(t) = f(t)h_2(\xi(t))$. If both $u_1(t)$ and $u_2(t)$ are identically zero, then the foregoing calculation shows that $(d/dt)h_3(\xi(t)) = 0$, and therefore $h_3(\xi(t)) = $ constant.

In order to get to the second part of the theorem, assume that the projection $g(t)$ of $\xi(t)$ is optimal and nonstationary. Both $h_1(\xi(t))$ and $h_2(\xi(t))$ are constant along $\xi(t)$, because $(a_i + h_i)(\xi(t)) = 0$, and each a_i is constant along $\xi(t)$ for $i = 1, 2$. Hence the total cost along $\xi(t)$ is given by

$$\int_0^T \sqrt{u_1^2 + u_2^2} \, dt = \sqrt{h_1^2 + h_2^2} \int_0^T \sqrt{f^2(t)} \, dt = \sqrt{h_1^2 + h_2^2} \int_0^T |f(t)| \, dt.$$

The initial and final configurations of $g(t)$ determine the value of $\int_0^T f(t) \, dt$, because $x_1(T) - x_1(0) = \int_0^T u_1(t) \, dt = h_1 \int_0^T f(t) \, dt$ and $x_2(T) - x_2(0) = h_2 \int_0^T f(t) \, dt$. It then follows that $\int_0^T \sqrt{f^2(t)} = \int_0^T |f(t)| \, dt$ is equal to the smallest value of $\int_0^T |\phi(t)| \, dt$ among all functions $\phi(t)$ that satisfy the constraint $\int_0^T \phi(t) \, dt = \int_0^T f(t) \, dt = c$. The minimal solution of this problem is constant, and the proof is finished.　∎

The regular extremal curves are described by the following:

Theorem 6 *Each regular extremal curve* $\xi(t) = (g(t), p(t))$ *is an integral curve of a family of Hamiltonians given by* $H_\lambda = \lambda((((a_1 + h_1)^2 + (a_2 + h_2)^2)^{1/2} - \frac{1}{2})^2 - \frac{1}{4})$, *with* λ *an arbitrary parameter. Moreover,* $H_\lambda(\xi(t)) = 0$ *for each* λ. *The control functions that correspond to* $\xi(t)$ *are given by*

$$u_1(t) = \lambda(a_1 + h_1)(p(t)), \qquad u_2(t) = \lambda(a_2 + h_2)(p(t)).$$

In particular, if $u_1^2 + u_2^2 \neq 0$, *then* $(a_1 + h_1)^2 + (a_2 + h_2)^2 = 1$.

Proof Assume that $(g(t), p(t))$ is a regular extremal curve generated by the control functions $u_1(t)$ and $u_2(t)$. Then it follows from the maximum principle that

$$-\sqrt{u_1^2(t) + u_2^2(t)} + u_1(t)(a_1 + h_1(p(t))) + u_2(t)(a_2 + h_2(p(t)))$$

$$\geq -\sqrt{v_1^2 + v_2^2} + v_1(a_1 + h_1(p(t))) + v_2(a_2 + h_2(p(t)))$$

for any (v_1, v_2) in $\mathbb{R}^2$ and almost all t. The function $f(v_1, v_2) = -(v_1^2 + v_2^2)^{1/2} + v_1 A + v_2 B$, with $A = (a_1 + h_1)(p(t))$ and $B = (a_2 + h_2)(p(t))$, is homogeneous. Therefore

$$\lambda f(v_1(t), v_2(t)) = f(\lambda v_1(t), \lambda v_2(t)) \leq f(u_1(t), u_2(t)),$$

and consequently $f(v_1, v_2) \leq 0$ for all (v_1, v_2) in $\mathbb{R}^2$.

If $A = B = 0$, then $u_1(t) = u_2(t) = 0$, and the conditions of the theorem are evidently satisfied. So assume that $A^2 + B^2 > 0$. Then $f \leq 0$ if and only if $f \leq 0$ on the unit circle $v_1^2 + v_2^2 = 1$. The maximum value of f on $v_1^2 + v_2^2 = 1$ is equal to zero and occurs for $v_1 = A/(A^2 + B^2)^{1/2}$ and $v_2 = B/(A^2 + B^2)^{1/2}$.

The maximality condition of the maximum principle then shows that $u_1(t) = \lambda A(t)$ and $u_2(t) = \lambda B(t)$, and consequently $(g(t), p(t))$ is an integral curve of

$$H_\lambda = \lambda\left(-\sqrt{(a_1 + h_1)^2 + (a_2 + h_2)^2} + (a_1 + h_1)^2 + (a_2 + h_2)^2\right).$$

H_λ can also be written as $\lambda(((a_1 + h_1)^2 + (a_2 + h_2)^2)^{1/2} - \frac{1}{2})^2 - \frac{1}{4})$, and the proof is now finished. ∎

Remark It is well known in the classic literature on the calculus of variations that a Lagrangian that is homogeneous with respect to the velocity variables does not satisfy the strong Legendre condition along an extremal curve, and consequently the extremal curve does not determine a unique Hamiltonian (Carathéodory, 1935). The nonuniqueness of the Hamiltonian in Theorem 6 is a manifestation of the same phenomenon; the extremal curves are determined only up to an arbitrary parametrization. For $\lambda = 1$, the nonzero extremal controls satisfy $u_1^2 + u_2^2 = 1$, and the corresponding extremal curves are parametrized by arc length. ∎

The Hamiltonian system defined by H_λ is completely integrable, for the following reasons. Let $H = \frac{1}{2}(a_1 + h_1)^2 + \frac{1}{2}(a_2 + h_2)^2$. Although H and H_λ are different functions, their Hamiltonian vector fields $\vec{H}$ and $\vec{H}_\lambda$, restricted to the hypersurface $H = \frac{1}{2}$ that contains the nonstationary extremal curves, satisfy

$$\vec{H}_\lambda = \lambda \vec{H}$$

Therefore the integral curves of $\vec{H}_\lambda$ and $\vec{H}$ contained in the surface $H = \frac{1}{2}$ differ only by parametrizations, and so their integrability properties are essentially the same.

For simplicity of exposition we shall work with H, rather than H_λ, keeping in mind that the extremal curves satisfy $(a_1 + h_1)^2 + (a_2 + h_2)^2 = 1$. Because H is a left-invariant function, it Poisson-commutes with each Hamiltonian defined by a right-invariant vector field on G. The translations on $\mathbb{R}^2$ are both left- and right-invariant, and therefore a_1 and a_2 are constants of motion. In addition to these integrals of motion, there is another integral that corresponds to a right-invariant vector field on $SO_3(R)$. This procedure accounts for three integrals of motion for H. The reader can easily check that the Casimir function $M^2 = h_1^2 + h_2^2 + h_3^2$ on $SO_3(R)$ is also a Casimir function on $E^2 \times SO_3(R)$. Together with H, these functions constitute five independent integrals of motion, all in involution with each other. Therefore, the Hamiltonian system $\vec{H}$ is completely integrable, and the remaining task is to find suitable coordinates in terms of which the extremal equations can be integrated by quadrature.

There is a system of coordinates naturally suggested by the geometry of the integrals of motion: Each extremal curve is contained in the intersection of the cylinder $(a_1 + h_1)^2 + (a_2 + h_2)^2 = 1$ with the sphere $M^2 = h_1^2 + h_2^2 + h_3^2$. Figure 14.4 describes distinct geometric types of intersections.

Fig. 14.4.

The geometric types of solutions depend on the location of the circle $(a_1 + h_1)^2 + (a_2 + h_2)^2 = 1$ in relation to the disc $h_1^2 + h_2^2 \leq M^2$ in the equatorial plane $h_3 = 0$. In the first case, the circle is contained in the interior of the disc; that is, $M > (a_1^2 + a_2^2)^{1/2} + 1$. In the second case, the circle is tangent to the boundary of the disc, $M = (a_1^2 + a_2^2)^{1/2} + 1$, and in the last case, $(a_1^2 + a_2^2)^{1/2} - 1 < M < (a_1^2 + a_2^2)^{1/2} + 1$, and the circle intersects the boundary of the disc (Figure 14.5).

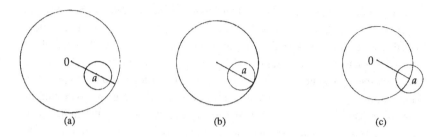

(a)   (b)   (c)

Fig. 14.5.

Each of these types of solutions can be naturally parametrized by an angle θ and a phase shift α defined as follows: α is an angle such that $\sin \alpha = a_1/(a_1^2 + a_2^2)^{1/2}$, and $\cos \alpha = a_2/(a_1^2 + a_2^2)^{1/2}$. We take $\alpha = 0$ if $(a_1^2 + a_2^2)^{1/2} = 0$. θ is defined through the formula $a_1 + h_1 = \sin(\theta + \alpha)$, and $a_2 + h_2 = \cos(\theta + \alpha)$.

Each extremal curve $(g(t), p(t))$ induces a movement of θ that obeys the following differential equation:

$$\frac{d\theta}{dt} = h_3(p(t)).$$

The argument is as follows:

$$\cos(\theta + \alpha)\frac{d\theta}{dt} = \frac{d}{dt}(a_1 + h_1(p(t))) = \{h_1, H\} = (a_2 + h_2)\{h_1, h_2\}$$
$$= h_3(p(t))\cos(\theta(t) + \alpha).$$

The differential equation for θ is the same as the equation for the mathematical pendulum whose total energy $E = \frac{1}{2}(M^2 - (1 + a_1^2 + a_2^2))$, as can be seen from the following calculation:

$$\left(\frac{d\theta}{dt}\right)^2 = (h_3(p(t)))^2 = M^2 - \left(h_1^2 + h_2^2\right)$$
$$= M^2 - \left(\sin(\theta + \alpha) - \sin\alpha\sqrt{a_1^2 + a_2^2}\right)^2$$
$$- \left(\cos(\theta + \alpha) - \cos\alpha\sqrt{a_1^2 + a_2^2}\right)^2$$
$$= M^2 - \left(1 + (a_1^2 + a_2^2)\right) + 2\cos\theta\sqrt{a_1^2 + a_2^2}.$$

It follows that

$$\frac{d\theta}{dt} = \pm\sqrt{2(E + A\cos\theta)}, \quad \text{with} \quad A = \sqrt{a_1^2 + a_2^2}.$$

The geometric types of solutions depicted in Figure 14.5 can now be classified in terms of the constants E and A, as follows: $E > A$ will be called the inflectional case, $E = A$ will be called the critical case, and $E < A$ will be called the noninflectional case. It follows that $E - A = \frac{1}{2}(M^2 - (1 + (a_1^2 + a_2^2)^{1/2})))$, and therefore these cases agree with the types of solutions described by Figure 14.5. In the noninflectional case the pendulum has sufficient energy to go around its top; in the critical case it reaches the top in infinite time; in the inflectional case there is a cutoff angle θ_c, and the pendulum oscillates between its extreme values $-\theta_c$ and θ_c. The choice of terminology for these cases is motivated by the movements of the center of the ball.

Recall that $dx_1/dt = u_1(t)$, and $dx_2/dt = u_2(t)$. Along the extremal curves,

$$\frac{dx_1}{dt} = a_1 + h_1(p(t)) = \sin(\theta + \alpha) \text{ and } \frac{dx_2}{dt} = a_2 + h_2(p(t)) = \cos(\theta + \alpha).$$

Let $\bar{x}_1 = x_1 \cos\alpha - x_2 \sin\alpha$, and $\bar{x}_2 = x_1 \sin\alpha + x_2 \cos\alpha$. Then

$$\frac{d\bar{x}_1}{dt} = \sin\theta(t), \qquad \frac{d\bar{x}_2}{dt} = \cos\theta(t), \qquad \frac{d\theta}{dt} = \pm\sqrt{2(E + A\cos\theta)}.$$

The foregoing equations coincide with the equations for the elastic problem of Euler, and therefore we have the following:

Theorem 7 *Each extremal path of the center of the sphere is also an extremal curve for the elastic problem of Euler.*

In the inflectional case, the geodesic curvature of the path traced by the center of the sphere changes its sign, whereas in the noninfectional case the sign is always the same.

The rotational kinematics of the sphere also undergo a bifurcating type of behavior as the system changes from one geometric type to another. The bifurcating characteristics are best described in terms of the changes in Euler angles along the extremal curves, as follows:

Let $R(t)$ denote the rotation matrix associated with an extremal curve. Then

$$\frac{dR}{dt} = R \begin{pmatrix} 0 & 0 & -u_1 \\ 0 & 0 & -u_2 \\ u_1 & u_2 & 0 \end{pmatrix},$$

with $u_1 = a_1 + h_1(p(t)) = \sin(\theta + \alpha)$ and $u_2 = a_2 + h_2(p(t)) = \cos(\theta + \alpha)$.

It is convenient to eliminate the phase shift α through an additional rotation

$$R_0 = \begin{pmatrix} \cos\alpha & -\sin\alpha & 0 \\ \sin\alpha & \cos\alpha & 0 \\ 0 & 0 & 1 \end{pmatrix}.$$

It follows that

$$\frac{d}{dt}R(t)R_0^{-1} = R(t)R_0^{-1}R_0 \begin{pmatrix} 0 & 0 & -u_1 \\ 0 & 0 & -u_2 \\ u_1 & u_2 & 0 \end{pmatrix} R_0^{-1} = \bar{R}(t) \begin{pmatrix} 0 & 0 & -\bar{u}_1 \\ 0 & 0 & -\bar{u}_2 \\ \bar{u}_1 & \bar{u}_2 & 0 \end{pmatrix},$$

with $\bar{R}(t) = R(t)R_0^{-1}$, $\bar{u}_1 = u_1\cos\alpha - u_2\sin\alpha$, $\bar{u}_2 = u_1\sin\alpha + u_2\cos\alpha$.
Hence $\bar{u}_1(t) = \sin\theta(t)$ and $\bar{u}_2(t) = \cos\theta(t)$.

Let ϕ_1, ϕ_2, ϕ_3 denote the Euler angles defined by $R(t) = e^{\phi_1 A_3}e^{\phi_2 A_2}e^{\phi_3 A_3}$,
with

$$A_1 = \begin{pmatrix} 0 & 0 & 0 \\ 0 & 0 & -1 \\ 0 & 1 & 0 \end{pmatrix}, \qquad A_2 = \begin{pmatrix} 0 & 0 & 1 \\ 0 & 0 & 0 \\ 1 & 0 & 0 \end{pmatrix}, \qquad A_3 = \begin{pmatrix} 0 & 1 & 0 \\ -1 & 0 & 0 \\ 0 & 0 & 0 \end{pmatrix}.$$

Recall that the rotational kinematics are given by

$$\frac{dR}{dt} = -\vec{A}_2(R)u_1 + \vec{A}_1(R)u_2.$$

It follows, by a calculation completely analogous to that used in Section 1.1,
that

$$\dot{\phi}_1 = -\frac{1}{\sin\phi_2}(\bar{u}_2\cos\phi_3 + \bar{u}_1\sin\phi_3),$$

$$\dot{\phi}_2 = \bar{u}_2\sin\phi_3 - \bar{u}_1\cos\phi_3,$$

$$\dot{\phi}_3 = \frac{\cos\phi_2}{\sin\phi_2}(\bar{u}_2\cos\phi_3 + \bar{u}_1\sin\phi_3).$$

Along an extremal curve $(g(t), p(t))$, the projection $p(t)$ on the dual of the Lie
algebra of $SO_3(R)$ can be identified with the matrix

$$P = \begin{pmatrix} 0 & h_3 & -h_1 \\ -h_3 & 0 & -h_2 \\ h_1 & h_2 & 0 \end{pmatrix}$$

via the trace form.

Then $R(t)P(t)R^{-1}(t)$ is a constant matrix Λ along each extremal curve,
and consequently $P(t) = R^{-1}\Lambda R$. It will be convenient to make a change of

variables in the matrix P that absorbs the phase shift α. Let $\bar{P}(t)$ denote the matrix $R_0 P(t) R_0^{-1}$. Then

$$\bar{P}(t) = \begin{pmatrix} 0 & h_3 & -\bar{h}_1 \\ -h_3 & 0 & -\bar{h}_2 \\ \bar{h}_1 & \bar{h}_2 & 0 \end{pmatrix},$$

with $\bar{h}_1 = (\cos\alpha)h_1 - (\sin\alpha)h_2$, and $\bar{h}_2 = (\sin\alpha)h_1 + (\cos\alpha)h_2$. Making use of the substitutions $h_1 = \sin(\theta + \alpha) - A\sin\alpha$ and $h_2 = \cos(\theta + \alpha) - A\cos\alpha$ in the foregoing expressions for $\bar{h}_1$ and $\bar{h}_2$ gives $\bar{h}_1 = \sin\theta$ and $\bar{h}_2 = \cos\theta - A$.

We shall further simplify the integration procedure by taking Λ to be equal to $-M A_3$, with M equal to the value of the Casimir function $h_1^2 + h_2^2 + h_3^2$. (Any antisymmetric matrix Λ can be rotated so that it becomes a scalar multiple of A_3). This choice of Λ naturally goes with the choice of Euler angles, because

$$\bar{h}_2 e_1 - \bar{h}_1 e_2 - h_3 e_3 = \hat{\bar{P}}(t) = -M\bar{R}^{-1} e_3 = -M e^{-\phi_3 A_3} e^{-\phi_2 A_2} e_3.$$

It then follows that

$$M\sin\phi_2 \cos\phi_3 = \bar{h}_2 = \cos\theta - A, \qquad M\sin\phi_2 \sin\phi_3 = \bar{h}_1 = \sin\theta,$$

$$M\cos\phi_2 = h_3 = \dot{\theta}. \tag{21}$$

The foregoing formulas will now be used to express the differential equations for ϕ_1, ϕ_2, ϕ_3 along the extremal curves in terms of the angle θ. To begin with,

$$\frac{d\phi_1}{dt} = -\frac{1}{\sin\phi_2}(\bar{u}_2 \cos\phi_3 + \bar{u}_1 \sin\phi_3) = -\frac{1}{M\sin\phi_2}\left(\cos\theta\,\frac{\bar{h}_2}{\sin\phi_2} + \frac{\bar{h}_1 \sin\theta}{\sin\phi_2}\right)$$

$$= -\frac{1}{M\sin^2\phi_2}(\cos\theta(\cos\theta - A) + \sin^2\theta) = \frac{A\cos\theta - 1}{M\sin^2\phi_2}.$$

Because $M^2\cos^2\theta = h_3^2 = (d\theta/dt)^2 = 2(E + A\cos\theta)$, the foregoing differential equation can also be written as

$$\frac{d\phi_1}{dt} = \frac{(A\cos\theta - 1)M}{M^2 - 2(E + A\cos\theta)}.$$

Continuing,

$$\frac{d\phi_2}{dt} = \bar{u}_2 \sin\phi_3 - \bar{u}_1 \cos\phi_3 = \frac{\bar{h}_1 \cos\theta}{M\sin\phi_2} - \frac{\bar{h}_2 \sin\theta}{M\sin\phi_2}$$

$$= \frac{1}{M\sin\phi_2}(\cos\theta \sin\theta - \sin\theta(\cos\theta - A)) = \frac{A\sin\theta}{M\sin\phi_2}.$$

The substitution $\sin \phi_2 = \pm(1/M)(M^2 - 2(E + A \cos \theta))^{1/2}$ gives $d\phi_2/dt = \pm(A \sin \theta)/(M^2 - 2(E + A \cos \theta))^{1/2}$. Finally,

$$\frac{d\phi_3}{dt} = \frac{\cos \phi_2}{\sin \phi_2}(\bar{u}_2 \cos \phi_3 + \bar{u}_1 \sin \phi_3) = \frac{(1 - A \cos \theta)\sqrt{2(E + A \cos \theta)}}{M^2 - 2(E + A \cos \theta)}.$$

The foregoing differential equations are valid under the assumption that

$$\sin \phi_2 = \pm\frac{1}{M}\sqrt{M^2 - 2(E + A \cos \theta)} \neq 0.$$

It follows that $\sin \phi_2 \neq 0$ for all A, with $0 \leq A < 1$, for the following reasons: The expression $M^2 - 2(E + A \cos \theta)$ is equal to $A^2 - 2A \cos \theta + 1$. Hence

$$(1 - A)^2 = A^2 - 2A + 1 \leq M^2 - 2(E + A \cos \theta) \leq A^2 + 2A + 1 = (1 + A)^2,$$

and therefore $\sin \phi_2 \neq 0$ for all A that satisfy $0 \leq A < 1$.

It remains to analyze the rotational kinematics of the sphere as the geometric type changes from the noninflectional case to the inflectional case, under the assumption that $0 \leq A < 1$.

3.2 Noninflectional solutions

Because $E > A$, there is no value of θ for which $(2(E + A \cos \theta))^{1/2}$ is equal to zero. Furthermore, $d\theta/dt = h_3(t) = \pm(2(E + A \cos \theta))^{1/2}$ implies that the sign of $d\theta/dt$ is determined by the initial value of h_3. If this value is positive, then $d\theta/dt = (2(E + A \cos \theta))^{1/2}$; otherwise, $d\theta/dt = -(2(E + A \cos \theta))^{1/2}$ for all values of θ.

In the first case, $\theta(t)$ is increasing, but in the second case, $\theta(t)$ is decreasing. In either case,

$$\frac{d\phi_1}{dt} = \frac{(A \cos \theta - 1)M}{M^2 - 2(E + A \cos \theta)} = \frac{(A \cos \theta - 1)M}{A^2 - 2A \cos \theta + 1} < 0,$$

and therefore $\phi_1(t)$ is a decreasing function of time.

The angles ϕ_1, ϕ_2, and ϕ_3 can also be parametrized by the angle θ, in which case

$$\frac{d\phi_1}{d\theta} = \pm\frac{(A \cos \theta - 1)M}{(A^2 - 2A \cos \theta + 1)\sqrt{2(E + A \cos \theta)}},$$

and therefore

$$\phi_1(\theta_2) - \phi_1(\theta_1) = \pm M \int_{\theta_1}^{\theta_2} \frac{(A \cos t - 1)\, dt}{(A^2 - 2A \cos t + 1)\sqrt{2(E + A \cos t)}}.$$

The remaining angles ϕ_2 and ϕ_3 are determined from the relations (21).

It then follows that

$$e^{\phi_2 A_2} = \begin{pmatrix} \cos\phi_2 & 0 & \sin\phi_2 \\ 0 & 1 & 0 \\ -\sin\phi_2 & 0 & \cos\phi_2 \end{pmatrix} \quad \text{and} \quad e^{\phi_3 A_3} = \begin{pmatrix} 1 & 0 & 0 \\ 0 & \cos\phi_3 & -\sin\phi_3 \\ 0 & \sin\phi_3 & \cos\phi_3 \end{pmatrix}$$

with

$$\cos\phi_2 = \pm\frac{1}{M}\sqrt{2(E + A\cos\theta)}, \quad \sin\phi_2 = \pm\frac{1}{M}\sqrt{A^2 - 2A\cos\theta + 1},$$

$$\cos\phi_3 = \pm\frac{(\cos\theta - 1)M}{\sqrt{A^2 - 2A\cos\theta + 1}}, \quad \sin\phi_3 = \pm\frac{M\sin\theta}{\sqrt{A^2 - 2A\cos\theta + 1}}.$$

These observations can be assembled in the following theorem.

Theorem 8 *If $E > A$, then both matrices $e^{\phi_2 A_2}$ and $e^{\phi_3 A_3}$ are periodic, with period equal to 2π. The rotation matrix $R(t)$ is periodic if and only if*

$$\phi_1(2\pi) - \phi_1(0) = M \int_0^{2\pi} \frac{(A\cos t - 1)\, dt}{(A^2 - 2A\cos t + 1)\sqrt{2(E + A\cos t)}} \tag{22}$$

is a rational multiple of 2π.

The rotational kinematics described by Theorem 8 can also be viewed as the motions on the torus T^2, with the coordinate ϕ_2 playing the role of the varying radius that changes from its maximal value $\cos^{-1}((A + 1)/M)$ to the minimal value $\cos^{-1}((1 - A)/M)$.

The limiting case $A = 0$ corresponds to particularly simple solutions. In such a case, $E = \frac{1}{2}(M^2 - 1)$, and $d\theta/dt = \sqrt{2E} = (M^2 - 1)^{1/2}$. Because the corresponding controls are given by $u_1 = \sin\theta(t)$ and $u_2 = \cos\theta(t)$, it follows that both u_1 and u_2 are constant for $E = 0$. The corresponding motions consist of paths along straight lines, with the sphere rolling with constant angular velocity $\omega = (\cos\theta)e_1 - (\sin\theta)e_2$.

For the values of $E \neq 0$, $\theta(t) = \sqrt{2E}t + \theta_0$. Then $dx_1/d\theta = \sin\theta/\sqrt{2E}$, and $dx_2/d\theta = \cos\theta/\sqrt{2E}$, and therefore

$$x_1(\theta) = -\frac{1}{\sqrt{2E}}(\cos\theta - \cos\theta_0) \quad \text{and} \quad x_2(\theta) = \frac{1}{\sqrt{2E}}(\sin\theta - \sin\theta_0).$$

Hence the center of the sphere moves along a circle passing through the origin whose radius is equal to $1/\sqrt{2E} = 1/(M^2 - 1)^{1/2}$.

The rotational kinematics are described by the angles ϕ_1 and ϕ_3, since $\phi_2 =$ constant. It follows from (22) that $\phi_1(2\pi) - \phi_1(0) = M \int_0^{2\pi} -dt/\sqrt{2E} = (-M/(M^2 - 1)^{1/2})2\pi$.

Therefore the motion is periodic if and only if $M/(M^2 - 1)^{1/2}$ is a rational number. Otherwise, the motion is dense on the torus defined by the angles ϕ_1, ϕ_3. Some typical noninflectional solutions are sketched in Figure 14.6.

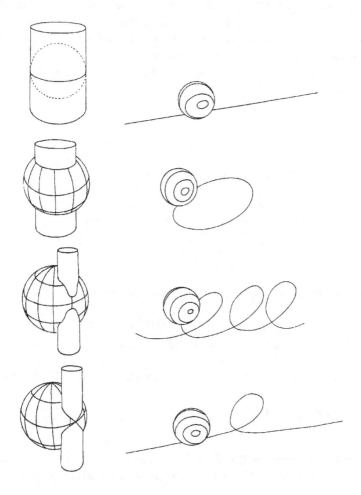

Fig. 14.6.

3.3 Inflectional solutions

Because $E < A$, the angle θ is confined to the interval $-\theta_c \leq \theta \leq 2\pi - \theta_c$, with $\theta_c = \cos^{-1}(E/A)$. As the pendulum swings from θ_c to $2\pi - \theta_c$, the quantity

$h_3(t)$ is positive, because $h_3(t) = d\theta/dt$. At the extreme values $\pm\theta_c$, $h_3(t) = 0$. Because

$$\frac{dh_3}{dt} = \frac{d^2\theta}{dt^2} = -A\sin\theta,$$

it follows that $h_3(t)$ reaches its maximum at $\theta = \pi$.

On the way back, $h_3(t)$ reaches its negative minimum value at $\theta = \pi$, and then the pattern repeats with each additional swing of the pendulum (Figure 14.7). It then follows from relations (21) that $\cos\phi_2 = (1/M)(2(E + A\cos\theta))^{1/2}$ as θ increases from θ_c to $2\pi - \theta_c$ and that $\cos\phi_2 = -1/M(2(E + A\cos\theta))^{1/2}$ as θ decreases from $2\pi - \theta_c$ back to θ_c. Therefore ϕ_2 is a periodic function of θ of period $4(\pi - \theta_c)$.

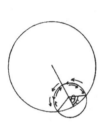 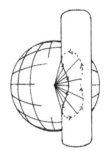

Fig. 14.7.

Whenever $\theta = \pm\theta_c$, $d\phi_3/dt = 0$, as can be seen from

$$\frac{d\phi_3}{dt} = \frac{(1 - A\cos\theta)\sqrt{2(E + A\cos\theta)}}{M^2 - 2(E + A\cos\theta)}.$$

Therefore, ϕ_3 oscillates between its extreme values $\phi_3^{\max}$ and $\phi_3^{\min}$, given by

$$\phi_3^{\max} = \sin^{-1}\frac{\sin\theta_c}{\sqrt{A^2 - 2A\cos\theta_c + 1}} \quad\text{and}\quad \phi_3^{\min} = -\phi_3^{\max}.$$

Theorem 9 *Suppose that $0 < E < A < 1$. Then both ϕ_2 and ϕ_3 are periodic functions of θ of period $4(\pi - \theta_c)$. The rotation matrix $R(t)$ is periodic if and only if*

$$\phi_1(2\pi - \theta_c) - \phi_1(\theta_c) = M\int_{\theta_c}^{2\pi - \theta_c}\frac{(A\cos t - 1)\,dt}{(A^2 - 2A\cos t + 1)\sqrt{2(E + A\cos t)}}$$

is a rational multiple of $2\pi - \theta_c$.

Theorems 8 and 9 reflect the qualitative changes in the rotational kinematics of the sphere as the parameter A decreases through the critical value $M = (1+A)$, with the angle ϕ_3 changing from an oscillatory pattern in the inflectional case to a rolling pattern in the noninflectional case. This bifurcating pattern can be described in slightly more picturesque language by saying that the sphere changes its motion from wobbling to rolling as A decreases through the critical value. Figure 14.8 depicts some typical inflectional solutions.

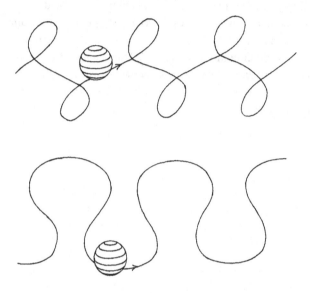

Fig. 14.8.

Notes and sources

The classic treatment of the equilibrium configurations of an elastic rod by Love (1927) does not contain much information about the details of integration of the Hamiltonian equations, apart from referring the reader to the literature of the heavy top, in accordance with Kirchhoff's theorem. As far as I know, Antman and Jordan (1974) were the first authors to deal successfully with the integration of the equations for the symmetric elastic problem. Holmes and Mielke (1988) verified the Hamiltonian equation for the elastic problem by using the Marsden-Weinstein reduction procedure and then proceeded to show the existence of chaotic orbits near the symmetric case.

The non-Euclidean model of the elastic problem for space forms is largely inspired by the papers of Bryant and Griffiths (1986), Langer and Singer (1984), and Jurdjevic (1995a). The present exposition follows my own paper (Jurdjevic, 1995a).

All of the literature on the Kowalewski top deals exclusively with the heavy top. The paper of Bobenko et al. (1989) is indicative of the mathematical attention that the original paper of Kowalewski generated. It may be that the Kowalewski elastic problem will shed new light on the mysterious integral of motion and provide a geometric explanation for its existence.

The rolling-sphere problem was first solved by Arthur and Walsh (1986) in response to the challenge posed by Hammersley (1983). The same problem was solved independently by Jurdjevic (1993) in a completely different context in connection with an optimal-control problem suggested by R. Brockett and L. Dai. The present exposition follows my own paper, even though Arthur and Walsh also used the maximum principle, which they applied to a control system described by quaternions.

References

Abraham, R., and Marsden, J. 1985. *Foundations of Mechanics*. Reading, MA: Benjamin/Cumming.

Agrachev, A. A., and Gamkrelidze, R. V. 1978. The exponential representation of flows and chronological calculus. *Matem. Sbornik* **109**:467–532.

Agrachev, A. A., Gamkrelidze, R. V. 1986. Symplectic geometry for optimal control. In *Nonlinear Controllability and Optimal Control*, ed. H. J. Sussmann, pp. 263–79. New York: Marcel Dekker.

Albrecht, F. 1968. *Topics in Control Theory*. Lecture Notes in Mathematics, no. 63. New York: Springer-Verlag.

Antman, S. S., and Jordan, K. B. 1974. Qualitative aspects of the spatial deformations of non-linearly elastic rods. *Proc. R. Soc. Edinburgh* 73A:85–105.

Arnold, V. 1976. *Methodes mathematiques de la mechanique classique*. Moscow: Mir Editions.

Arthur, A. M., and Walsh, G. R. 1986. On the Hammersley's minimum problem for a rolling sphere. *Math. Proc. Cambridge Phil. Soc.* **99**:529–34.

Bellman, R., Glicksberg, I., and Gross, O. 1956. On the "bang-bang" control problem. *Quart. Appl. Math.* **14**:11–18.

Blaschke, W. 1930. *Differential-geometrie*. Berlin: Verlag von Julius Springer.

Bliss, A. G. 1925. *Calculus of Variations*. La Salle, IL: American Mathematical Society.

Bobenko, A. I., Reyman, A. G., and Semenov-Tian-Shansky, M. A. 1989. The Kowalewski top 99 years later: a Lax pair, generalizations and explicit solutions. *Comm. Math. Phys.* **122**:321–54.

Bogoyavlenski, O. 1984. Integrable Euler equations on $SO(4)$ and their physical applications. *Comm. Math. Phys.* **93**:417–36.

Bonnard, B. 1981. Controlabilité des systèmes bilinéaires. *Math. Systems Theory* **15**:79–92.

Bonnard, B., Jurdjevic, V., Kupka, I. K., and Sallet, G. 1982. Transitivity of families of invariant vector fields on semi-direct products of Lie groups. *Trans. Am. Math. Soc.* **271**:525–35.

Boothby, W. 1975. *An Introduction to Differentiable Manifolds and Riemannian Geometry*. New York: Academic Press.

Boothby, W., and Wilson, E. N. 1979. A transitivity problem from control theory. *SIAM J. Control* 17:212–21.

Borisov, V. F., and Zelikin, M. I. 1993. Regimes with increasingly more frequent switchings in optimal control problems. *Proc. Steklov Inst. of Math.* 1:95–185.

483

Brockett, R. W. 1972. Systems theory on group manifolds and coset spaces. *SIAM J. Control.* **10**:265–84.

Brockett, R. W. 1981. Control theory and singular Riemannian geometry. In *New Directions in Applied Mathematics*, ed. P. J. Hilton and G. S. Young, pp. 11–27. New York: Springer-Verlag.

Brunovsky, P. 1976. Local controllability of odd systems. *Banach Center Publications* **1**:39–44.

Brunovsky, P., and Lobry, C. 1975. Contrôlabilité bang-bang, contrôlabilité différentielle et perturbations des systèmes non-lineaires. *Anali di Math. Pura et Applicata* **4**:93–119.

Bryant, R., and Griffiths, P. 1986. Reductions for constrained variational problems and $\frac{1}{2} \int k^2 ds$. *Am. J. Math.* **108**:525–70.

Bushaw, D. 1958. Optimal discontinuous forcing terms. In *Contributions to the Theory of Nonlinear Oscillations*, vol. 4, ed. S. Lefschetz, pp. 29–52. Princeton University Press.

Carathéodory, C. 1935. *Calculus of Variations*. Berlin: Teubner. Reprinted 1982. New York: Chelsea.

Cartan, E. 1958. *Leçons sur les invariants integraux*. Paris: Hermann.

Chow, W. L. 1939. Über Systeme von linearen partiellen Differentialgleichungen erster Ordinung. *Math. Ann.* **117**:98–105.

Clarke, F. H. 1983. *Optimization and Non-smooth Analysis*. New York: Wiley.

Coddington, E. A., and Levinson, N. 1955. *Theory of Ordinary Differential Equations*. New York: McGraw-Hill.

Dirac, P. A. M. 1950. Generalized Hamiltonian dynamics. *Can. J. Math.* **1**:129–48.

Dubins, L. E. 1957. On curves of minimal length with a constraint on average curvature and with prescribed initial and terminal positions and tangents. *Am. J. Math.* **79**:497–516.

Dugundji, J., and Granas, A. 1982. *Fixed Point Theory*, vol. 1. Monografie Matematyczne, Tom 61. Warszawa: Polska Academia Nauk, Institut Matematyczny.

El Assoudi, R., Gauthier, J. P., and Kupka, I. K. in press. Controllability of right invariant systems on semi-simple Lie groups. *Ann. Inst. H. Poincaré.*

Elliott, D. L. 1971. A consequence of controllability. *J. Diff. Equations* **10**:364–70.

Fuller, A. T. 1963. Study of an optimum nonlinear system. *J. Electronics Control* **15**:63–71.

Fuller, T. 1960. Relay control systems optimized for various performance criteria. In *Automation and Remote Control*, vol. 1, ed. J. F. Coates et al., pp. 510–19. London: Butterworth.

Gardner, R. B., and Shadwick, W. F. 1987. Feedback equivalence of control systems. *Systems and Control Letters* **8**:463–5.

Goldstine, H. H. 1980. *A History of the Calculus of Variations, from the 17ᵗʰ through the 19ᵗʰ Century*. New York: Springer-Verlag.

Grasse, K. A. 1985. A condition equivalent to global controllability in systems of vector fields. *J. Diff. Equations* **56**:263–9.

Griffiths, P. A. 1974. On Cartan's method of Lie groups and moving frames as applied to uniqueness and existence questions in differential geometry. *Duke Math. J.* **41**: 775–814.

Griffiths, P. A. 1983. *Exterior Differential Systems and the Calculus of Variations*. Boston: Birkhäuser.

Guckenheimer, J., and Holmes, P. 1986. *Non-linear Oscillations, Dynamical Systems and Bifurcations of Vector Fields*. New York: Springer-Verlag.

Hammersley, J. M. 1983. Oxford commemoration ball. In *Probability, Statistics and Analysis*, pp. 112–42. *London Math. Soc.* lecture notes, ser. 79.

Hardy, G. H., Littlewood, J. E., and Polya, G. 1934. *Inequalities*. Cambridge University Press.

Haynes, G. W., and Hermes, H. 1970. Non-linear controllability via Lie theory. *SIAM J. Control* 8:450–60.

Hermann, R. 1962a. Some differential geometric aspects of the Lagrange variational problem. *Illinois J. Math.* 6:634–73.

Hermann, R. 1962b. The differential geometry of foliations II. *J. Math. Mech.* 11:303–16.

Hermann, R. 1963. On the accessibility problem in control theory. In *International Symposium on Non-linear Differential Equations*, ed. J. P. La Salle and S. Lefschetz, pp. 325–32. New York: Academic Press.

Hermann, R., and Martin, C. 1982. Lie and Morse theory for periodic orbits of vector fields and matrix Riccati equations. I: General Lie theoretic methods. *Math. Systems Theory* 15:277–84.

Hermes, H. 1974. On local and global controllability. *SIAM J. Control Opt.* 12:252–61.

Hermes, H., and La Salle, J. P. 1969. *Functional Analysis and Time Optimal Control*. New York: Academic Press.

Hilgert, J., Hofmann, K. H., and Lawson, J. D. 1989. *Lie Groups, Convex Cones and Semigroups*. Oxford University Press.

Hirsch, M., and Smale, S. 1974. *Differential Equations, Dynamical Systems and Linear Algebra*. New York: Academic Press.

Holmes, P., and Mielke, A. 1988. Spatially complete equilibria of buckled rods. *Arch. Rat. Mech. Anal.* 101:319–48.

Holmes, R. B. 1975. *Geometric Functional Analysis and Its Applications*. New York: Springer-Verlag.

Hörmander, L. 1967. Hypoelliptic second order differential equations. *Acta Math.* 119:147–71.

Jakubczyk, B., and Respondek, W. 1980. On linearization of control systems. *Bull. Acad. Polonaise Sci.* 23:517–22.

John, F. 1982. *Partial Differential Equations*. New York: Springer-Verlag.

Jurdjevic, V. 1993a. The geometry of the ball–plate problem. *Arch. Rat. Mech. Anal.* 124:305–28.

Jurdjevic, V. 1993b. Optimal control problems on Lie groups: crossroads between geometry and mechanics. In *Geometry of Feedback and Optimal Control*, ed. B. Jakubczyk and W. Respondek. New York: Marcel Dekker.

Jurdjevic, V. 1995a. Non-Euclidean elastica. *Am. J. Math.* 117:93–125.

Jurdjevic, V. 1995b. Casimir elements and optimal control. In *Geometry in Non-linear Control and Differential Inclusions*, ed. B. Jakubczyk, W. Respondek, and T. Rzezuchovski, pp. 261–75. Warczawa: Banach Center Publications.

Jurdjevic, V., and Kogan, J. 1989. Optimality of extremals for linear systems with quadratic costs. *J. Math. Anal. App.* 143:87–107.

Jurdjevic, V., and Kupka, I. K. 1981a. Control systems subordinated to a group action: accessibility. *J. Diff. Equations* 39:186–211.

Jurdjevic, V., and Kupka, I. K. 1981b. Control systems on semi-simple Lie groups and their homogeneous spaces. *Ann. Inst. Fourier* 31:151–79.

Jurdjevic, V., and Kupka, I. K. 1985. Polynomial control systems. *Math. Ann.* 272:361–8.

Jurdjevic, V., and Sallet, G. 1984. Controllability properties of affine systems. *SIAM J. Control Opt.* 22:501–8.

Jurdjevic, V., and Sussmann, H. J. 1972. Control systems on Lie groups. *J. Diff. Equations* 12:313–29.

Kalman, R. E. 1960. Contributions to the theory of optimal control. *Bol. Soc. Matem. Mexicana (Ser. 2)* 5:102–19.

Kalman, R. E. 1963. Mathematical description of linear dynamical systems. *SIAM J. Control (A)* 1:152–92.

Kalman, R. E., Ho, Y. C., and Narendra, K. S. 1962. Controllability of linear dynamical systems. *Contrib. Diff. Equations* 1:186–213.

Kelly, H., and Edelbaum, T. 1970. Energy climbs, energy turns, and asymptotic expansions. *J. Aircraft* 7:93–5.

Kitapcu, A., Silverman, L. M., and Willems, J. C. 1986. Singular optimal control; a geometric approach. *SIAM J. Control Opt.* 24:369–402.

Kogan, J. 1986. *Bifurcation of Extremals in Optimal Control.* Lecture notes in mathematics no. 1216. New York: Springer-Verlag.

Kostant, B. 1958. A characterization of classical groups. *Duke Math. J.* 25:107–23.

Kowalewski, S. 1889. Sur le probléme de la rotation d'un corps solid autour d'un point fixe. *Acta Mathematica* 12:177–232.

Krener, A. 1974. A generalization of Chow's theorem and the bang-bang theorem to non-linear control problems. *SIAM J. Control Opt.* 12:43–51.

Krener, A. 1977. The high order maximum principle and its applications to singular extremals. *SIAM J. Control Opt.* 15:256–93.

Kučera, J. 1966. Solution in large of control problem: $\dot{x} = (A(1 - u) + Bu)x$. *Czech. Math. J.* 91:600–23.

Kupka, I. K. 1987. Geometric theory of extremals in optimal control problems. I: The fold and the manifold case. *Trans. Am. Math. Soc.* 299:225–45.

Kupka, I. K. 1990. The ubiquity of Fuller's phenomenon. In *Non-linear Controllability and Optimal Control,* ed. H. J. Sussmann, pp. 313–50. New York: Marcel Dekker.

Kwong, M. N., and Zettl, A. 1992. *Norm Inequalities for Derivatives and Differences.* Lecture notes in mathematics no. 1536. New York: Springer-Verlag.

Langer, J., and Singer, D. 1984. The total squared curvature of closed curves. *J. Diff. Geometry* 20:1–22.

La Salle, J. P. 1954. Basic principle of the "bang-bang" servo. *Bull. Am. Math. Soc.* 60:154.

Lee, E. B., and Markus, L. 1967. *Foundations of Optimal Control Theory.* New York: Wiley.

Lefschetz, S. 1963. *Differential Equations: Geometric Theory.* New York: Wiley Interscience.

Lie, S. 1888. *Transformations Gruppen,* 3 vols. Leipzig: Teubner. Reprinted 1970. New York: Chelsea.

Lions, P. L. 1982. *Generalized Solutions of the Hamilton-Jacobi Equations.* Research notes in mathematics. Boston: Pittman Advances Publishing Program.

Lobry, C. 1970. Contrôlabilité des systèmes non linéaires. *SIAM J. Control Opt.* 8:573–605.

Lobry, C. 1974. Controllability of non-linear systems on compact manifolds. *SIAM J. Control Opt.* 1:1–4.

Love, A. E. H. 1927. *A Treatise on the Mathematical Theory of Elasticity,* 4th ed. New York: Dover.

McShane, E. J. 1967. Relaxed controls and variational problems. *SIAM J. Control Opt.* 5:438–85.

Marchal, C. 1973. Chattering arcs and chattering controls. *J. Optim. Theory App.* 11:441–68.

Mayne, D. Q., and Brockett, R. (eds.) 1973. *Geometric Methods in System Theory*. Proceedings of the NATO Advanced Study Series. Dordrecht: Reidel.

Nagano, T. 1966. Linear differential systems with singularities and applications to transitive Lie algebras. *J. Math. Soc. Japan* **18**:394–404.

Neimark, J. I., and Fufaev, N. A. 1972. *Dynamics of Non-holonomic Systems*. Translations of mathematical monographs, vol. 33. Providence: American Mathematical Society.

Poincaré, H. 1892. *Les méthodes nouvelles de la mechanique celeste*. Paris: Gauthier-Villars.

Poincaré, H. 1901. Sur une forme nouvelle des equations de la mechanique. *Comptes Rendus des Sciences* **132**:369–71.

Pontryagin, L. S., Boltyanski, V. G., Gamkrelidze, R. V., and Mischenko, E. F. 1962. *The Mathematical Theory of Optimal Processes*. New York: Wiley.

San Martin, L. A. B., and Tonelli, P. A. 1994. Semigroup actions on homogeneous spaces. *Semigroup Forum* **14**:1–30.

Shayman, M. A. 1986. Phase portrait of the matrix Riccati equation. *SIAM J. Control Opt.* **24**:1–65.

Sontag, E. D. 1990. *Mathematical Control Theory: Deterministic Finite Dimensional Systems*. New York: Springer-Verlag.

Spivak, M. 1970. *A Comprehensive Introduction to Differential Geometry*, vol. 1. Berkeley: Publish or Perish, Inc.

Stefan, P. 1974. Accessible sets, orbits and foliations with singularities. *London Math. Soc. Proc., ser. 3* **29**:699–713.

Stefani, G. 1991. On maximum principles. In *Analysis of Controlled Dynamical Systems*, ed. B. Bonnard, B. Bride, J. P. Gauthier, and I. Kupka, pp. 373–82. Boston: Birkhäuser.

Sternberg, S. 1964. *Lectures on Differential Geometry*. Englewood Cliffs: Prentice-Hall.

Sussmann, H. J. 1973. Orbits of families of vector fields and integrability of distributions. *Trans. Am. Math. Soc.* **180**:171–88.

Sussmann, H. J. 1976. Some properties of vector fields not altered by small perturbations. *J. Diff. Equations* **20**:292–315.

Sussmann, H. J. 1987. Regular synthesis for time-optimal control of single input real-analytic systems in the plane. *SIAM J. Control Opt.* **25**:1145–62.

Sussmann, H. J. 1988. Why real analyticity is important in control theory. In *Perspectives in Control Theory*, ed. B. H. Jakubczyk, K. Malanovski, and W. Respondek, pp. 315–40. Boston: Birkhäuser.

Sussmann, H. J., and Jurdjevic, V. 1972. Controllability of non-linear systems. *J. Diff. Equations* **12**:95–116.

van der Schaft, A. J. 1987. Equations of motion for Hamiltonian systems with constraints. *J. Physics, ser. A, Math. Gen.* **20**:3271–7.

Warga, J. 1972. *Optimal Control of Differential and Functional Equations*. New York: Academic Press.

Wolf, J. A. 1984. *Spaces of Constant Curvature*, 5th ed. Wilmington, DE: Publish or Perish, Inc.

Wonham, W. M. 1974. *Linear Multivariable Control, a Geometric Approach*. Lecture notes in economics and mathematical systems. New York: Springer-Verlag.

Young, L. C. 1969. *Lectures on the Calculus of Variations and Optimal Control*. Philadelphia: Saunders.

Index

489